Art of the Dogfight
Volume 2

Jim Wilberg, M.D.

For Anne Marie, who kept the faith.

Acknowledgements
For our aviation books, please see our website at: **www.aeronautbooks.com**.

I am looking for photographs of the less well-known German aircraft of WWI to complete this series. For questions or to help with photographs you may contact me at **jherris@me.com**.

ISBN: **978-1-953201-26-3**

Design and layout: Jack Herris
Cover design: Aaron Weaver
Digital photo editing: Jack Herris

Aeronaut Books

Books for Enthusiasts by Enthusiasts
www.aeronautbooks.com

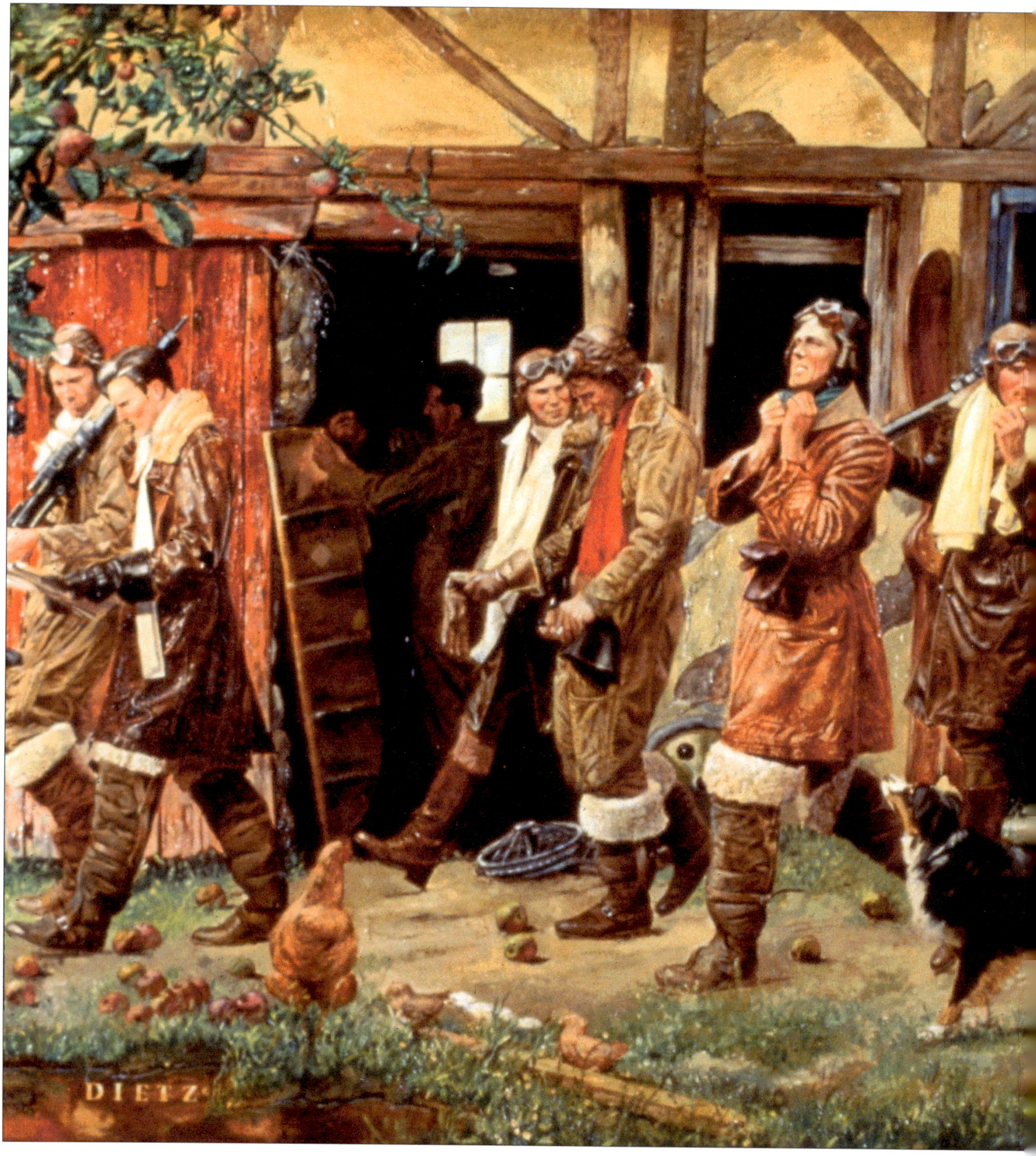

Above: Jim Dietz's two paintings, *Bonne Chance* (front endpaper) and *C'est le Guerre* (back endpaper) are wonderfully evocative pieces depicting the start of and the return of a sortie by a British squadron in the First World War. The details are superb, and for the observant reader, a story is visually told.

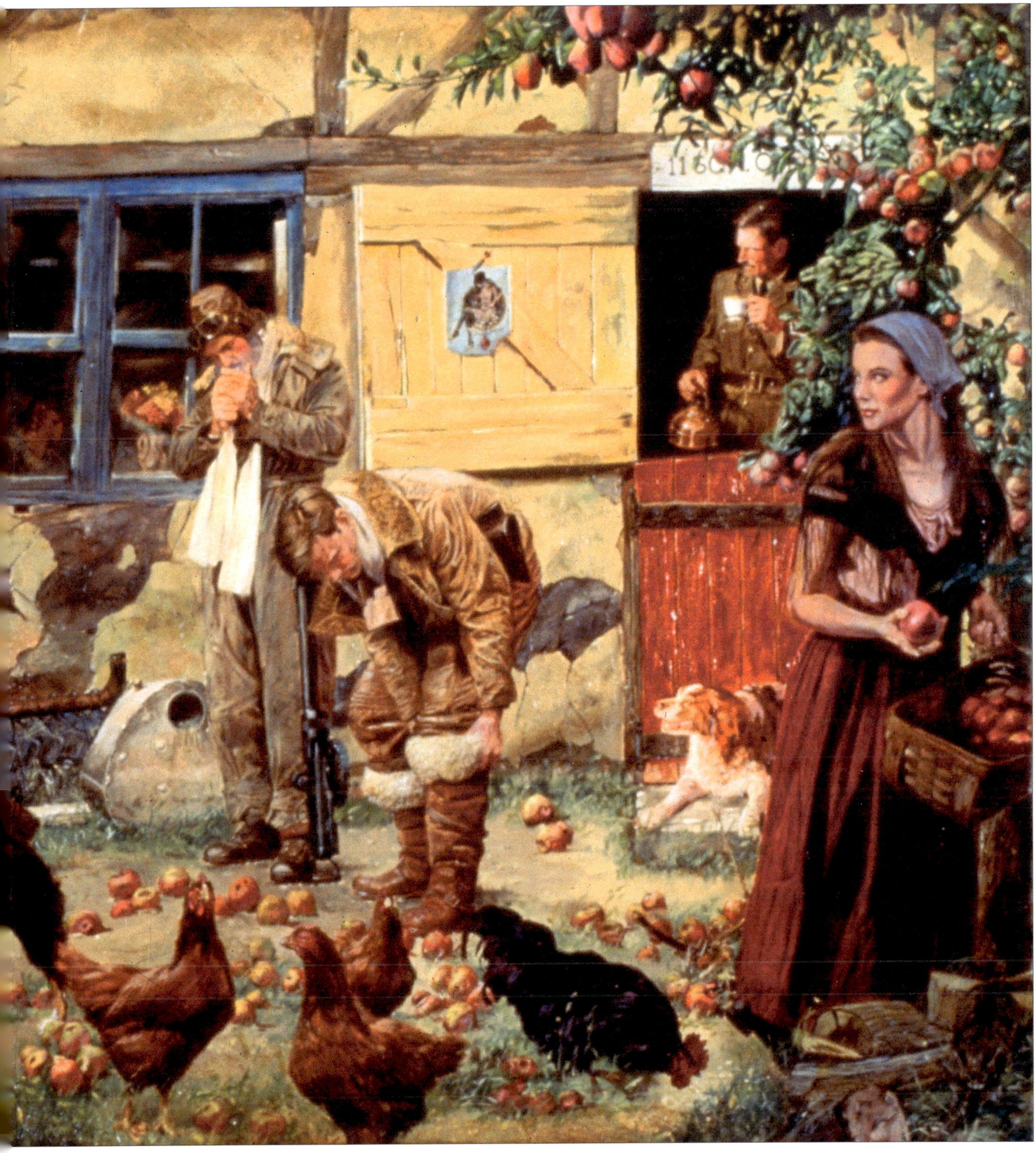

"Every man who went aloft was marked for death
Sooner or later
Once his wheels had left the ground"

Anthony Fokker

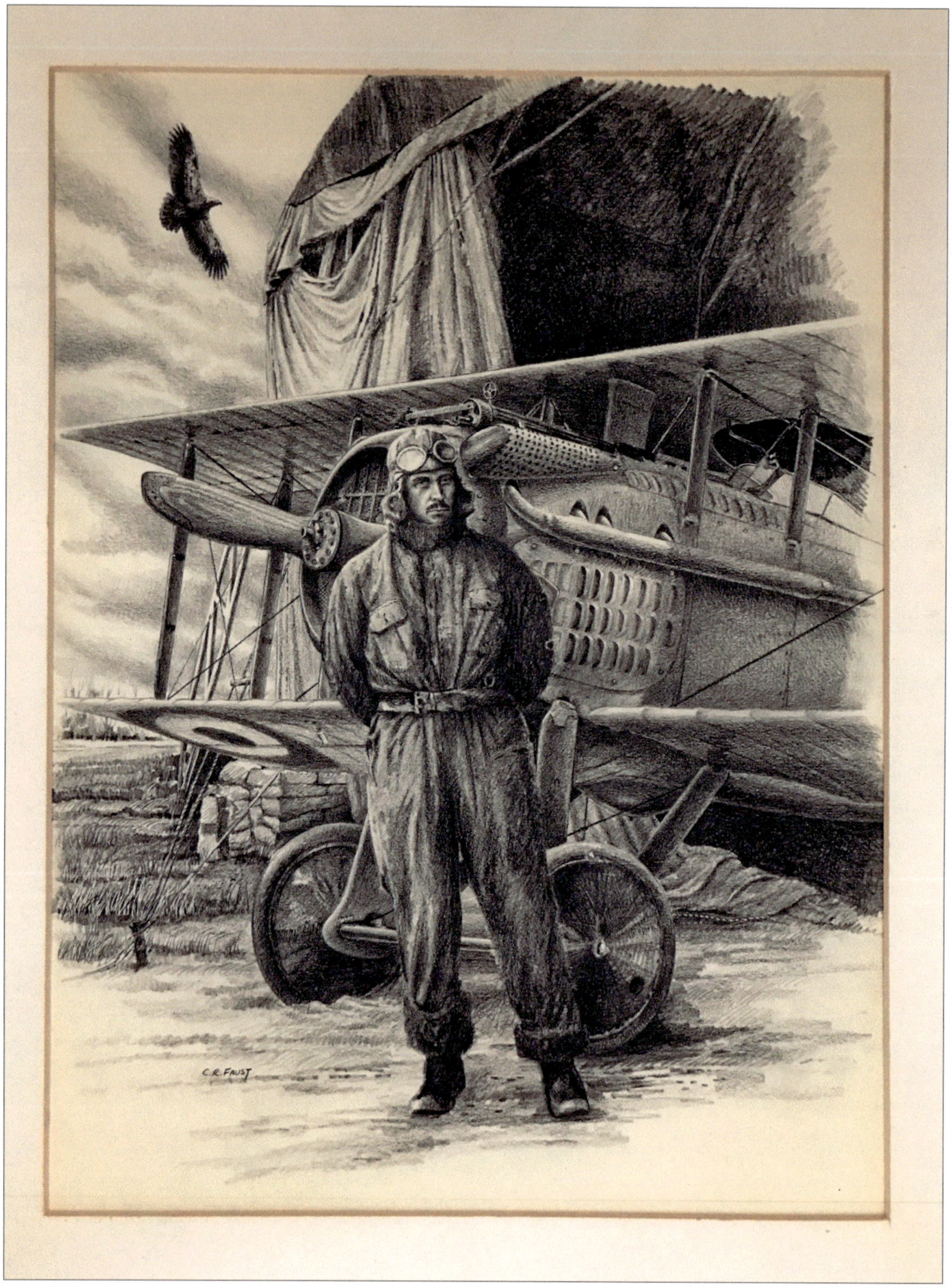

The Hawk by Charles Faust

Introduction Volume 2

This is the second of a four volume series, *Art of the Dogfight*. The four volumes actually represent one continuous body of work and should be recognized as such. The reader is referred to Volume 1 for the acknowledgements and introduction for the series and to the end of Volume 4 for a section comprised of brief biographies of the artists whose work is included in this series. All inquiries should be sent to jherris@me.com.

Art of the Dogfight Volume 2 Table of Contents

Cover Painting: Bolt From the Blue

My friend, renowned aviation artist Jim Dietz created this painting especially for the cover of ***Art Of The Dogfight***. It depicts an offensive patrol flight of Sopwith Camels from B flight, 10 Squadron RNAS engaging a kette of Pfalz DIII fighters of Royal Prussian Jasta 30 in Feb or March, 1918. This painting exemplifies the author's concept of "visual history" in that the aircraft are portrayed as accurately as possible, and that research would indicate that these two fighter formations were indeed opposite from each other on the Western front at the time depicted, thanks to research by Greg van Wyngarden and Bruno Schmalling.

It is the extensive collaboration between the aviation historians and the aviation artist that gives rise to the creation of visual history for us to enjoy and learn from.

Hawker V.C.

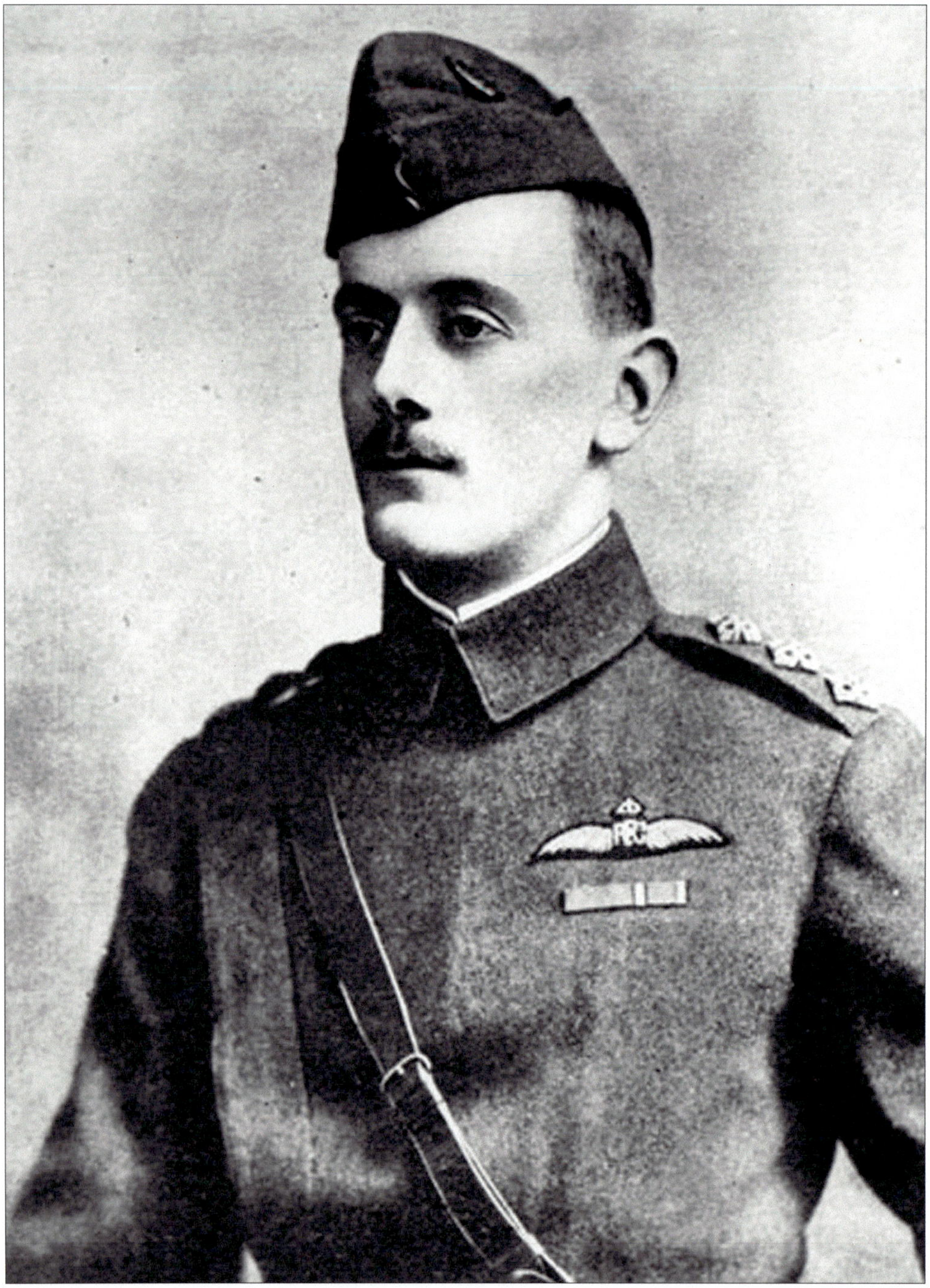

Left: The indomitable Lanoe Hawker, RFC.

24 Squadron, R.F.C., arrived in France in February, 1916 as the first-single seat fighter squadron ever to be deployed by the R.F.C. It was lead by its most charismatic pilot and most indomitable fighter at the time, Major Lanoe Hawker,V.C. The squadron would do much to reverse the tide of German aviation supremacy on the Somme Front, the so-called "Fokker Scourge".

Emblematic of the spirit in the squadron, developed by Hawker and fostered by his frequent exploits and inexhaustible courage, on the eve of the Battle of the Somme (July, 1916) the pilots in the 24th noticed a sheet of paper pinned to the notice board outside of their commander's office:

"Tactical orders by O.C. No.24 Squadron, Royal Flying Corps.

ATTACK EVERYTHING

L.G. Hawker, Major, Commanding No.24, RFC"

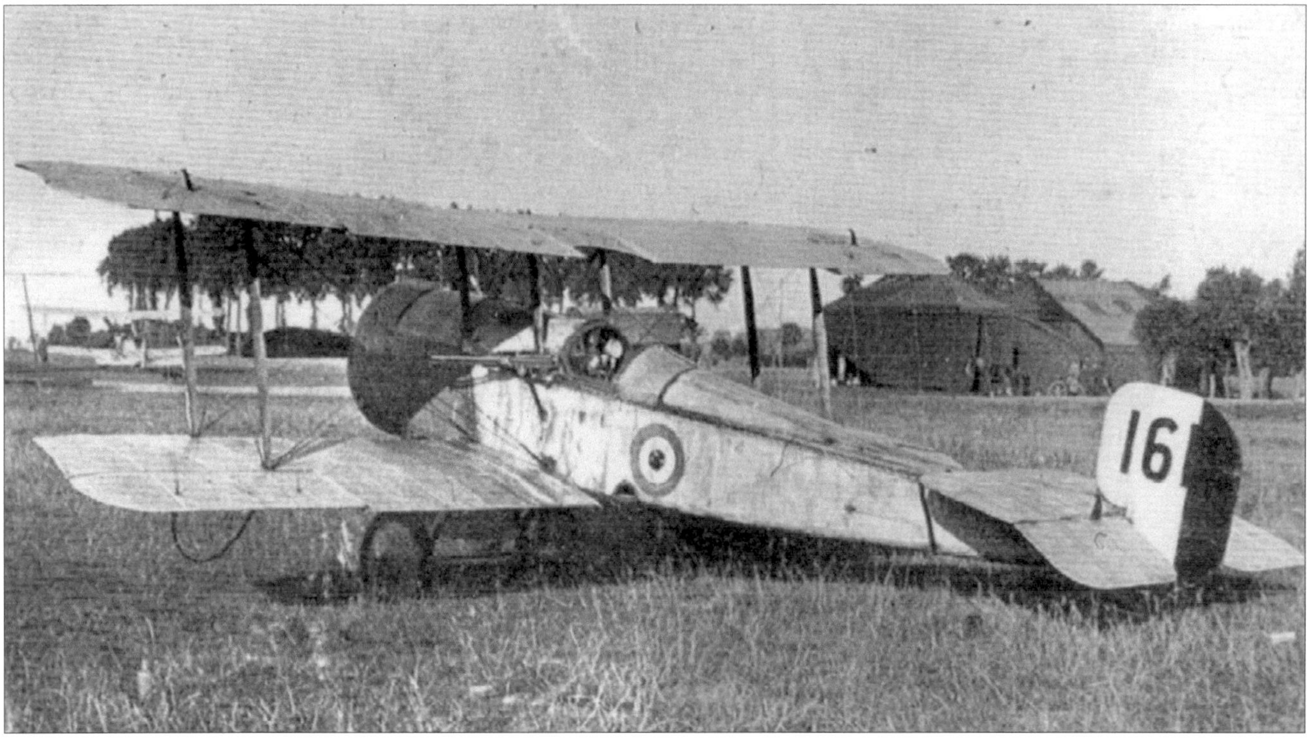

Above: "I have a beautiful little toy", Bristol Scout 1611, in which Hawker won the first V.C. for air-to-air combat. Note the Lewis machine gun mounted to fire at a 45 degree angle so as to miss the aircraft's propeller.

Hawker's father had been a career officer in the Royal Navy and at age 15 young Lanoe seemed destined to follow in his father's footsteps as he entered the Britannia Naval College at Dartmouth. However, he became quite ill while there and some residual medical effects from that illness forced Hawker to abandon his idea of a naval career.

He then began studies which would ultimately end in his placement into the Royal Engineers. However, during the course of these studies both Lanoe and his brother Tyrrel "caught" the flying bug and both joined the Royal Aero Club in 1910. By March, 1913 Lanoe had earned his Royal Aero Club certificate. In June, 1914 Hawker received orders to report to the Central Flying School. On 11th August Hawker had flown his first solo flight in the military. World events would move the pace of things along for the fledgling aviator as Britain was preparing for war. Hawker passed out of the CFS in early October 1914.

Lanoe Hawker was a 23 year old when he first arrived in France as a member of No.6 squadron,

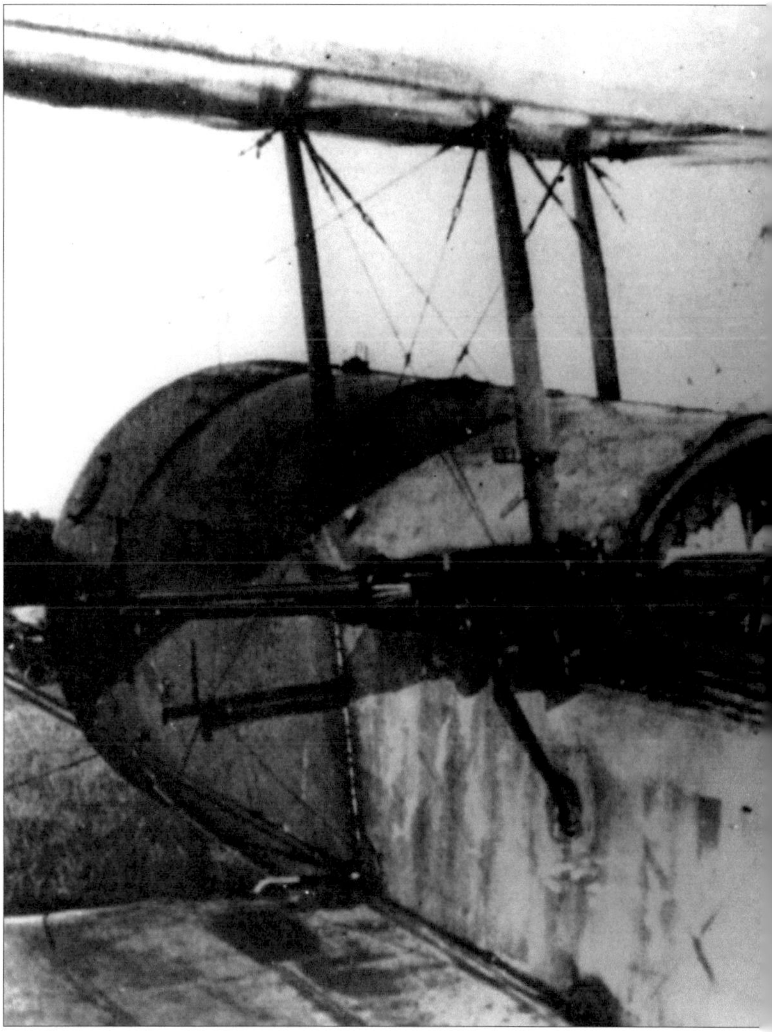

Right: A close up shows the Lewis gun mounted on the side of the fuselage. Hawker had to fly his aircraft and shoot at his opponent at a 45 degree angle, not an easy task at all.

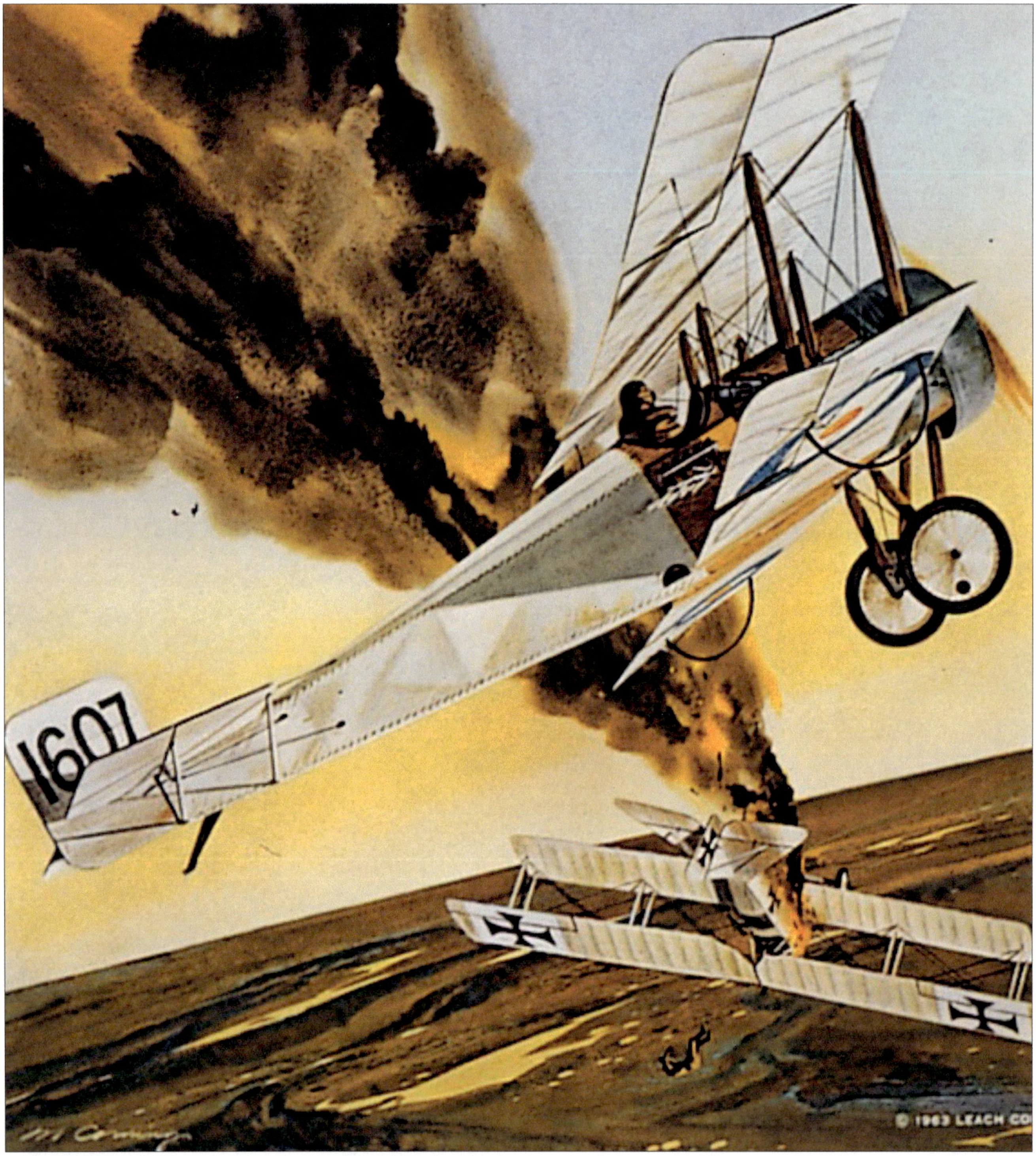

Above: Hawker earned his Victoria Cross by downing two German observation aircraft and damaging a third in a single sortie while flying Bristol Scout 1611. Merv Corning

R.F.C. in October of 1914. The unit was originally equipped with both Henri-Farman pusher aircraft as well as B.E.2a tractor aircraft.

As with all of the early aviators destined to make a mark as the pioneers of fighter aviation, Hawker was thoroughly displeased with not being able to retaliate against his enemies as he flew reconnaissance missions. He decided to carry a revolver with him and on 31 October:

"I met a German biplane yesterday afternoon and fired six shots at him with my revolver- no good, of course, but he took fright and made for home." [*Hawker V.C.*, Hawker, 1965]

Hawker was not only an aggressive pilot, he was

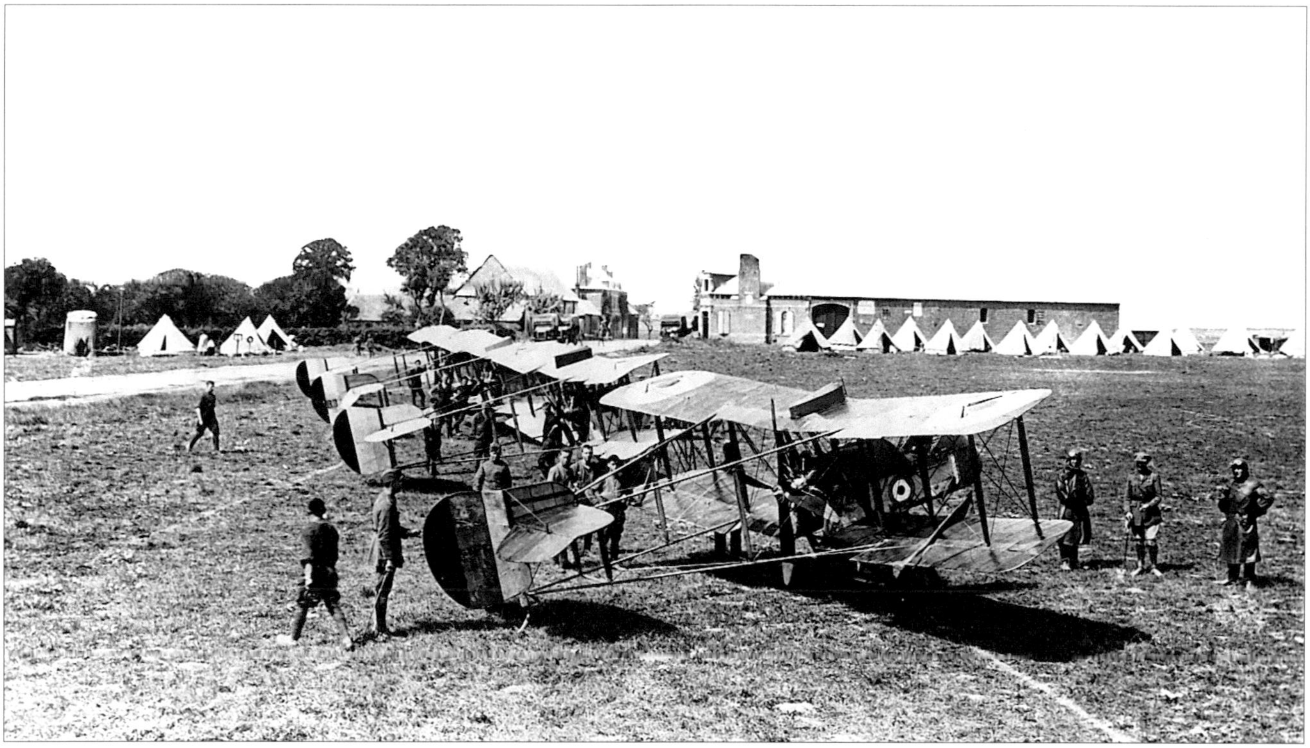

Above: This photo is of No.32 squadron RFC, while stationed at the aerodrome at Vert Galand: although the aerodrome for 24 squadron would have been quite similar in many aspects. Note the "Bell" tents which would be for the enlisted men of the unit, the officers would likely have been billeted in a house or houses nearby: possibly seen in the back of the photo.

Below: Capt. Herman W. Von Poellnitz, commander of B flight, 32 sqdn. taking off at the height of the Battle of the Somme, note the streamers on the wing struts indicating a flight leader.

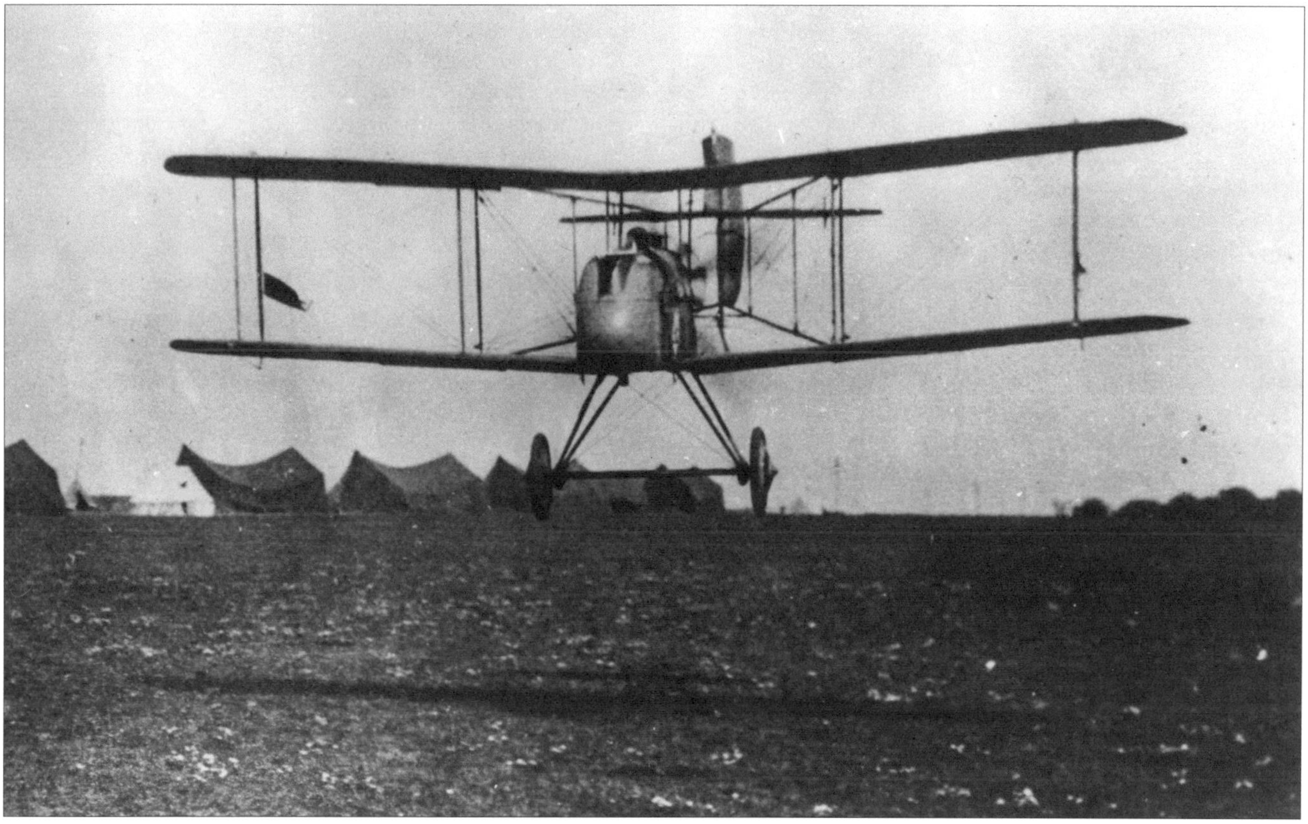

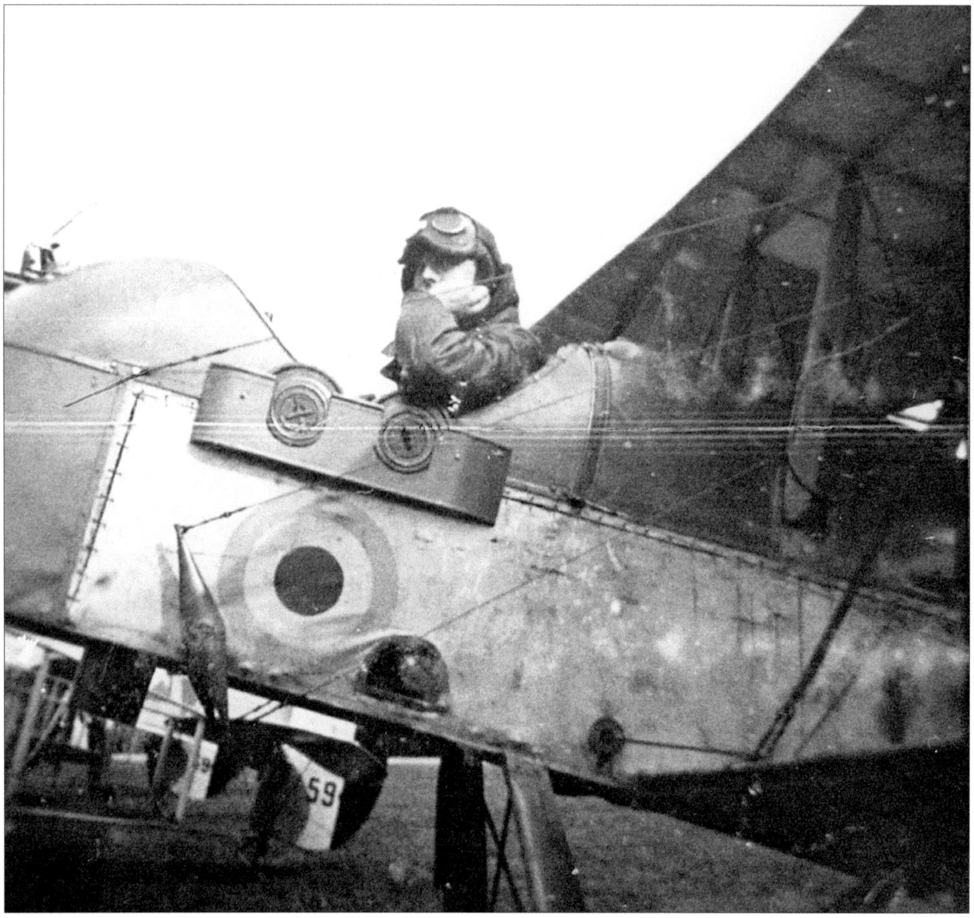

Left: Sitting in the cockpit reveals the openness that a DH 2 pilot had to contend with, Lt. Muir of No. 24 Sqdn. awaiting take off: note the extra ammunition drums for the Lewis gun on the side of the fuselage.

a very skilled pilot and was recognized as such by his commanding officer, Major Gordon Shephard.

Therefore, Hawker was selected as the pilot from his squadron for a mission to attack the Zeppelin base at Gontrode on 18 April, 1915. Despite intense anti-aircraft fire he pressed home his attack and dropped his three bombs on the target (without effect): Hawker received the DSO for his determined and daring attacks during this mission.

Hawker was wounded by ground fire during a reconnaissance mission on 24 April. However Hawker insisted on continuing to fly during the Second Battle of Ypres despite the fact that he would have to be carried to and from his aircraft since the wound made walking too painful!

As the intensity of the battle wound down, Hawker was finally sent home for an extended leave of three weeks to recover.

Hawker would return to the front in late May after three weeks sick leave at home. He was sent to collect a Bristol Scout C (No.1611) from St.Omer during the first week of June. At this time the R.F.C. was beginning to "distribute" small, fast scout planes to its squadrons with the squadron commanders to determine the pilot who best demonstrated the aggressive and fighting spirit to use these aircraft to engage the enemy.

There was no doubt, Hawker was the best choice in No. 6 squadron. Being an engineer, Hawker (along with his mechanic E.J. Elton) rigged a mounting for a Lewis gun to be fired at an approximately 45 degree angle so as not to damage the propeller blades since the British did not have a synchronization mechanism yet. To fly the nimble, rotary-powered aircraft in one direction so as to shoot at your enemy in a completely different direction would be beyond the skills of most pilots: Hawker was certainly not like most pilots.

Hawker was completely re-energized now:

"I have a beautiful little toy, a new Bristol Scout that goes at 80 and climbs 5 or 600 feet a minute! I'm having a machine-gun fitted to see how they like it...My little beast is a beauty and make no mistake. Was up twice this morning and worried a German each time but they don't wait as a rule. However, it is an art that requires practice and I hope to get proficient soon." [*Hawker VC*, Hawker, p. 94 and p.96]

During the evening of the 25th July, 1915 Hawker attacked several German reconnaissance aircraft from Fl.Abt. 3 and managed to down two of them in the action. He would be awarded the

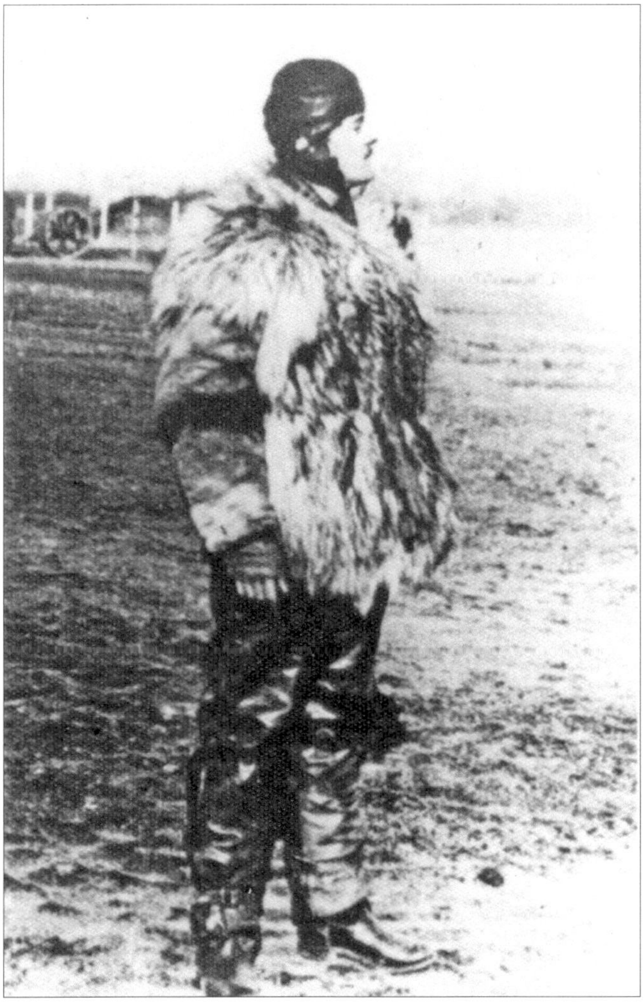

Above: Hawker getting ready for a flight in his DH2. Note the heavy bearskin suit and the tall boots needed to stay warm in the very open cockpit of the pusher aircraft.

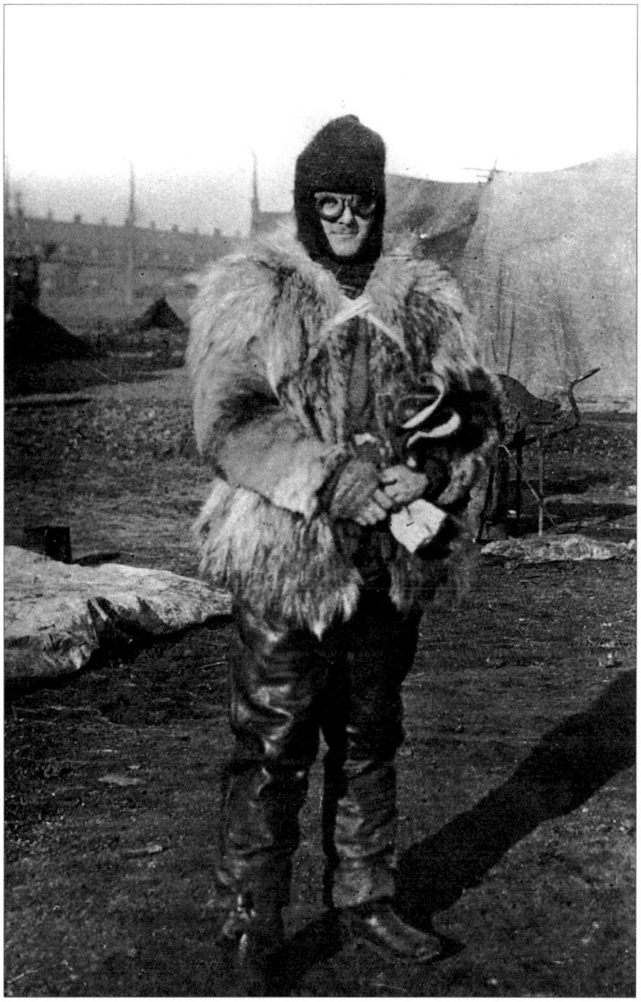

Above: Hawker now completely suited up and ready for the flight.

Victoria Cross for this action (as well as his "untiring work"), the first V.C. that was awarded for air-to-air combat. Hawker was still required to fly his share of reconnaissance missions, flying his Bristol Scout was deemed "extra" for him.

No.6 squadron had by this time received several FE2a aircraft which were to fly escort to the squadron's BE2 aircraft as they did reconnaissance or artillery spotting missions. It was while flying one of the FE2s on 11 August that Hawker(and his gunner,Lt. Clifton) scored another double victory.

To summarize his combat report from that mission:

I. The F.E. attacked and drove off an Albatros
II. The F.E. was then attacked by two Halberstadt scouts and an Albatros, after a 20 minute fight they all "retired"
III. The F.E. then attacked an Aviatic which "left in a nose dive"
IV. The F.E. was attacked by a very fast scout monoplane, "I turned the F.E. very sharp and succeeded in facing it"

Note: ammunition used : Lewis gun 360 rounds, Rifle 20 rounds

One is left with the impression that Hawker was flying and firing a rifle while Lt. Clifton managed the Lewis gun. The Aviatik biplane and the Fokker E-type monoplane were both seen to crash for confirmed victories. Which of course was the second double victory for Hawker in less than a month!

In fact, unknown to either Lanoe or his brother Tyrell (an officer in a nearby artillery unit), Tyrell was actually watching his brother battle the Fokker from a telescopic sight in his observation post.

Hawker was flying reconnaissance missions, escort missions, "hunting missions" in his Bristol Scout and was also always utilizing his engineering skills in improving his squadron's aircraft. He was also clearly worn out. Hawker was sent home for a brief leave in August 1915, but was back at No.6

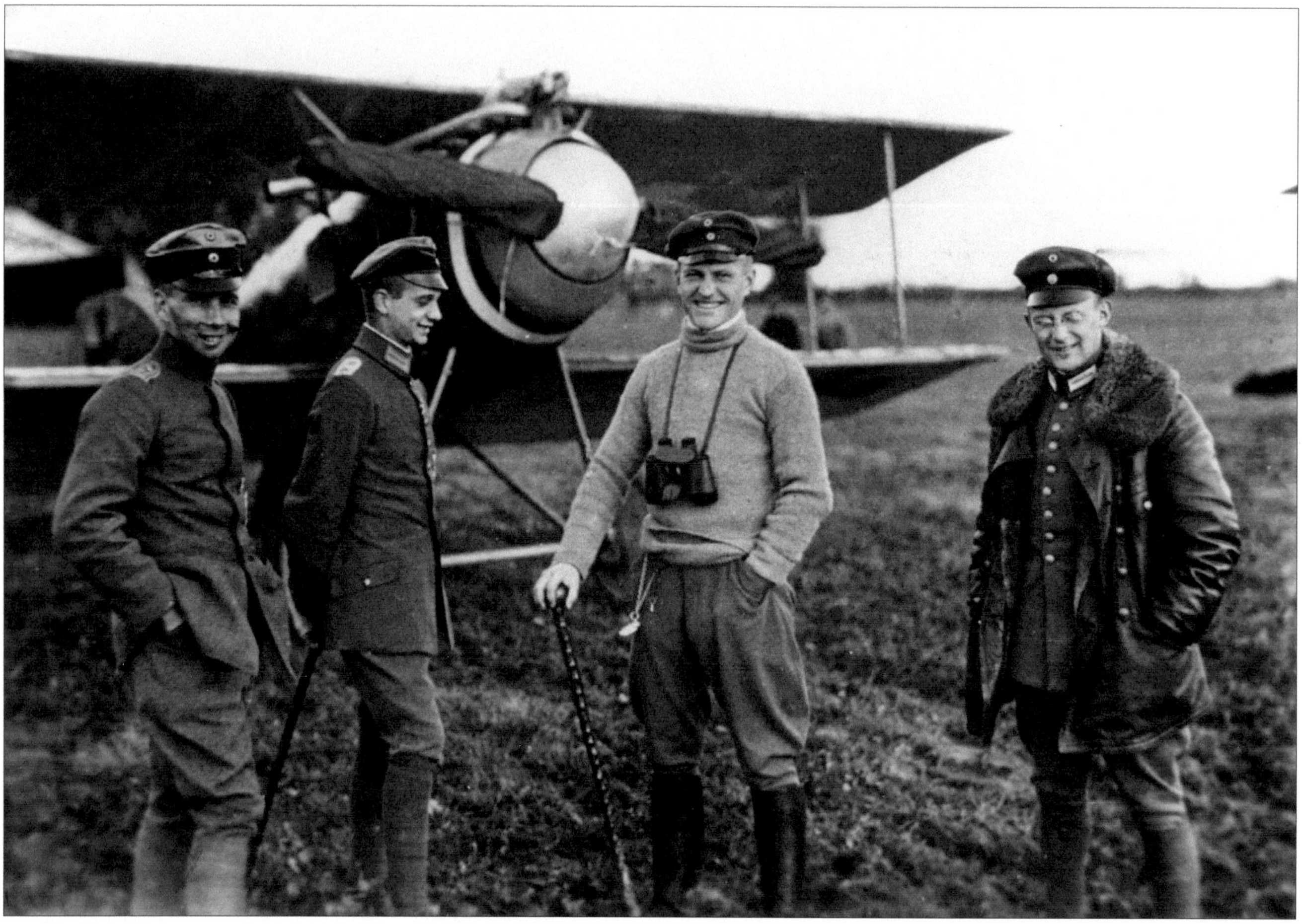

Above: Richthofen in a cheerful mood and relaxed dress in front of his new Albatros D.II fighter-note the white line around the nose, his personal marking at the time. At the far left is Obltn. Stefan Kirmaier who became the leader of Jasta 2 after Boelcke's death. Kirmaier himself was shot down and killed by J.O. Andrews of 24 squadron on the 22nd of November, one day before the epic dogfight between Hawker and Richthofen.

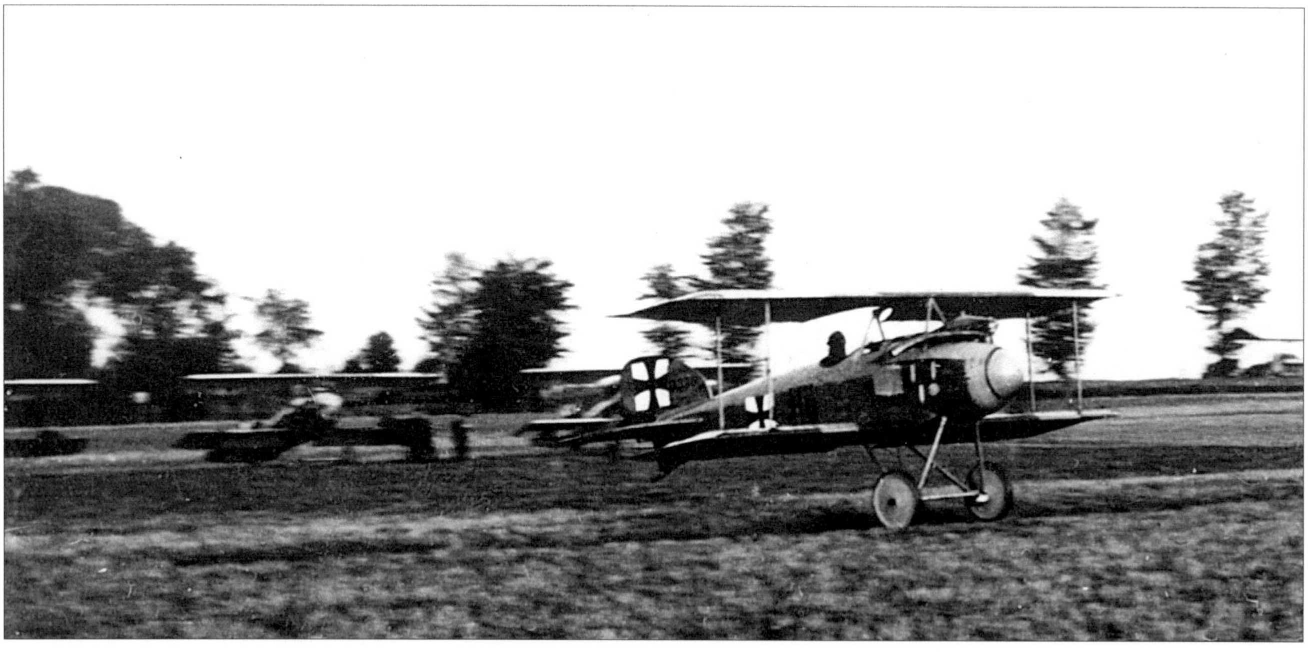

Above: An Albatros D.I fighter of Jasta 2 takes off; the pilot Ltn. Hohne.

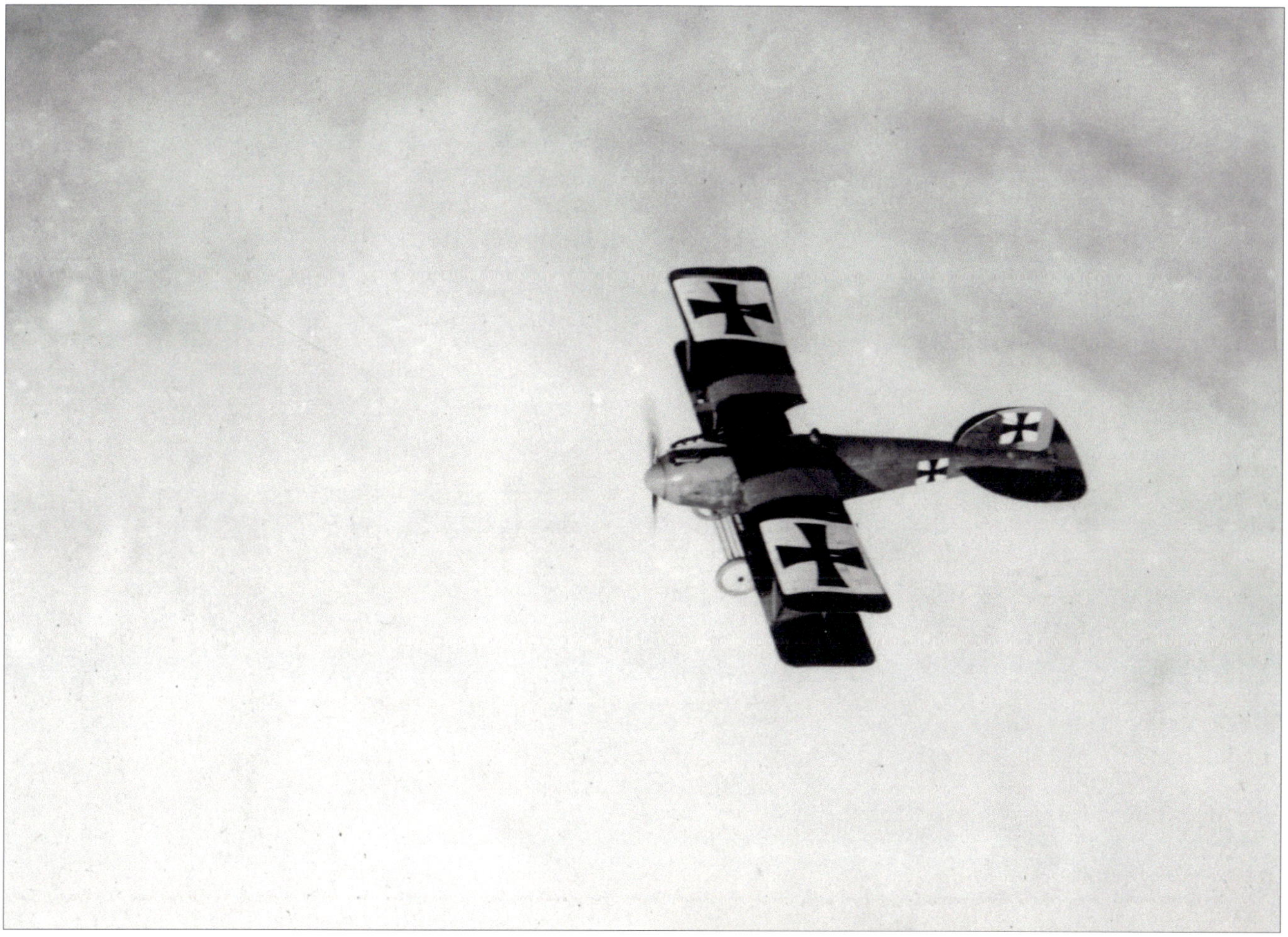

Above: An Albatros D.II fighter in flight; this view would have been what Hawker saw as he battled Richthofen.

squadron before the month was done. However, the leave seemed to be inadequate as demonstrated in his letter of 3rd September 1915:

"It is almost impossible to write about the 'deep things' as no words can adequately express emotions we don't quite understand...If you want to help that way, don't try and write, it is so easily done by prayer...I am not quite as well as I should like, and if I don't get better soon I shall apply for a home job. Only I am very keen on completing a full year out here, a thing no one else had done...This must be a short letter as I am tired of writing." [*Hawker, V.C.*, Hawker, p.112–113]

Although his commanding officer had trouble sparing such an experienced and talented officer, Gordon Shephard could see the strain and emotional burdens being carried by Hawker, who by this time was the last remaining pilot of the original group who had flown from England to France the year before.

Hawker's time at home would not be too extended. Commanding officer of the R.F.C., Hugh Trenchard had taken note of the French strategy

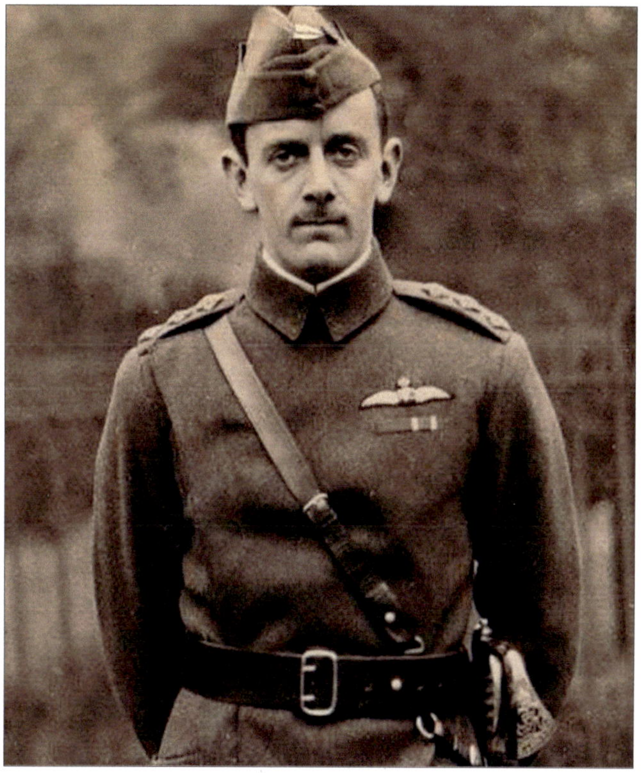

Above: "Who stands if freedom fall? Who dies if England live?"

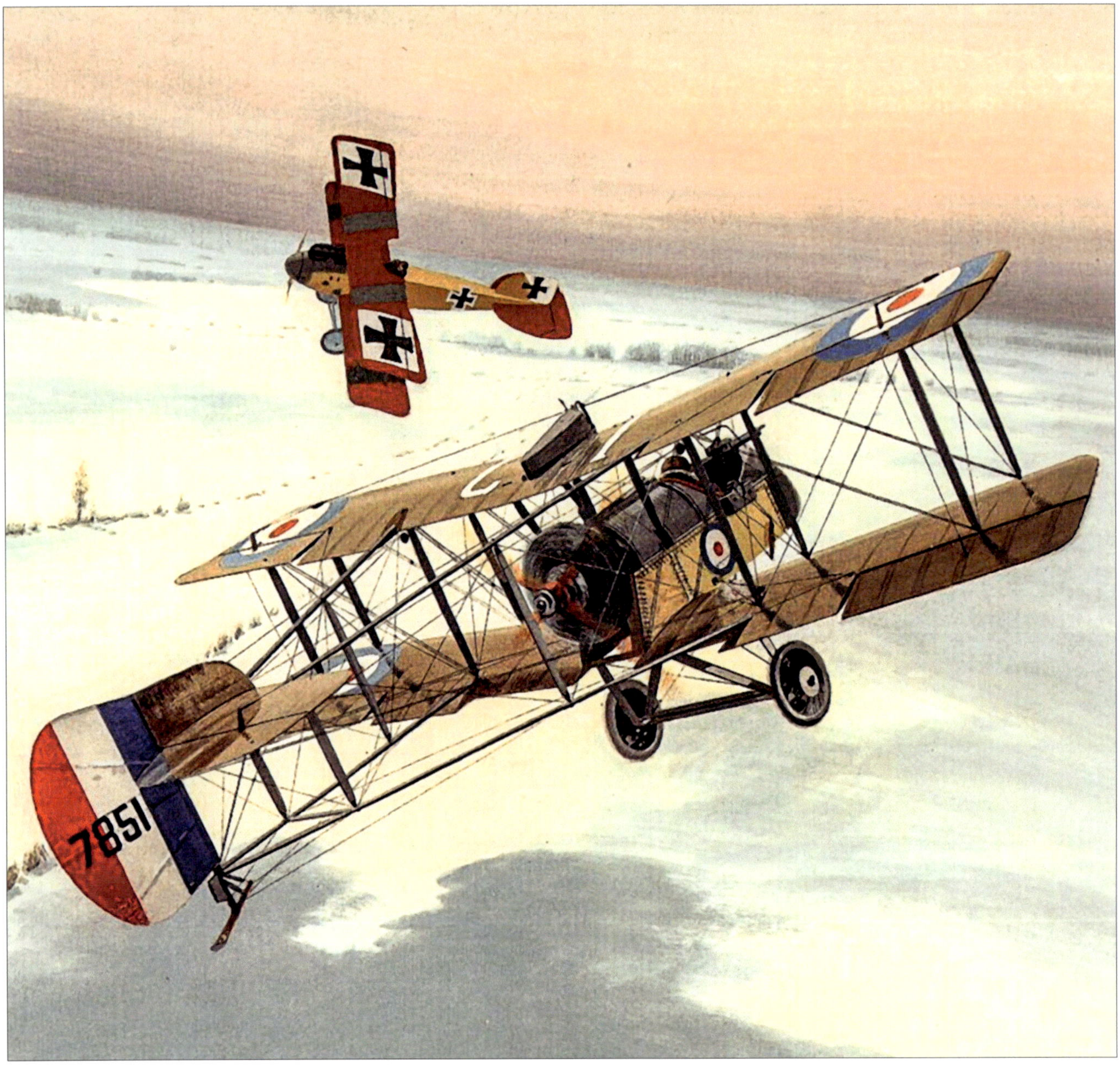

Above: One of the most famous air duels in all of the First World War. Note the lower left wing in this painting, it has a roundel on it which indicates it is actually a top wing: the DH 2 wings were all interchangeable. Brian Knight

of forming specific fighter squadrons and therefore ordered now Major Hawker to take over command of the first, single seat fighter squadron of the R.F.C., No. 24.

The squadron was to be equipped with the new DH 2 single seat, pusher aircraft (its forward firing Lewis gun did not need to be synchronized as the engine was located behind the pilot).

At the very end of September, Hawker was ordered to take command of No. 24.

The squadron would not deploy to France until February of 1916 after it had come to strength and the pilots learned how to fly their new fighters. The DH 2 had a reputation for being hard to control and that if the pilot entered into a spin, it was likely to be fatal. Hawker solved that worry by flying a DH 2 into all sorts of spins and maneuvers over the field as his pilots looked on. He then taught them too, how to handle the DH 2 and so the pilot's concerns were effectively eliminated.

Hawker also designed a 97 round canister for the Lewis gun (doubling its capacity) and he also designed special flying clothes for his squadron because of the profoundly cold cockpit of the DH 2 (including the coveted Fug boots).

No. 24 squadron arrived at the Somme front

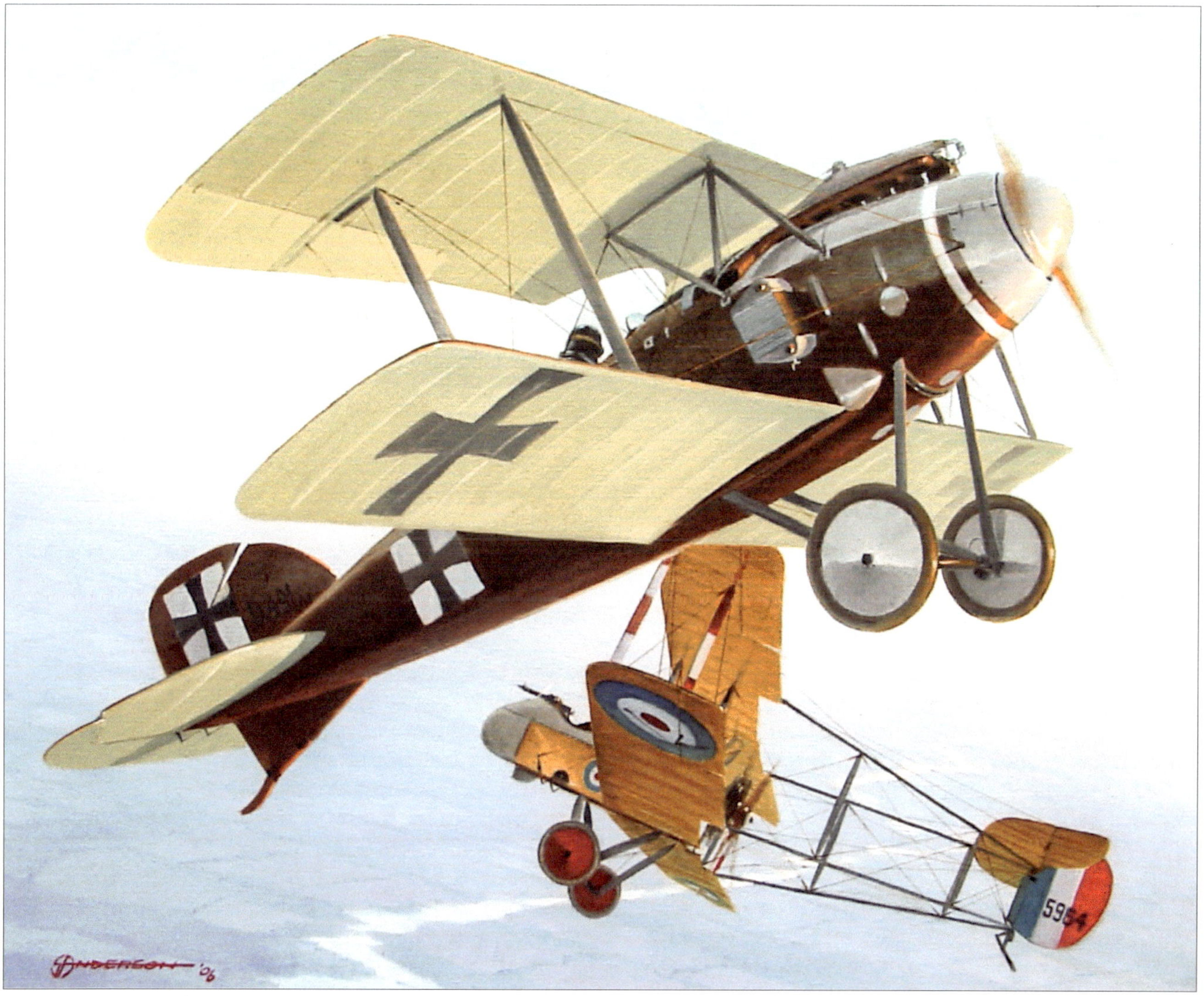

Above: The battle between Hawker and Richthofen ultimately came down to Richthofen having a faster aircraft with superior climb rate that he could use to gain altitude and an aircraft with two machine guns with 1,000 rounds of ammunition. Note the white band, Richthofen's personal marking. Steve Anderson

in February 1916 and Hawker's men quickly made their presence known. The Germans had achieved air superiority over the front since Hawker had been sent home to rest, primarily because of the Fokker monoplanes now equipped with a synchronized machine gun leading to what became known as the Fokker Scourge.

Hawker's squadron was joined by two other units equipped with the DH 2 fighter (No. 29 and 32), these units together would fly aggressively, confidently and would re-establish the control of the air above the Western Front for the British.

By April, 1916 the situation had completely changed in favor of the British and the DH 2 was proving more than a match for the Fokker monoplane series. A typical air battle for the time occurred on 25th April when DH 2's from No. 24 squadron were escorting a B.E.2 flying a reconnaissance mission when the formation was set upon by a flight of Fokkers. THe DH 2's were able to drive off the Germans without a loss to themselves and forced one of the Germans down: Max Immelmann himself.

The onset of the Somme offensive in July 1916 would ratchet up the air combats between the British and the Germans to a new intensity. Although Hawker and other squadron commanders were forbidden to fly, he still would insert himself into a sortie from time to time: in order to know what his men were facing in the air.

Immelmann, the Eagle of Lille, fell on 18th June likely destroying his own propeller due to a malfunctioning synchronization gear. This loss left Oswald Boelcke the leader of the German fighter pilots and acknowledged master of combat flying. The German high command re-organized its units, again copying the French, and formed jagdstaffeln (hunting units, usually referred to as Jastas). Boelcke was placed in command of Jasta 2 and found that his unit was essentially opposite of Hawker's No.24 squadron.

Thus, throughout the summer and into the fall of 1916 the pilots lead by Hawker and Boelcke often encountered each other in the skies over the Somme battlefield.

In September 1916, Boelcke would tally Jasta 2's efforts as follows:

- 186 front line combat sorties
- 69 combats
- 25 confirmed victories

Hawker would write in a letter from 6th September 1916:

" How time does fly! September already, and the days getting shorter and shorter thank goodness... I wish I could give details of our work, as really it's very interesting, but that would hardly do...I have just been compiling our records, and find we have shot down 28 up to the end of August, and a further four on Sept.2nd....No wonder the Hun is a little shy down here!"[*Hawker.V.C.*, Hawker, p.206]

Despite the enormous losses on the ground suffered by the British army, the R.F.C. had been diligently pursuing the doctrine of its leader Brigadier-General Hugh Trenchard in flying over the front lines and bringing the battle in the air behind the German lines. Continually taking the offensive and by sheer will and incredible effort, the R.F.C. was able to maintain nearly complete air superiority over the battlefront.

However, indomitable spirit alone cannot make up for inferior equipment and while the British were still flying their DH 2 fighters, the Germans were supplied with the new Albatros DI and DII fighters in mid-September. The Albatros had a 160 hp Mercedes inline engine with a maximum speed of 108 mph (almost 20 mph faster than the DH2), and was armed with two machine guns, each with 500 rounds of ammunition.

Still, Hawker's men fought on and brought the battle to the Germans, with the most experienced pilots still able to hold their own by utilization of their combat experience and the better maneuverability of their DH 2 to good advantage.

Men such as J.O. "Jock" Andrews who was A flight leader took to heart their commanding officer's orders and often, quite simply: attacked everything.

On 26th October Oswald Boelcke collided with another of his jagdfliegers while in a dogfight with two members of No.24 squadron and perished in the subsequent crash.

Oblt. Stefan Kirmaier took over command of Jasta 2, but he was shot down and killed by "Jock" Andrews on the 22 November 1916.

Therefore, the most experienced and successful fighter pilot in Jasta 2 took over de facto command, Manfred von Richthofen.

J.O. Andrews was scheduled to lead "A" flight on a defensive patrol during the afternoon of the 23rd November 1916. One of the planned members of that flight was a young pilot scheduled for leave the next day, Hawker took advantage of that fact and utilized it so that he filled the spot for his young pilot. Thus, Andrews set off leading four DH 2 fighters flown by himself, Robert Saundby, John Cruth and his commander, Major Lanoe Hawker.

Cruth would be forced out of the patrol by engine troubles, leaving Andrews to lead the other two DH 2s towards a pair of German Roland two-seaters at approximately 6,000 ft near Bapaume. Andrews led his flight down to attack them, however during the first firing pass Andrews noted patrols of German fighters above him and so used his speed from the diving pass to regain altitude. Saundby followed him, but Hawker pressed on against the Rolands as they turned east with Hawker now vulnerable from attack above.

The German fighters were from Jasta 2 and led by von Richthofen. They quickly dove upon the disadvantaged Hawker. Andrews and Saundby now dove upon the Germans to try to come to the aid of their squadron commander.

In the ensuing dogfight, Andrews was able to make several firing passes at von Richthofen and hit his machine several times. However, Andrews was not able to drive him away before he himself was hit during the combat and forced to dive away with his aircraft on fire and trailing smoke.

Likewise, Saundby did what he could before his aircraft was also hit numerous times and also had to dive away trailing smoke.

This left Hawker alone with the German fighters, all but von Richthofen seemed to clear away- likely so as to not accidentally ram into one another as had recently happened to Boelcke. Hawker and von Richthofen turned and jockeyed for position. However, the prevailing winds were from the west and were slowly pushing the two farther into German held territory.

Likely running out of fuel, Hawker finished a zoom to gain some altitude and then turned and dove to try to regain Allied lines. Richthofen followed, intermittently firing at the zig-zagging British fighter. Richthofen would write in his combat report that he fired 900 rounds in the fight!

Nearing the trenches, and almost escaping, Hawker was cut down by a single round that instantly killed him. He crashed just inside the German lines, Richthofen would later praise Hawker as the "English Boelcke". Richthofen would also remove the Lewis machine gun from Hawker's aircraft and mount it in his quarters as a souvenir from his 11th victory.

As Boelcke would inspire future German aces with his bravery and the skills that he taught his fledgling eagles, so to would Hawker live on as an inspiration to future RFC fighters such as Ball, McCudden and others:

"There is but one task for all,
One life for each to give.
Who stands if freedom fall?
Who dies if England live?"

[*Hawker V.C.*, Hawker, p.217]

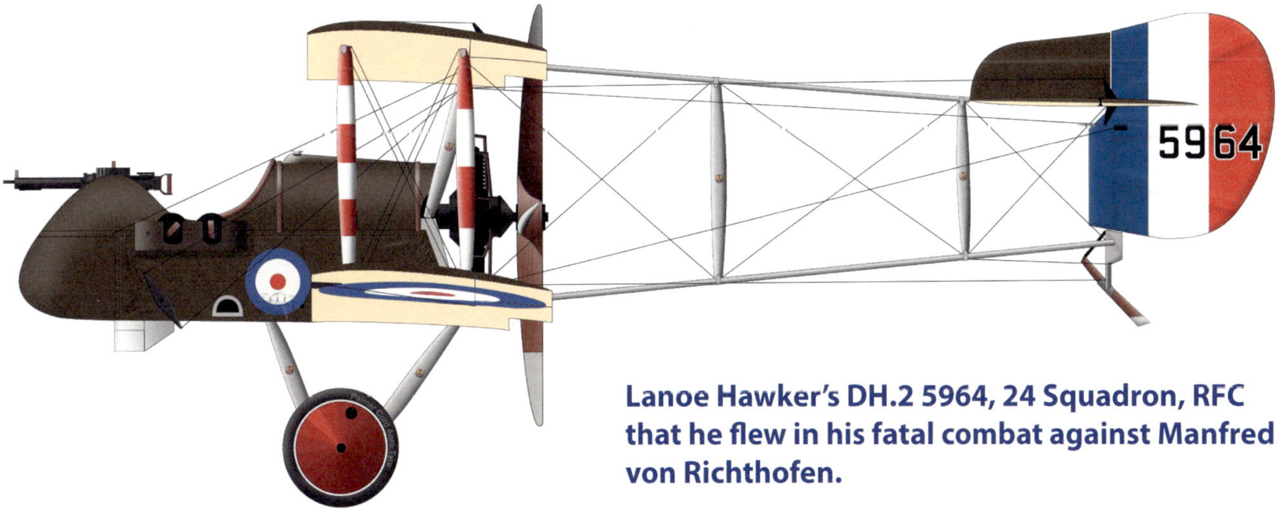

Lanoe Hawker's DH.2 5964, 24 Squadron, RFC that he flew in his fatal combat against Manfred von Richthofen.

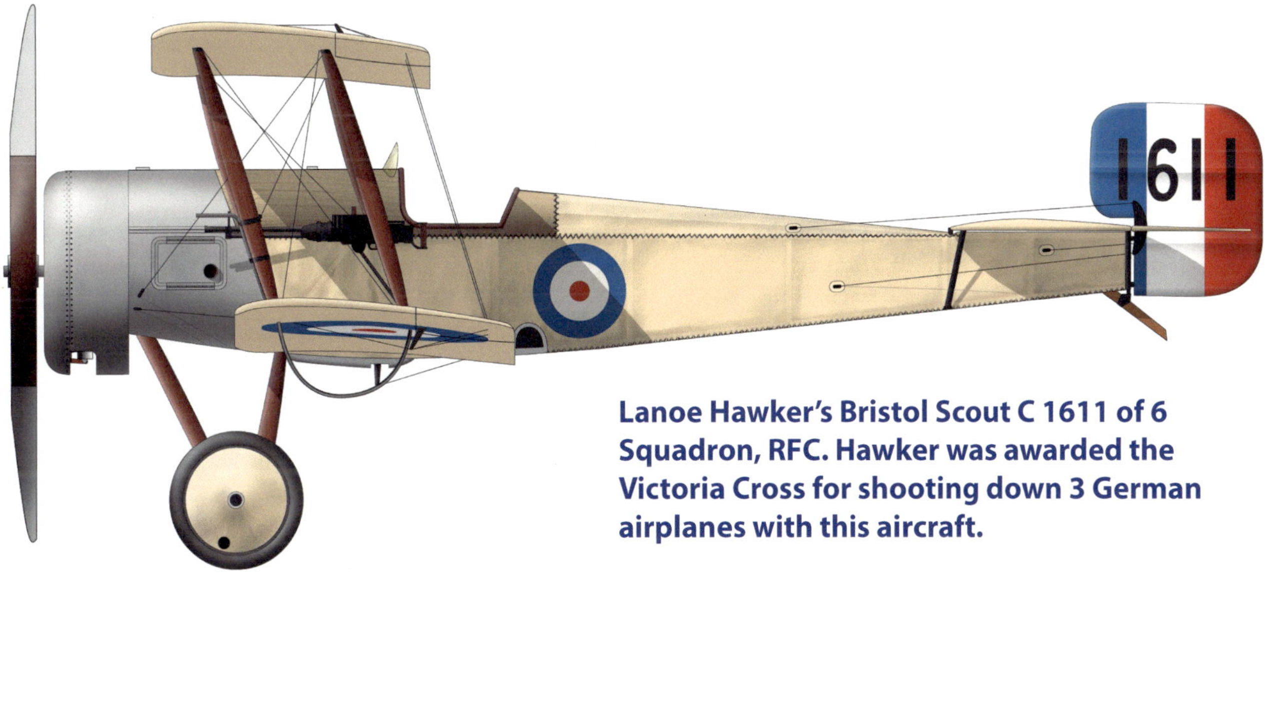

Lanoe Hawker's Bristol Scout C 1611 of 6 Squadron, RFC. Hawker was awarded the Victoria Cross for shooting down 3 German airplanes with this aircraft.

Lufbery of the Lafayette Escadrille

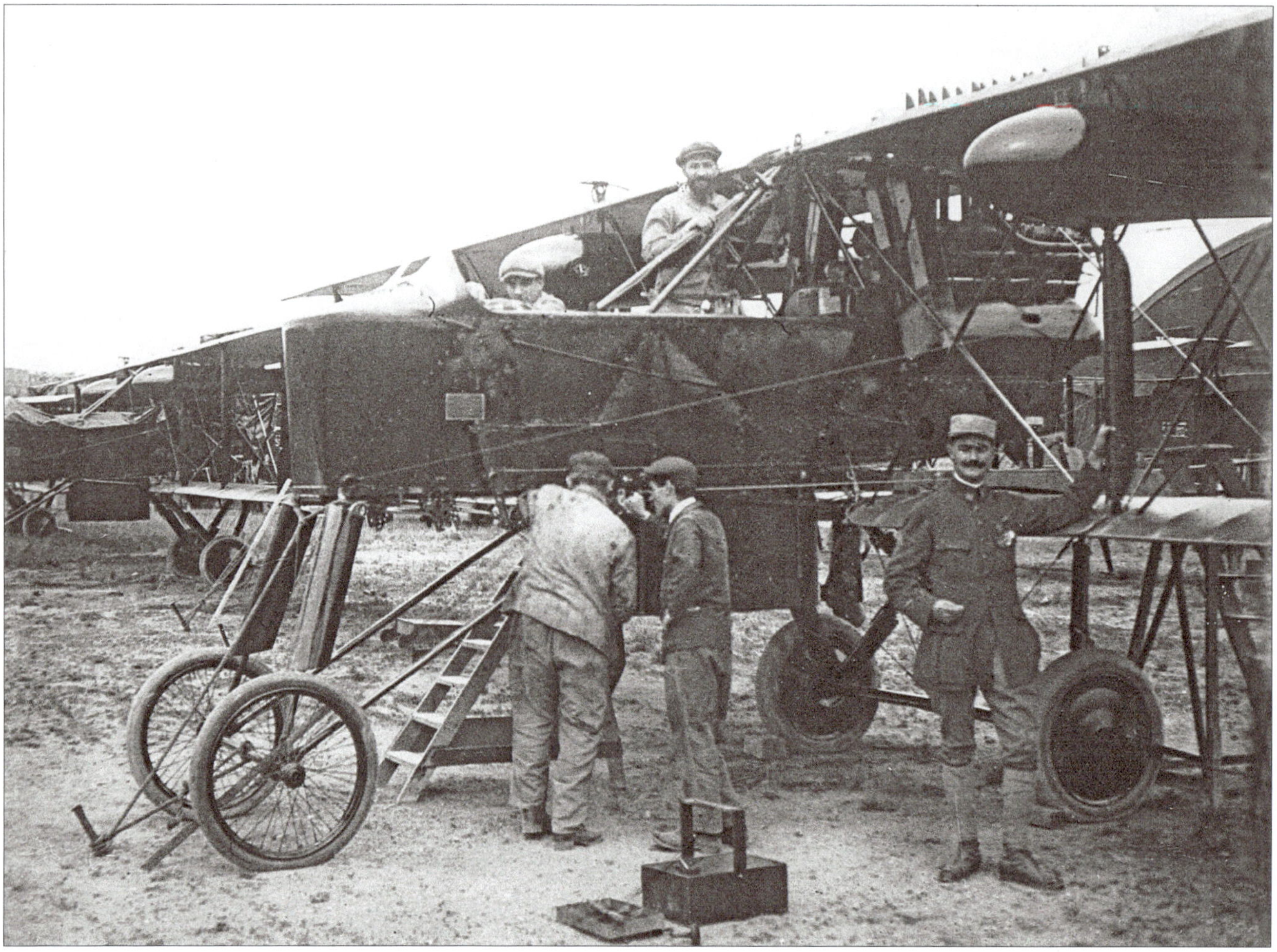

Above: William Thaw beside the wing of a Voison night bomber.

William Thaw was born in Pittsburgh, Pennsylvania to a father who was a wealthy businessman and a mother who was an heiress to her family's fortune. He was thus lucky enough to attend some of the finest schools in America and was at Yale University for his second year of study in 1911 when he left to pursue aviation.

HIs family's wealth provided him with the funds needed to learn to fly and earn his wings in 1913 at the Curtiss School of Aviation at Hammondsport, New York. He then purchased a Curtis flying boat and entertained the social elite by giving them rides during that summer while residing at his family's "cottage" in Newport, Rhode Island.

Thaw made a name for himself when he made a flight in New York City in which he flew under the four East River bridges and then flew around the Statue of Liberty. In the spring of 1914 he went to Europe and earned the Aero Club of France F.I. Hydro License No.2 He earned the sobriquet of "playboy of the Riviera" after he brought his plane there and again flew the wealthy for joy rides along the coast.

With the outbreak of war Thaw presented himself as a volunteer to France and offered his services as a known, trained seaplane pilot. A grateful France accepted his aircraft, but turned down Thaw as a pilot.

Undaunted, William Thaw immediately enlisted into the French Foreign Legion as an infantryman, however Thaw never wavered in his desire to fly for France. He met several other American volunteers in the Foreign Legion and rather quickly, in October 1914, Thaw developed an idea of forming an all-volunteer, all-American escadrille to fly for France and for the cause of "humanity".

Thaw himself was able to persevere and joined Escadrille D.6, commanded by Lt. Brocard and was

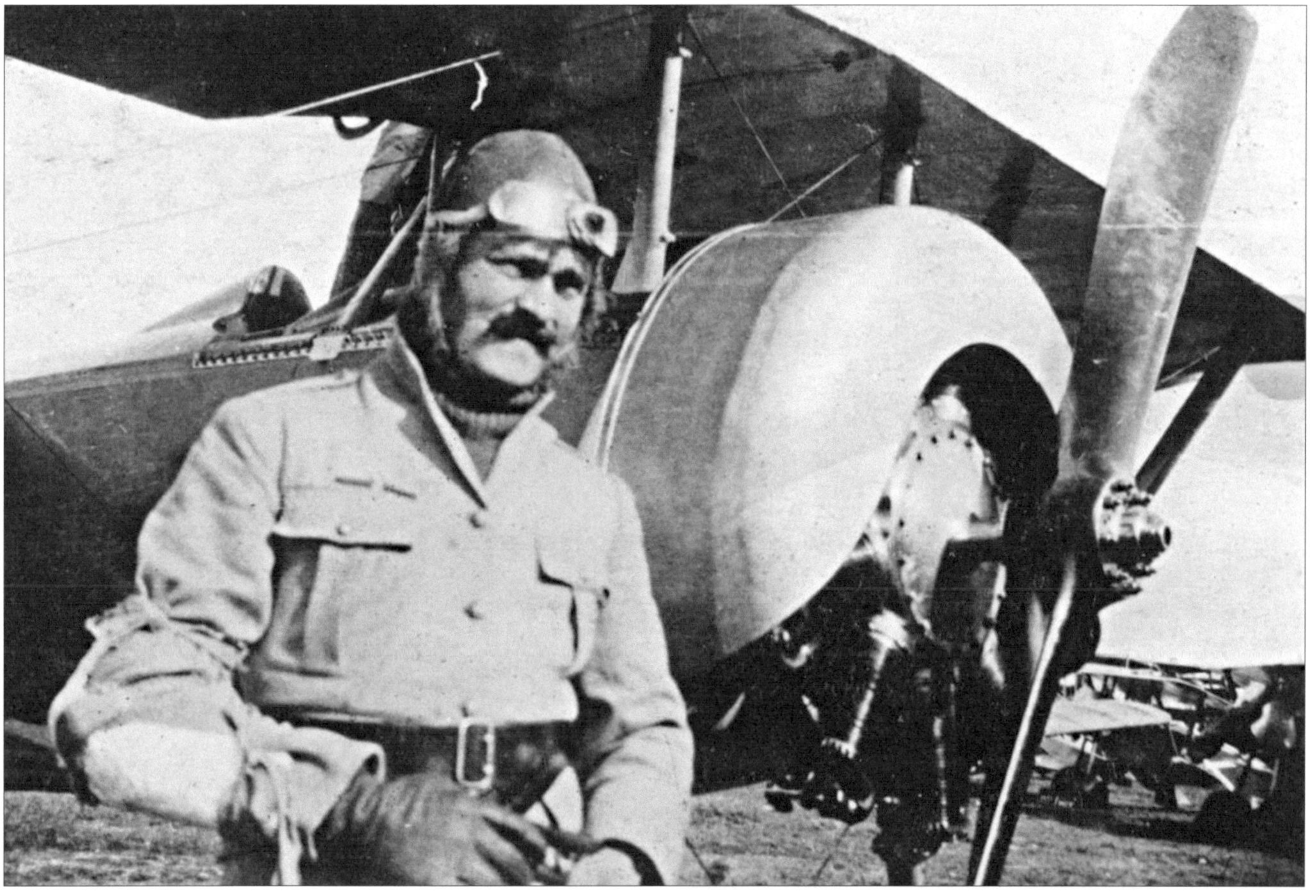

Above: Thaw preparing for a flight, note the bandages on his right arm-likely dating this photo to sometime not long after his dogfight in which he was shot down behind enemy lines.

Right: Thaw returned from a leave in Paris with a new squadron mascot, a lion cub from the Paris zoo, he was given the name whiskey by the pilots.

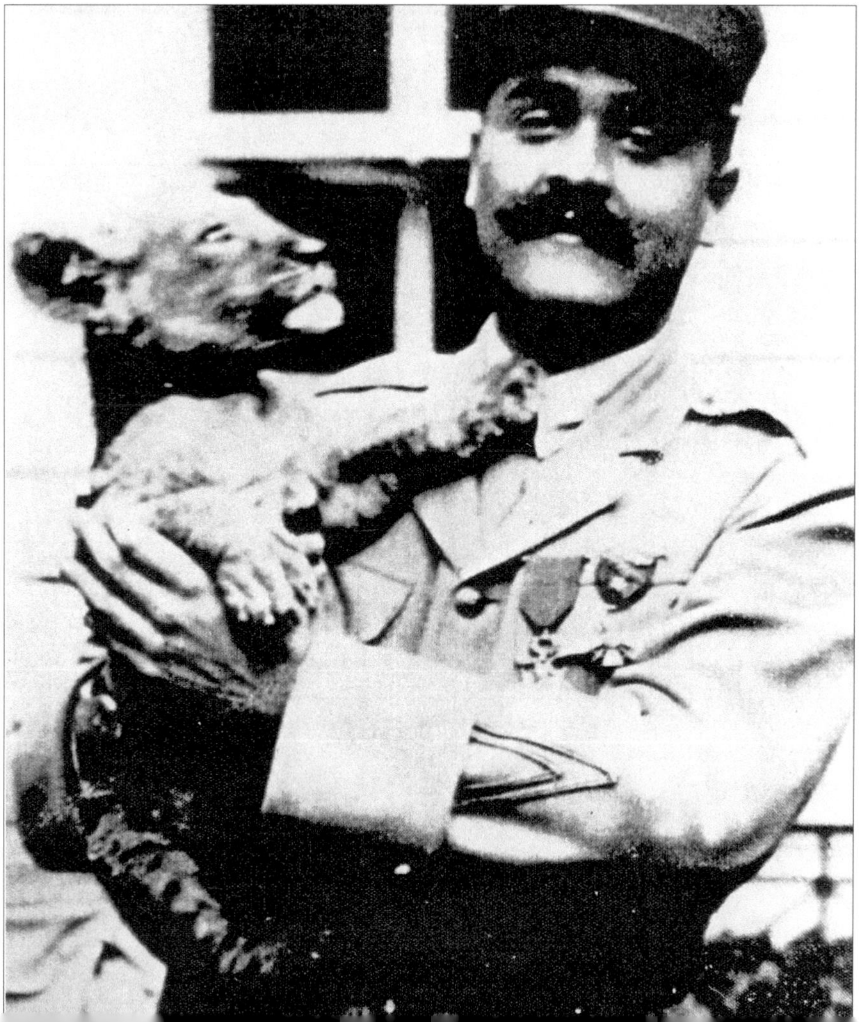

therefore the first American to serve at the front lines as a pilot in the French Air Force. He would be transferred to Escadrille C.42, and served with enough distinction there to earn the Croix de Guerre with three palms. In March, 1915 Thaw became the first American to receive his brevet militaire(no. 714).

Thaw was proving by his service and his courage that the idea of an American flying and fighting for France was potentially a good idea. Unknown to Thaw, there was another American who

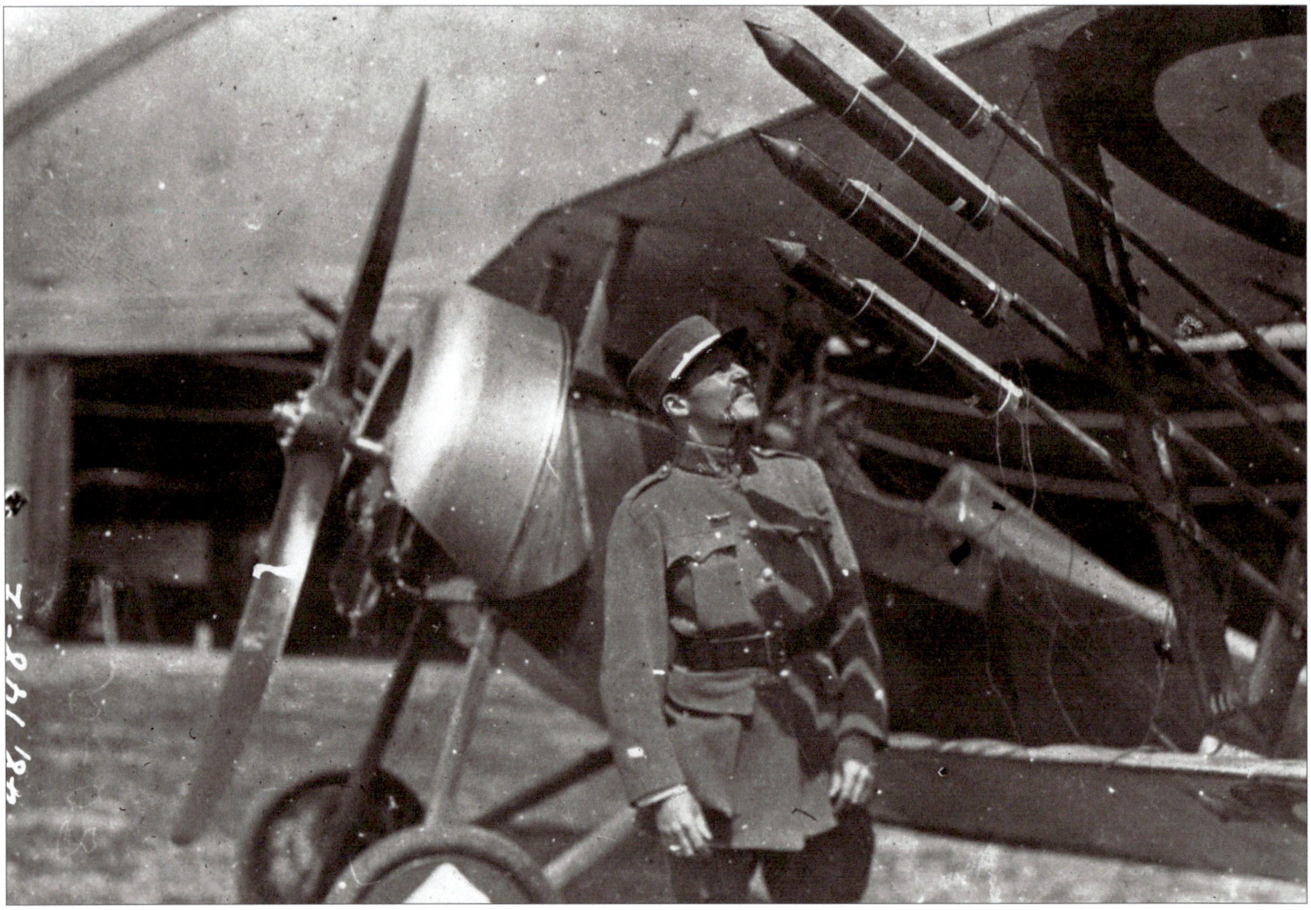

Above: Norman Prince examines the Le Prieur rocket system attached to a Nieuport fighter.

Below: Thenault, commander of N.124, called this a "tragic photograph". From L to R: McConnell, Rockwell, Thenaualt, Prince, and Chapman in April, 1916 at the beginning of the escadrille. Within a few months, all four American pilots would be dead.

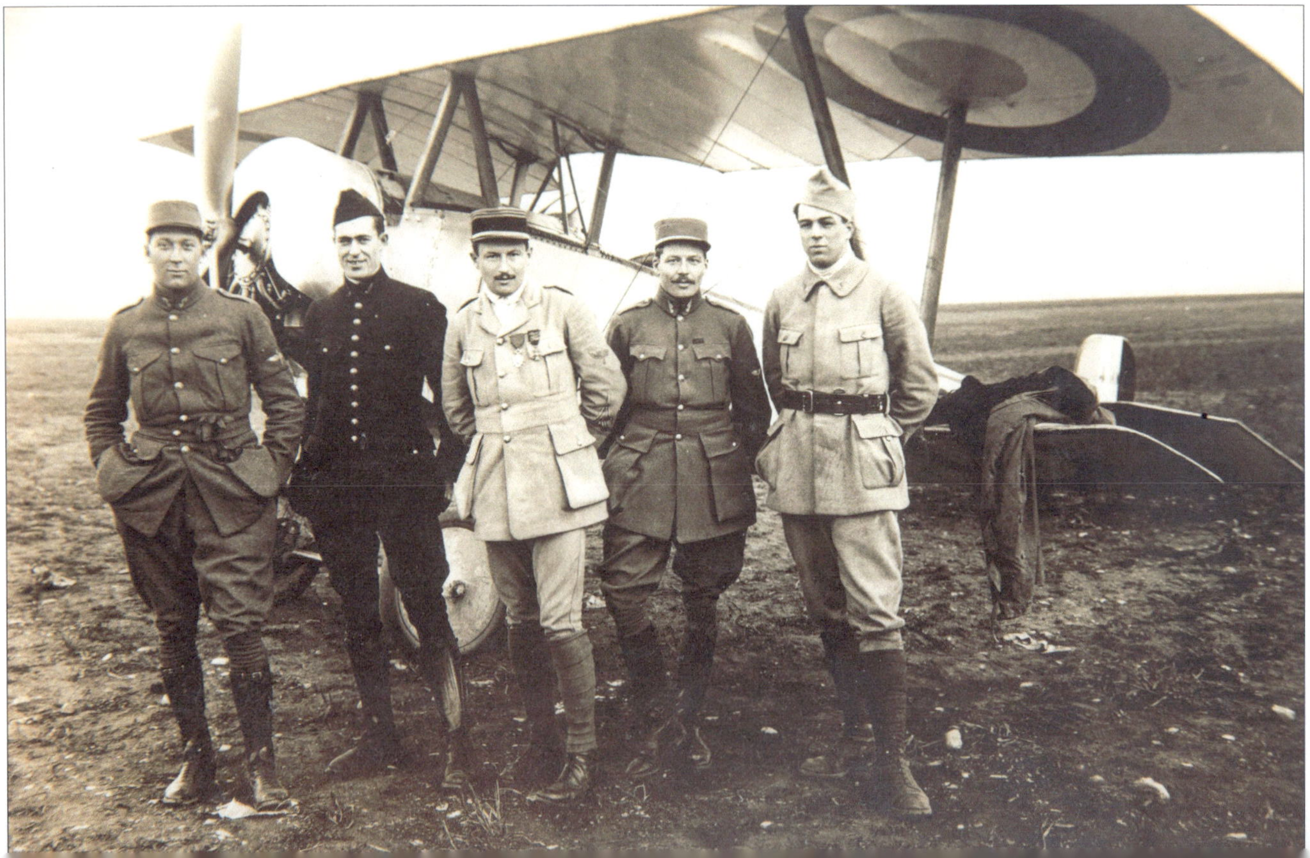

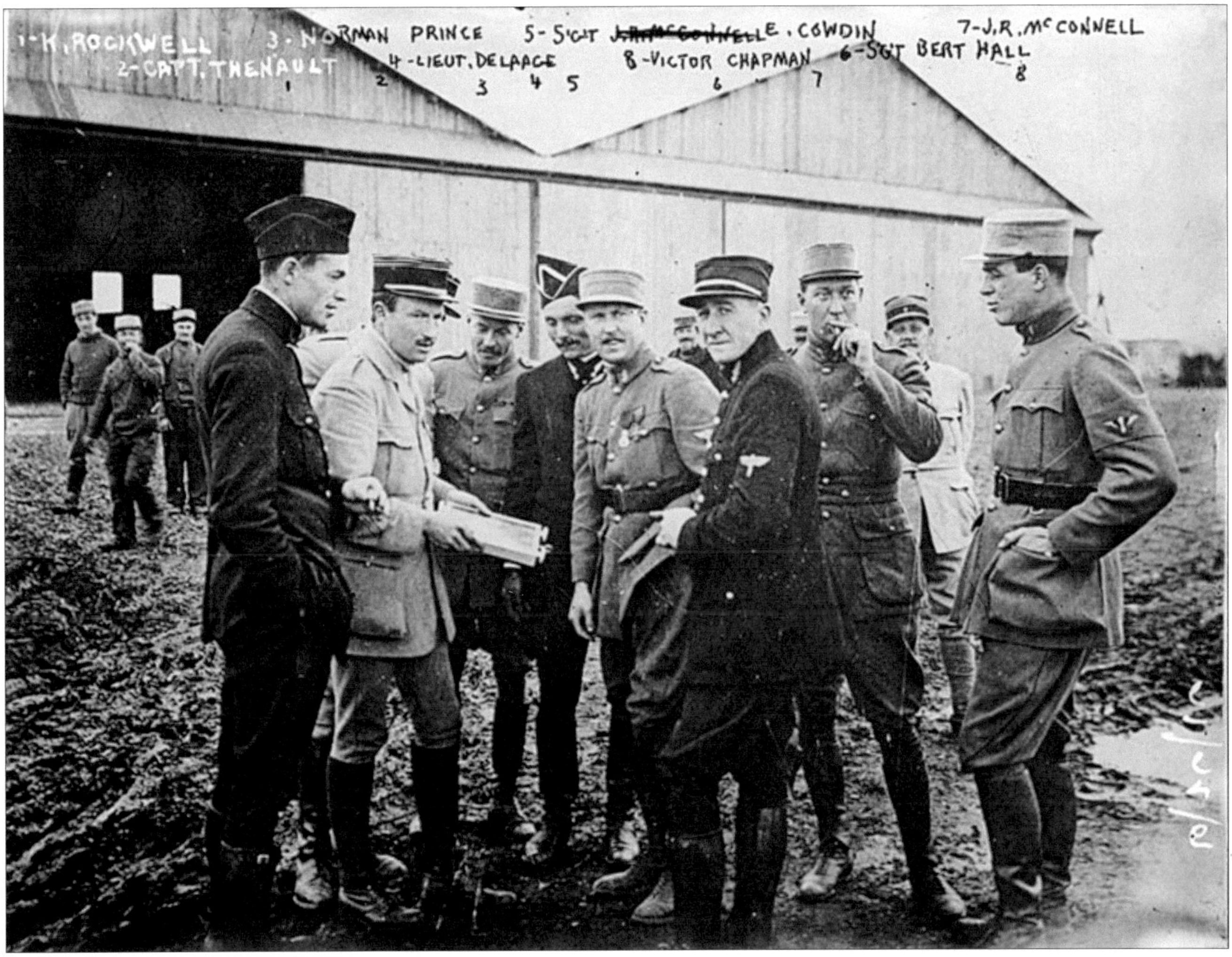

Above: A photo that shows the original seven members of the escadrille N.124, L to R: Rockwell, Thenault, Prince,Lt. De Laage(French second in command),Cowdin,Hall,McConnell, Chapman. Thaw is directly behind Thenault. Luxeuil, May, 1916.

also fervently desired to help make an escadrille of American volunteers a reality.

Norman Prince was born in Prides Crossing, Massachusetts: the son of an extremely wealthy investment banker and owner of the enormous Chicago's Union Stockyards. In addition to homes in the United States, the Prince family had homes in Biarritz and Pau, France. Prince's father was the "master of the hounds" in Pau and his son considered France a second home.

Norman Prince graduated from Groton, Harvard and then Harvard Law(1911) before re-locating to Chicago to practice law.

However, he had caught the flying bug and under an alias, since his father did not approve, earned his Aero Club of America certificate(no.55). He sailed to France in January, 1915 to offer his services to France, but it was not until May 1915 that he was assigned to V.B. 108 along with another American volunteer Elliot C. Cowdin.

Prince, with the help of his father, would eventually be able to bring pressure in the "right" places, because of his family's wealth and high-level connections in order to bring about the American volunteer squadron.

Initially, the French authorities were very reluctant to allow such a squadron to be formed, American authorities were equally skeptical because of America's official neutrality status. Pro-German groups in America protested any hint of America actually choosing a side in the war and avowed neutralists were worried about getting dragged into a European war.

In December, 1915 Thaw, Prince and Cowdin sailed back to the USA while on a 30 day leave from their duties. The trio were besieged by the press and were feted as brave adventurers and heroes. The fact that so much attention was paid to them and that

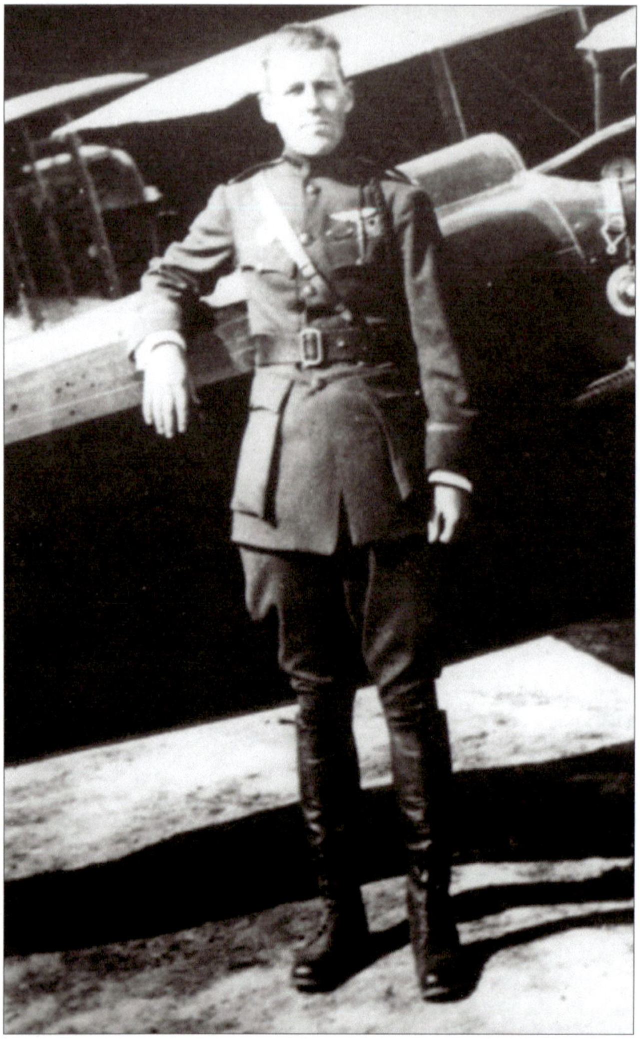

Above: David Endicott Putnam of Jamaica Plains, MA was a descendant of a Revolutionary War hero. He left Harvard College to fight for France. He would initially fly in combat from December 1917 with MS 156, then Spa. 38 and finally as a flight leader in USAS 139 squadron, With 13 confirmed victories he was the second highest scorer of the LFC. Like Lufbery, he likely had many more victories, perhaps over 30 that were unconfirmed. He wrote in a letter to his mother: "Mother, there is no question about the hereafter of men who give themselves in such a cause. If I am called upon to make it, I shall go with a grin of satisfaction and a smile."

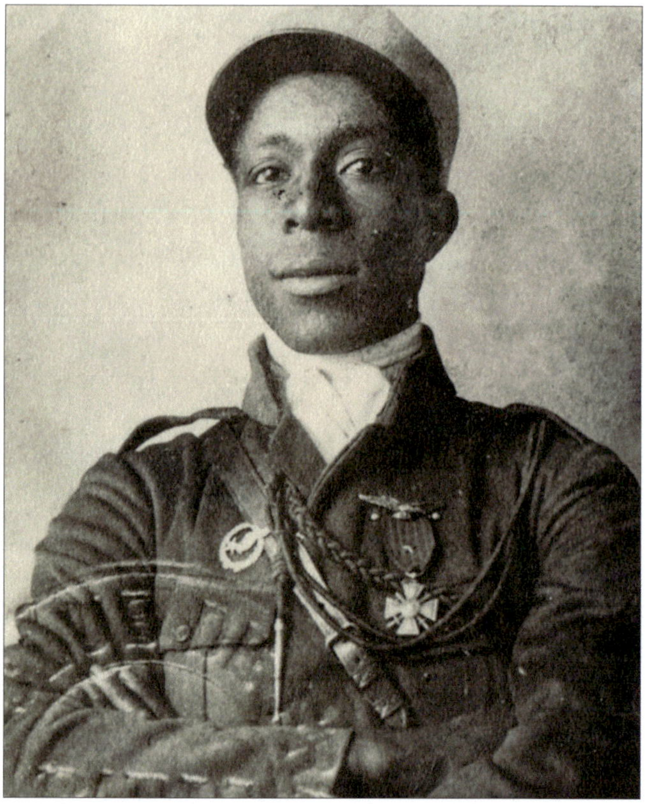

Above: Perhaps the most unusual member of the LFC in WWI was Eugene Bullard who became the first black military aviator in history when he earned his brevet and joined Spa. 93 His French comrades called him "l'Hirondelle noire" the Black Swallow. Shamefully, Dr. Edmond Gros, who was the head of the LFC wrote a letter that discouraged the USAS from accepting Bullard because he was a black American. Further, Gros denied Bullard's certificate of LFC membership after the war because of his bigoted views.

the attention was almost uniformly in a favorable light was not lost upon the French authorities.

Both Thaw and Princes's fathers began to take up the cause to help convince the French authorities to develop an American escadrille and both, especially the elder Prince, brought considerable weight to the efforts. Prince himself was offering some of his lands in France and significant amounts of money to the French cause.

In addition, his friends such as Vanderbilt and Morgan brought hundreds of millions of dollars to the French side of the war and were critical in moving the slow machinery of the authorities to make things happen. They also combined to set up generous amounts of funds that would be allocated for the American volunteer aviators who would be serving.

A system was established in which a candidate's passage from America was paid for and the volunteer flyer would receive a generous monthly allowance once he was officially enrolled in the aviation service. There was also "bonus" money paid for every confirmed aerial victory and for every military decoration that was awarded to the aviator.

Upon arrival in France, a very wishful candidate would be examined by Dr.Edmund Gros and if the physical exam was passed, the candidate would then officially join the French Foreign Legion via a special form which allowed for aviation service

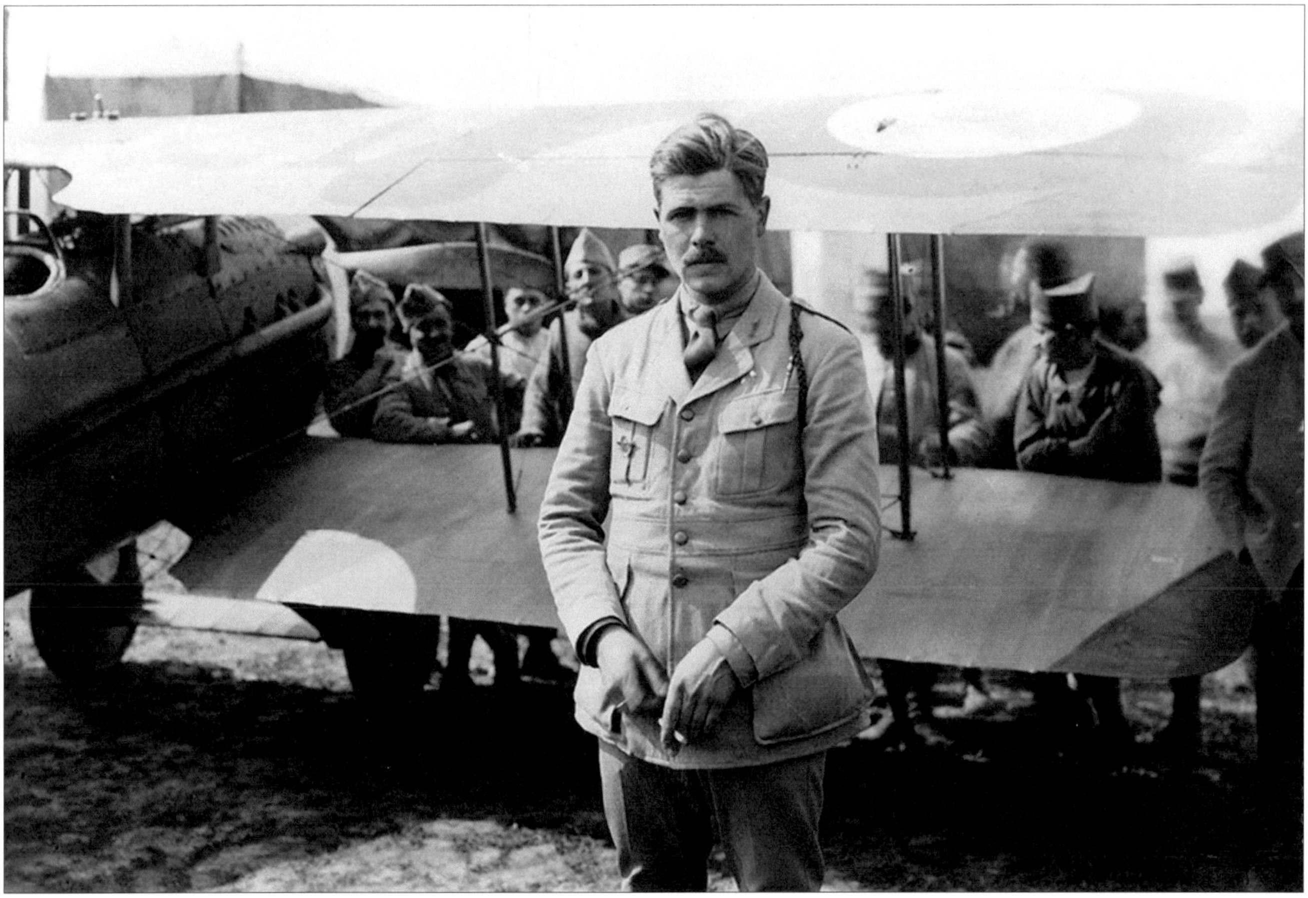

Above: Another common way for American volunteers to make their way into the LFC was via the American Ambulance Service. Frank Baylies of New Bedford, MA was such an individual. He served in the AAS from Oct 196 until May 1917. He then transferred to the Air Service and after training was assigned to the famous Spa 3 escadrille. He would score 12 victories between November 1917 until his death in June 1918. When members of the LFC were being transferred to the USAS, he declined a captaincy and instead accepted a 2LT. rank in the USAS so that he could stay and fight in Spa3.

only. In other words, if the candidate was not able to complete aviation training for whatever reason, that person would not be obliged to serve in the infantry or any other branch of the Foreign Legion.

By June 1916, the Franco-American Flying Corps was up and running with more than enough cash on hand to handle all its expenses. Eventually this group would become the Lafayette Flying Corps. Although final, exact numbers vary depending on the source approximately 250 Americans would join this corps.

Of these aviators, 179 men served on the Western Front and fully 25% of them would die in the course of their service to France. Of these volunteer aviators only 38 would serve in the actual escadrille Lafayette. The others would serve in French units, all of the aviators were considered part of the Lafayette Flying Corps.

Another important contribution made by the LFC was that 128 of its members would transfer to the USAS once America entered the war and provided the nascent air corps with a good number of experienced leaders to help the fledglings prepare for combat.

In the beginning though, there would be only seven pilots joining N.124, escadrille Americaine, in April, 1916: Kiffin Rockwell, Norman Prince, Elliot Cowdin, James McConnell, Victor Chapman, Bert Hall and William Thaw.

The squadron's name would change several times, initially complaints were made by German authorities regards the name and the official neutrality of the United States. Therefore the squadron was re-named in November 1916 to escadrille de Volontaires. However the aviators thought this name was far-too bland and so in December of 1916 ,N.124 was given its final name of escadrille Lafayette(in recognition of the famous Frenchman who came to the aid of the United States during its war for independence).

Many, if not most of the young Americans who would serve in the Lafayette escadrille or the

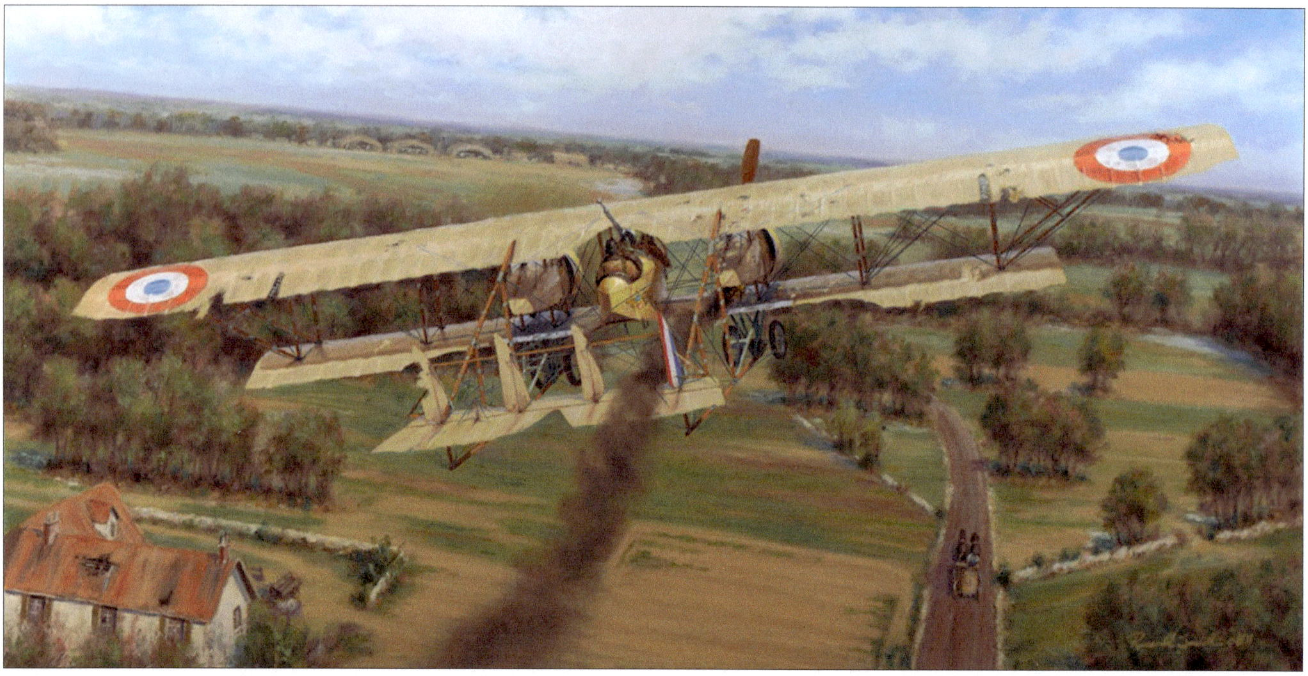

Above: William Thaw manages to bring his seriously damaged Caudron observation aircraft to the home airfield of his escadrille, C.42. Russell Smith

Lafayette Flying Corps were from a unique and privileged group of individuals. They were college graduates or came directly from college campuses at a time when only a minority of American males could attend college. In addition, most of the early volunteer pilots were from the east coast and a majority of them from Ivy League colleges.

Thus, the overall impression of the average squadron member was that of a young well-educated man from a privileged background.However, even at the very beginning there were exceptions to that observation. And the greatest fighter pilot to emerge from the escadrille Lafayette would certainly not fit that description.

Upon his return to France, Sgt. William Thaw returned to his escadrille C.42 until sometime in early February, 1916. He then left to join the newly formed unit of American volunteers. He did not come alone, the commander of C.42- Capt. George Thenault- came along as well, presumably because of a suggestion for the post by Thaw.

Thaw would join in the squadron's first combat patrol on 13th May,1916 which was led by Sgt. Kiffin Rockwell. The following week Thaw was able to shoot down a Fokker E-type, but on a second patrol that same day was engaged by a numerically superior force and Thaw was momentarily forced to disengage because of a jammed machine gun. He was set upon by several German fighters and had his left arm broken by machine gun fire. His gas tank had been punctured, so that gas poured out over him and as the gas evaporated caused his skin to burn. Thaw was able to crash-land his aircraft behind enemy lines. His wound did not keep him away from "his" escadrille for very long.

Thaw had been promoted to a full Lieutenant rank, however it was the force of Thaw himself, the strength of his character which led him to be the unofficial American commander of the escadrille which was, of course, officially led by its French commander Capt. Thenault(whom Thaw knew quite well).

Edwin Parsons(fighter ace and member of N.124) would later write:

"Most of the time he was in practical, if not nominal, command and handled with unfailing tact and good humor one of the toughest jobs ever thrown on the shoulders of one man. It is reasonable to suppose, since he seemed to be doing all right before there was any such thing, that he could have gotten along without the escadrille. But most certainly the escadrille could never have gotten along without Bill. Many times he must have been seething inside, but he never showed it. He made a brilliant success of a job where most men would have gone mad."

The pre-eminent escadrille Lafayette historian, Dennis Gordon, had this to say of Thaw:

"William Thaw was revered by the pilots who served with him. And for good reason. He had an innate ability to lead men....He led the best way possible, by winning the pilots' devotion and respect.

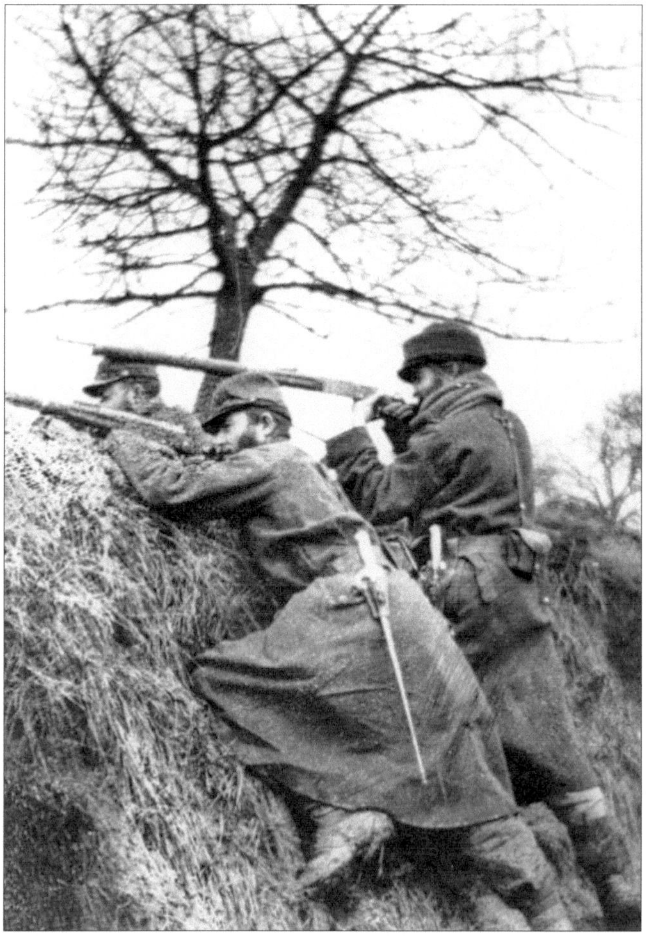

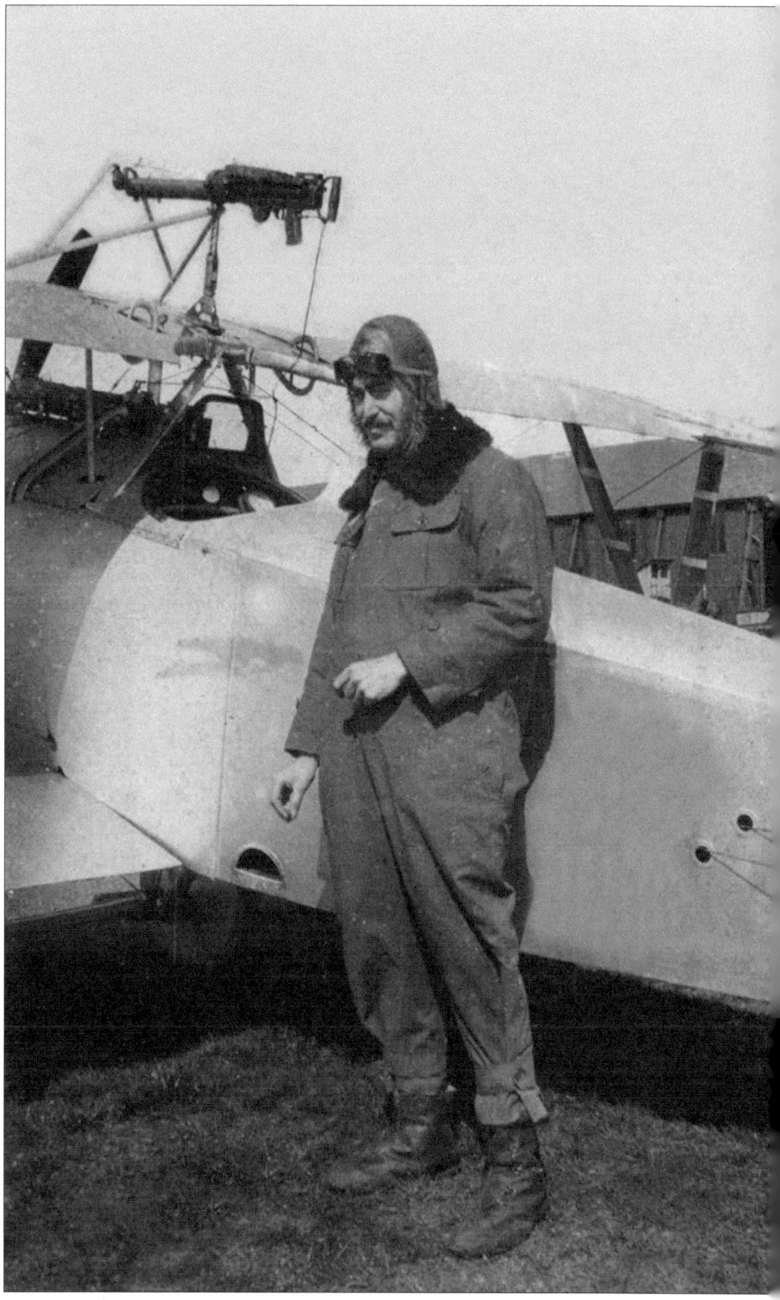

Above: American volunteers in the Foreign Legion fight in the Aisle trenches, December 1914. L to R: Dennis Dowd, Kiffin Rockwell, Charles Trinkard. All three would eventually transfer to the Lafayette Flying Corps, and all three would be killed in the war while serving France.

Right: Kiffin Rockwell ready for a flight in his Nieuport 17 fighter.

Thaw never asked them to do anything which he was unwilling to do himself. And his heroic record bore this out."

Unlike Thaw, the other founder of the escadrille, Norman Prince would not survive the war. He also did not seem to generate the universal affection that Thaw did amongst his comrades. In fact, Prince would have an ongoing bitter dispute with Kiffin Rockwell.

Still, Prince was very active in the initial life of the escadrille and managed to score four official victories before his death following a landing accident in which his undercarriage caught on unseen telegraph wires and Prince was severely injured in the crash: he succumbed to complications of those injuries.

Kiffin Rockwell was an idealistic young man who perfectly fit the established mold for the dashing, heroic fighter pilot. He had attended the Virginia Military Institute before he enrolled at Washington and Lee University. Upon the outbreak of war, he and his brother Paul sailed to Europe to fight, he believed that the cause of France was the "cause of humanity". He joined the French Foreign Legion and fought in the trenches.

Both he and Paul would be wounded, Paul's wound was such that he could no longer serve in the military and Kiffin's wound was such that he could no longer serve in the infantry, however he could carry on fighting if he transferred into the air service.

As stated earlier, Rockwell was given the

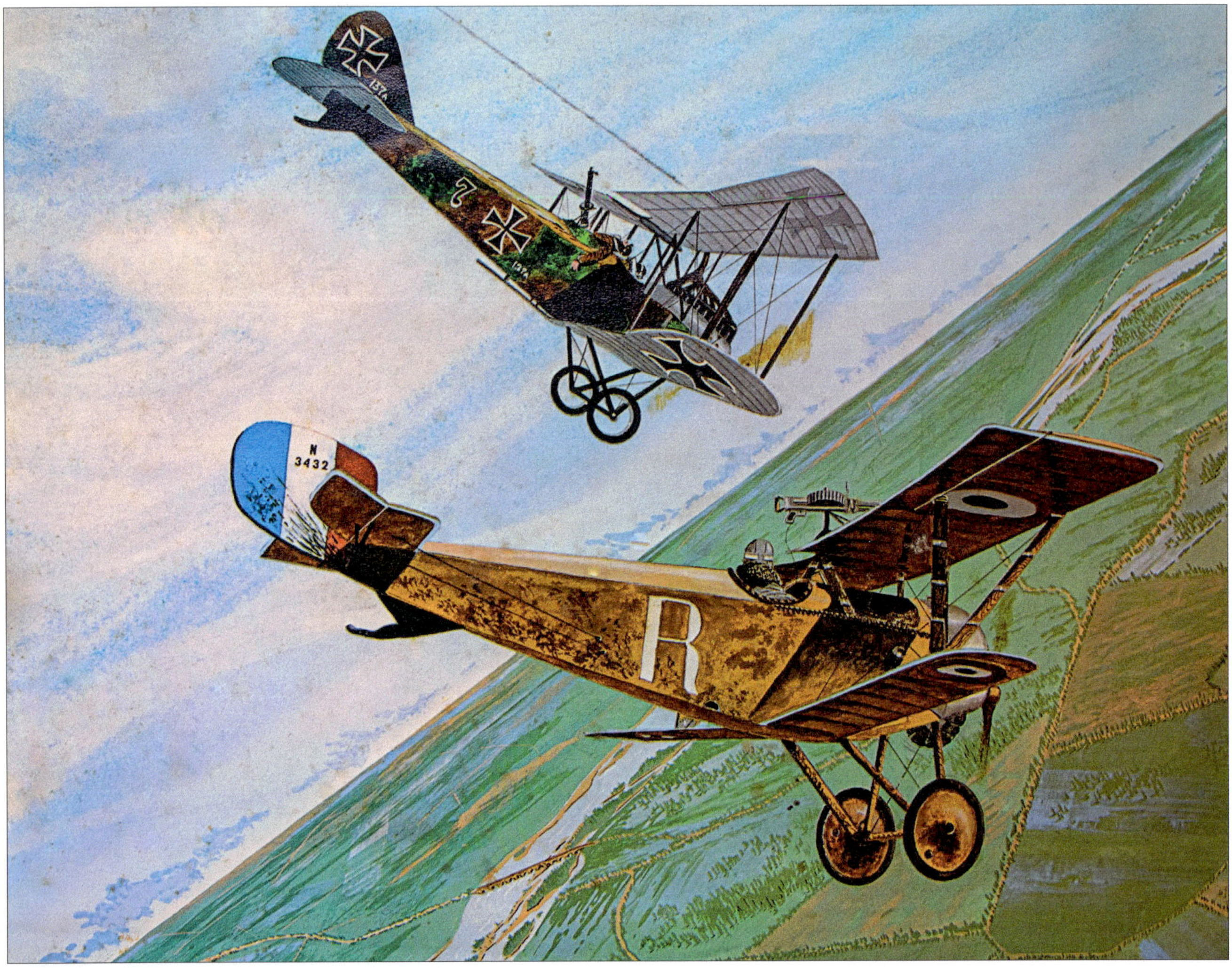

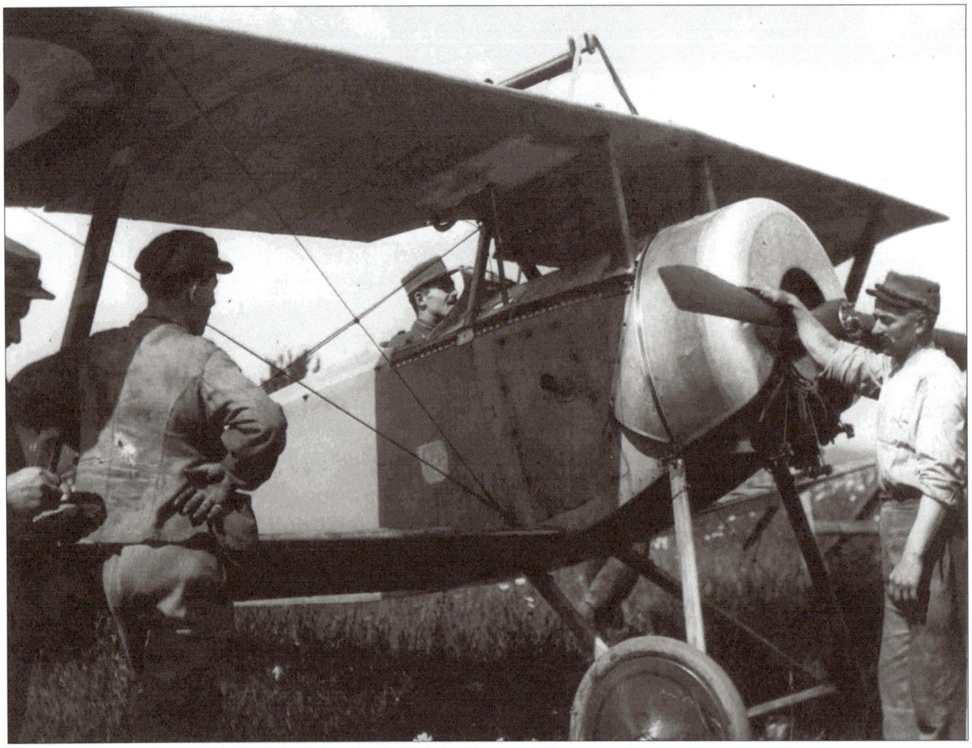

Above: Kiffin Rockwell scores the escadrille's first officially confirmed victory after only four shots. Robert Carlin

Left: Rockwell in the cockpit of a Nieuport 11 fighter working with mechanics to adjust the Lewis gun.

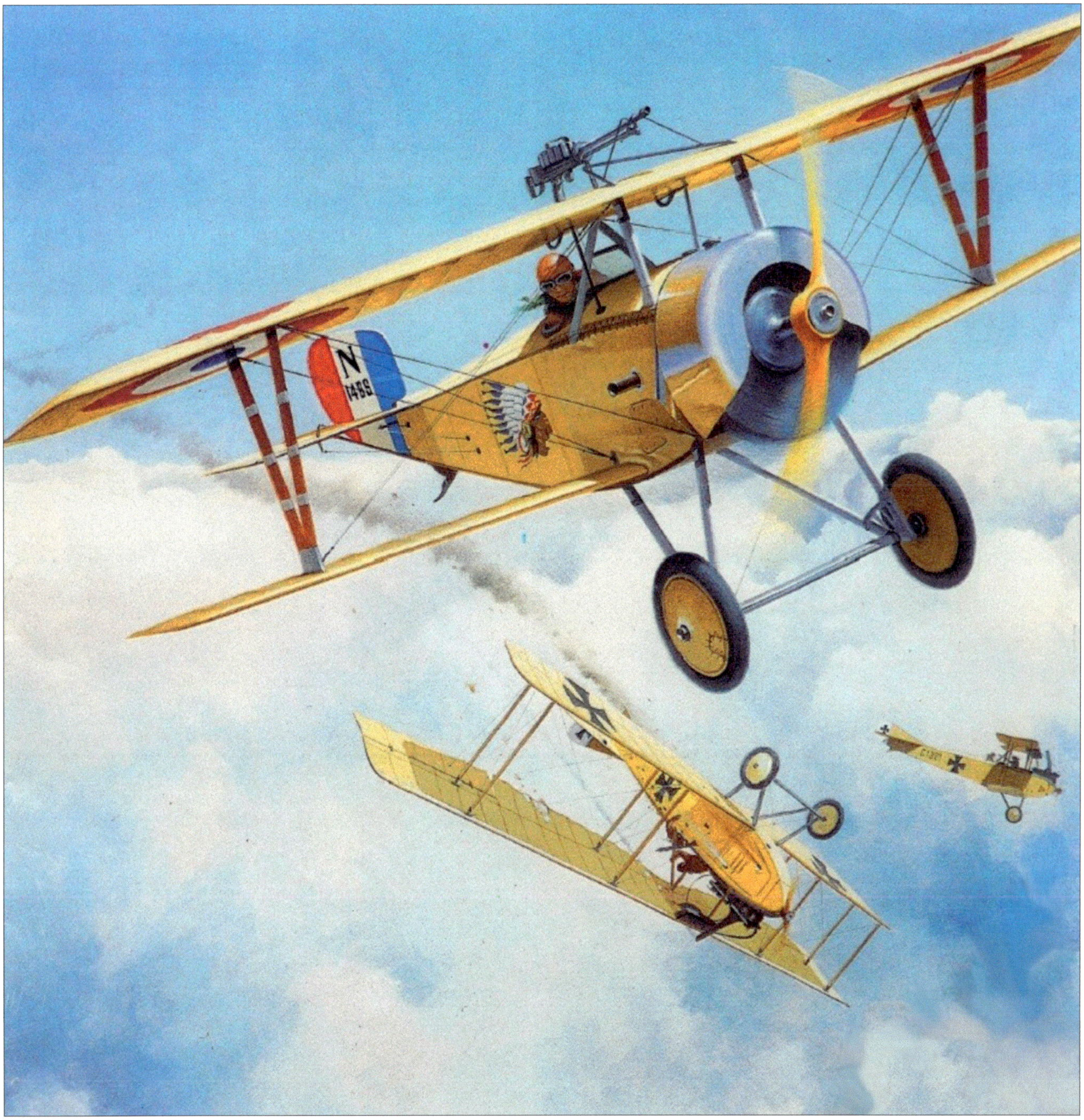

Above: Lufbery would often ignore odds and was an aggressive fighter in the sky. Here he downs a German two-seater from a flight of several. Brian Knight

honor to lead the first official combat patrol of the escadrille and then five days later, during his fourth patrol on 18 May 1916, Rockwell scored the first official victory of the escadrille. He closed in very close to a German observation aircraft and after only four shots, he had managed to kill both the pilot and observer. The aircraft crashed behind the Allied lines so that confirmation had already been called in to the airfield before Rockwell had returned. Rockwell seemed to always be in the air. In July he recorded 40 combats and in August another 34 combats. According to the historian Dennis Gordon:

" Rockwell continued to prove an excellent aviator. His comrades saw him as the soul of the squadron and greatly admired his idealism, his enthusiasm and the fearlessness he had demonstrated in the air on numerous occasions as he closed to within feet of his adversary before he tripped his machine gun."

However, Kiffin Rockwell was not happy and

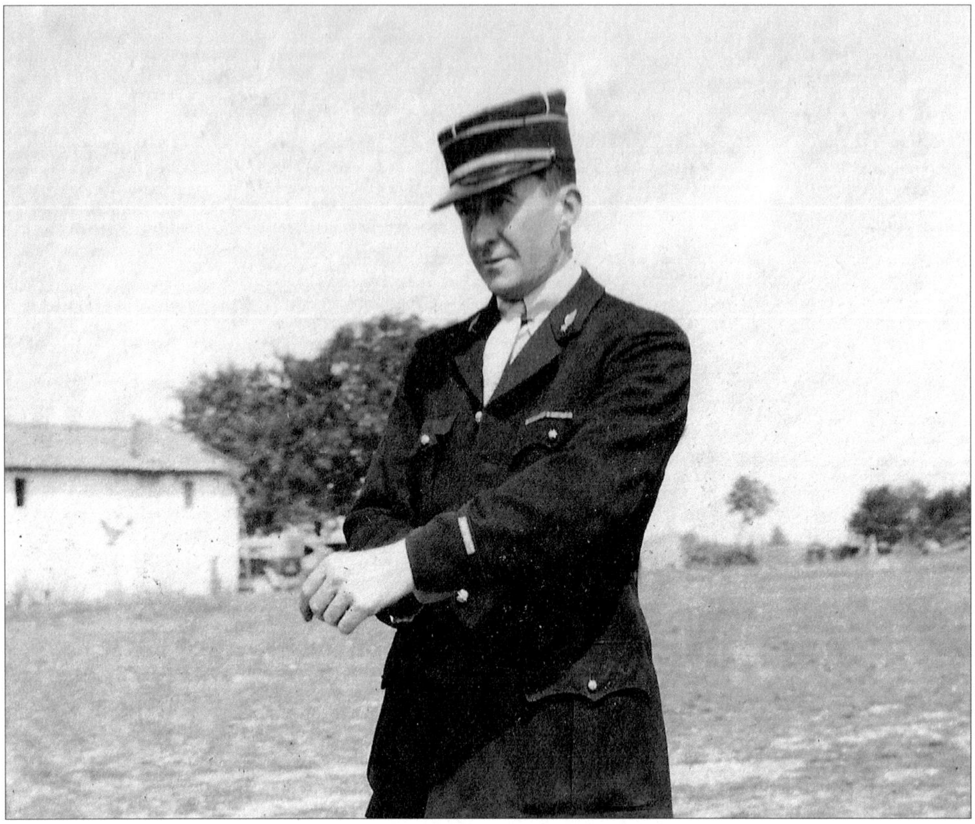

Left: Bert Hall did not fit in well with his comrades because of their dispirit backgrounds.

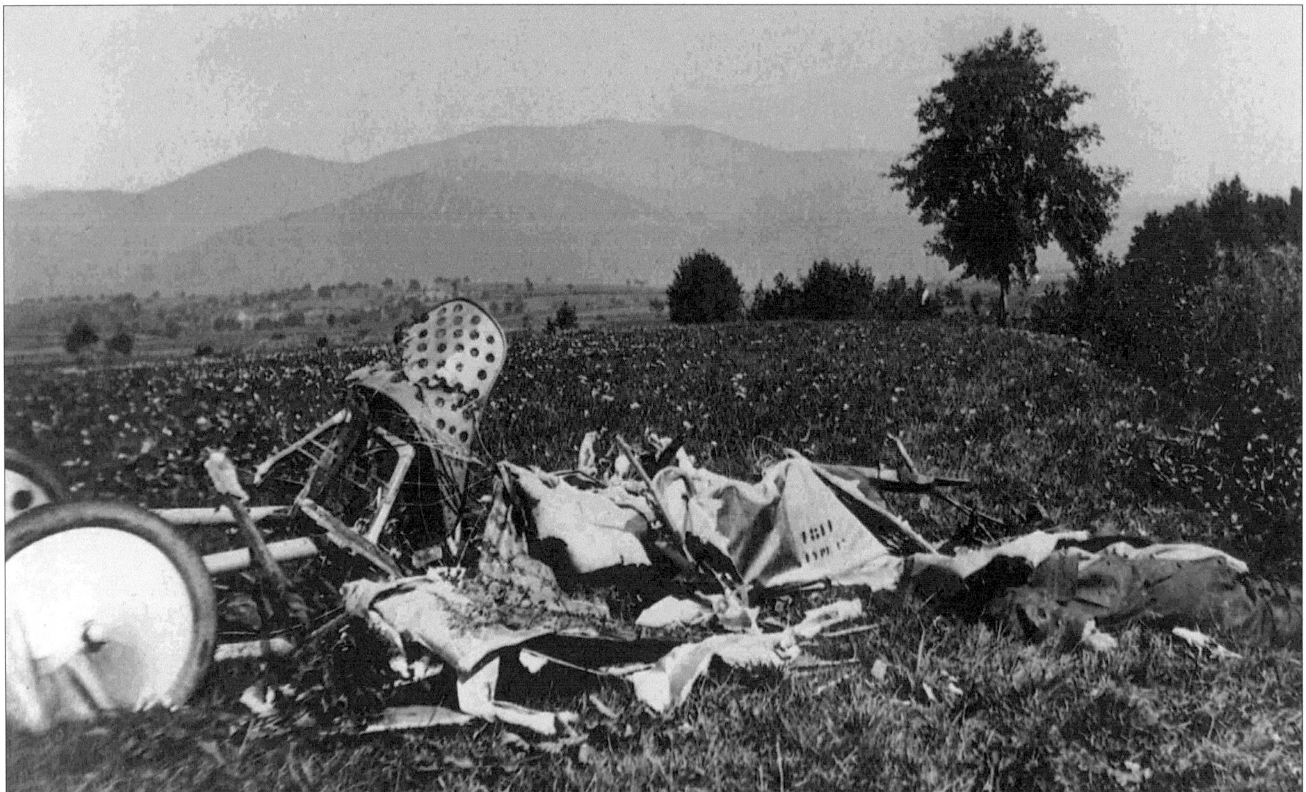

Above: Rockwell's fatal crash after he was shot down almost exactly four months after leading the unit's first combat sortie.

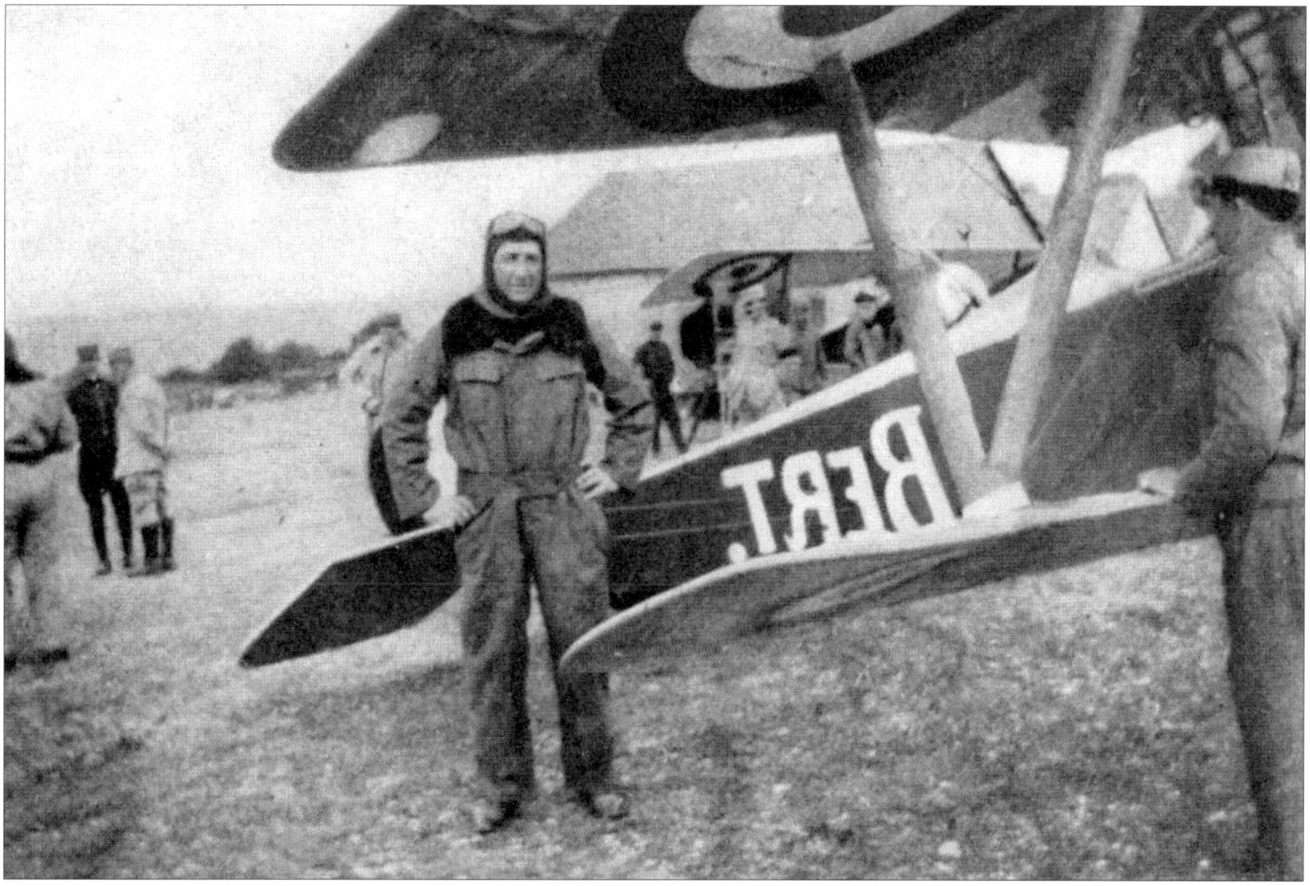

Above: Hall has just returned from a flight in his Nieuport which had his name BERT spelled forward on one side and backwards on the other side.

Below: Rockwell mounts the Nieuport fighter of his friend Bert Hall. Note the 'H' on the fuselage.

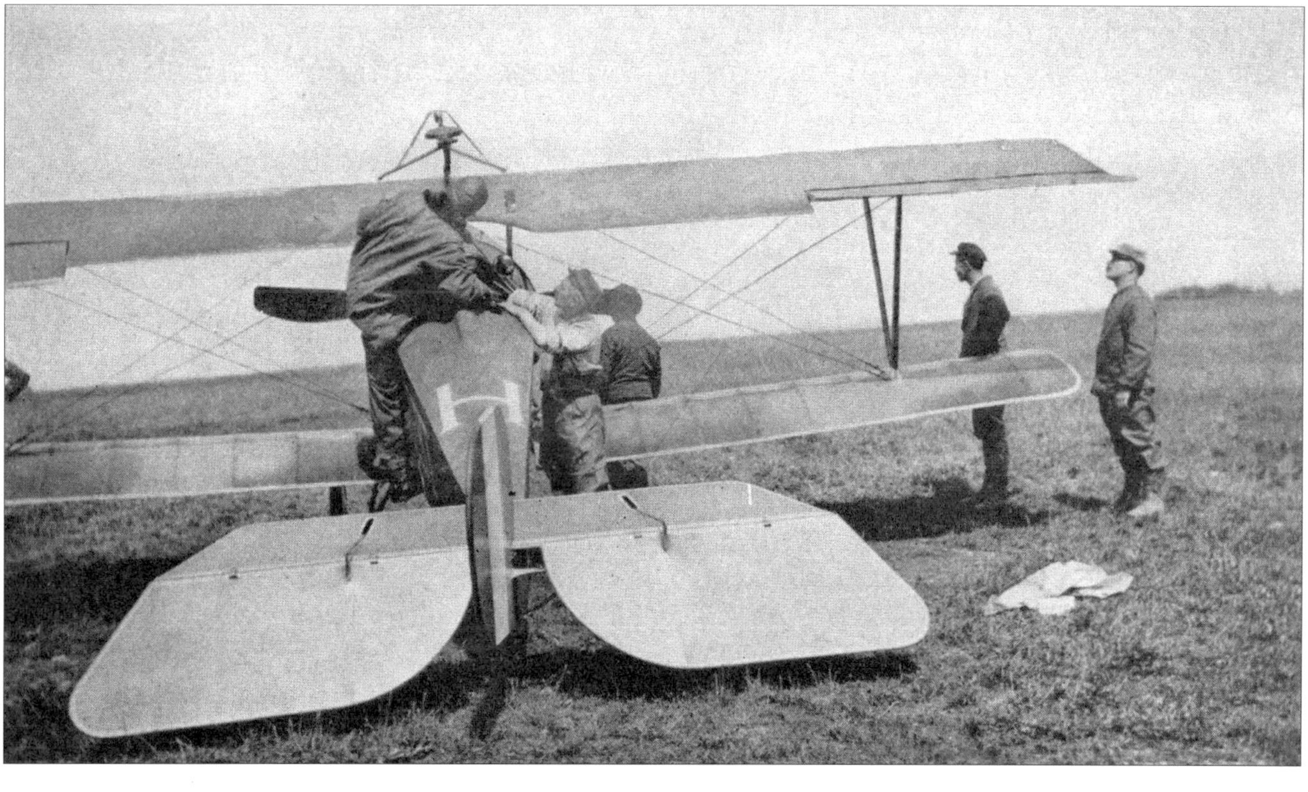

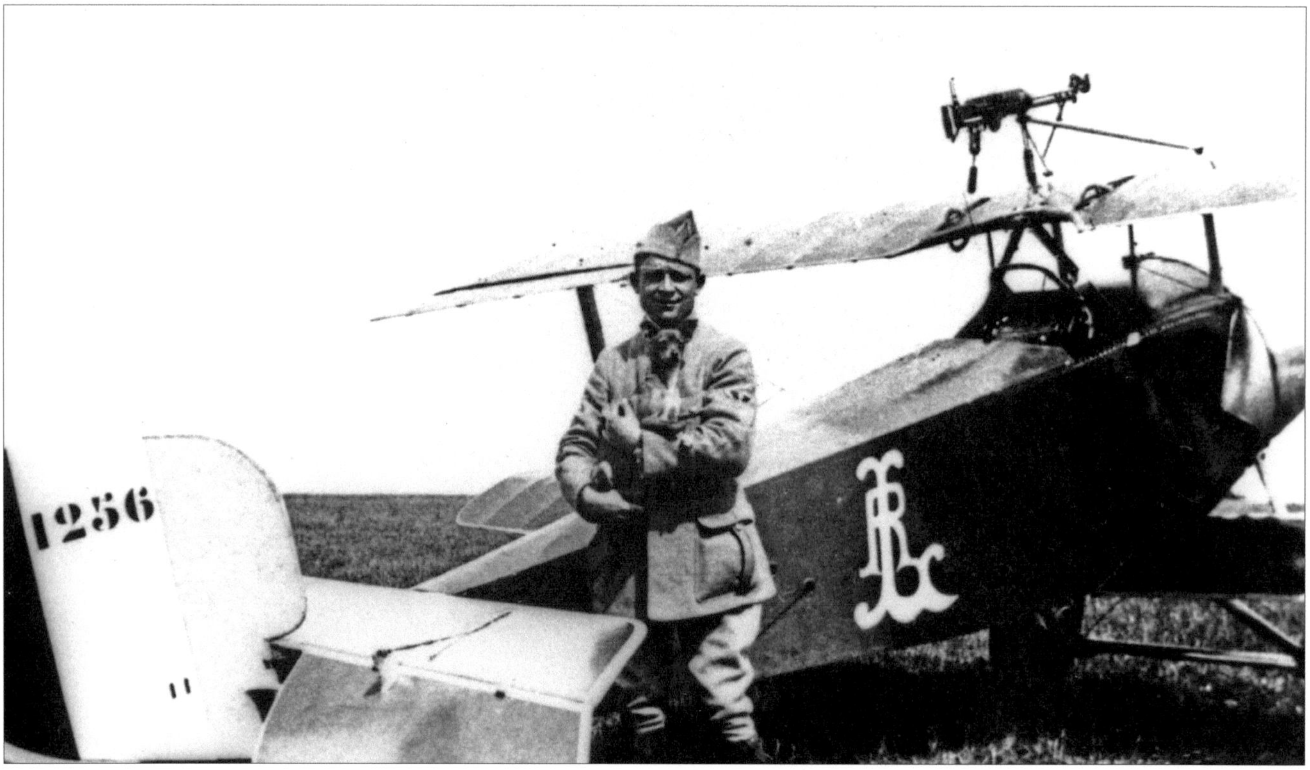

Above: Raoul Lufbery holds a dog of one of his comrades; there were numerous pets to be found on any WWI airfield. Note the stylized 'R and L' on his fighter.

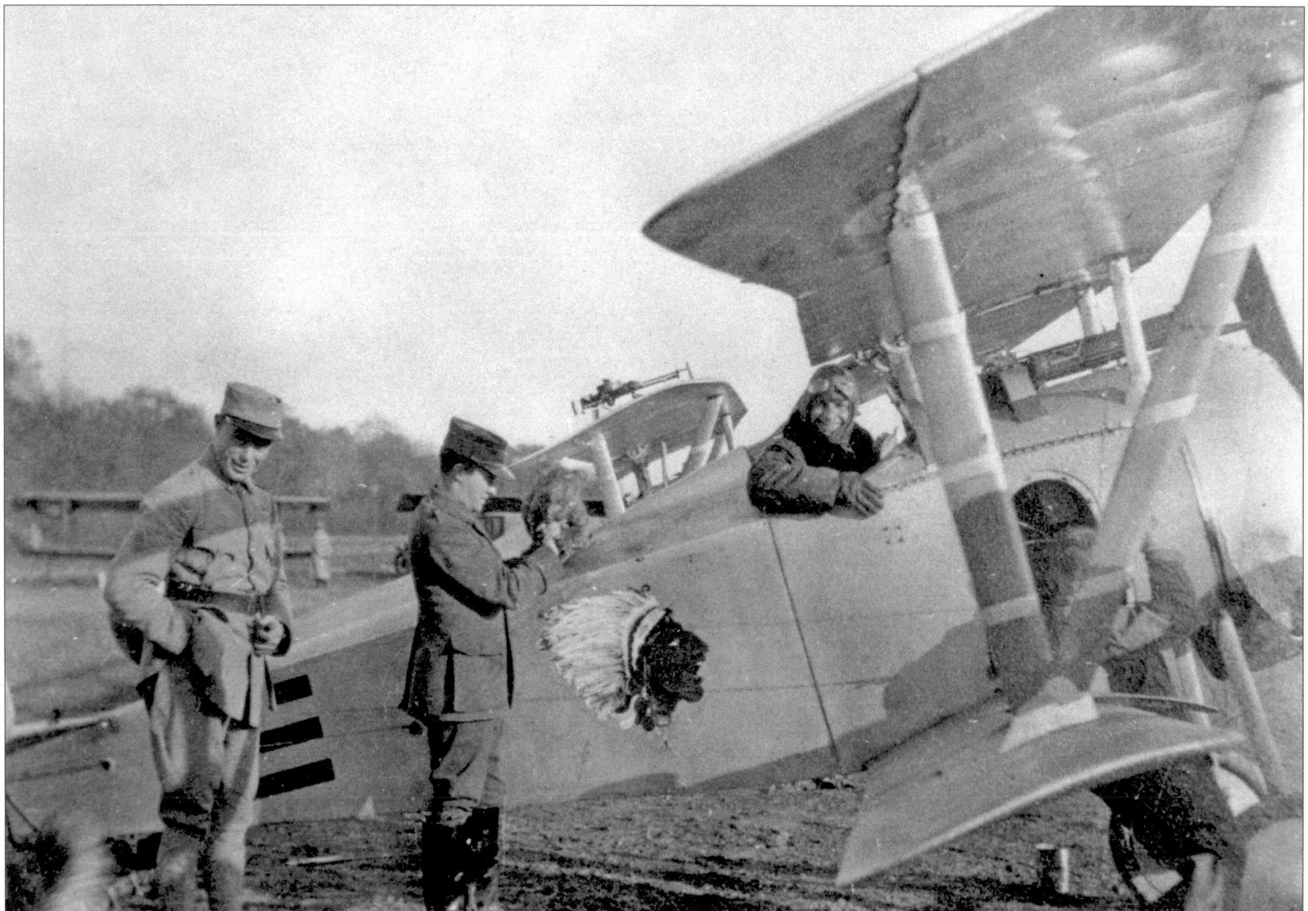

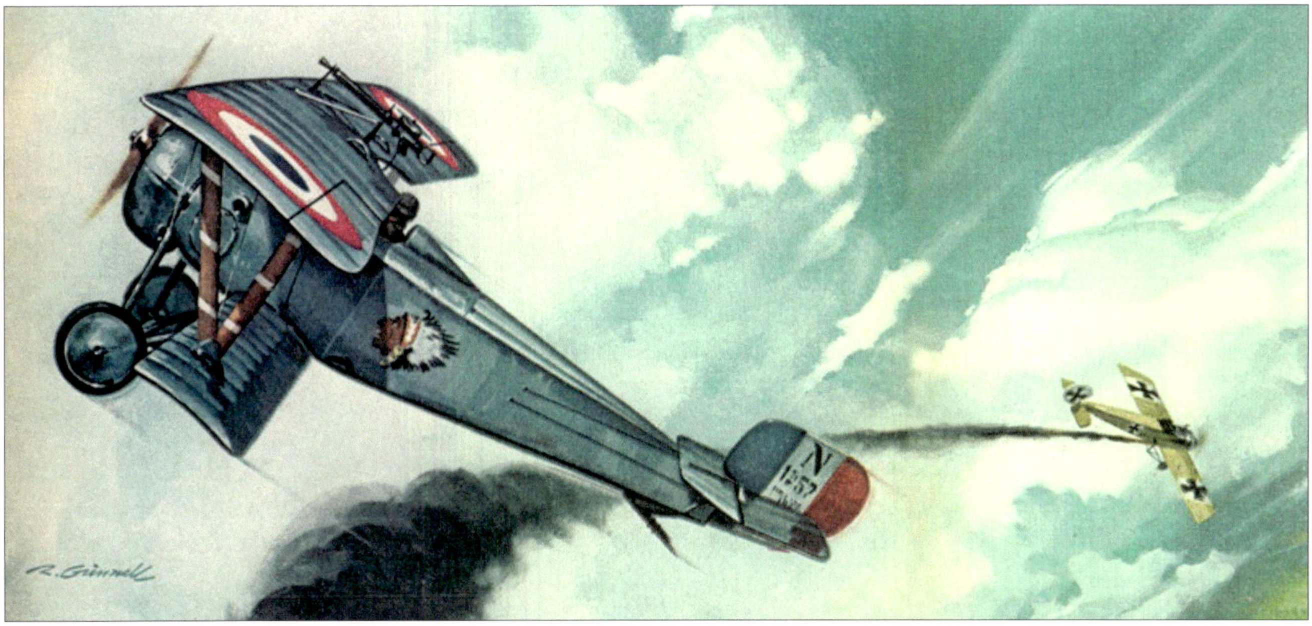

Above: Lufbery and his comrades battled German Fokker fighters over the Verdun battlefield. Roy Grinnell

Facing Page, Below: Lufbery about to take off for a combat sortie gets a send off from the Lafayette's unique mascot, the lion cub Whiskey. Note the original escadrille marking of a Seminole Warrior profile.

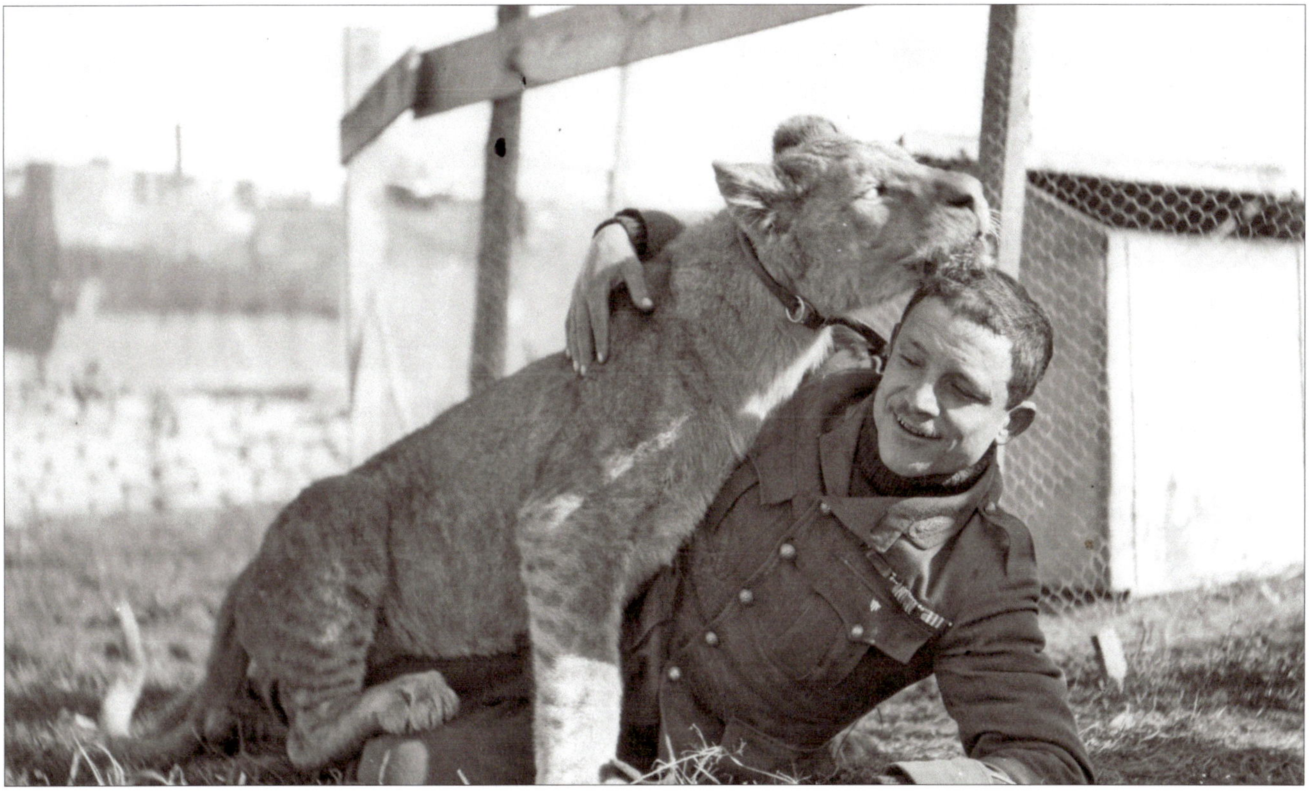

Above: The lion cub grew, but remained playful and was most fond of Lufbery,who called him his "Whiskey-man."

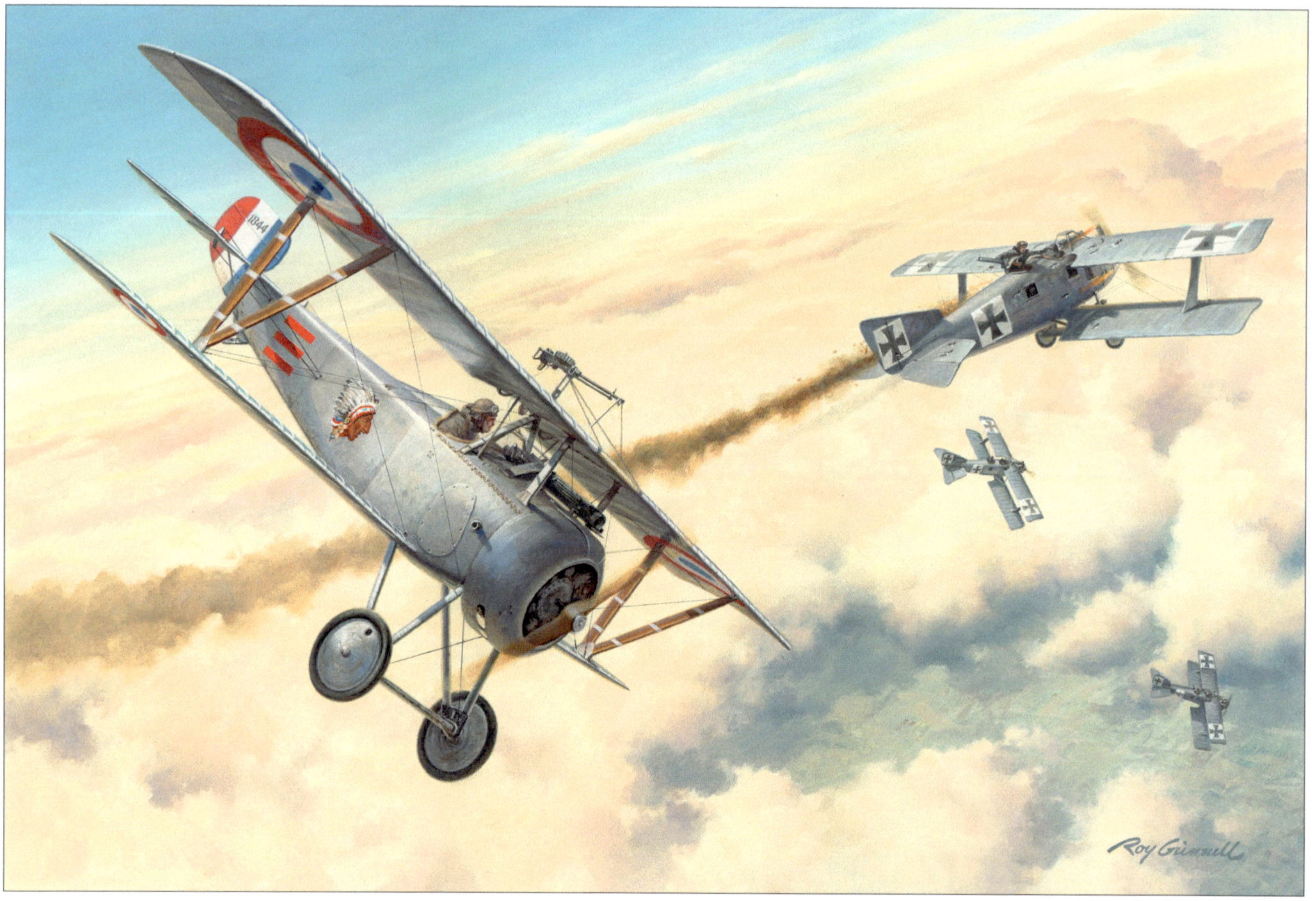

Above: Lufbery scores his fifth confirmed victory while the escadrille served as escorts to the sixty Allied bombers in the Oberndorf Raid on October 12, 1916. Roy Grinnell.

Right: Lufbery, somewhere in Paris while on leave from the front.

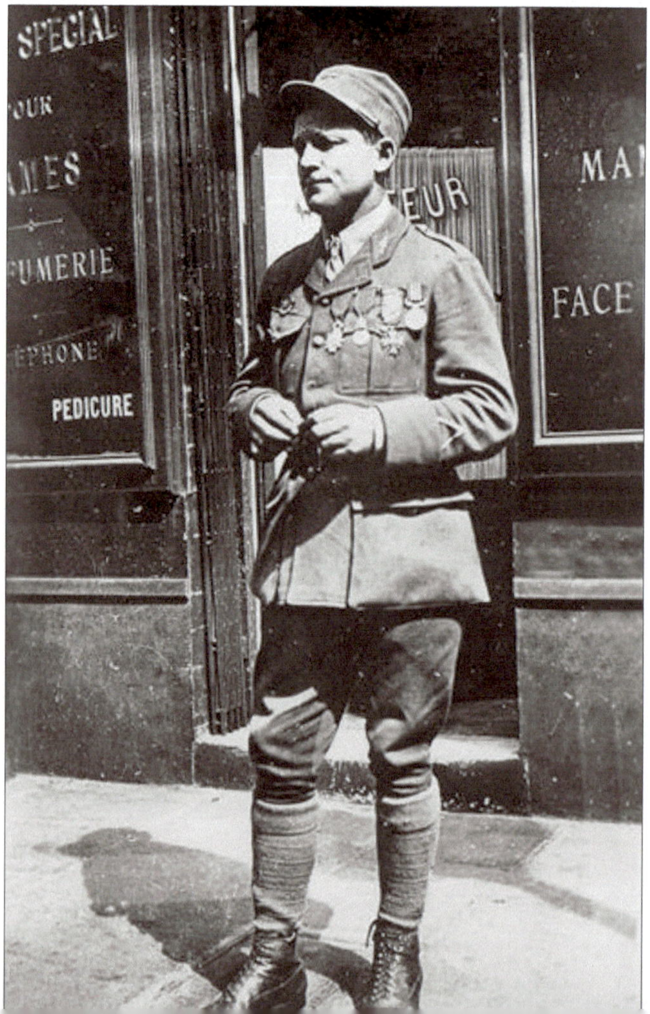

"felt that his efforts [were] not appreciated" by his commander. In addition, he felt that Norman Prince was receiving undo attention and praise. At one point he even requested a transfer to another escadrille, but he was convinced to stay.

September 1916 found the escadrille re-assigned from the Verdun front, back to their initial posting at Luxeuil.

Thus, on the 23rd of that month, Rockwell found himself patrolling in the same sky in which he had scored his initial victory only four months prior. He was patrolling with his comrade Raoul Lufbery, but Lufbery had to force land due to trouble with his machine gun. Rockwell carried on by himself and soon spied another German two-seater aircraft. As before, Rockwell closed in for the attack and held his fire until he was in very close range. Tragically, a bullet struck Kiffin in his chest immediately killing him and his Nieuport fighter was seen to dive straight into the ground at full speed.

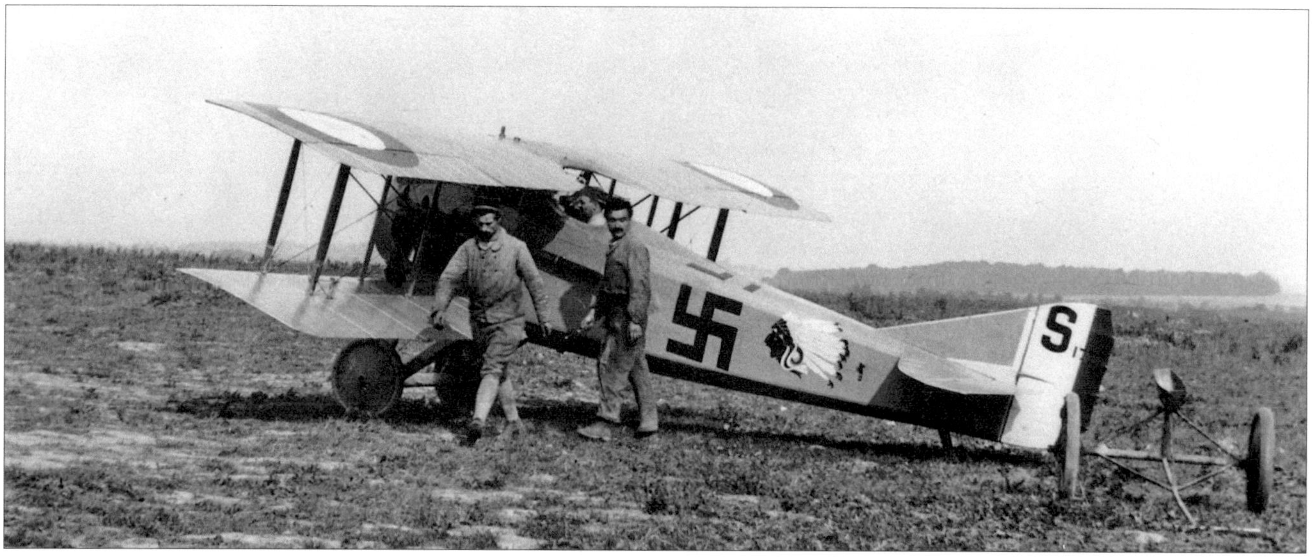

Above: A well-known Spad VII fighter, S.1777 with Didier Mason in the cockpit. Many pilots would jump at the chance to fly one of Lufbery's planes because of his meticulous care for the engine and aircraft. This Spad would have 10 different pilots fly it in a total of 89 patrols between June and December 1917 – which was an exceptionally long front-line career for the aircraft. Of the ten pilots, only Lufbery engaged in combat during these flights and managed two victories: revealing that a successful fighter pilot was one who would engage the enemy and not merely fly the mission.

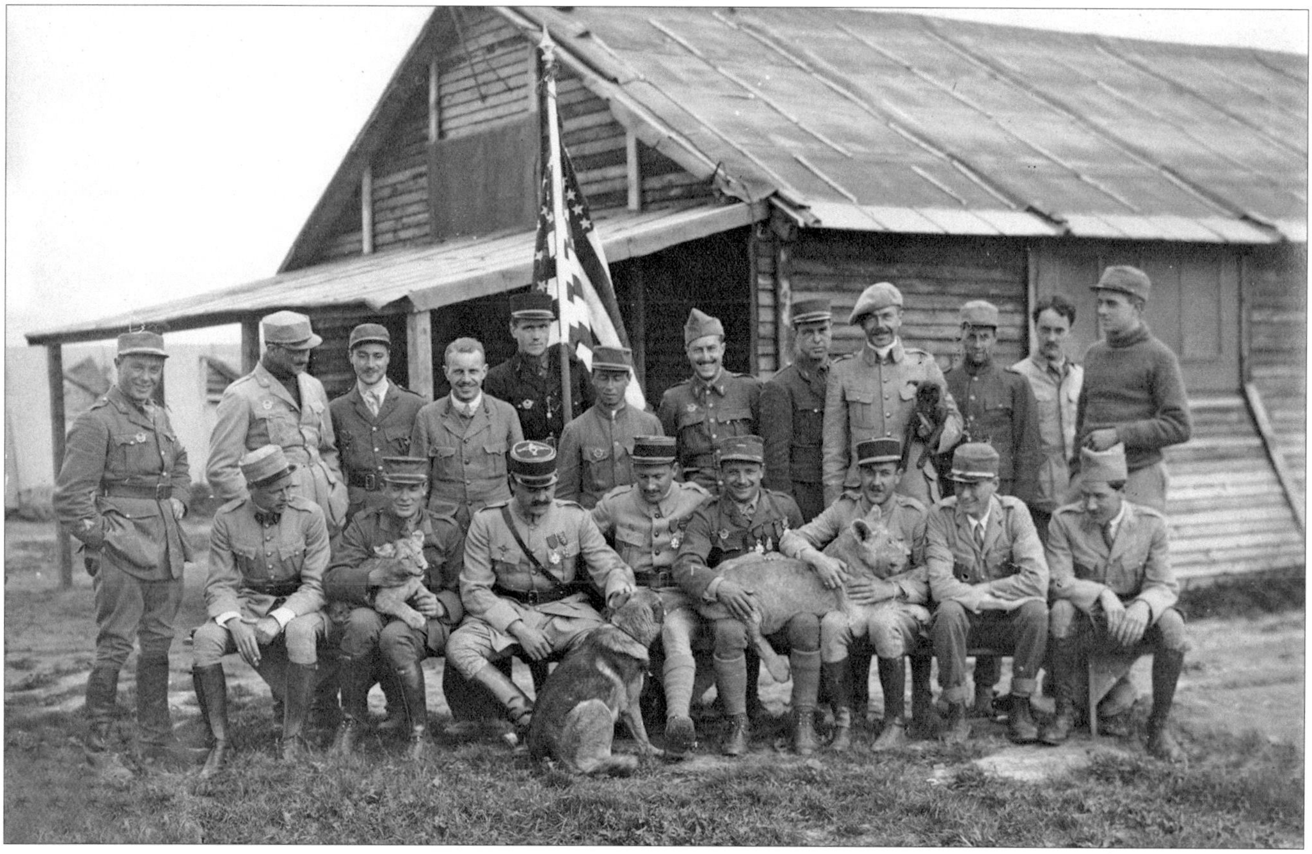

Above: A ceremony was held at the airfield at Chaudun on July 7, 1917 when the escadrille received the gift of a handmade American flag. The escadrille was at its peak and the proud members posed for a photo taken by the ever present Paul Rockwell (Kiffin's brother): L to R:sitting- Dudley Hill, Didier Mason (holding the second lion cub mascot Soda), William Thaw, George Thenault, Raoul Lufbery and Chouteau Johnson(holding Whiskey), Stephen Bigelow and Robert Rockwell: standing:Robert Soubiran, James Doolittle, Courtney Campbell, Ted Parsons, Ray Bridgman, William Dugan, Douglas MacMonagle, Walter Lovell, Harold Willis, Hank Jones, Antoine de Maison-Rouge and David Peterson.

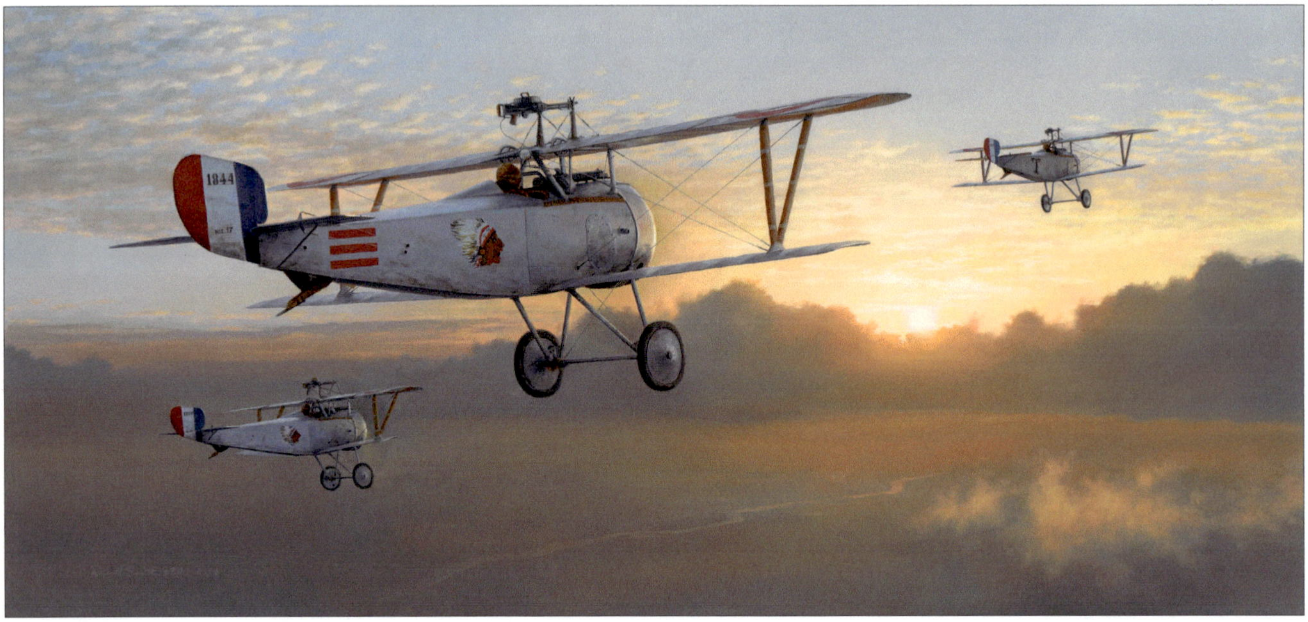

Above: William Thaw leads Raoul Lufbery (foreground) and Dudley Hill in a patrol at Dawn in the autumn of 1916. Russell Smith

Although Kiffin Rockwell perfectly fit the perceived mold of the volunteer American fighter pilot, one of his closest friends in the escadrille (he called him Bertie), Weston "Bert" Hall did not. Bert Hall was born on a farm in Missouri who, throughout his life, seemed to walk a fine line between being an unsavory fellow or merely a "lovable old rogue."

The outbreak of World War I found Hall working as a taxi driver in Paris. He was in the first group of 43 Americans who enlisted in the French Foreign Legion on 21st August 1914. In the infantry, he served with Thaw, Bach and Rockwell. Thanks to the assistance of Thaw, Hall was ordered to transfer to aviation in December 1914.

He began his frontline service flying two-seater, observation aircraft in escadrille M.S.38 along with his fellow American James Bach. In October 1915, the unit was supplied with the new Nieuport aircraft and so Hall and Bach were among the first twenty of France's Nieuport pilots. He flew with that unit until January 1916 when he was sent to Avord for pursuit pilot training.

After finishing the course, he then became an instructor and from there Hall reported for duty with the escadrille Americaine as one of its founding seven members. Although Hall did not fit in regards either the social status or the educational status of his comrades, Hall clearly did not shy away from his duty:

- 22 May- scored his first victory over an Aviatik observation aircraft for which Hall received the Medal Militaire and the Croix de Guerre with palm (N.124's second victory)
- 23 July- scored his second victory over a Fokker E-type fighter
- 28 August- scored his third confirmed victory and at this point only Lufbery had more victories with four

Despite his apparent valor and success, Hall found himself on the wrong side of most of his fellow escadrille pilots, Kiffin Rockwell being the exception. Although it is not certain exactly why this was so, speculation by historians is that Hall would often win at cards and would take home a lot of money form his well-endowed fellow pilots who suspected him of cheating. Overall, his ungentlemanly behavior and background conspired against Hall.

When Rockwell was shot down and killed, his only friend in the escadrille was now gone,Hall realized he had to do something. He requested a transfer to another squadron, which was granted. Hall reported to escadrille N.103 in early November 1916. He proceeded to score his fourth official victory on 26 November for which another palm was added to his Croix de Guerre and this citation was written:

"Clever, energetic, and courageous pilot, full of spirit. Daily attacking enemy planes at very short distance...The following day, after a combat held quite near, returned with his machine hit by several shots, also a shot in his helmet."

Clearly, Bert Hall was not only performing his duty, he was doing so energetically and quite bravely.

He left N.103 in December 1916 with orders to

report to the under secretary of war in Paris. It is here that Hall's story becomes very contradictory to how he had been serving. He went missing and wrote a very self-serving book that was full of embellishments and fabrications.

He supposedly transferred to a French aviation unit in Romania, but there are some historians who feel he made his way back to America and sat out the rest of the war. After the war, Hall was married four times, was involved in several business enterprises including the film industry. But his most adventurous times involved selling arms and aircraft to a Chinese warlord from whom he apparently stole a lot of money. His later, checkered life seems to have thrown a dark shadow over his early, respectable service to France in the escadrille.

Another member of the escadrille who clearly did not fit the mold young men like Rockwell, Thaw and Prince filled was Raoul Lufbery. Unlike Hall, Lufbery would have a distinguished military career and in fact Lufbery would rise to become the best fighter pilot of the escadrille and America's first fighter pilot ace.

Lufbery was born in France to an American father and a French mother. His mother died before Raoul had reached his first birthday. Five years later his Father re-married and moved to America. Raoul and his two brothers were left behind with relatives in France. In 1906 he and one of his brothers traveled to America to see their father, not aware of the fact that their father had recently set sail to France.

Lufbery wanted adventure and so joined the U.S. Army where his unit was sent to the Philippines. Lufbery was chosen as his unit's top marksman and before he completed his tour of duty, Lufbery became a naturalized US citizen. After his tour was finished, Lufbery continued to chase adventure and slowly traveled west until in 1912 he met Marc Pourpe in South-east Asia. Pourpe was a well-known French aviator who was touring the world. Lufbery became his chief mechanic. In 1914, Lufbery accompanied Pourpe as he flew the first airmail route from Cairo to Khartoum. A journey of 1,250 miles that took the two eight days.

With the outbreak of war, Pourpe immediately returned to France to offer his services and was assigned to escadrille N.23. Lufbery enlisted as well and continued in his role as Pourpe's mechanic. After flying over 80 combat hours in 30 sorties, Pourpe was killed in a crash of his aircraft while exiting clouds after a combat mission on 2nd December 1914.

Lufbery was devastated and quickly applied for transfer to become a pilot himself.

Later, one of Lufbery's escadrille comrades Edwin "Ted" Parsons had this to say:

"I only know one certain thing about him, Raoul Lufbery flew, fought and died for revenge."

From October 1915 until May 1916 he flew bombing and observation missions in V.B. 106. Then, at his own request he was sent back for training to become a pursuit pilot. After completion of his training, Lufbery joined N.124 as its eighth member. Now, Lufbery could truly pursue his course of vengeance against the Germans for the loss of his very close friend.

In Lufbery there would be a consolidation of the traits for a successful fighter pilot: keen eyesight to see the enemy first, superior marksmanship to hit his target and the inner ferocity to close with his enemy in a fearless manner.

Lufbery had another advantage as his years as Pourpe's mechanic taught him about his aircraft and the engines that powered them, in fact fellow escadrille member Edward Hinckle wrote this:

"Lufbery was a wonderful mechanic and his plane was always the best in the Escadrille. Anyone would rather have a second-hand Lufbery machine than a new one, anytime."

Aviation historian and escadrille Lafayette expert Dennis Gordon wrote this of Lufbery:

" Lufbery lived to fight. With no concept of fear, the prospect of death could not dissuade him from his missions. And during his relentless hunts he often took as much as he gave, many times returning to the squadron's airfield, his riddled plane a testimony to his fierce combats. Only one of the frequent attacks of rheumatism could keep him out of the sky, attacks which at one point became so severe that Lufbery required hospitalization. Often he flew in great pain, his joints and muscles half crippled."

At the end of the war, Lufbery would have 17 official victories, however his comrades knew that he had many more than that and many thought he had more than twice as many. His frequent combats were behind the German front lines and so often could not be confirmed by the requisite witnesses from the ground required by the French system.

However, Lufbery could care less about the confirmation of his score as he continued to seek his revenge. Carl Dolan had personally witnessed Lufbery down three German planes in a single sortie, however he received credit for only one. When Dolan asked Lufbery if he was bothered by the lack of confirmed victories Lufbery replied: "What the hell do I care. I know I got them."

Lufbery accepted a rank of Major in the newly formed USAS after America entered the war and managed to eventually become assigned to the first

U.S. pursuit squadrons, the 95th and the 94th, as a combat instructor. He would prove invaluable to the new, green pilots as he taught them their new craft as only an experienced front line fighter pilot could.

On 19th May 1918, anti-aircraft batteries began firing at a high altitude German observation aircraft that headed directly for the airfield of the American squadrons. One of the pilots who were on stand by, took off to intercept the German but likely because of his inexperience was unable to effectively engage the enemy.

From the ground, Major Lufbery could only watch the unsuccessful air combat and he decided to take measures into his own hands. Unfortunately, his own personally cared-for fighter was not in service at that moment and so Lufbery took off in another machine without having been able to check its guns.

From the ground, the other members of the unit could watch as Lufbery dove to the attack only to have to verve away after several shots could be heard: likely to deal with a jammed machine gun.

Finally, once again it seemed that Lufbery dove close in to the tail of the enemy Rumpler aircraft. The bystanders were shocked to see flames erupt from Lufbery's Nieuport fighter and could see him quickly dive away in the hopes of putting out the fire.

Then, they saw Lufbery exit the aircraft at about two thousand feet from the ground. Some of the people there thought that he had jumped from the aircraft towards a small stream that was below him. Some thought he jumped to escape the flames and some thought he fell from the aircraft by accident. There can never be certainty as to what actually happened.

The crowd rushed to where the greatest American fighter pilot of the First World War had fallen to find that the local villagers had already covered the body in fresh flowers cut from their gardens.

His commander from the escadrille Lafayette, Capt. Georges Thenault later wrote of Lufbery:

"His Spad was always the highest and every day he won new victories. He seemed to hardly care about having them confirmed. Calmly he reigned as a sovereign lord in his chosen element and beat down his foes to accomplish his duty and not for the sake of glory."

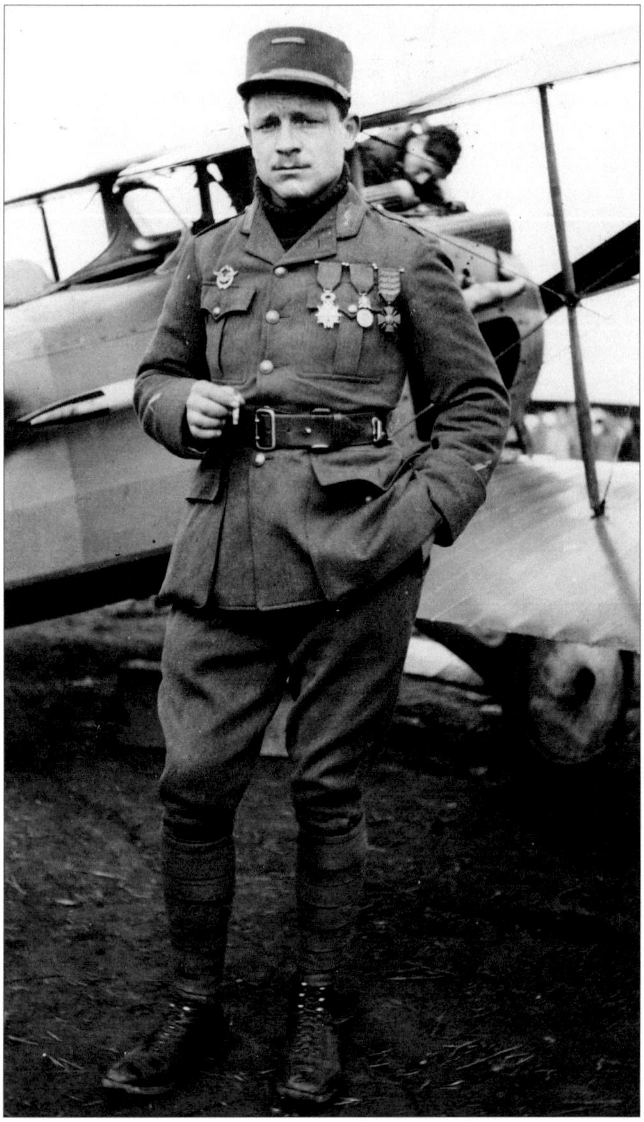

Above: The incomparable Raoul Lufbery.

Kingston On Thames

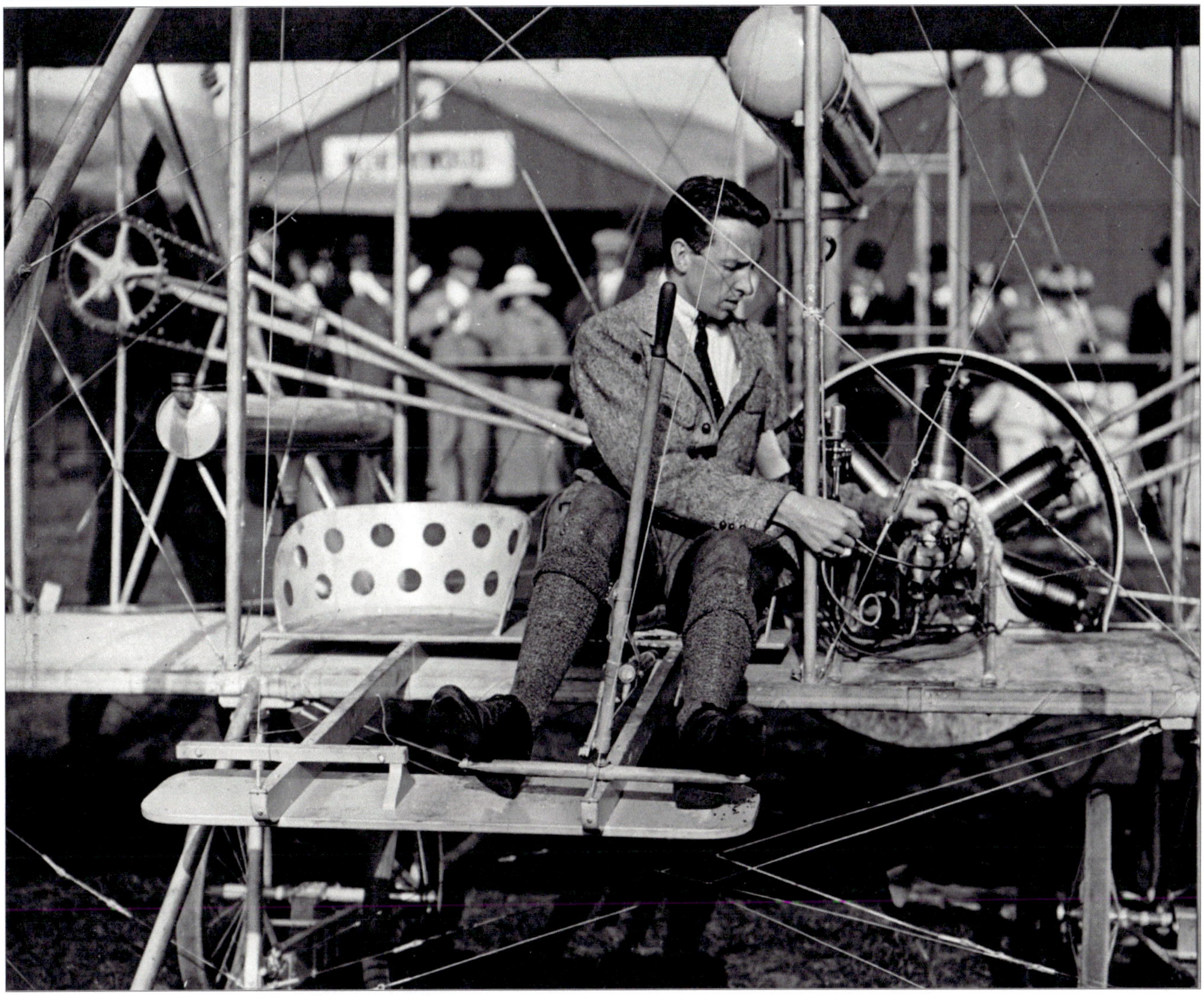

Above: Thomas Sopwith became a renowned aviator before the war and was known to improve the capabilities of the various aircraft designs that he flew.

On the 18th January 1888 Thomas Octave Murdoch Sopwith was born, the eighth child and only son to a prosperous family. At his birth, his father was 50 years of age and being the first son and heir, young Tom was likely considered a special child.

However, such good fortunes does not always shield one from tragedy. At the age of ten, Tom was with his father on a shooting holiday on the Isle of Lismore when a gun that was in his lap accidentally discharged and killed his father. A sadness that he apparently felt for the remainder of his life, which would be an incredibly successful life for the next ninety years.

Sopwith would be associated with many of the most famous and effective aircraft designs for the Royal Flying Corps and Royal Air Force in the First World War. His designs would be pitted against the designs of Tony Fokker over the Western Front throughout the war. Like Fokker, Sopwith did not excel in the traditional schools at the time. He ended up graduating from Seafield Park Engineering College where he devoted himself to learn about engines and the new automobiles that were beginning to be seen throughout London and its environs.

In 1906, with a childhood friend, Sopwith opened up a showroom in London selling Rolls Royce autos.

He began to embrace motor racing, sailing and motor boat racing. His favorite pursuit at the time

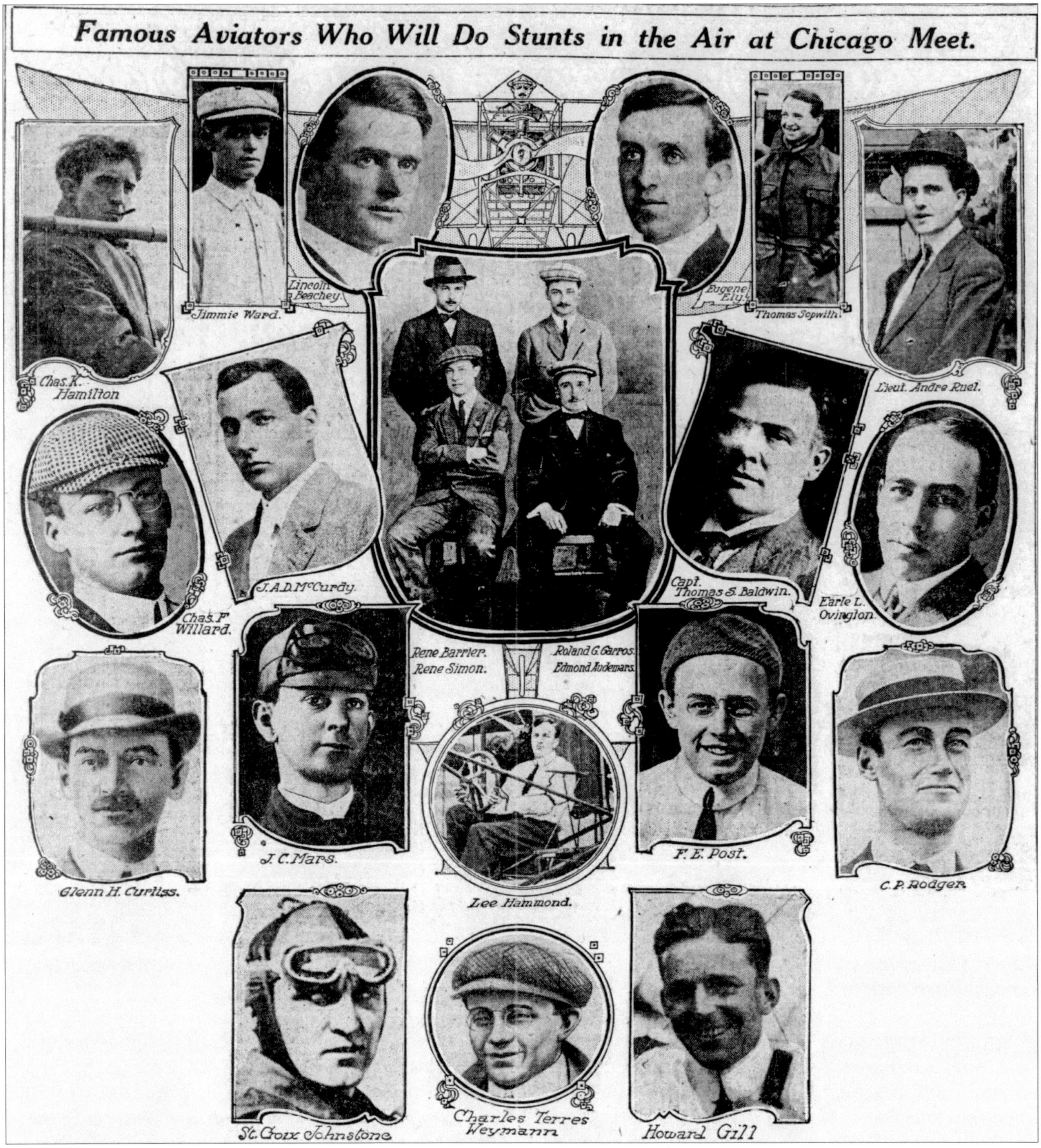

Above: In 1911, Sopwith won approximately 4,000 dollars in prize money at the air show in Chicago, Illinois. He returned to England and used that money to help purchase the building that would become his aircraft company, located in Kingston Upon Thames.

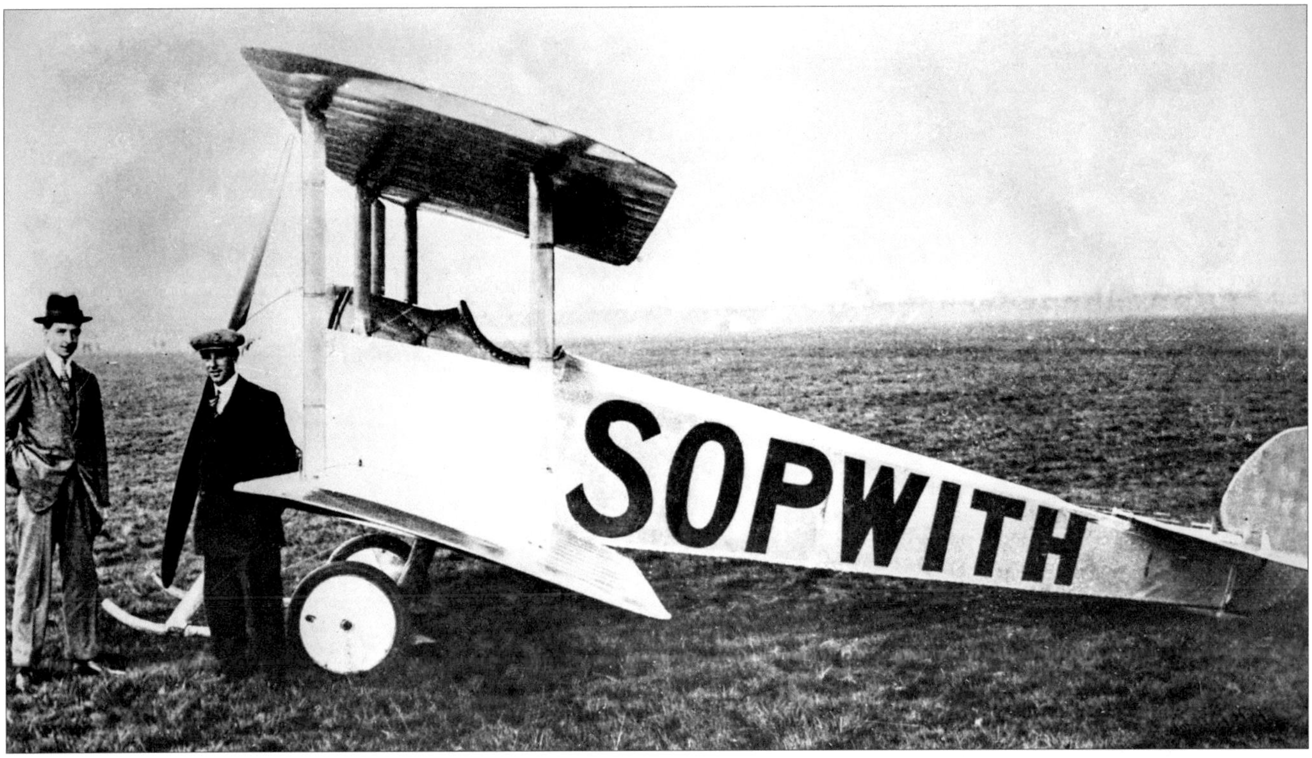

Above: The design that brought fortune to the Sopwith company was the Sopwith side-by-side, most of the future Sopwith designs can be seen in this aircraft's lines and dimensions already. Sopwith on the left, Harry Hooker on the right.

was ballooning. In that same year, Sopwith and a friend bought a second hand balloon from the Short brothers and became avid aeronauts. For the next few years Sopwith often flew his balloon on journeys around southern England.

His love of flying began at this time. In 1910 he met a supremely talented mechanic, Fred Sigrist and the duo would stay together from then on. That same year Sopwith made his first flight in an aeroplane and quickly bought his first aircraft. He taught himself to fly and on 21st November 1910 earned Aviator's License #31. Sopwith would prove to be a gifted pilot and with Sigrist as his mechanic constantly improving the aircraft/engines of Sopwith's aircraft, Sopwith would quickly become a very famous early aviator.

On the 18th December 1910 (less than one month after earning his license) Sopwith won the 4,000 pound Baron de Forest prize for the longest non-stop flight from England to continental Europe. He became the sixth aviator to fly across the English Channel during his 169 mile flight that lasted for 3hours and 40 minutes: an incredible feat for the time.

He had caught, like so many of his contemporaries, the flying bug. Sopwith also had a knack for showmanship and embraced the fame that aviators could earn during those years before the war.

On 1st February 1911 he flew to WIndsor Castle and landed on its East Lawn: there he was photographed being greeted by his King and the King's sons. Many young men from nearby schools also found their way to this exciting display. One of those young men was 12y/o Sydney Camm, who would eventually be employed by Sopwith and who would become a very successful aircraft designer with the Hawker Hurricane amongst his famous designs.

Shortly after that successful promotion, Sopwith and Sigrist headed to America in order to take the marvel of his flying on tour:

" Sopwith's enterprise left the Americans breathless. Here was a modest, self-effacing young man of the English upper set who was highly organized, very competitive and not averse to a little self-promotion." Alan Bramson

In August 1911 Sopwith won several events at the Chicago Aviation Meet and earned a little more than $4,000 in prize money. He then took that money and returned to England to establish the Sopwith School of Flying in 1912.

One of his pupils would be Hugh Trenchard, who would be later acknowledged as the "father" of the Royal Air Force. Another of his pupils was a young Australian named Harry Hawker who would prove to an outstanding pilot. Hawker would become the chief test pilot of the Sopwith Aviation Company

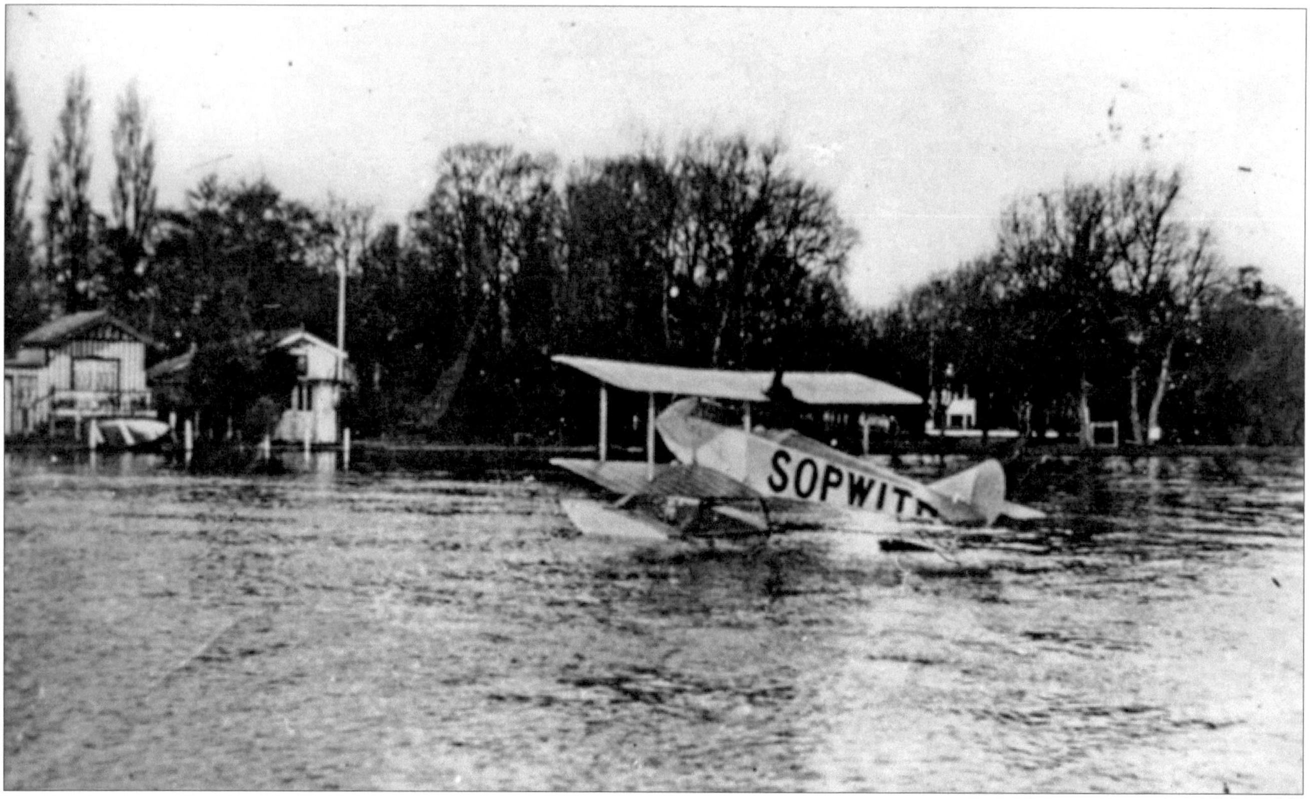

Above: Testing the Sopwith Schneider entry in England. Note the pilot standing in the cockpit to help get his aircraft safely back to shore.

Below: C. Howard Pixton flew this aircraft to win the 1914 Schneider Cup race and set a speed record of 92 mph for a floatplane. This victory lead to a substantial order from the Royal Navy.

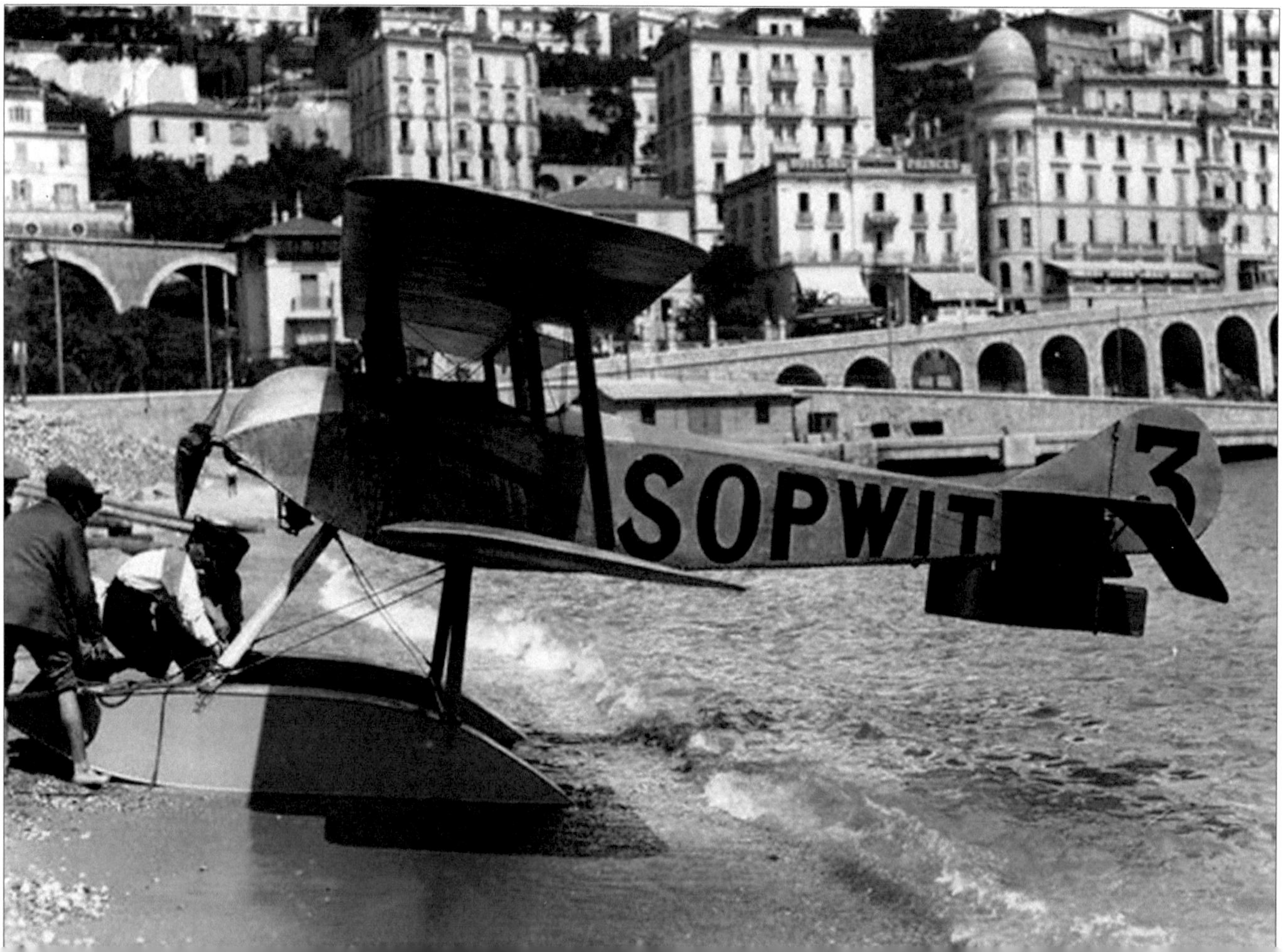

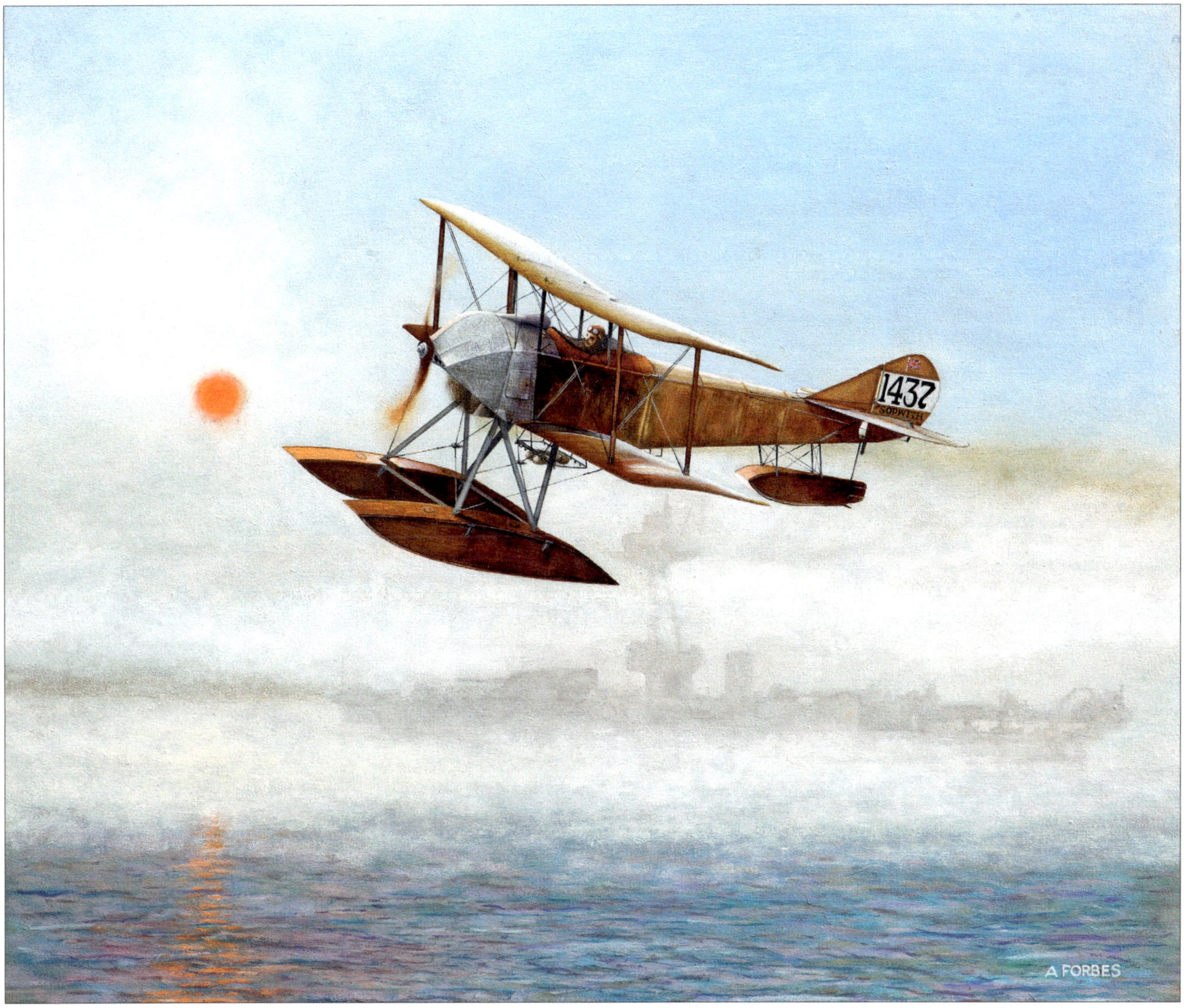

Above: A Sopwith Baby takes flight in the early morning mist in the Dardanelles. Alan Forbes

and a much-admired, close friend of Tom Sopwith.

Sopwith attended the military trials on Salisbury Plain in 1912 and was troubled by the inadequate British aircraft that took part in the exercise. He compared them to the far superior designs he saw while at the Rheims Aviation Meeting at the end of 1911. He was induced to build better aircraft for the British military. Therefore the trio of Sopwith, Sigrist and Hawker started the Sopwith Aviation Company Ltd. in March, 1914.

One of their first designs, known as the Sopwith Hybrid was successfully sold to the Royal Navy for 900 pounds. Using a portion of that money, Sopwith purchased a disused roller skating rink that was 16,000 square feet in size located in a village about 12 miles from central London, Kingston-on-Thames.

The next design proved to be truly outstanding for its time. The Sopwith SS (side-by-side) became more widely known as the Sopwith Tabloid. The aircraft was trim and fast , with a unique cockpit for two occupants who would sit next to each other instead of the typical arrangement. The plan was that Hawker would take an example with him to fly exhibitions and give flying lessons on an extended trip to his native Australia.

In the meantime, an example was modified in order to participate in the 2nd Schneider Trophy Race to be held at Monte Carlo in April 1914. The Tabloid did not simply compete, it won the race with such an outstanding performance that many of the assembled pilots, such as the Frenchman Roland Garros and the American William Thaw, did not even fly the course.

Its pilot, C. Howard Pixton, set a speed record of 92 mph for seaplanes. The performance was also noted by the Royal Navy which placed an order for

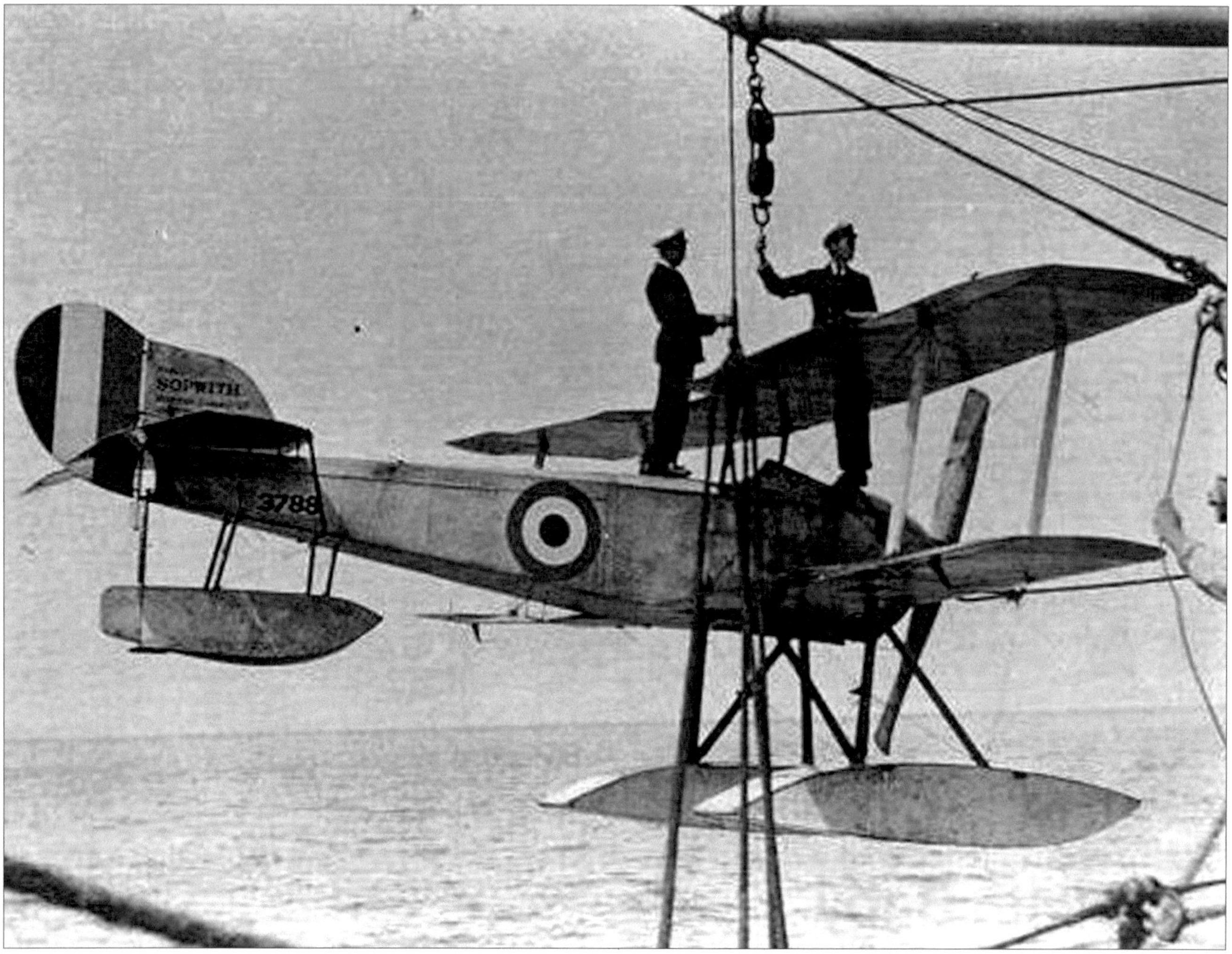

Above: The Sopwith Schneider and Baby floatplanes were made by several aviation firms with some slight differences based on the manufacturer. These floatplanes saw extensive service in RNAS units in Italy, Egypt and the Aegean Sea.

130 aircraft.

This order would be the beginning of the success that Sopwith held steady throughout the war and at some points during the war Sopwith was the only British aircraft firm that was producing successful combat aircraft. The Sopwith 1 ½ Strutter, Pup, Triplane, Camel and Snipe would all become proven fighter aircraft, at times filling absolutely essential roles that no other aircraft was capable of fulfilling in the Royal Flying Corps.

Starting with the Pup, Sopwith aircraft put the pilot, fuel and engine in close proximity in order to maximize maneuverability. The Sopwith Camel took this idea to the extreme and became renowned as a very difficult aircraft to master, but once mastered did prove to be a formidable foe to its German adversaries. Over 5,700 Camels were built and the type was credited with approximately 2,700 confirmed victories making it the single most successful type of fighter aircraft in the war. At the end of the war, over 50% of British fighter aircraft on the Western Front were Sopwith fighters.

Tom Sopwith, at the ripe old age of 30 y/o found himself managing the largest aircraft company in Britain. The Sopwith company had designed 32 types of aircraft during the war, with the company (and designated sub-contractors) producing over 16,200 airplanes. In fact, most of the Sopwith designed aircraft in the war would be produced in other aircraft production facilities. For example, in 1917 the Sopwith Company itself manufactured 849 aircraft while other companies produces a total over 2,900 aircraft of his design in the same year.

An incredible achievement by any measure, even more special when considering he had accomplished all of these feats within a mere eight years of having taught himself how to fly.

However, the success achieved in wartime rapidly faded during the peacetime that followed. The Royal AIr Force was drastically being reduced

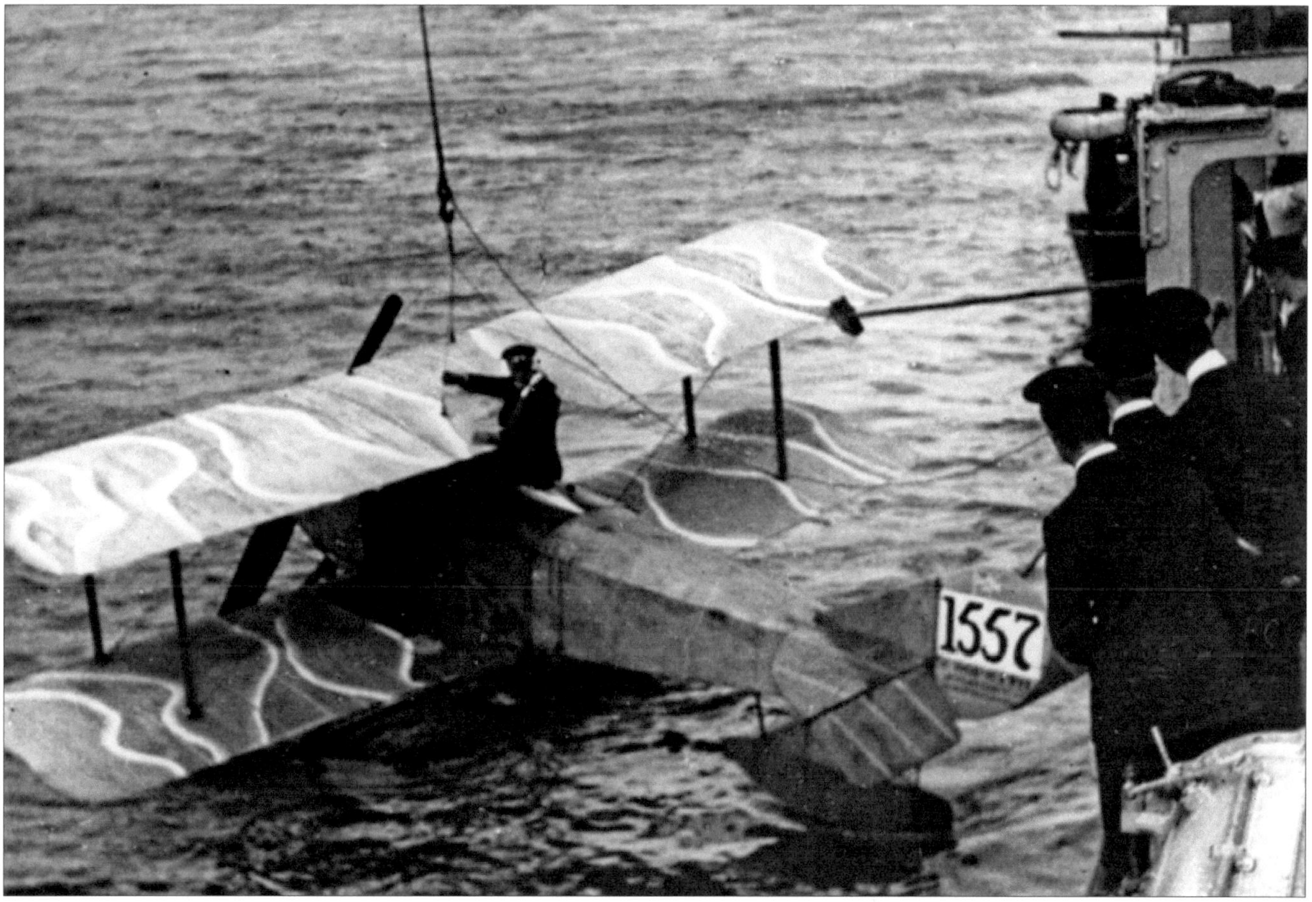

Above: Some of the variations were also in the color schemes used on the aircraft.

in strength and numbers and the civilian flying numbers did not make for too many orders. The British government desperately needed cash because of its huge debt that accumulated during the war. The government pressed for prompt payment of delayed taxes, which forced Sopwith to close his aircraft company.

Soon however, in November 1920 a new company, H.G. Hawker Engineering Company Ltd. was formed with the trio of Sopwith, Sigrist, and Harry Hawker together once again. Hawker would die tragically in an aircraft crash in July, 1921.

Sopwith and Sigrist carried on, and in the late 1920's hired Sydney Camm as chief aircraft designer. By the early 1930's, the nimble and sleek biplanes that he designed, the Hawker Hart and Hawker Fury were the mainstays of the RAF. At one point, Sopwith's new company, Hawker Siddeley were responsible for 80% of the aircraft in the RAF inventory. Camm would continue to design outstanding aircraft such as the Hawker Hurricane, Tempest and Typhoon.

Carrying on even after WWII, the company produced the Hawker Hunter jet and the Hawker Siddeley Harrier, the principle fighter aircraft of the Falklands war.

Sopwith celebrated his 100th birthday at his beloved home, Compton Manor. Although he had lost his sight by this advanced age, he was brought out onto the terrace of his home as a flyover in his tribute was performed. Initially a Sopwith Pup flew by, followed by a Hawker Hurricane and finally by a Hawker Harrier jump jet: signifying Sopwith's incredible successes as well as his longevity in the history of aviation.

He would die just after celebrating his 101st birthday. A few years before , when he was "only" 96 y/o, Sopwith had been interviewed for a BBC television feature:

"The last question asked of this pioneer pilot, engineer, sportsman and industrialist on the grand scale was 'what do you consider is the reason for your outstanding success?', to which the great man replied: 'Pure luck.'" A. Bramson

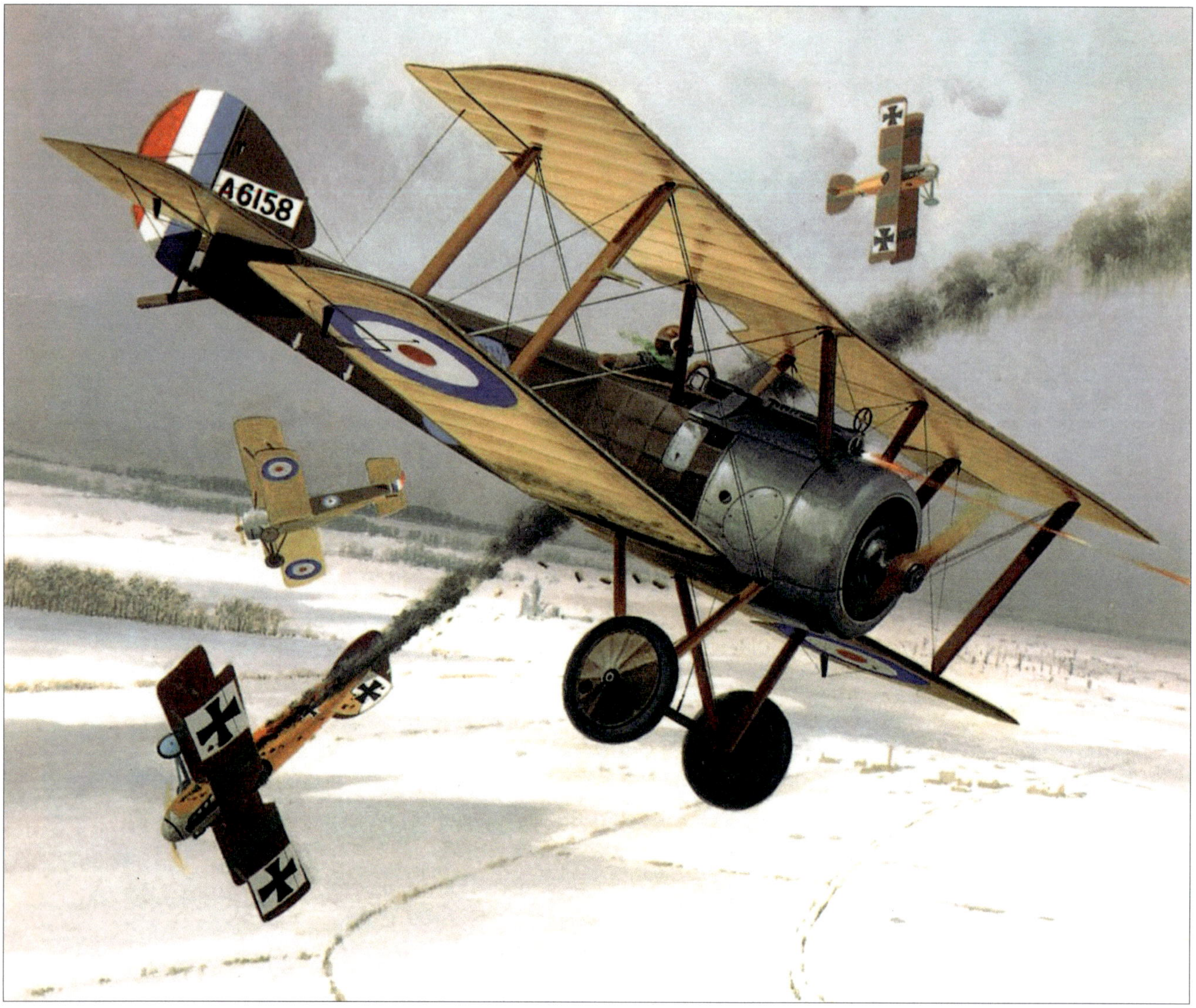

Above: The Sopwith Pup was the first Sopwith fighter on the Western Front that could hold its own and better the German Albatros fighters it was facing. Brian Knight

Below: The beloved Sopwith Pup, per the pilots who flew it the aircraft was a "delight".

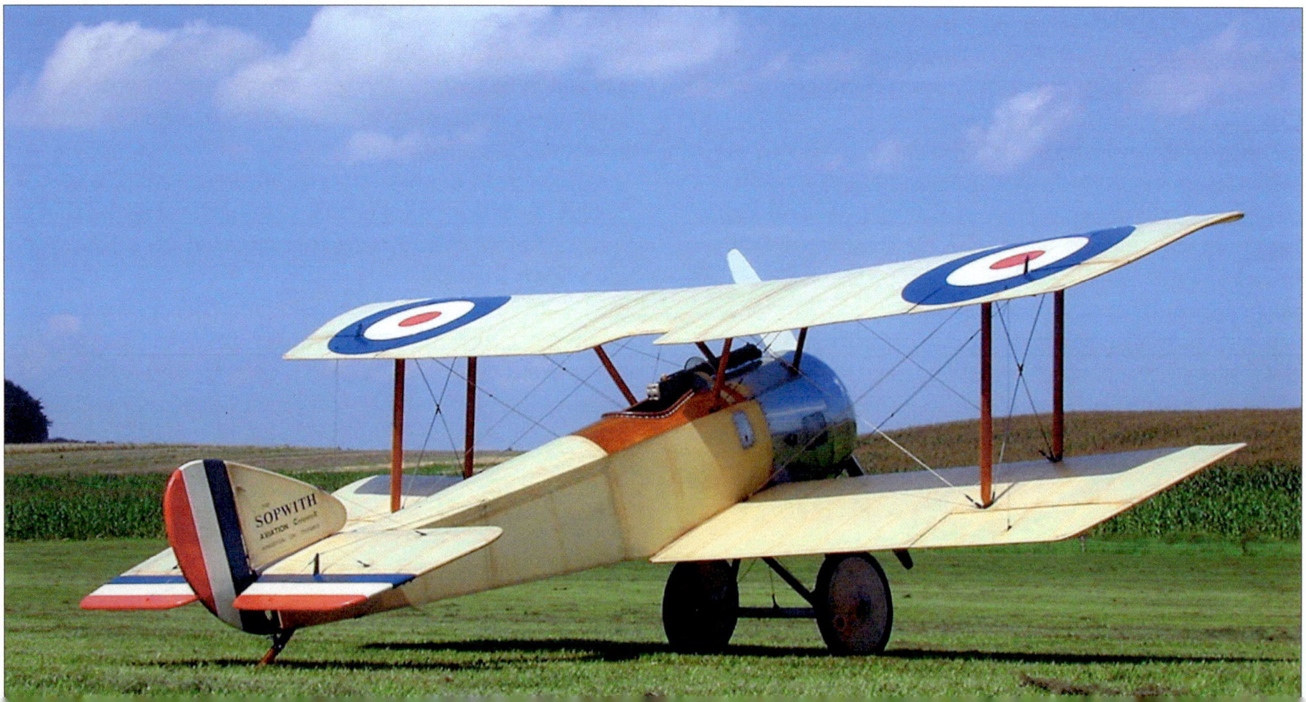

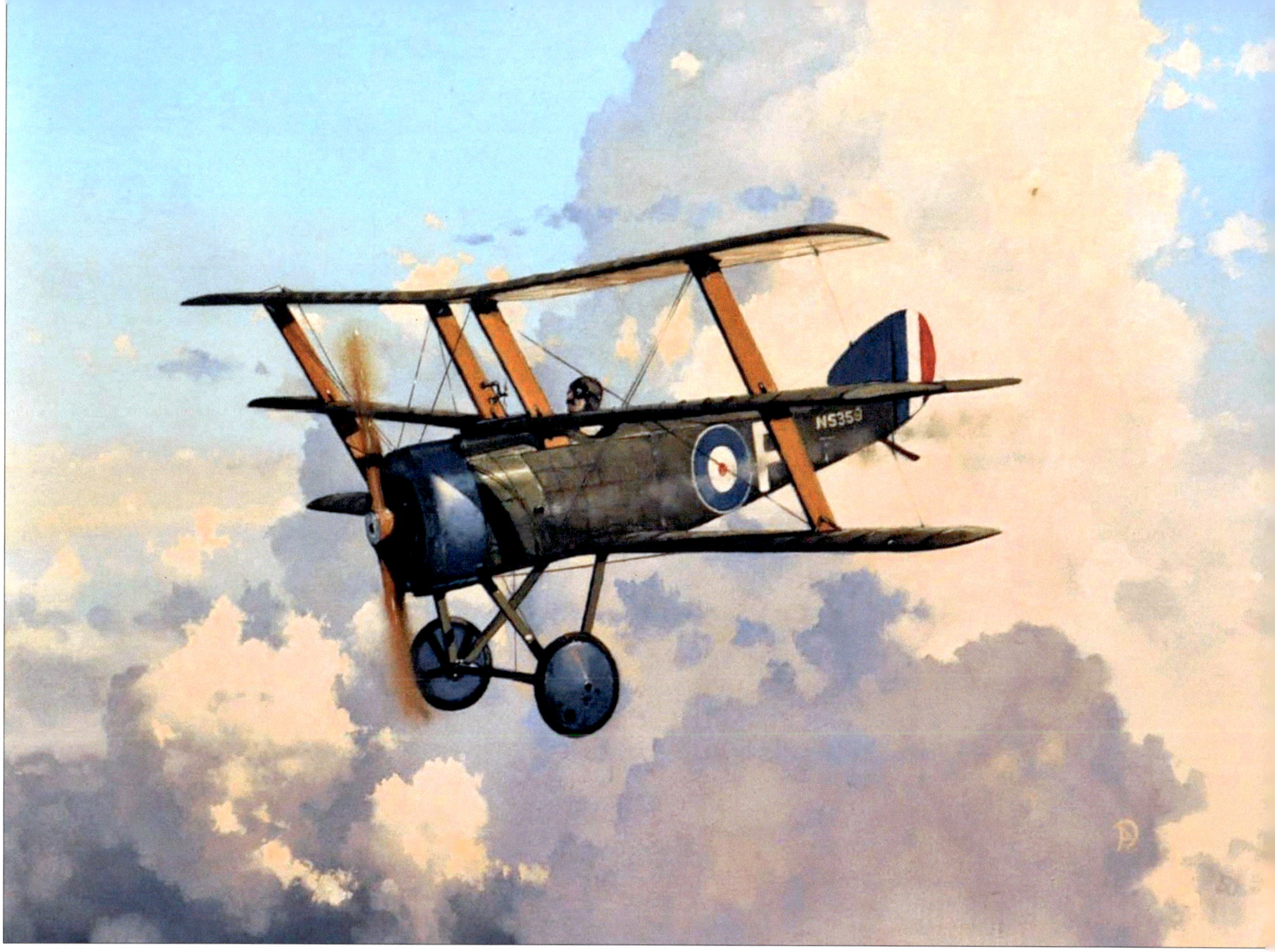

Above: The Sopwith Triplane. Andrew Dillon

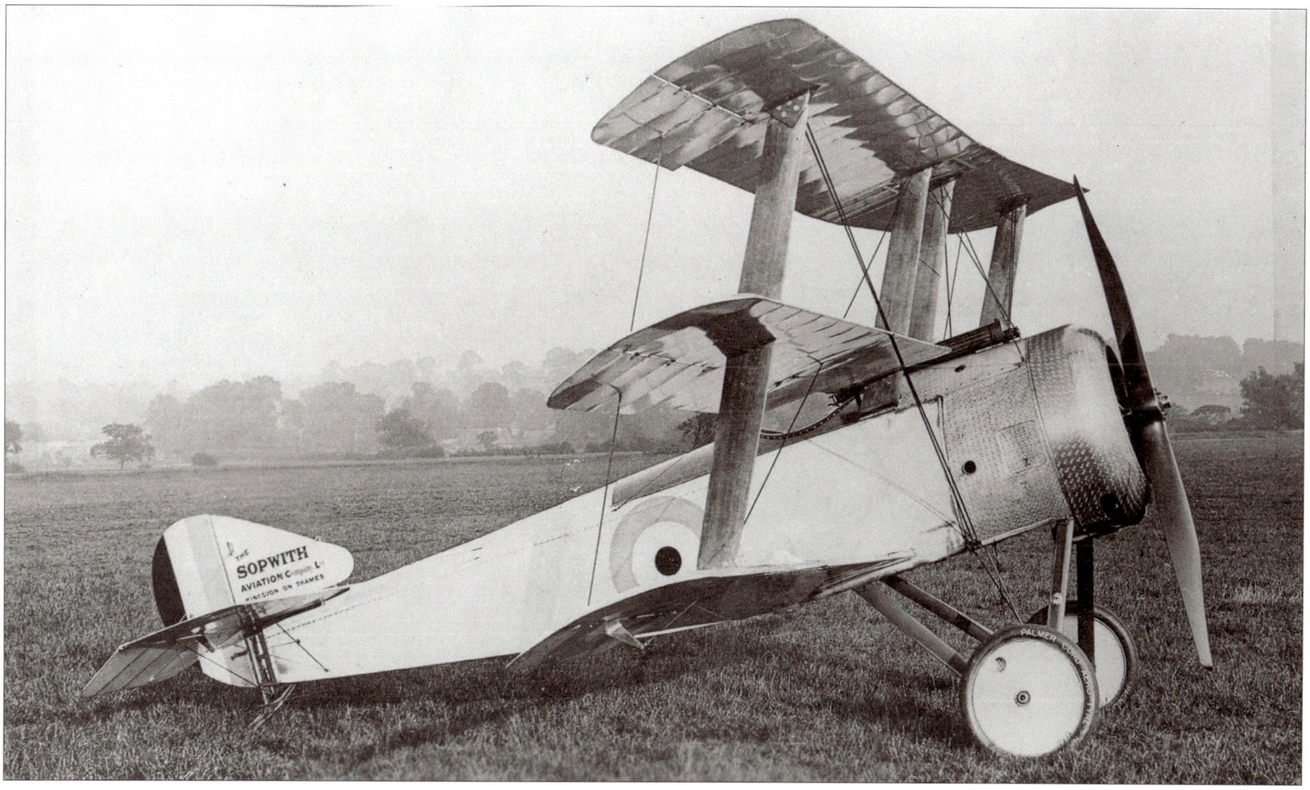

Above: An unusual design to help improve rate of climb and pilot vision was the Sopwith Triplane.

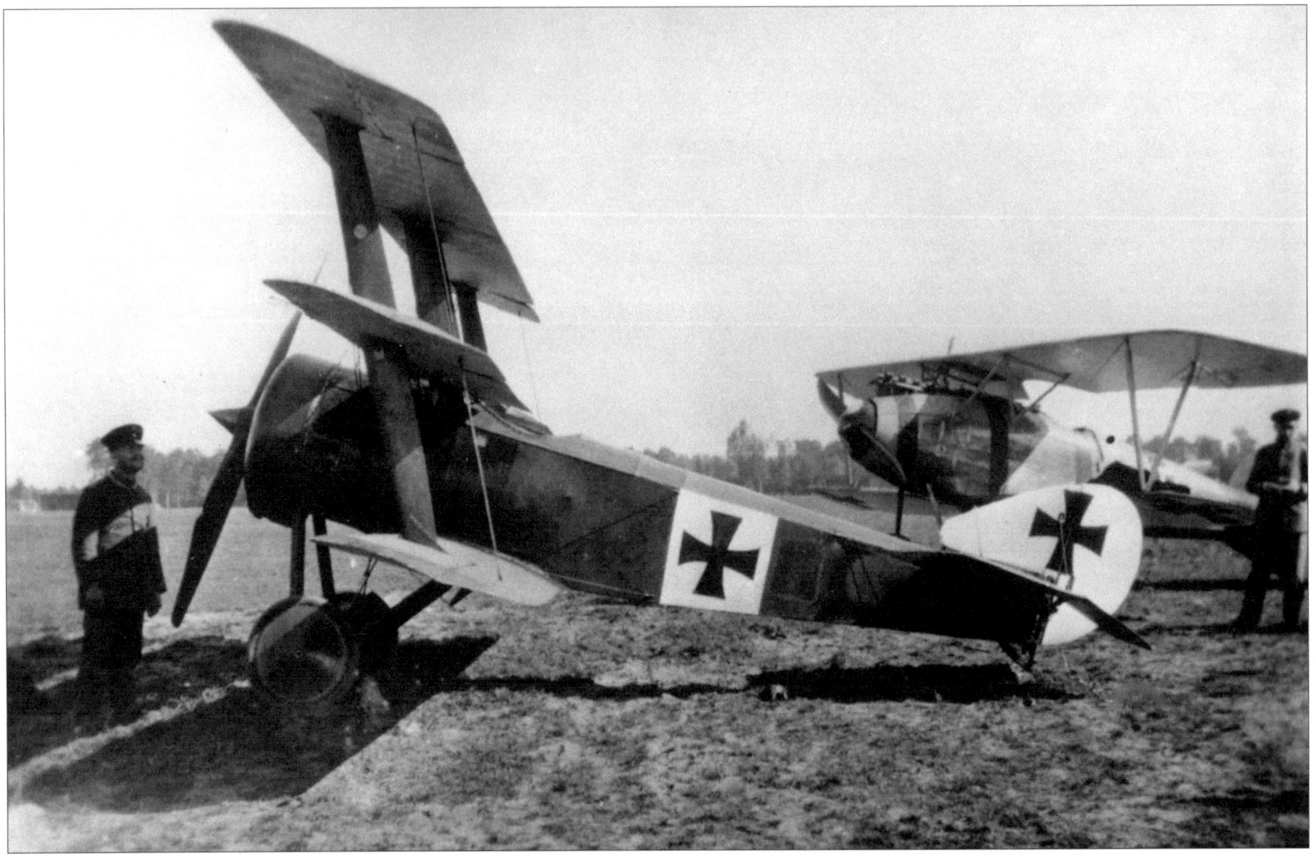

Above: Several examples of captured triplanes were flown to evaluate the type's performance. The evaluation was positive, such that in a very short time Anthony Fokker would have a triplane of his own design.

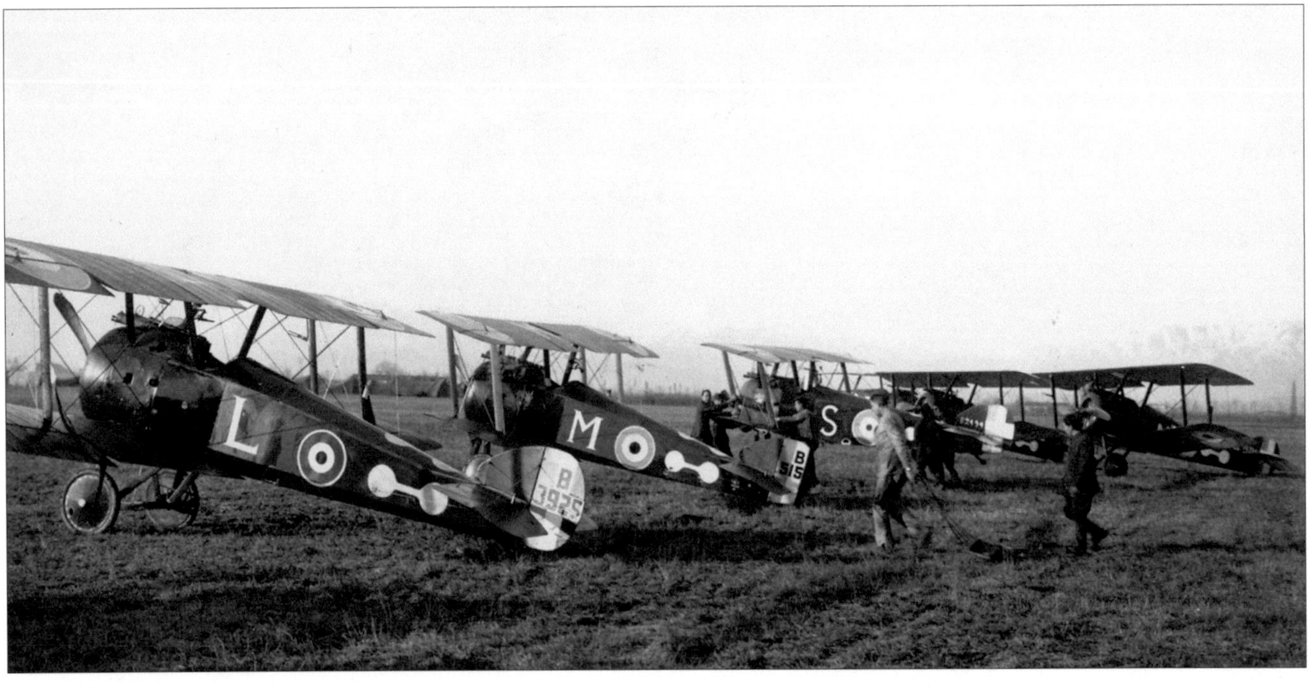

Above: The Sopwith Camel became the king of fighter aircraft in WWI, having been credited with the most enemy aircraft shot down by any single type of design in the war. The Camel served everywhere in the war, these Camels are of No. 455 squadron part of the RAF force in Italy.

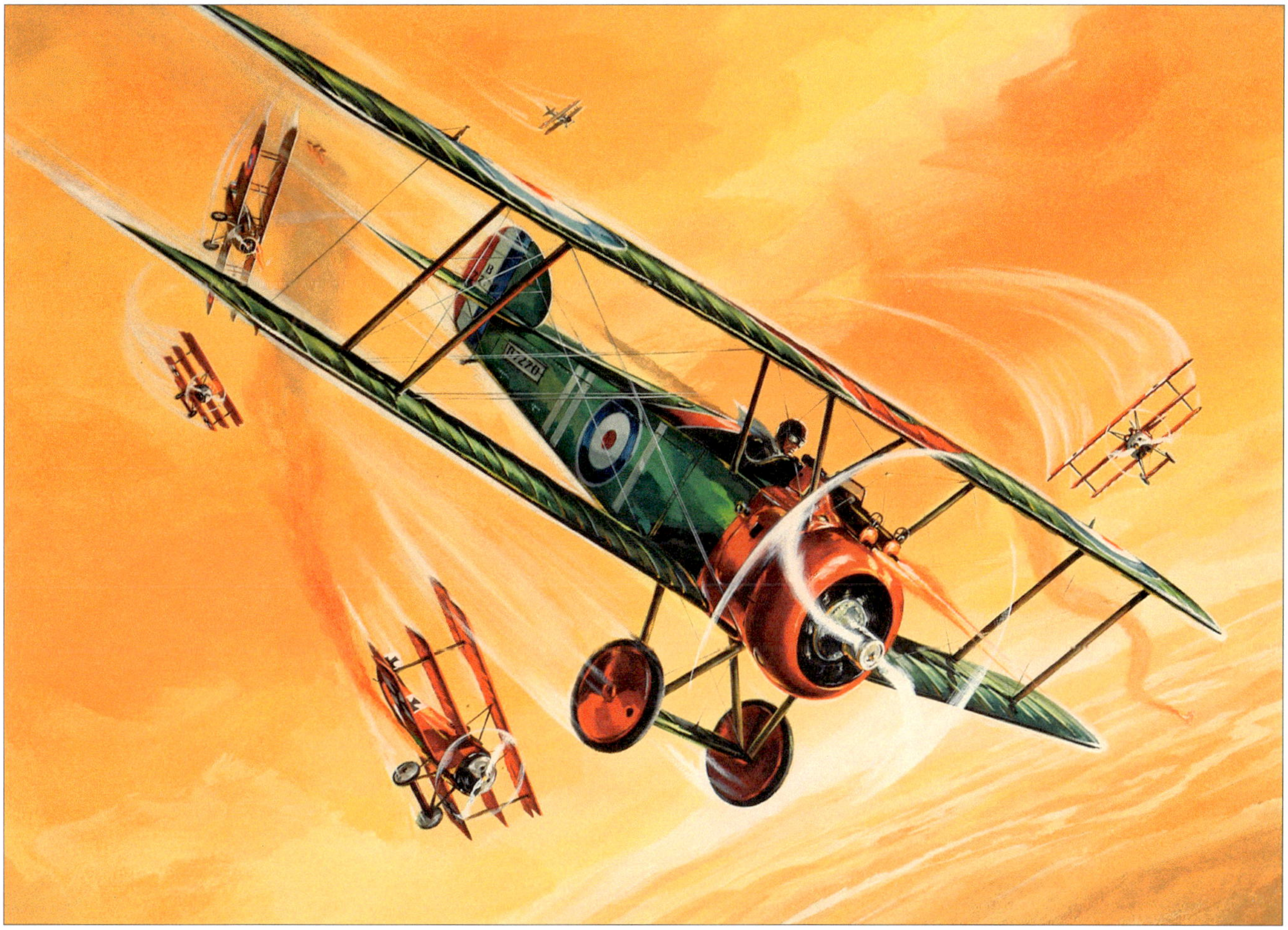

Above: Sopwith Camel in the middle of a melee or dogfight. Jack Leynwood.

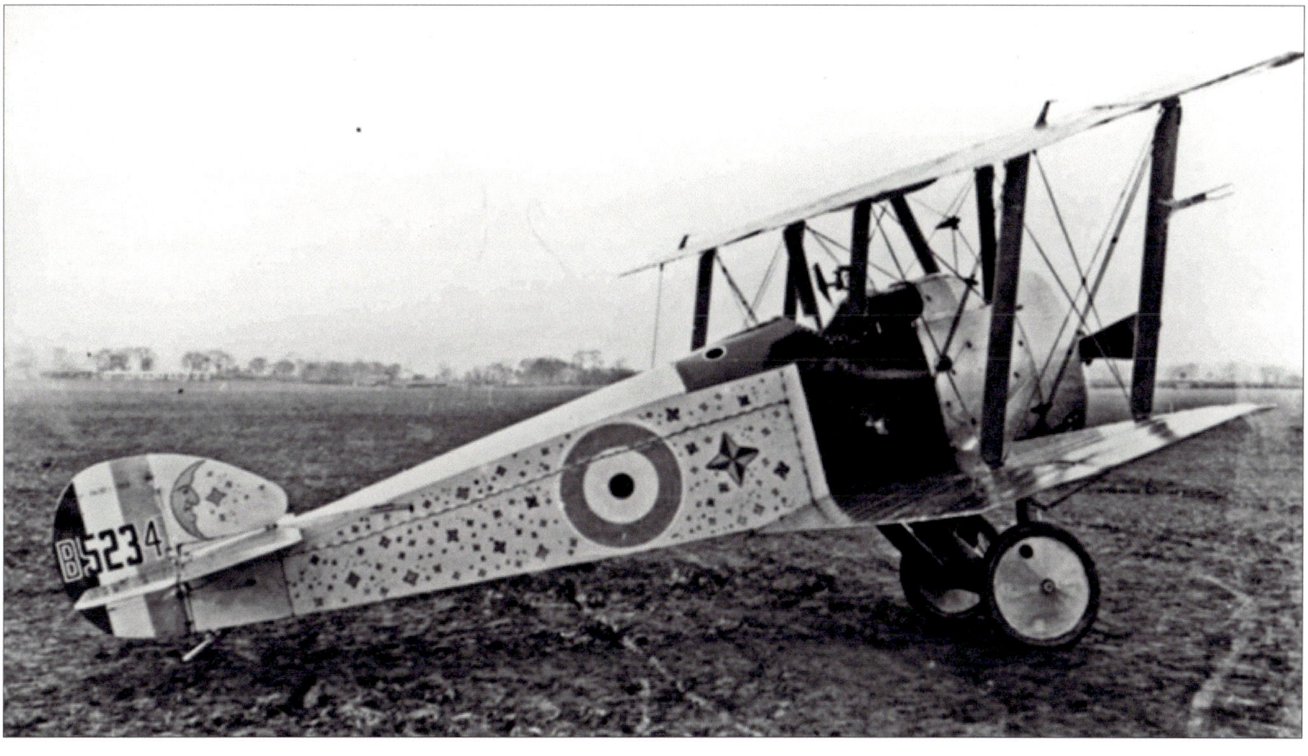

Above: Some Camels received elaborate decorations, especially ones flown on the home front in England.

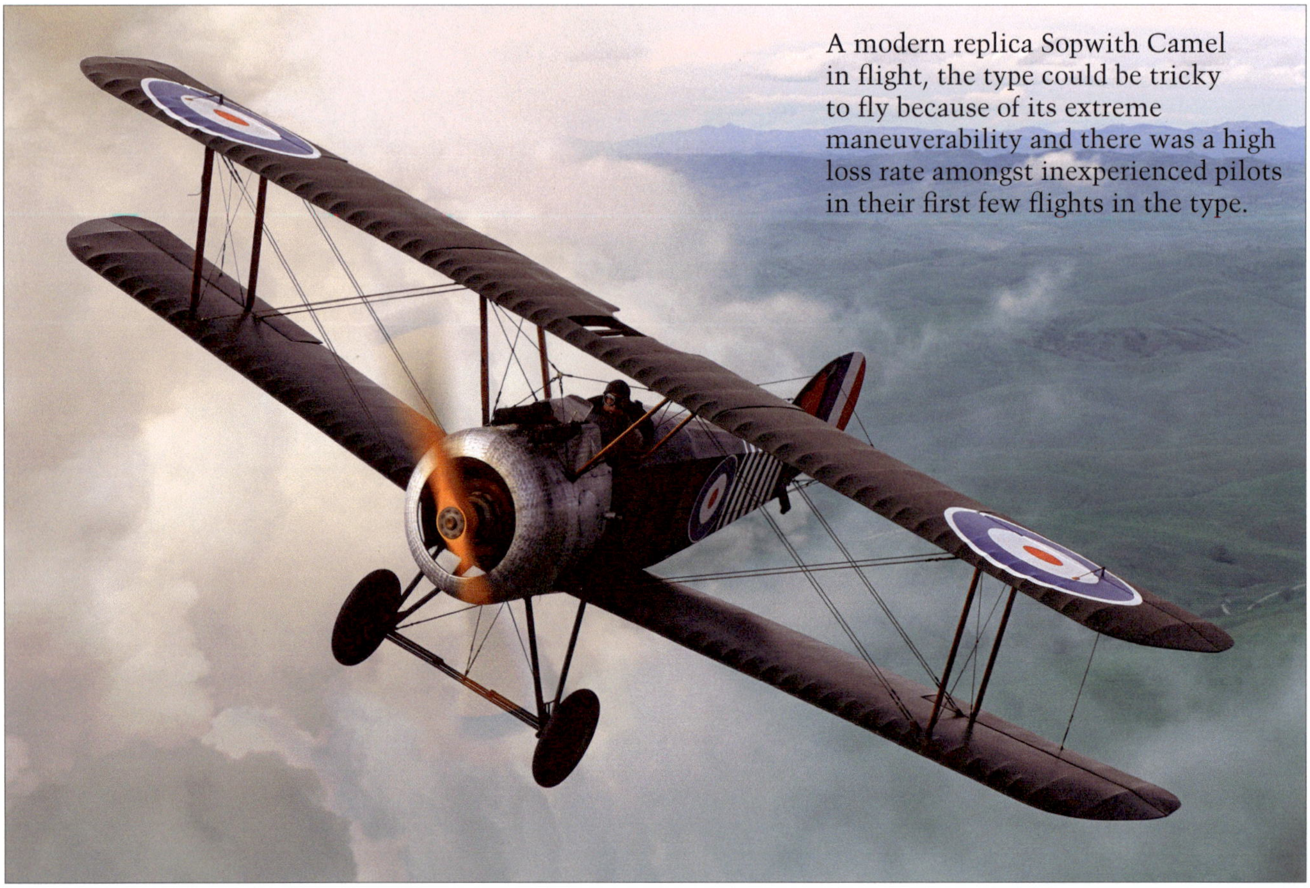

A modern replica Sopwith Camel in flight, the type could be tricky to fly because of its extreme maneuverability and there was a high loss rate amongst inexperienced pilots in their first few flights in the type.

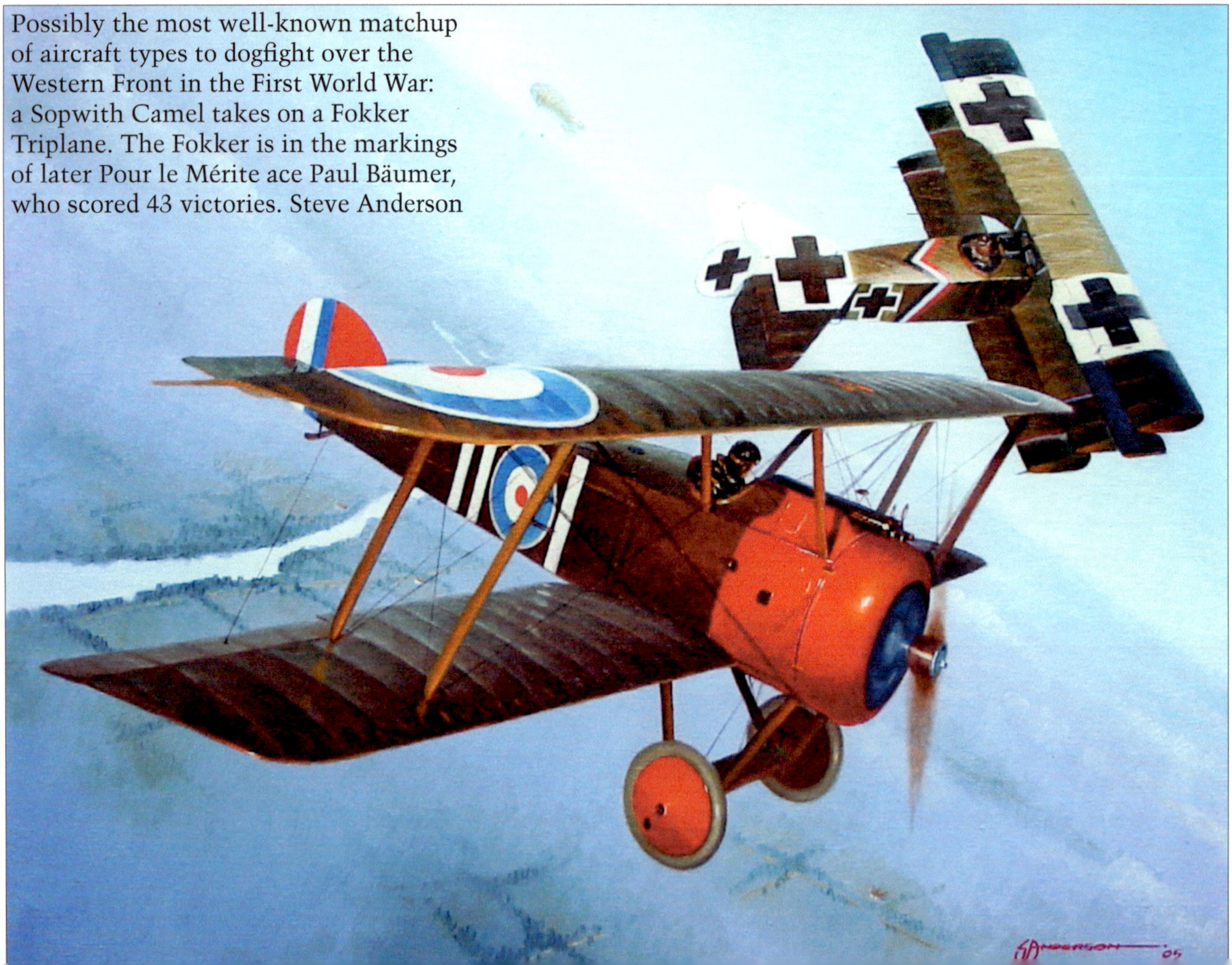

Possibly the most well-known matchup of aircraft types to dogfight over the Western Front in the First World War: a Sopwith Camel takes on a Fokker Triplane. The Fokker is in the markings of later Pour le Mérite ace Paul Bäumer, who scored 43 victories. Steve Anderson

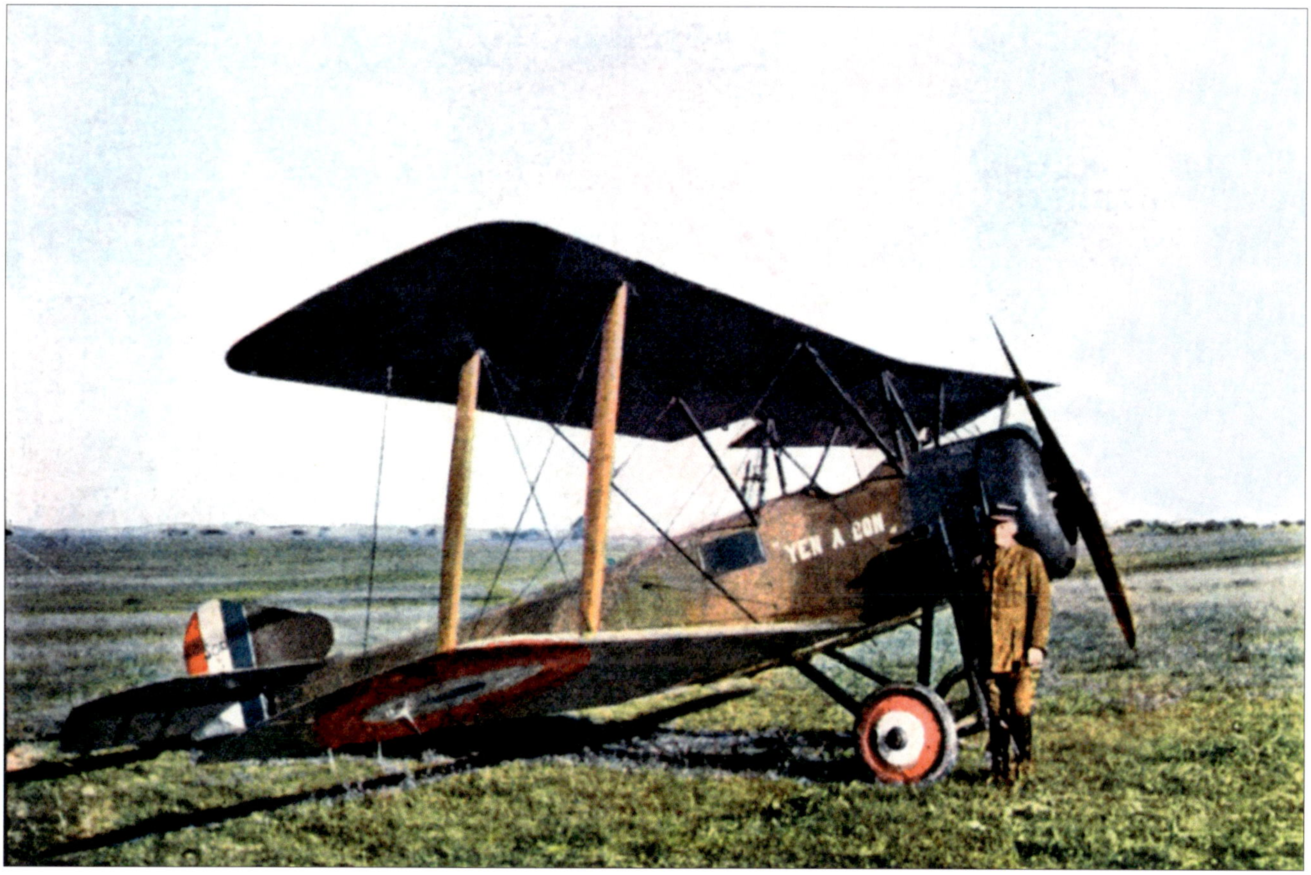

Above: An original color photo from WWI.The type was fitted with either a Clerget or Le Rhone rotary engine. The Sopwith 1½ Strutter was used as a single seat light bomber and two-seat reconnaissance plane by the French.

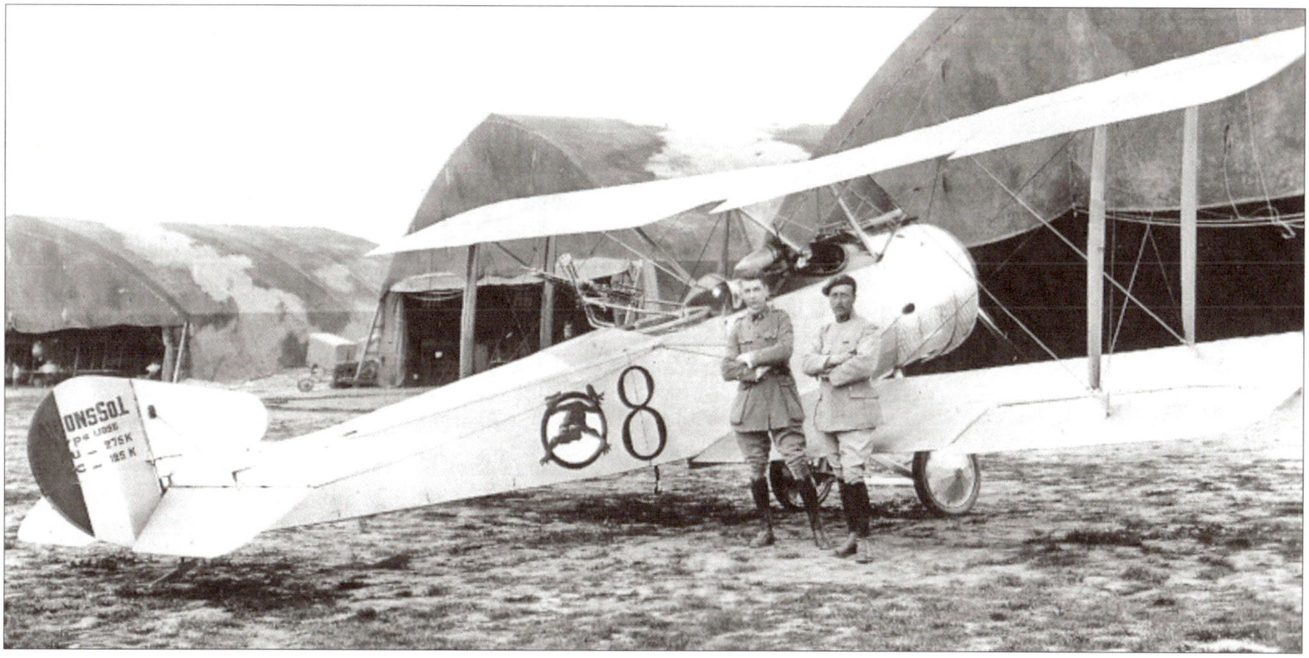

Above: Of the 5,700 Sopwith 1½ strutters built in the war, 4,200 were built under license in France. Here is another example of a French-built Sopwith, which was also used for reconnaissance.

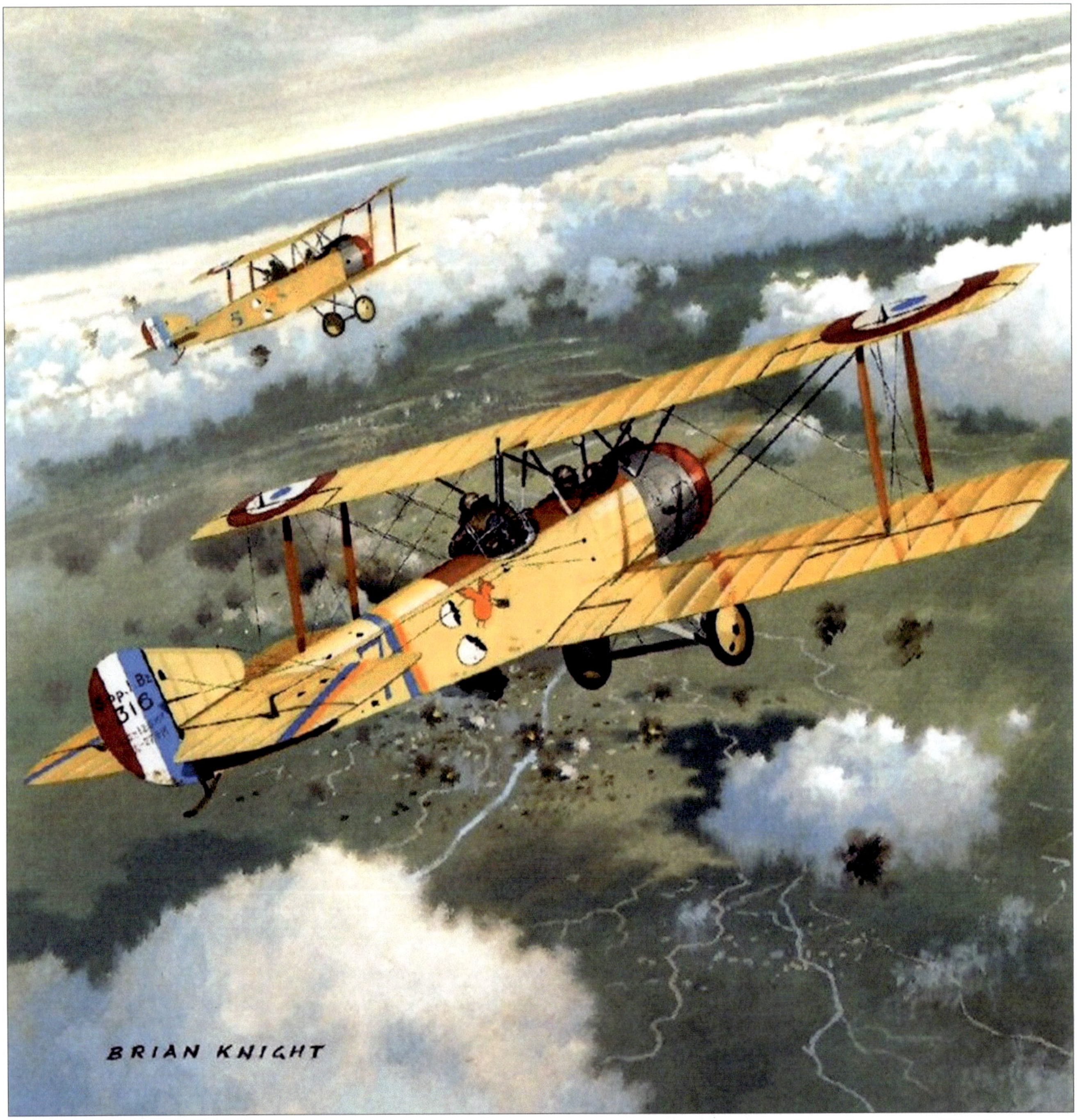

Above: A Spowith 1½ Strutter of the French Air Service. These aircraft were of a reconnaissance unit. The Sopwith type 9700 (RNAS designation) was also known as the Sopwith two-seater (RFC designation); however, the majority of the design was actually built in France for the French Air Service.

Wolfchen

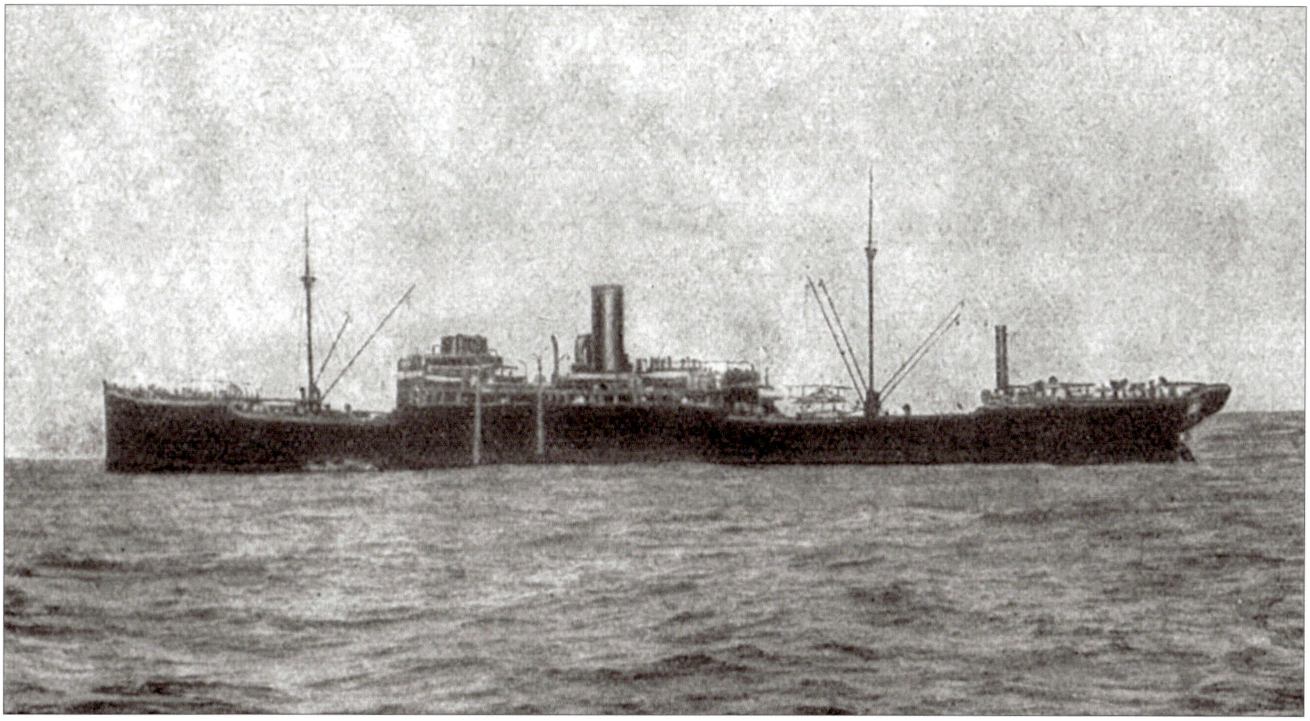

Above: The German raider SMS *Wolf*, note the seaplane just aft of midships.

One of the more remarkable naval stories of the First World War involved the incredible voyage of the German raider S.M.S. Wolf. Its captain, Karl Nerger, would earn his PLM award by successfully sinking 30 ships for a total of 110,000 tons during his epic 444 days at sea. He directed his ship from Kiel harbor on 30th November 1916 and would return to that same harbor on 24th February 1918 after sailing through parts of the Atlantic, Pacific and Indian oceans as well as much of the South Seas.

What makes his story also unique is that on board his merchantman converted to being a commerce raider was a Friedrichshafen 33 two-seat reconnaissance aircraft that was nicknamed the Wolfchen (little wolf). The seaplane would prove invaluable to Nerger in a variety of roles and circumstances, significantly contributing to the successes he achieved on his voyage and equally important, often would keep him safe from being discovered and sunk himself by Allied ships prowling the waters in an attempt to stop him.

His mission was to mine several key harbors and trade routes as well as to capture prize vessels. The fact that Merger was able to keep S.M.S. Wolf out to sea for so long was in large part due to his utilization of coal and other key booty (such as food and spare parts) found on many of his victims.

It had already been a very long journey and he had already accounted for ten ships (six by mines and four by capture) when, in the last week of May 1917, Nerger brought his ship into the protected harbor at Sunday Island (also called Raoul Island). There the S.M.S. Wolf dropped its anchor for the first time since leaving Kiel in February.

Nerger planned to lay mines off of Australia and New Zealand before proceeding North to weave through the waters around the island of New Guinea and the Dutch East Indies so as to return to the Indian Ocean to resume looking for more prizes.

First, however, a complete engine overhaul would be needed and numerous repairs made throughout the ship to help restore its optimal seaworthiness. After several weeks of hard work, the engines were still only partially functional when a lookout spotted an approaching ship on the horizon.

Even from a distance Nerger could spot the freighter's radio aerial quite clearly, and if he could see the Allied ship, it could see him. However, the ship was currently out of range of sending a signal. But soon, by nightfall it would be able to send the sighting of the German commerce raider to Samoa.

Knowing that with only one boiler fully

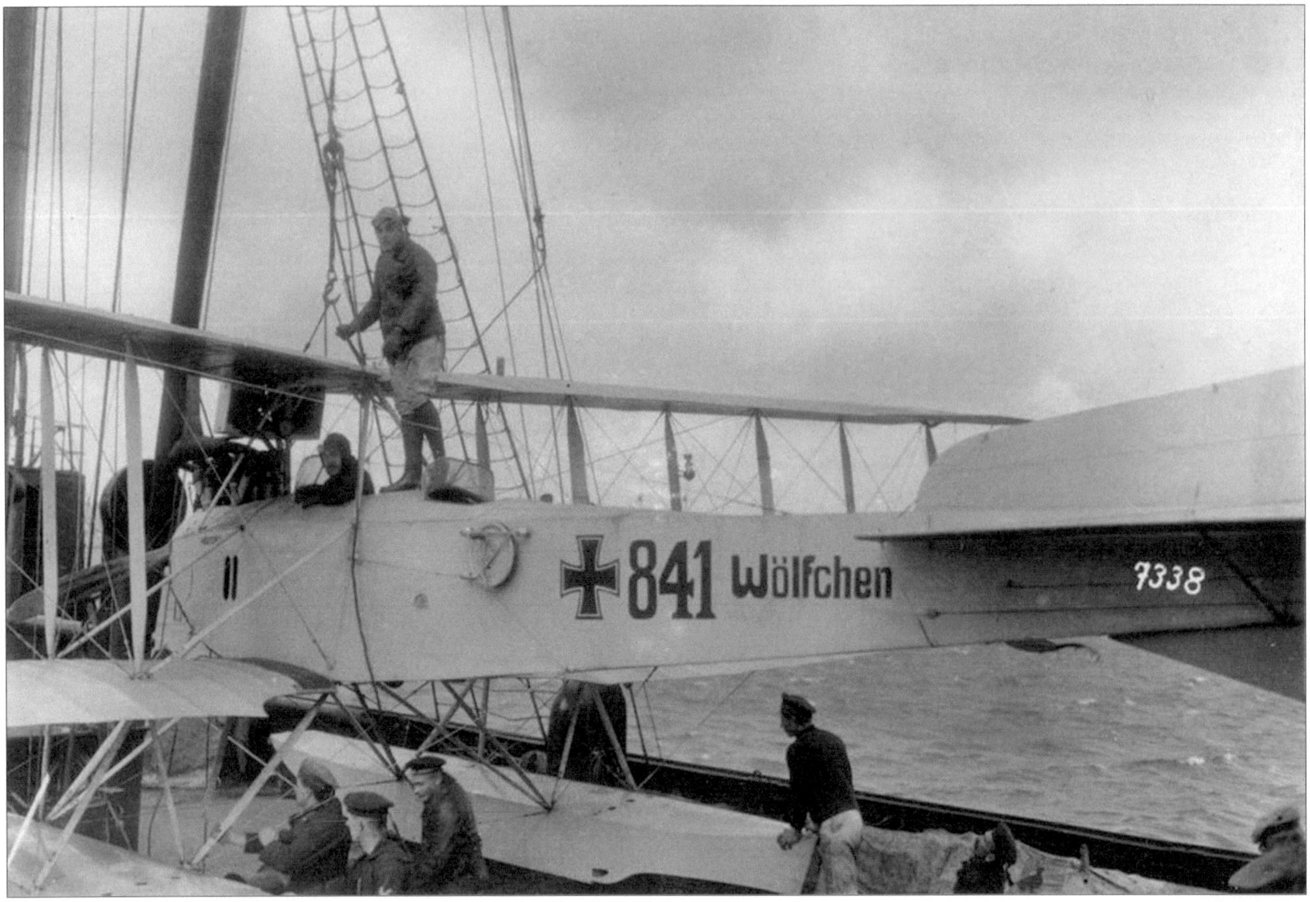
841 Wölfchen
7338

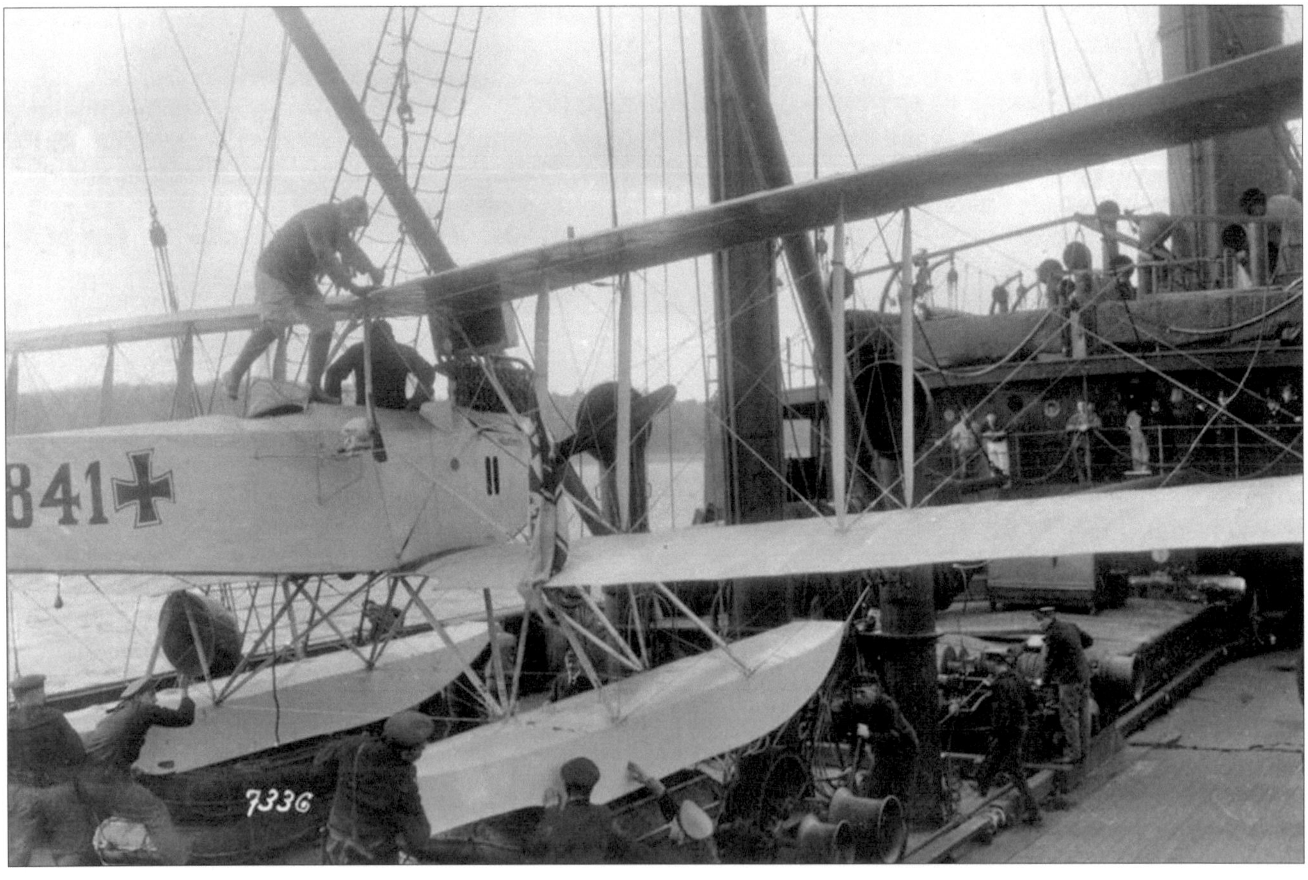
841
7336

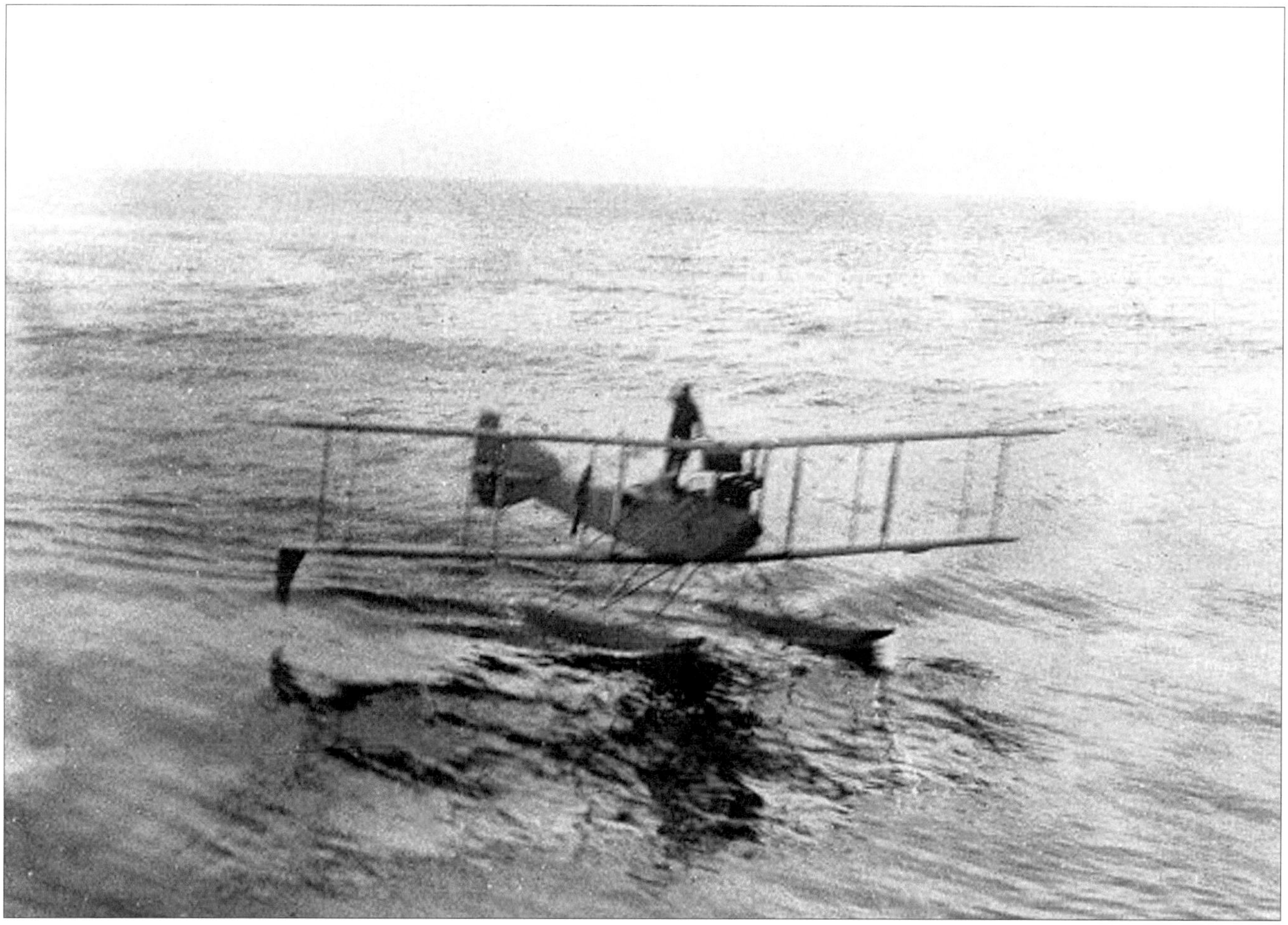

Above: A photo shows the approach of the *Wolfchen* after completion of a sortie.

Facing Page, Above: The seaplane was a Friedrichshafen FF 33 HFT variant which meant that the observer sat in the rear cockpit, which can be noted here because of the radio transmitter antenna reel seen beside the rear cockpit. Fabeck is in the front cockpit and Stein is standing to help with the launch of the aircraft.

Facing Page, Below: A view from the other side during the launch process of the *Wolfchen*.

operational, he could never catch the other ship, Nerger ordered the aircrew, pilot Paul Fabeck and observer/radio operator Matthaus Stein airborne to intercept the ship and try to bluff it into surrender as Nerger brought the S.M.S. Wolf from its anchorage.

The ship that the Germans had spotted was the cargo ship Wairuna, a ship that had left New Zealand just two days before, bound for San Francisco. The captain, Harold Saunders, had dismissed any concerns about the ship now weighing anchor and coming out from the small harbor, since he had not been issued any warnings about the possibilities of a hostile raider before embarking:

"So Saunders was entirely unprepared when, within twenty minutes of his first mate's remark, a biplane lifted off from the sea near the black freighter and headed towards them, droning so low overhead that it nearly clipped their masts as it dropped a weighted bag onto the Wairuna's deck. The bag was retrieved by a crewman and brought up to the bridge, where Saunders opened it and found the Wolf's customary note of greeting. On a second pass, the Wolfchen added emphasis by dropping a bomb, which exploded off the bow and Sauders hastily brought his ship to a halt. Well out of radio range of any land, he watched helplessly as the black raider cast off from Sunday Island and approached at slow speed with its guns visible." Richard Guilliatt

The Wolfchen had not only prevented an enemy ship from escaping and likely sounding the alarm about having seen another, suspicious ship, it had provided 1,150 tons of coal onboard the Wairuna for the taking by Nerger to replenish his ship's supply.

Only two weeks later, Wolfchen repeated its success by flying up to and then circling overhead of a four-masted American ship, the Winslow. Having dropped a message onto the deck, the ship reefed its sails and the S.M.S. Wolf had another

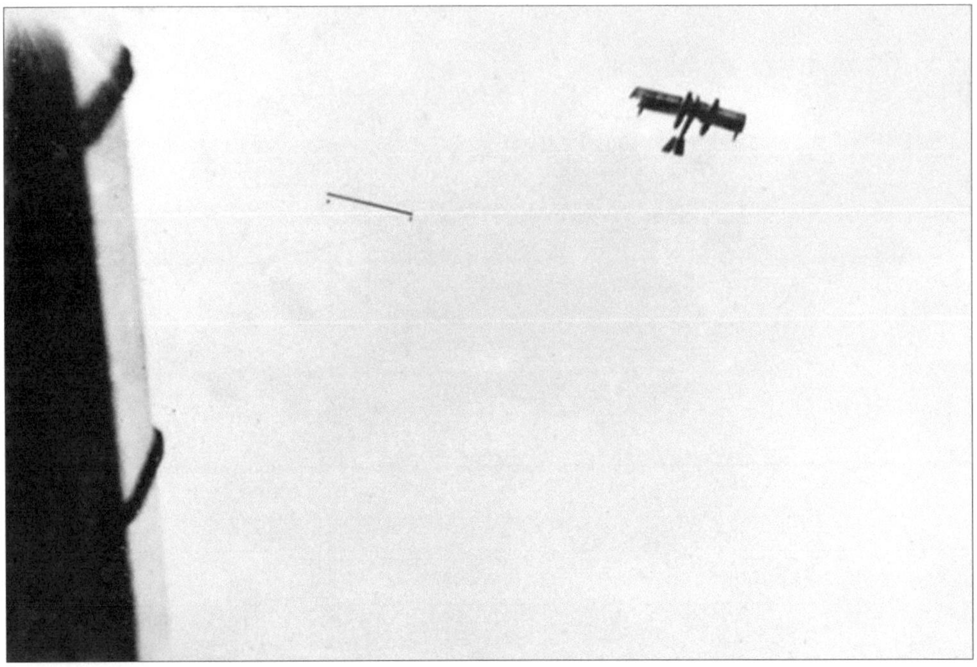

Left: *Wolfchen* above the SMS WOLF, this view would have been how the crews of the Allied ships would have first suspected that they were in trouble.

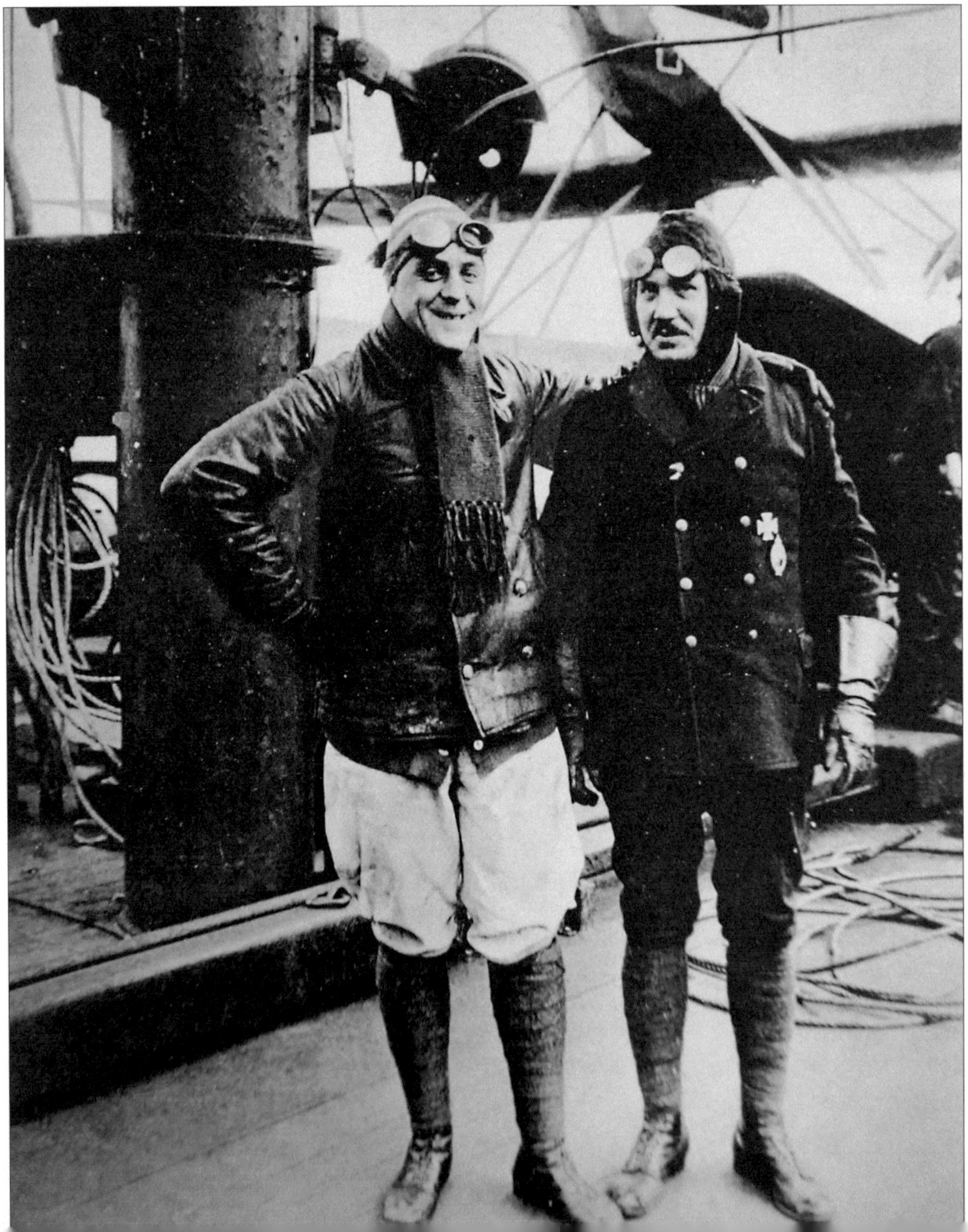

Left: Leutnant Flieger Alexander Stein on the left and Oberflugmeister Paul Fabeck on the right.

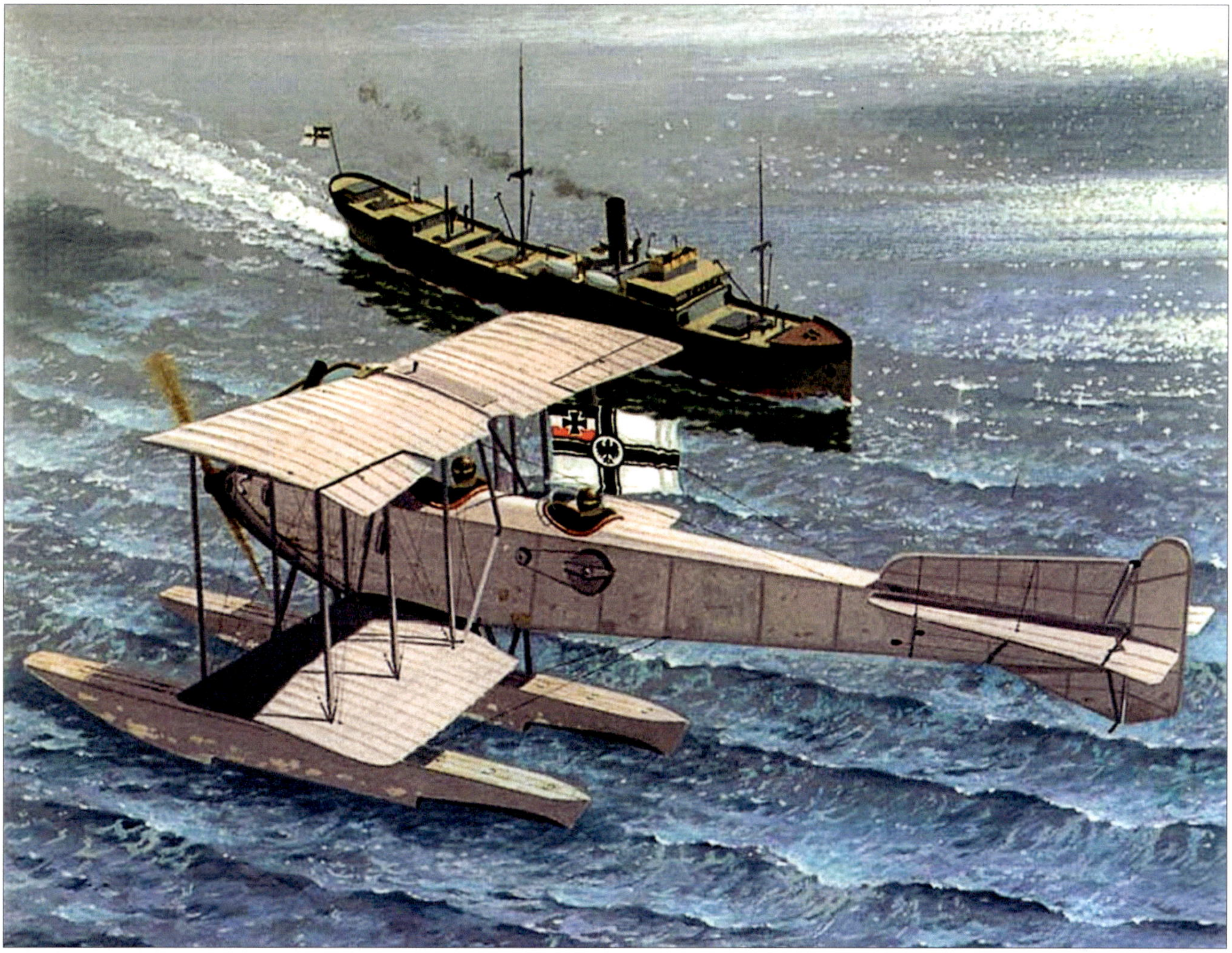

Above: Likely a depiction of the *Wolfchen* after its extensive repairs, note the lack of insignia on the aircraft, necessitating the German Naval Ensign from a starboard strut. Brian Knight

prize. Especially critical was the finding of 1,500 boiler bricks which were invaluable to reline the overworked boilers of the German ship.

Another American ship, the Beluga was captured on the 9th of July in a similar fashion. However, shortly after its capture bad luck struck:

"...the Wolfchen came in to land on a heavy swell and suddenly crashed nose-first into the ocean. The impact flipped the plane vertically onto its propeller, splintering its pontoons, tearing gashes in its threadbare fabric, and Forcing Fabeck and his copilot to climb the inverted tail as sharks began circling." Richard Guilliatt

With his recent successes, Nerger was worried that an increased number of navy warships had been sent out into the area looking for him. Without his eyes in the sky, he left the main trade routes, keeping him safe from capture but also making it very unlikely that he would claim another prize.

It took about three weeks to repair the seaplane and a lot of ingenuity. Wooden wing struts were fashioned from captured tea chests, spare sail canvas was stretched over the fuselage and the salt water soaked motor was finally brought back to working order.

However, by August and with the Wolfchen repaired, Nerger was feeling ready again. Nerger intercepted several radio transmissions which indicated that a very desirable target was at hand.

Nerger calculated that the Australian ship, Matunga, would have to pass in a narrow strait between the east coast of New Guinea and the Solomon Islands if it was to reach Rabaul- which he knew was its intended destination based on the intercepted messages.

He therefore positioned himself in a secluded area near that strait and sent the Wolfchen aloft to find his next prize.

When the German seaplane dropped its usual message onto the deck of the Matunga, and a warning shot was fired from the German raider, the helpless Australian freighter stopped all engines.

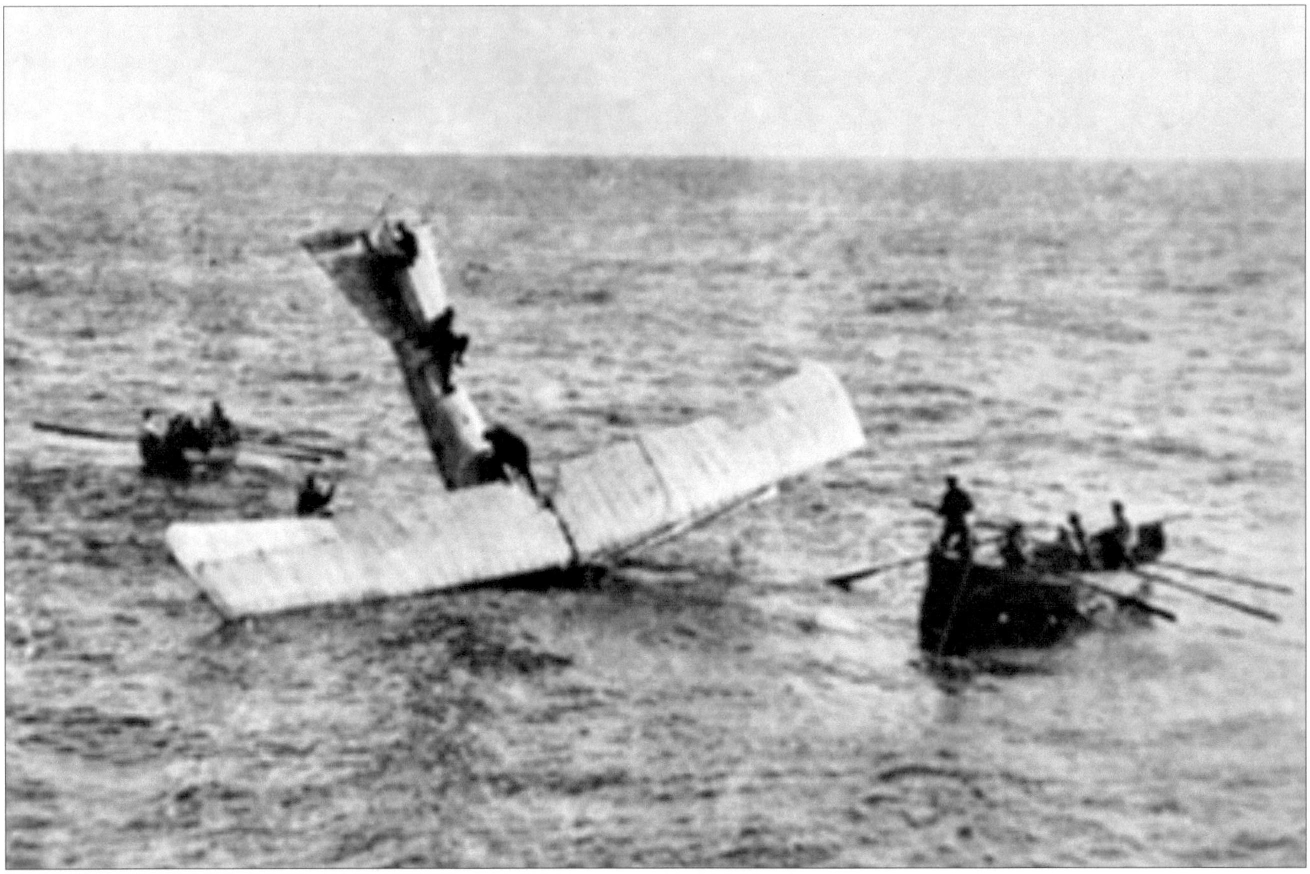

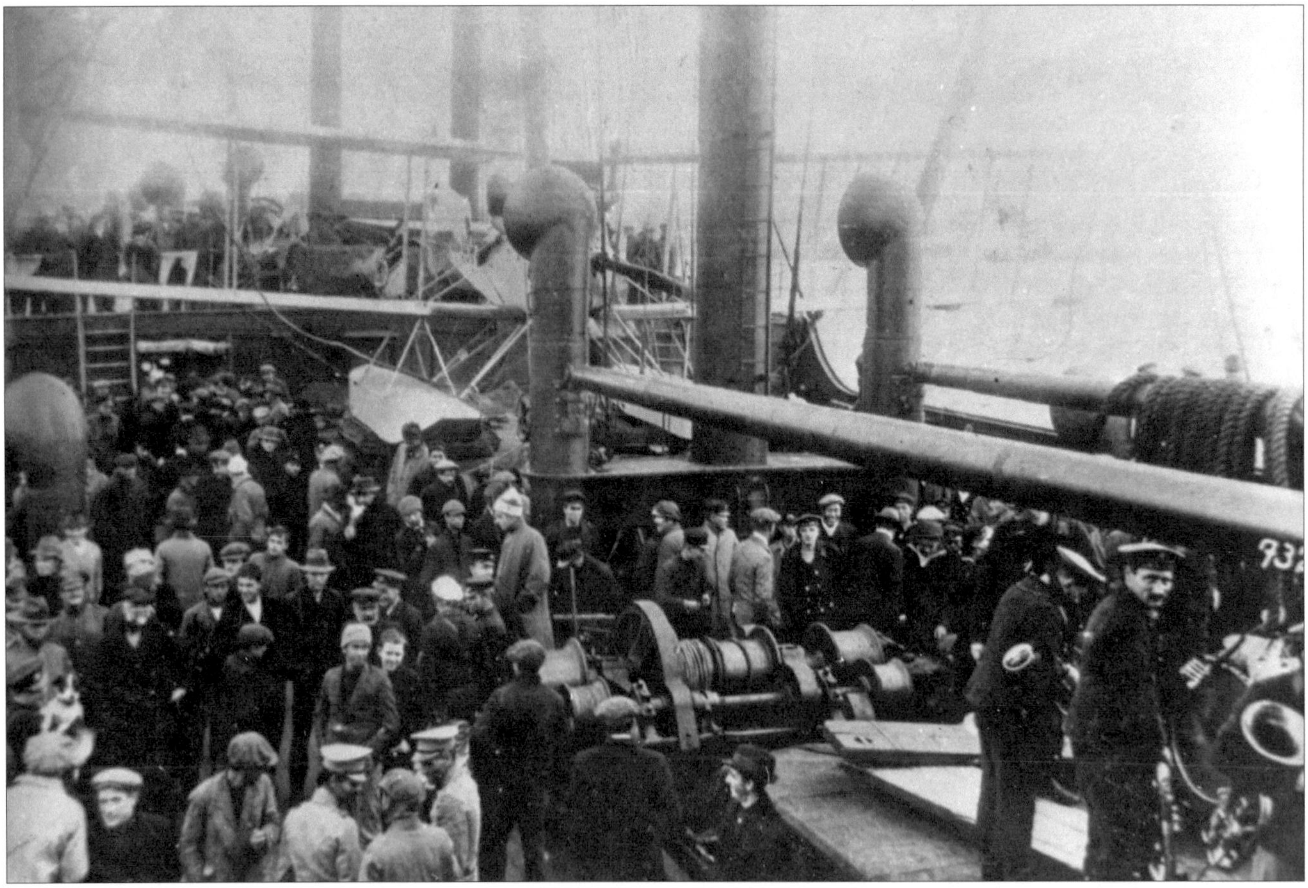

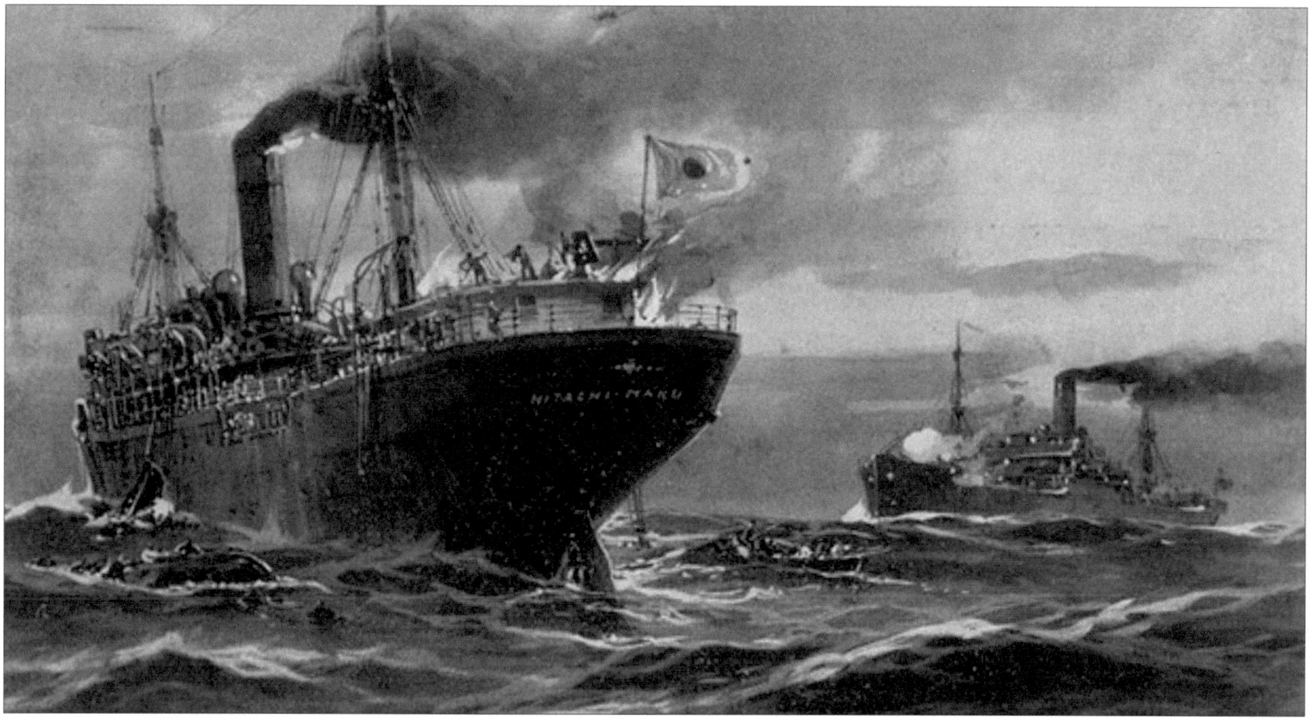

Above: The SMS *Wolf* had its only active combat engagement on 25th September 1917 when it battled the Japanese *Hitachi Maru*. Captain Nerger was especially pleased to capture this vessel because it allowed him to transfer almost all of his prisoners onto this ship.

Facing Page, Above: While returning from a flight in July 1917, the *Wolfchen* tipped up onto its nose. Stein and Fabeck can be seen climbing towards the tail of the aircraft to re-distribute weight as well as to stay away from the water since there were sharks sighted. Captain Nerger sailed to a safe place to anchor and did not resume his cruise until the aircraft was repaired into flyable condition: indicating how important he felt that *Wolfchen* was to his mission.

Facing Page, Below: One of the biggest problems faced by the German raider was how to care for the captured crews of the Allied ships – seen here crowded on the aft deck just below the *Wolfchen*.

Another load of coal for the Germans, as well as a large amount of food: including meat and fruits.

Now the *Wolfchen* performed another vital role as it flew ahead and scouted to make certain that there were no enemy ships near the German raider and its new prize as they sailed towards finding a safe harbor in order to transfer the stores from one ship to the other.

Throughout the voyage, the Wolfchen performed invaluable roles, which allowed the German raider to succeed at avoiding capture and significantly aided its ability to be at sea for so very long.

At various times, the Wolfchen would fly from its mother ship to one of the prize ships to inform the German crew that had been placed onto the captured ship when and where to meet with the S.M.S. Wolf: which allowed precise coordination and radio silence for the German raider.

The use of this new tool in its arsenal, the seaplane, clearly made for a more successful and prolonged cruise for the German raider than would have otherwise been possible. It is also likely that the use of its aircraft led to fewer casualties in the S.M.S. Wolf 's engagements because the "usual" practice of having the Wolfchen fly to the intended target and drop its warning message would allow the ship to stop engines and surrender before a shot was even fired.

This tactic however did not work as well as possible on the 25th September 1917. The Germans were about to capture the largest ship of its entire voyage (larger ships were credited to the S.M.S. Wolf, but these had been sunk by mines laid near harbors).

The Hitachi Maru was a 6,700 ton mail-passenger ship of the Nippon Yusen Kaisha line. It had left Yokohama, destined for London. There were over 40 travelers on the ship, in addition to its crew. The ship also had provisions to last for its long voyage, which included a very generous supply of alcohol of a variety of spirits. But most importantly, if captured it would allow Nerger to put many of his swelling numbers of captured people off from S.M.S. Wolf: which could barely accommodate the now several hundred people, including women and one

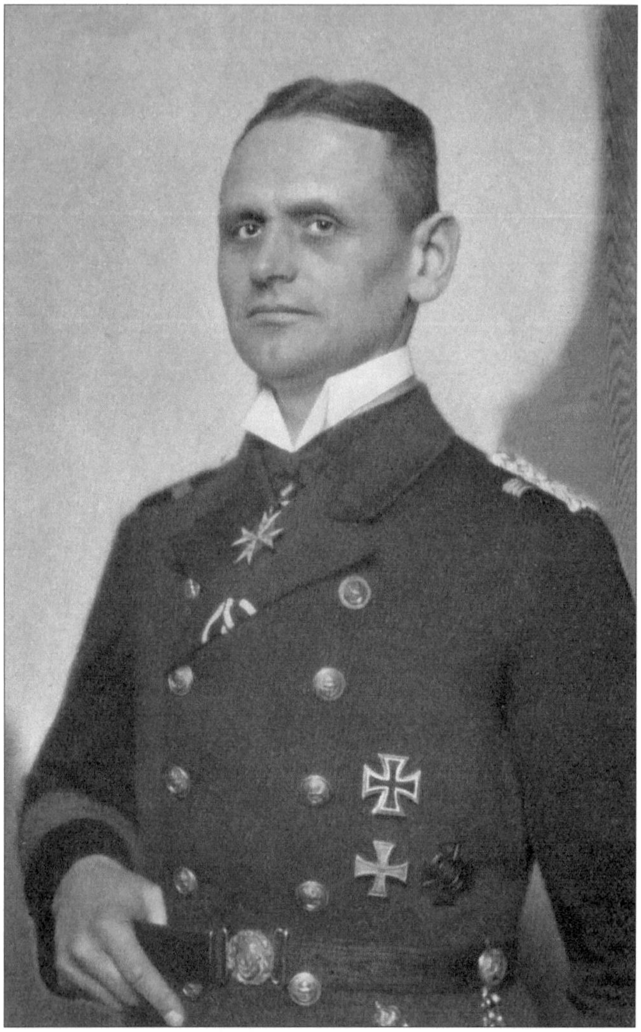

Above: A fine portrait of a proud privateer, Captain Karl Nerger shows his recently awarded Pour le Merite.

child, Juanita Cameron (she was the daughter of the captain of the Beluga).

Following the well-practiced drill, the Wolfchen flew overhead to drop a message upon the Maru and for good measure dropped a bomb close by to the ship. The S.M.S. Wolf ran up its battle ensign and fired a warning shot past the Maru's bow for good measure.

However, the Japanese commander had no intention of not engaging the German ship. A small gun had been installed on the Maru, which was now manned and Lt.-Commander Shiraishi ordered it to fire at the well-armed German raider fast approaching his ship.

For the first time in its now ten month voyage, Nerger ordered his men to open fire upon a ship that still had people on board(they had shot at prize ships in order to sink them after the people had been removed).

Facing Page, Above: It is very unlikely that the SMS *Wolf* would have been as successful if it was not for the efforts of its two naval aviators Stein (L) and Fabeck.

Facing Page, Below: Chart depicting the extraordinary cruise of the SMS *Wolf*, the longest by any warship of any nation in the First World War, a cruise not likely possible without its aircraft in support.

The resulting broadsides severely tore apart the stern section of the Japanese ship, where the gun was located. As more sailors rushed to try to man the gun, another salvo from the German raider effectively put the Japanese gun out of commission.

The end result was never in doubt, after Nerger placed a small, prize crew on board the Maru, the two ships sailed towards a safe anchorage in the Maldives. There the Germans worked for days to retrieve what was wanted from the Maru and after making repairs to the damaged ship, Nerger transferred many of his prisoners onto the Japanese ship where they found their new accommodations nearly luxurious compared to the "hell hole" shared space below decks of the German ship. Nerger was aided in his plans to make this new prize into a ship to accommodate any prisoners who might still be captured in future actions because he knew that the Wolfchen could fly back and forth if needed to coordinate the movement of the ships while still maintaining strict radio silence, which was crucial to the raider's survival.

During his mission, Nerger tried to treat the various people he captured as humanly as possible, with the limited resources available to him. Over time, it was not surprising that the humanity inherent in people led to interactions with the crew and their captives.

The young girl, Juanita Cameron became a favorite of the crew and she was given freedom to roam the ship and was always welcomed at the officers table for supper. The Wolfchen's pilot, Paul Fabeck became her favorite. He had a daughter at home of about the same age and no doubt saw in the little girl traits of his own daughter, far away. As an example of their friendship, on Christmas eve 1917, Juanita was expressing great doubts that Father Christmas could ever find her as she was now "lost at sea on a strange ship". Fabeck located several toys and gifts amongst the items captured from other ships and wrapped them. When he then presented them to the young girl, he told her the story of how he came across Father Christmas while flying in the sky in the Wolfchen. He told her he was given these gifts with the expressed wish of Father Christmas to bring them to her.

Ironically, when the S.M.S. Wolf finally returned

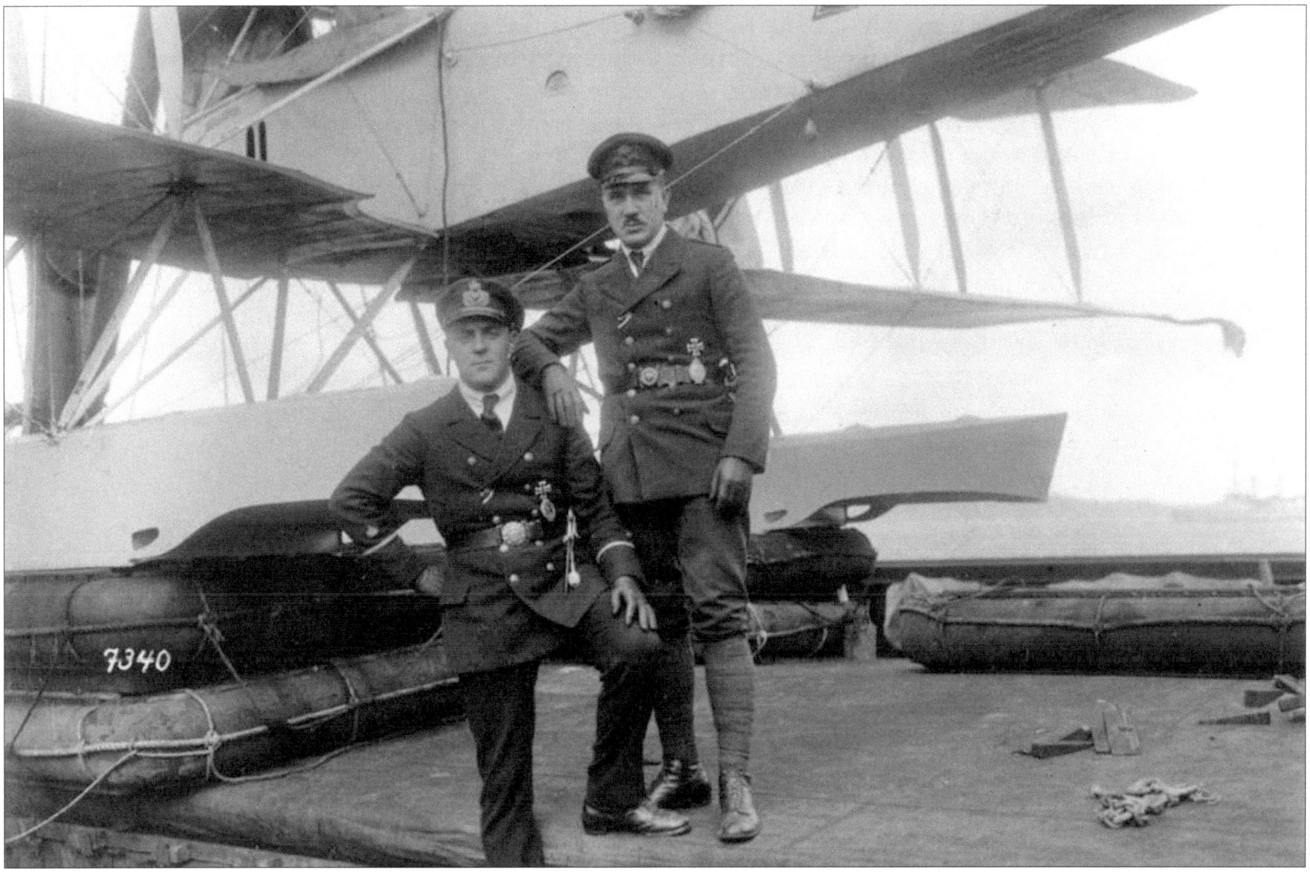

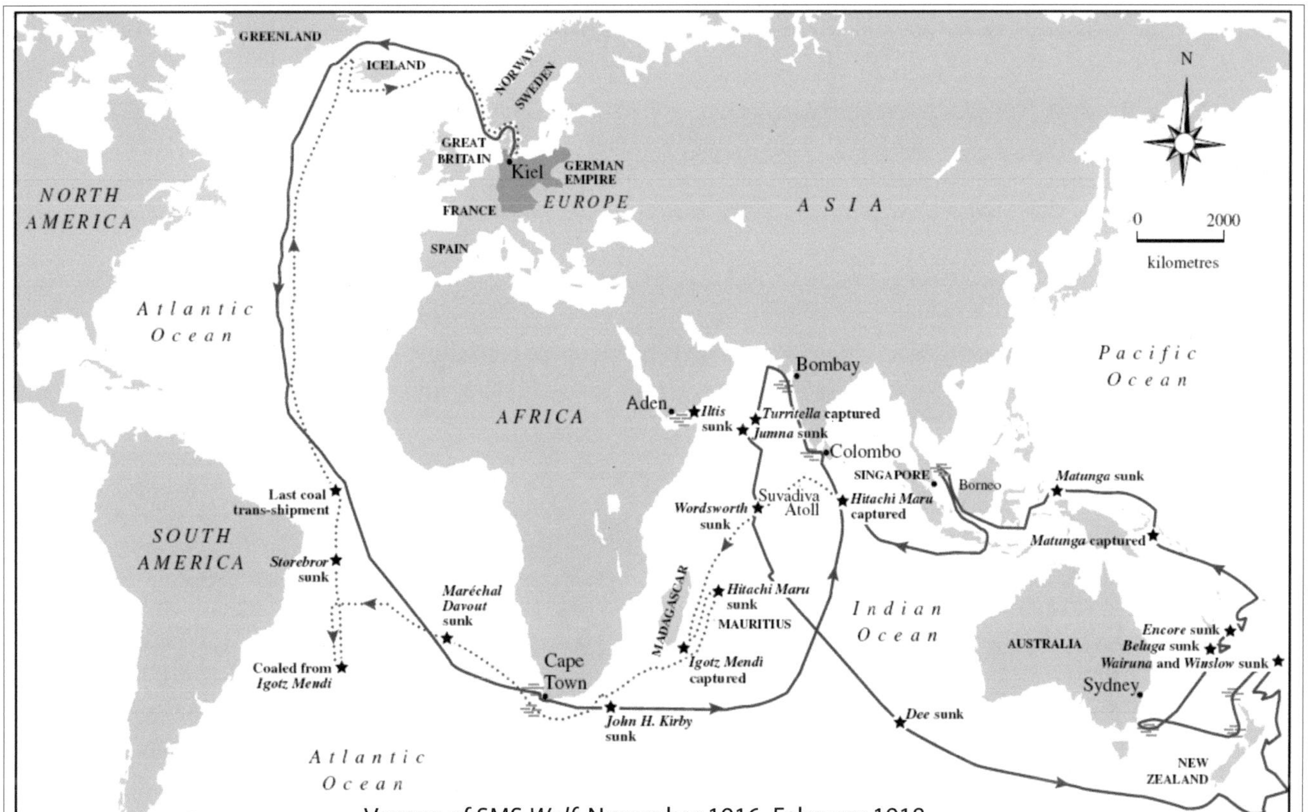

Voyage of SMS *Wolf*, November 1916–February 1918

to the waters outside of Kiel harbor and signaled to be guided into the inner harbor, Nerger was informed that everyone there thought he had already been long lost at sea. Work crews were sent out to paint the German raider and make it look respectable. New uniforms were brought out for the crew as well and after several days of preparation, the ship was deemed ready for its return to the harbor where the epic journey had all started.

The cruise of the S.M.S. WOLF was the longest voyage of any single warship in the First World War. Making the successful cruise even more unbelievable was that it was accomplished without any support of any kind- the armed cruiser was in enemy waters the entire voyage and could never hope to enter any friendly harbor for repairs or assistance until it made its journey back to its home port.

The only support the ship had to rely upon was its Wolfchen, which no doubt was a significant factor in the overall success of the German raider.

Nerger proudly sailed his ship into Kiel harbor past an honor guard of ninety warships, having completed his record setting 444 days mission. On the rear deck of the ship, a newly repaired and brightly painted Wolfchen prominently sat.

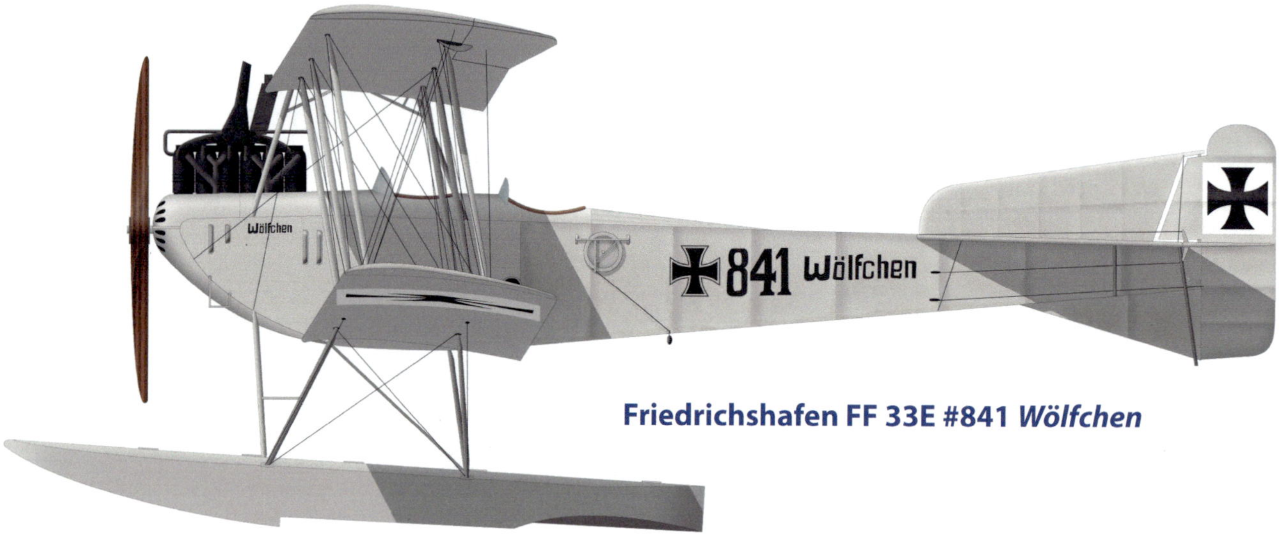

Friedrichshafen FF 33E #841 *Wölfchen*

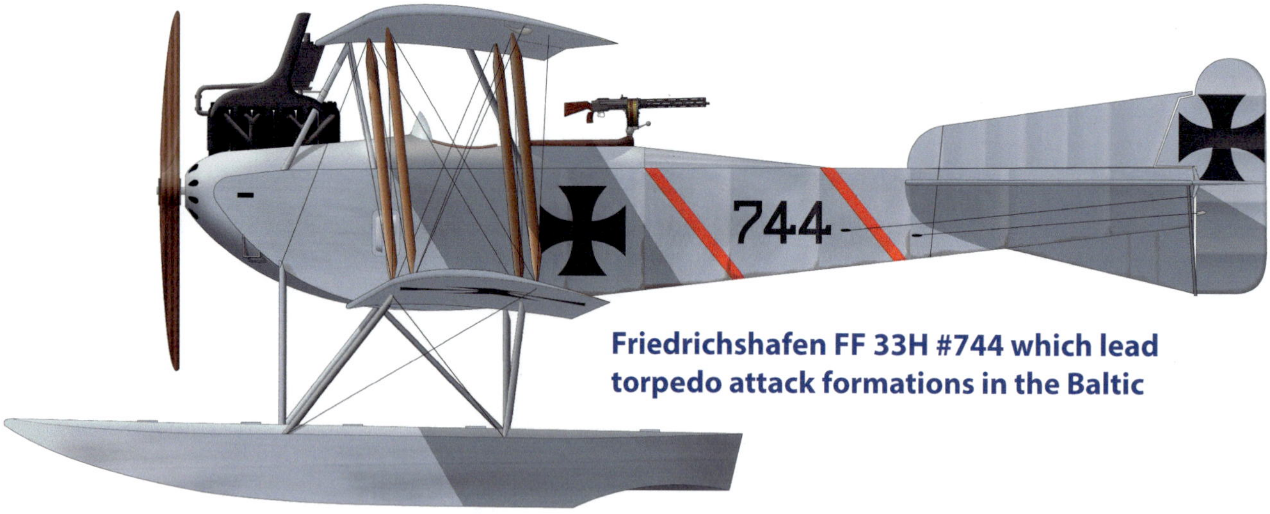

Friedrichshafen FF 33H #744 which lead torpedo attack formations in the Baltic

USAS Scores

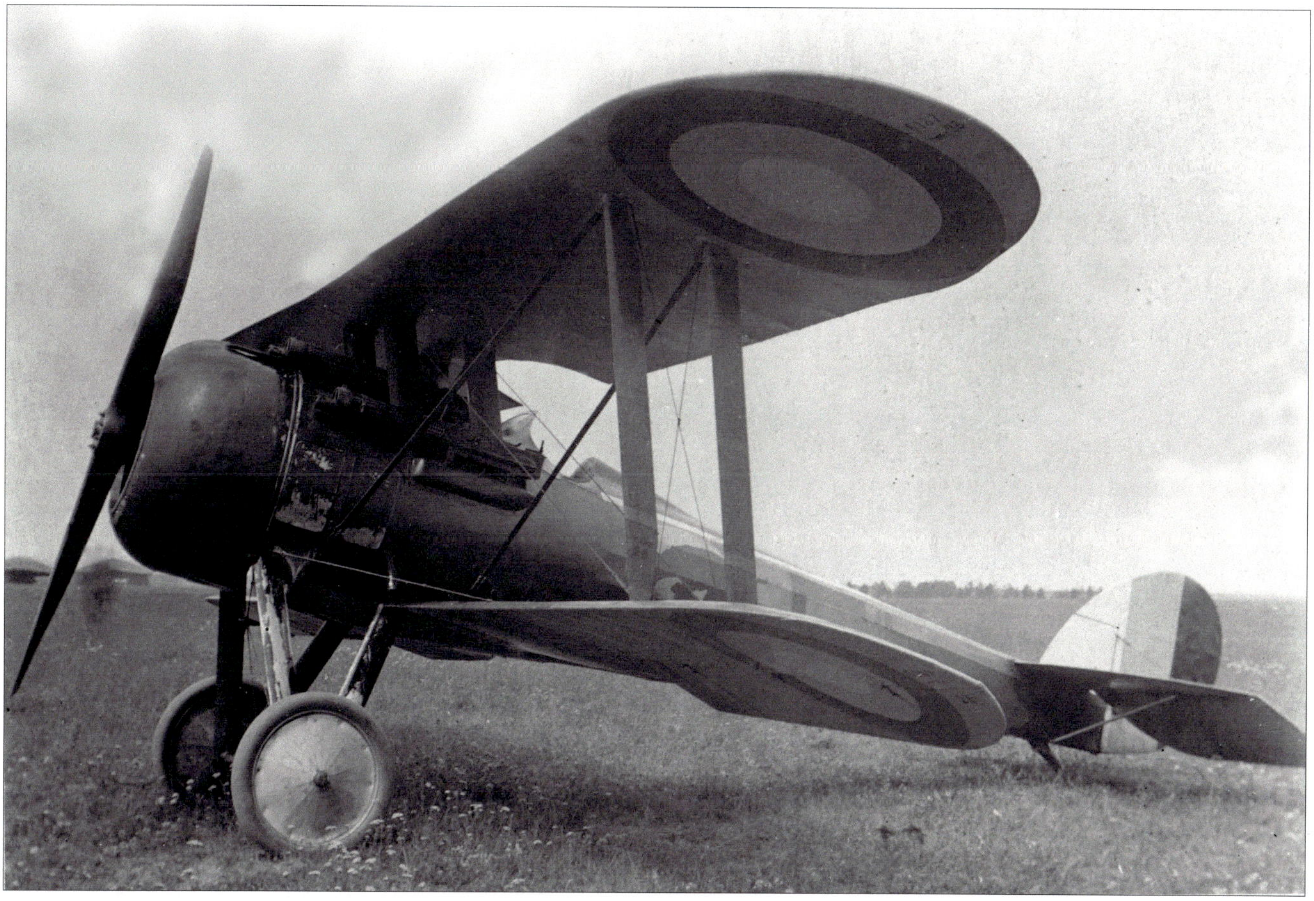

Above: The Nieuport type 28 was the first American fighter plane. It was initially delivered without any machine guns, obviously limiting its capability as a fighter.

Germany anticipated that the United States would enter the war at any moment. Therefore the Kaiserschlact (Kaiser's Battle) was initiated in the last week of March 1918 with the purpose of breaking through the Allied lines on the Western Front to force a peace initiative before the Americans could join in the actual fighting. Despite the United States formally declaring war against Germany on 8th April 1918, the Germans did not have to initially worry much about American air power.

The few airplanes that the U.S. military had were woefully inadequate and incapable of front line service. There were very few aviators in any case. However, the Americans did bring an unbridled enthusiasm and a firm conviction to win the war with them.

Major Raynal Bolling, an aviation reserve pilot was chosen to head a commission with the express purpose of finding airplanes for the fledgling United States Air Service (USAS). It was initially believed that the vast American manufacturing capability would be brought into action to make thousands of aircraft in license agreements from European firms and saturate the skies over the battlefields.

Rather quickly, such thoughts were recognized as being foolish. The time it would take to re-tool the factories, the time to produce the aircraft and then the time to ship them to Europe would mean that the aircraft designs would likely be outmoded by the time they reached the battlefront. The Americans looked to France (with plenty of cash) to purchase examples of their latest designs. However, the French were also working to fulfill their Air Service's demands as well.

The Nieuport aircraft company was founded by Edouard Nieuport before the war and was famous for having developed several successful designs that won some of the prestigious pre-war competitions. Edouard died in a fatal crash flying one of his designs and the firm was taken over by his brother Charles.

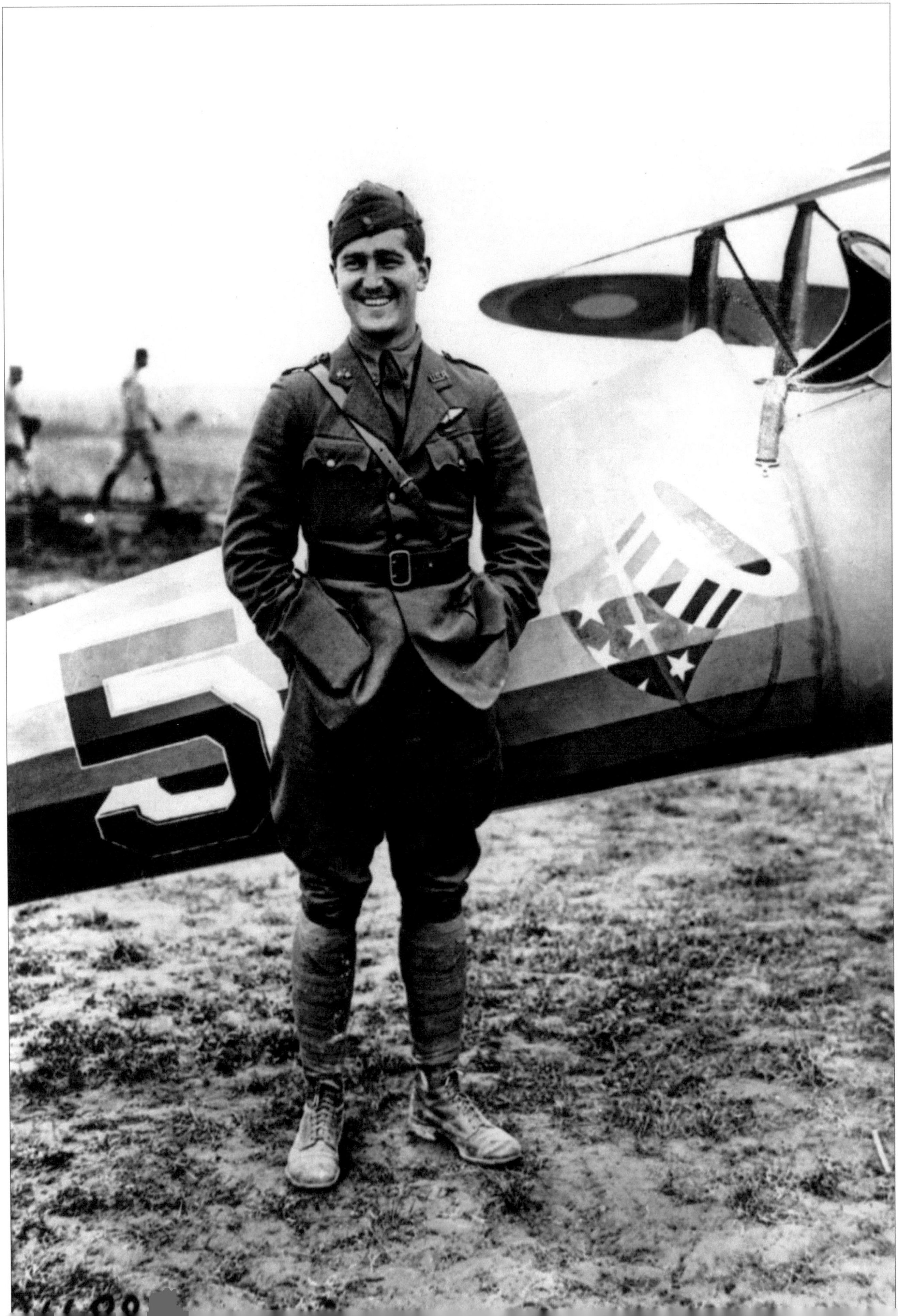

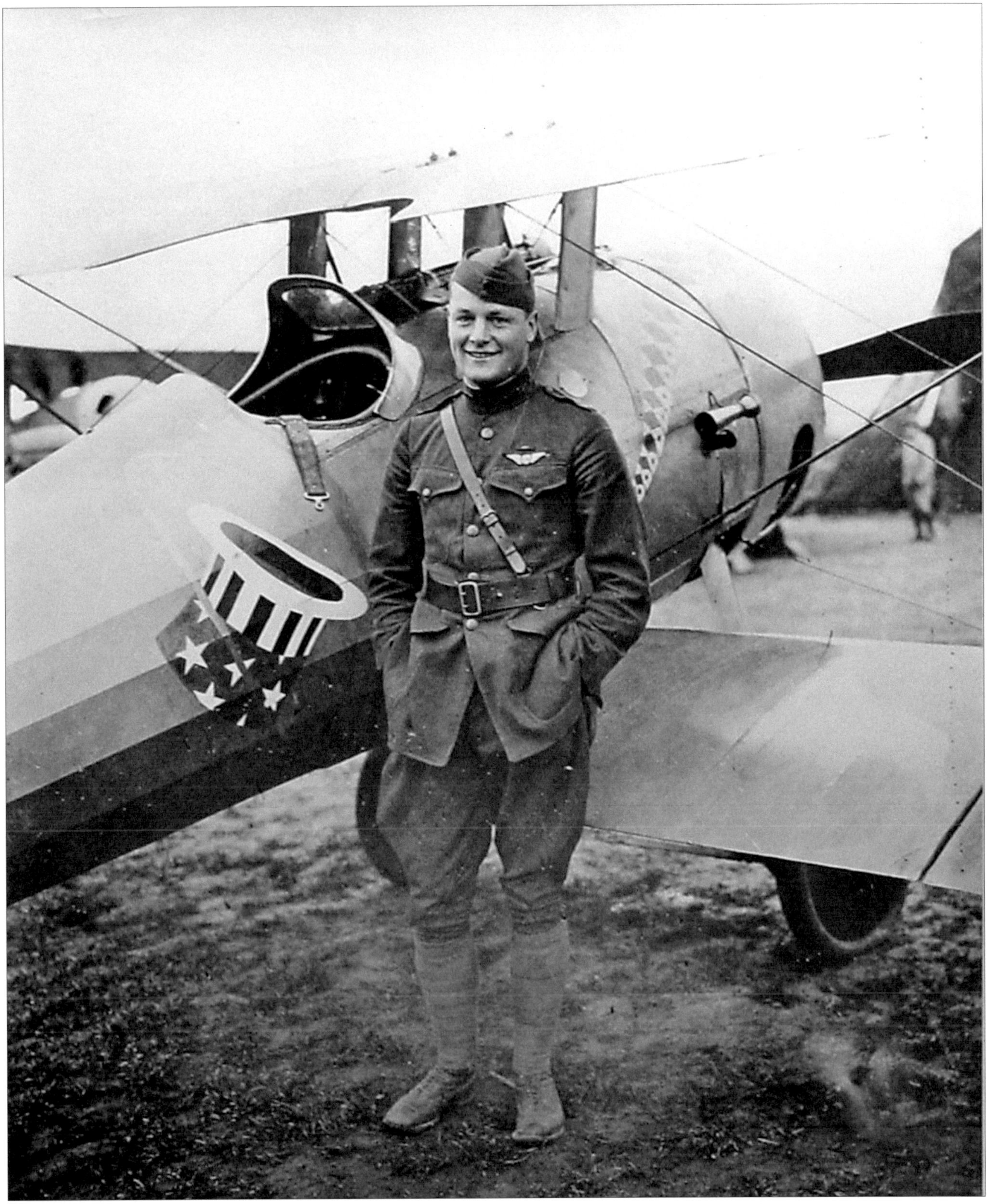

Above: Lt. Alan Winslow stands beside a Nieuport 28 fighter of the 94th pursuit squadron USAS. This aircraft was assigned to Lt. John Wentworth as can be determined by Wentworth's personal decoration of a snake that wrapped around the cowling.

Facing Page: Lt.Douglas Campbell beside his Nieuport 28; the squadron's insignia was based on the thought of Uncle Sam throwing his hat into the ring.

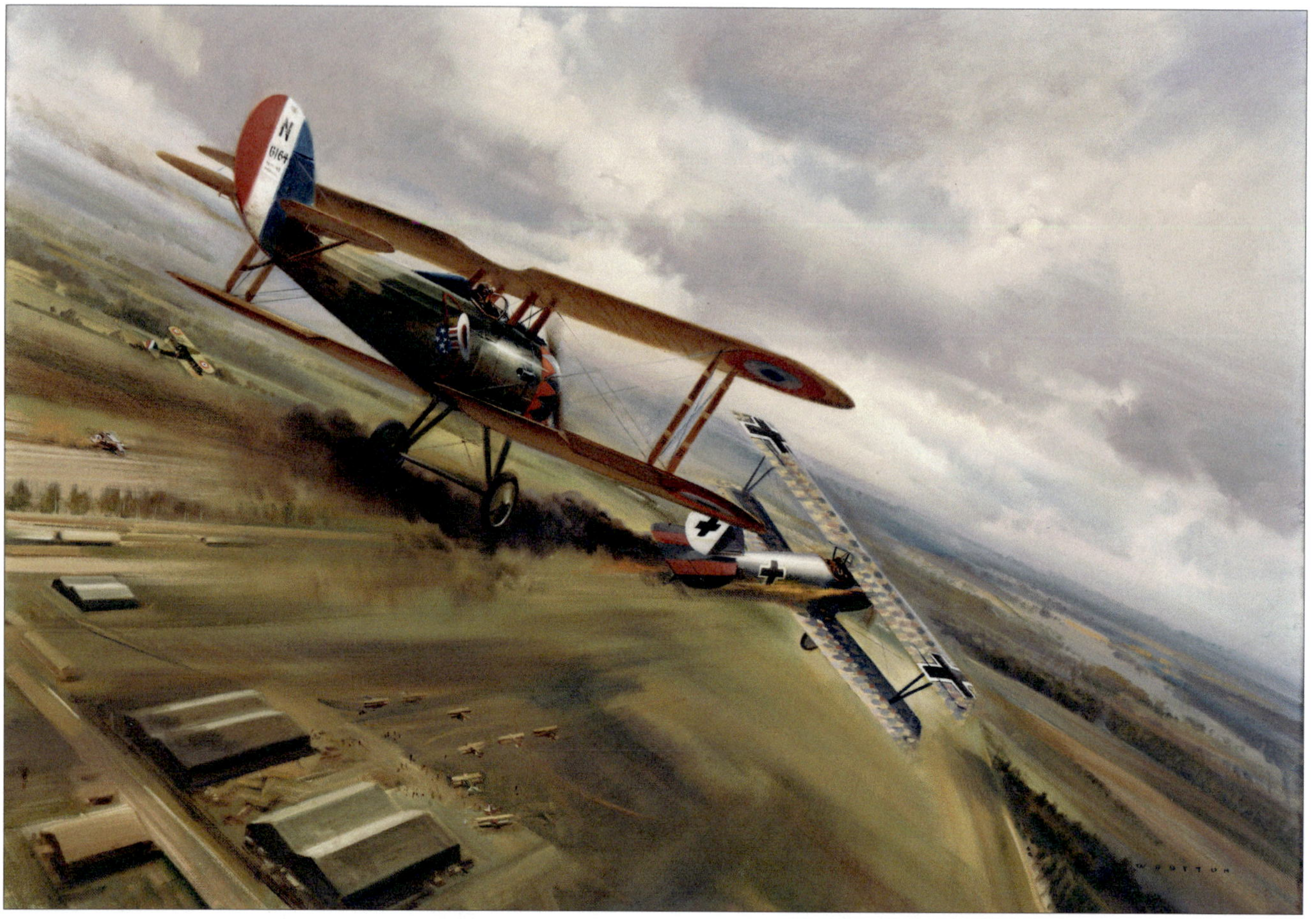

Above: Winslow is in the process of bringing down Vzfw. Wronieki in his Albatros D.Va. Frank Wootton

He also died in an aircraft accident in 1913. These well-publicized deadly accidents dealt a blow to the Nieuport company.

However, the company had a brilliant technical director responsible for aircraft design, Gustave Delage. Under his direction, a new type of Nieuport began to take shape for the 1914 air racing season. With the outbreak of the war, the type would be tweaked to conform to military desires for biplane designs. Delage incorporated a sesquiplane design because it would be lighter, faster and give the pilot superb vision compared to the more traditional biplane designs of the time.

The N.10 was the initial design, but this was a two-seater and so a smaller version, the N.11 "Bebe" was built as a single seat pursuit aircraft. These types would be so successful that Nieuport aircraft would eventually serve in every Allied country's air service and would bring the Nieuport company sales from 285,000 francs in 1914 to 26.4 million francs in 1916! Eventually, even the German aircraft companies copied the sesquiplane design which led to the well-known Albatros and Pfalz fighters.

The Nieuport fighters were built around a rotary engine which was known to have a superb power to weight ratio for the light aircraft of the day. However, rotary engines were inefficient and needed servicing quite often. The rotary engine also had an inherently limited capability of power output and so in the end, the design had an inescapable limitation that could not be ignored.

As lessons were learned from combat experiences, the need for more robust aircraft with greater speed and the ability to withstand greater G-forces was crucial. Thus, the Spad fighter designs with their Hispano-Suiza powerful inline engines eclipsed the Nieuport company offerings by early 1918.

Therefore as Major Bollings began to search for an aircraft type for the new American fighter units, he was drawn to the Spad XIII. However, the French wanted to use these aircraft for their own units. In a recent design competition, the Nieuport 28 had lost out to the latest variation of the Spad fighter. In fact, the Nieuport company had already moved on and was already involved in designing the Nieuport 29 built around the Hispano-Suiza engine (this aircraft would become the standard French fighter throughout the 1920's).

As Bollings was trying to find a fighter aircraft

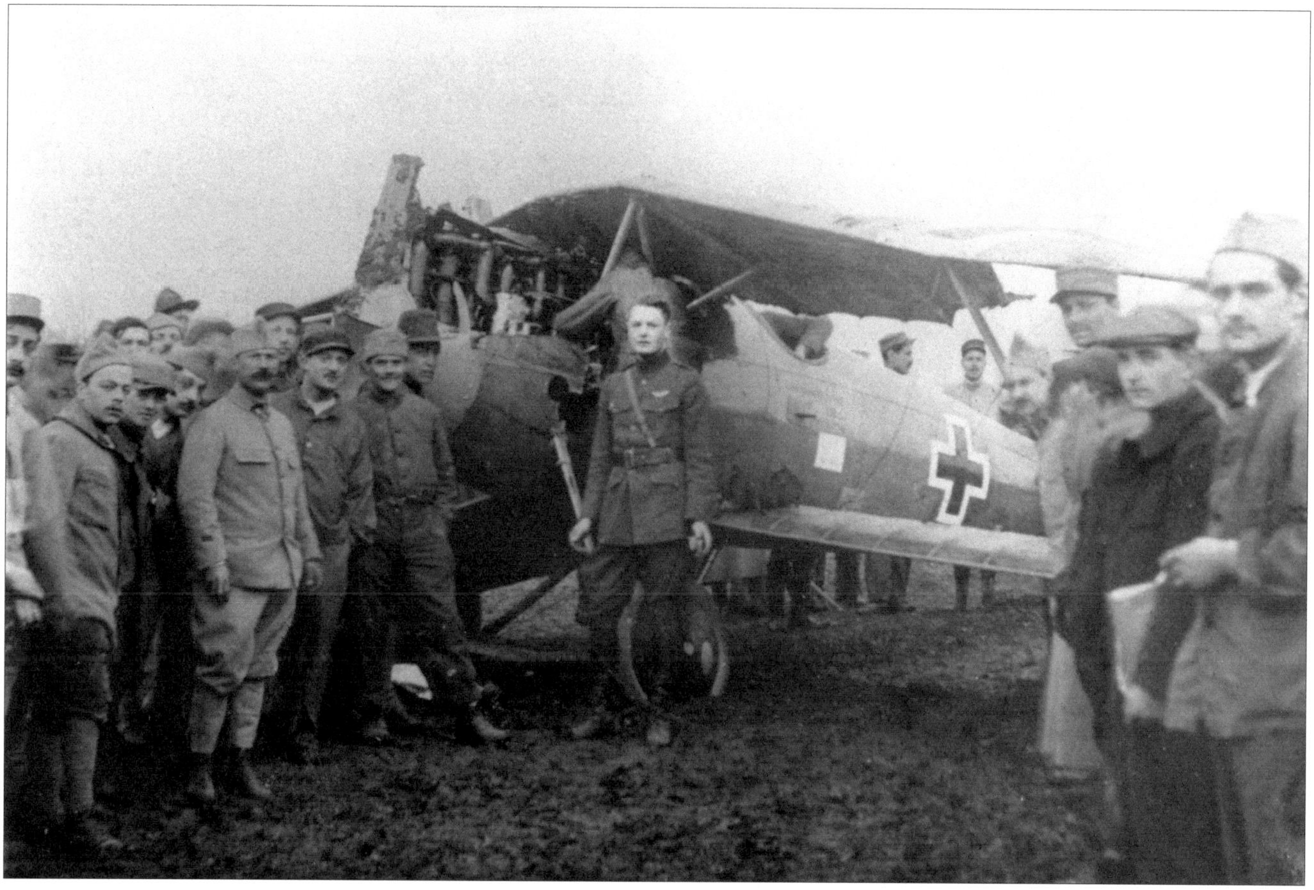

Above: Alan Winslow beside the Albatros fighter that he brought down. Its pilot was a Polish national, Antoni Wronieki, who was actually in the process of defecting. Wronieki would join a French/Polish unit before the end of the war.

Right: Winslow (L) and Campbell (R) received the Croix de Guerre for their victories.

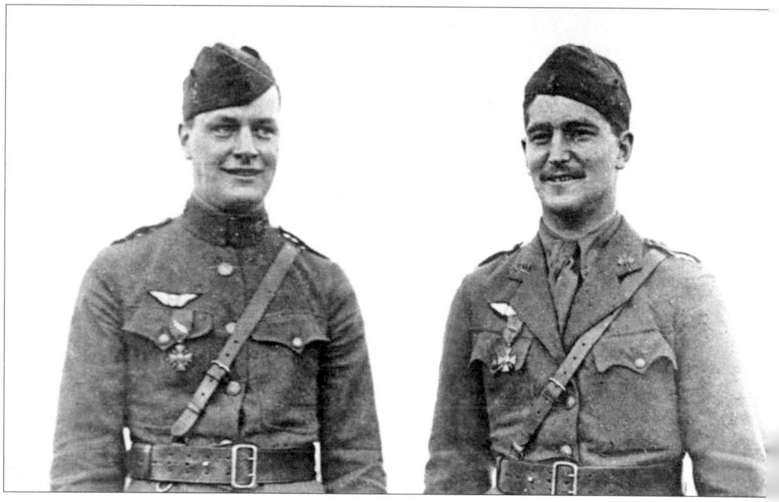

for the fledgling USAS,it happened that the Nieuport company was not otherwise producing any other types of aircraft, even though it was already fully prepared to build the N.28 type. In addition, since the N.28 design had a rotary engine, the requests for the production of the Nieuport 28 by the Americans would not be in jeopardy since the French Spad fighters were utilizing the inline Hispano-Suiza engine. It was primarily for these reasons that the first American fighter plane was produced by France and that the aircraft was the Nieuport 28.

What France could also offer to the USAS was utilization of their numerous aviation training facilities and instructor pilots. The camps would have to be expanded to accommodate the influx of the eager young Americans. Almost all of the aircraft used in training would be rotary powered Nieuport types and therefore the American pilots would enjoy the benefit of being trained on similar equipment that they would soon be flying in combat.

Hiram Bingham, a professor at Yale before the war, was placed in charge of recruiting and training of the new American eagles. In his superb book, The Unsubstantial Air, Samuel Hynes described Bingham's beliefs:

"This ideal American pilot-wellborn, well educated, athletic, patriotic, and honorable in all his doings- wasn't invented by Bingham; he was simply endorsing a type that already existed in upper and

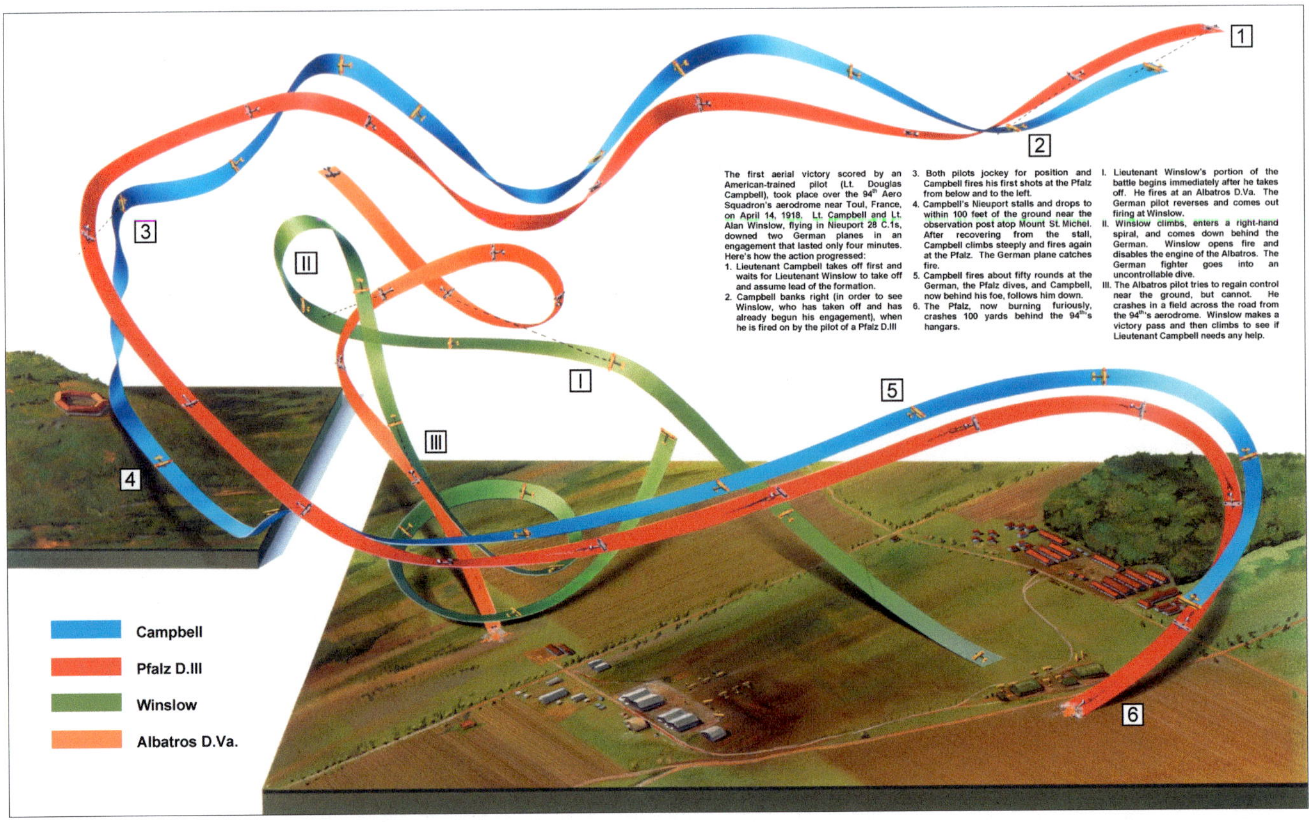

Above: Diagram of the aerial combat of 14 April 1918 that resulted in the first victories for the USAS. Ted Hamady

upper-middle-class American society. The ideal figure was there in the Yale men Bingham taught... By accepting the assumption that such young men were the best possible material for combat pilots, the Service made a class distinction: flying- American combat flying- would be an occupation for gentlemen." (p.18–19)

Thus, the first American fighter pilots literally maintained a collegial-type of rivalry as they went off to France. Many of them already knew each other, or had heard of each other, or had played against each other on the sporting fields of the elite schools back home. They went through training together and they were all eager to make the "first team".

The first unit to be deployed to the front lines, the first to the front, was the 95th aero squadron. When the young pilots presented themselves to their aerodrome at Villeneuve-les-Vertus they had nothing to do, since they had no aircraft! Finally, a handful of the pilots went directly to the Nieuport factory to pick up several of the first N.28 types to roll from the factory,

They flew the aircraft back to their base, eager to begin their contributions to the cause. Another problem arose in the fact that there were no machine guns to arm the fighters.

Undaunted, on the 15th March 1918 Sumner Sewall, Charles Woolley and Dick Blodgett piloted their unarmed N.28 aircraft over the front lines on the first combat mission of a USAS fighter squadron.

The men of the 95th lorded this "first" over their compatriots of the 94th aero squadron at the same airfield, who did not yet even have aircraft; armed or otherwise!

However, the members of the 95th were ordered to go back to the gunnery school at Cazeaux and turn their aircraft over to the members of the 94th. The arrival of some used Vickers machine guns followed and soon the Nieuport fighters had one machine gun each(the design called for two guns).

Thus, the stage was set for the events of the morning of 14th April 1918. At 0940 hrs. Forward observers noted two low flying German Aircraft headed in a Southerly direction(flying behind the Allied lines). The observers notified the closest airfield was at Gengoult where the 94th aero squadron was now based. Ltn. Alan Winslow and Ltn. Douglas Campbell were on standby and within 10 minutes had their N.28 fighters airborne. Literally as the two took off, they flew directly into the two German fighters and immediately engaged them. During the air combat the aircraft were never higher than 200 meters and in fact, in retrospect it is likely that at least one of the two German fighters was in the process of landing!

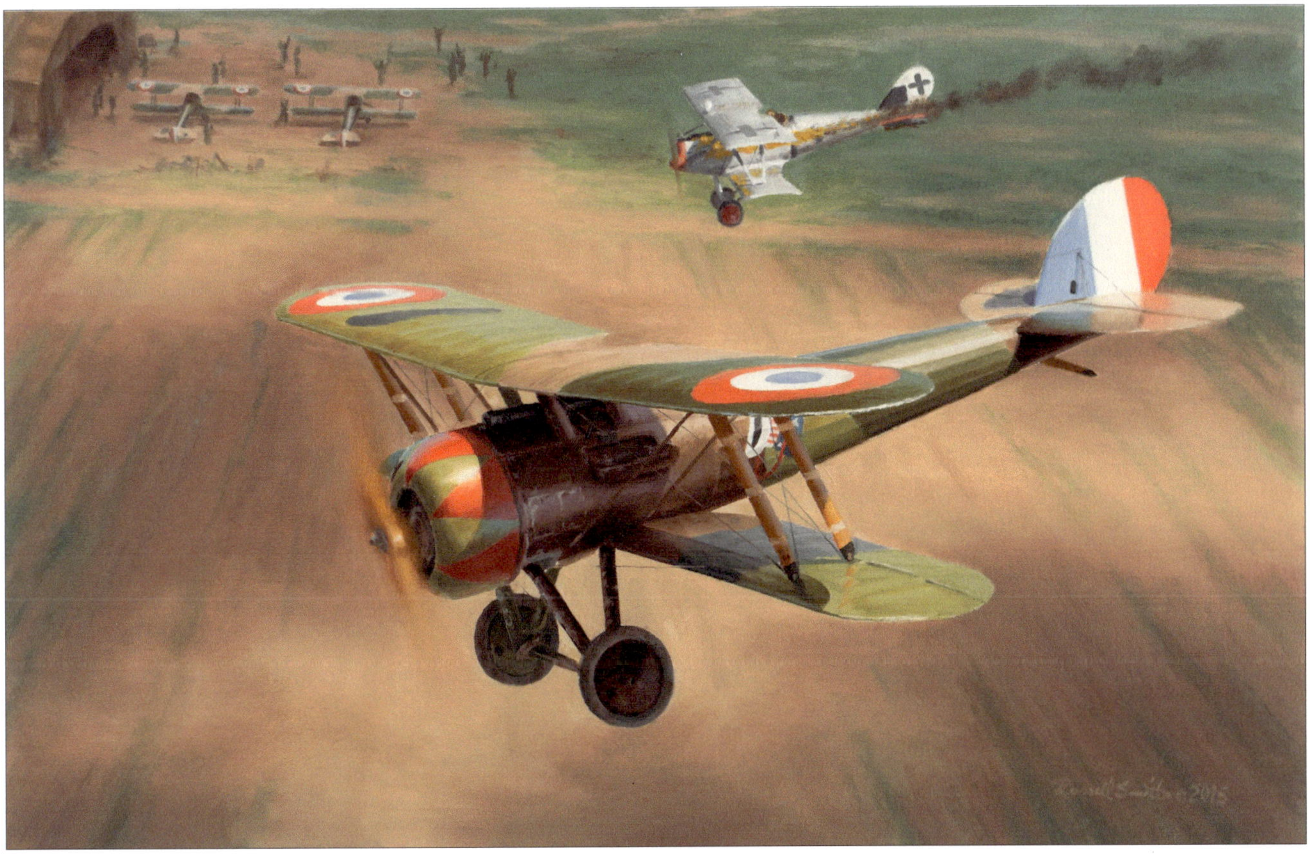

Above: Campbell brought down Unteroffizier Simon in his Pfalz D.IIIa in flames. Simon would perish from his wounds. Russell Smith

Winslow was an experienced pilot, who had previously flown in the Lafayette Flying Corps, and he quickly latched onto the German Albatros D.Va fighter and brought it down into what seemed to be a controlled crash landing on the aerodrome.

Campbell took only a few more minutes to gain an advantage over his adversary who was flying a Pfalz DIII.a fighter. Campbell had brought himself underneath the German fighter and when he fired into the Pfalz, it caught fire and crashed closely behind the 94th squadron hangars.

Campbell and Winslow quickly landed at their airfield: later Winslow would recall the scene:

"Mind you, the flight took place only three hundred meters up, in full view for all on the ground and the near-by town: and it took place directly above our Aviation field. Furthermore, mine dropped about one hundred yards to the right and Doug's one hundred yards to the left of our fields. These are remarkable facts, for one of our Majors[Major Raoul Lufberry, highest-scoring member of the Lafayette Escadrille], how, with the French army since 1915, has shot down seventeen machines, never had one land in France-and here we go, right off the bat and stage a fight over our aerodrome and bring down two Huns right on it. It was an opportunity of a lifetime-a great chance.

When we landed, only our respective mechanics were left in the 'drome' to help us out of our flying clothes. The whole camp was pouring out, flying by on foot, bicycles, side cars, automobiles: soldiers, women, children, majors, colonels, French and American- all poured out of the city. In ten minutes several thousand people must have gathered. "Doug" and I congratulated each other, and my mechanic, no longer military, jumping up and down, waving his hat, pounded me on the back instead of saluting and yelled:'Damn it! That's the stuff, old kid!' Then Campbell and I rushed to our respective Hun wrecks."

The Americans had scored, and all that was missing was pulling down the goalposts!

There is an interesting twist to this story. The pilot of the Albatros brought down by Winslow was of Polish descent, Vizefeldwebel Antoni Wronieki. He survived his crash with only minor bruises. Wronieki had been born in a part of Poland that had been amalgamated into Prussia at the end of the 18th century. He stated that he had been a pilot on the Western Front for two years, but had not been promoted since he was Polish. Wronieki had been leading the two aircraft further and further behind

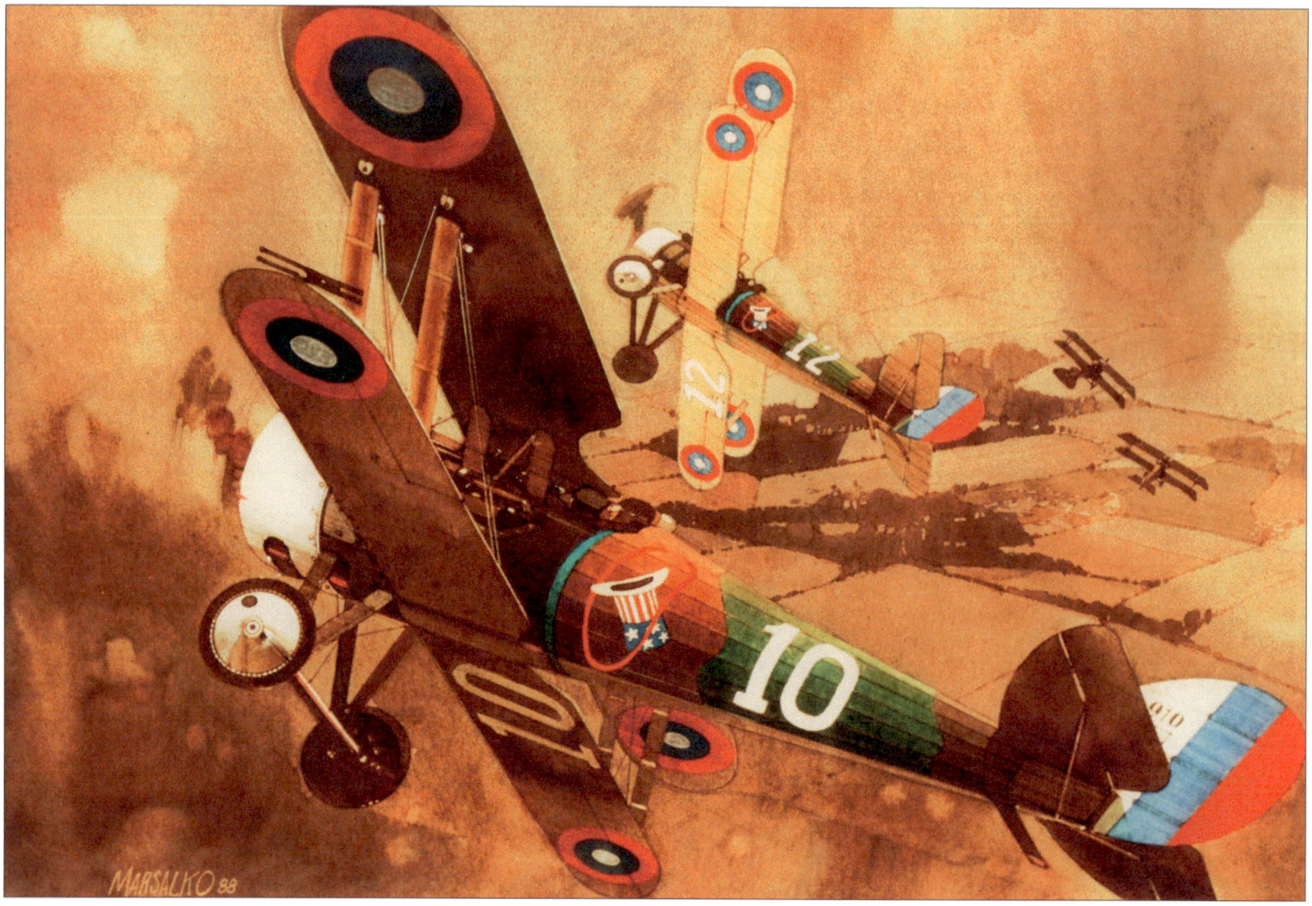

Above: This painting was done many years ago; it shows the first USAS dogfight and the aircraft involved as was first thought to be the markings of the aircraft. Subsequent research by aviation historians allowed the more accurate "visual history" as depicted by Wootton and Smith. This is a more stylized image by William Marsalko

the French lines so that he could defect. He saw the airfield and was indeed bringing his aircraft to a landing when the Americans struck.

The pilot of the Pfalz DIII.a fighter was Unteroffizier Simon, who died from his burns and other injuries shortly after he had crashed. Simon had actually fired at and engaged Campbell, therefore it was not likely that he was defecting like Wronieki.

However, in the official report the two enemy pilots were switched to indicate that Wronieki had perished in the burning Pfalz fighter. This allowed Wronieki to quickly volunteer to fly in the French Air Service under an assumed name,Wroblewski (wrobel is sparrow in Polish).

After the war Wronieki made his way to the newly independent Poland and joined its newly formed air force. In the early 1930s, Wronieki joined the Polish spy service (he was fluent in German and French) and had to be extricated from Berlin when it was felt that he had been betrayed.

In the Second World War, Wronieki made his way to France after Poland was overrun by Germany and from there to England. Squadron leader Wronieki died of natural causes while serving in the RAF.

Du Doch Nicht!

Right: Udet poses in front of his Fokker Eindecker fighter at Habsheim 1916.

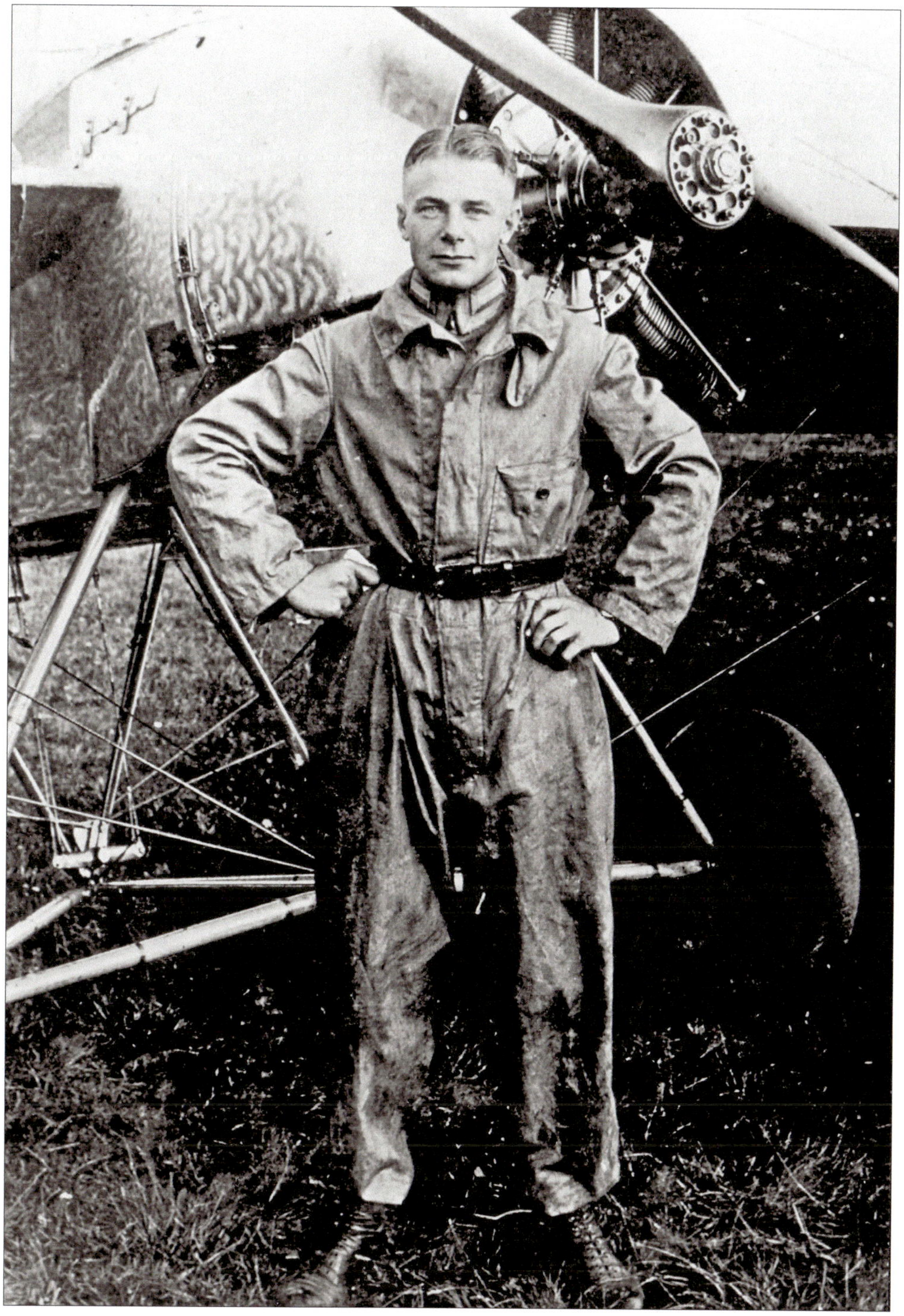

The German slang phrase that Ernst Udet adorned his aircraft's empennage with, Du Doch Nicht!, has been interpreted in several ways. My favorite version is "You and who else!". Udet's confidence, bluster and humor are all contained in that short phrase. At the time that his Fokker DVII fighter carried his challenge to any and all Allied fighter pilots, Udet was a successful, experienced and highly skilled fighter ace who would end the war as Germany's leading, surviving ace with 62 victories: his final tally would place him second overall, behind only Manfred von Richthofen.

However, the legendary Udet was much more than the second most successful German ace of

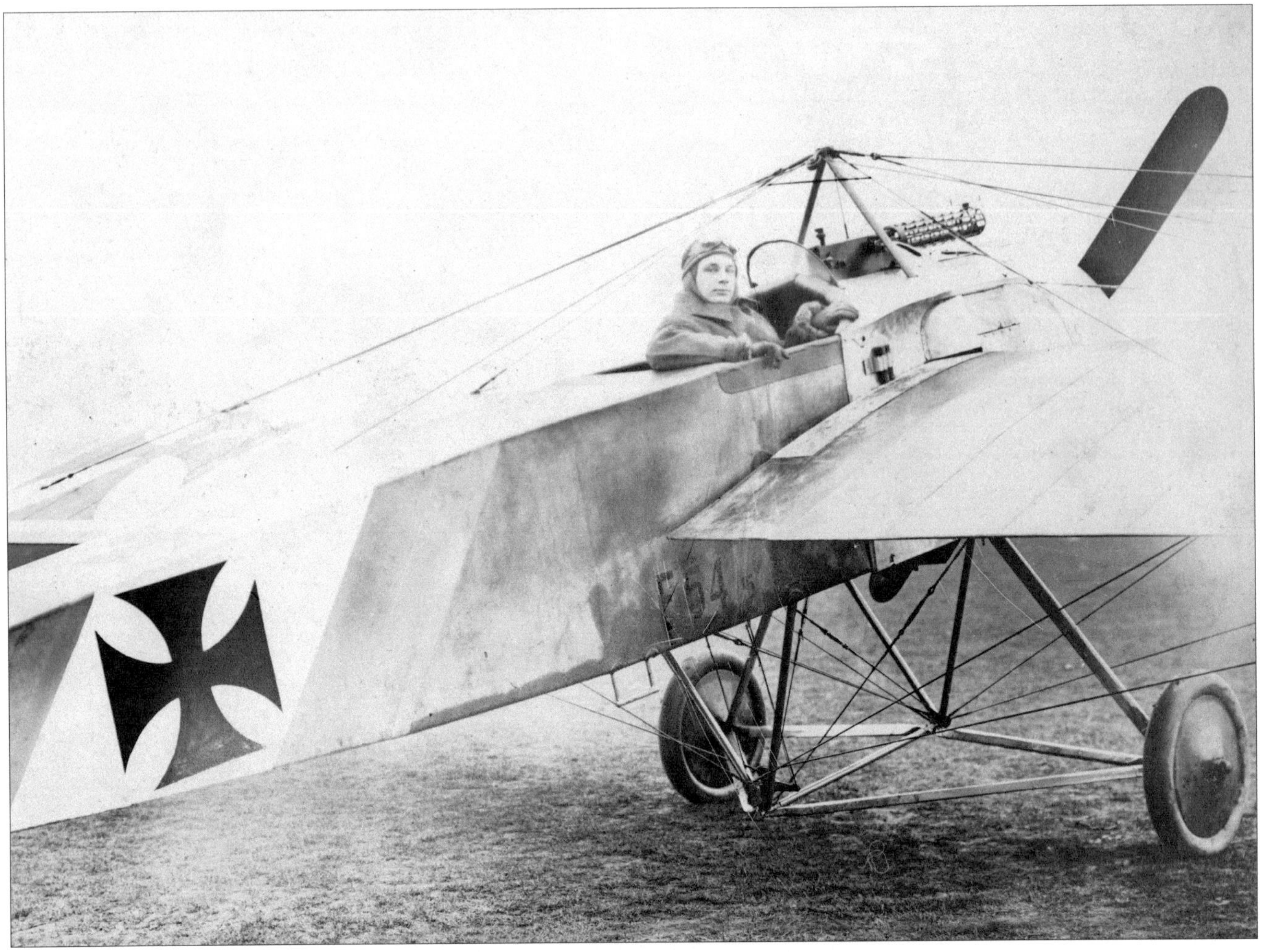

Above: Udet poses in the cockpit of Fokker E.I 54/16 after obtaining his first victory on 18 March 1916.

the war. He was arguably the most flamboyant character to emerge from the ranks of First World War aviation. Traveling the world after the war to demonstrate his phenomenal flying skills in movies and at aviation races and meets certainly helped to cement his "name recognition" everywhere. His signature maneuver of picking up a kerchief from off the ground utilizing his wingtip never failed to suitably impress the crowd.

He was fun loving, carefree and always an entertaining guest at any dinner or party. Udet's personality was likely what most people thought of whenever they thought about what the typical, daring aviator would be like. He led an incredible, adventurous life up until its catastrophic end.

He was born in Frankfurt am Main on the 26th April 1896. His father was an engineer and a very successful one who moved the family to Munich shortly after Ernst was born. Ernst would develop an intense interest in aviation, which led him to form with two friends the Aero-Club Munchen in 1909 at the age of thirteen! He built several model gliders, and even attempted to fly one of his own design, without much success. His love of aviation was such that the test scores from his studies were lower than desired by his father. However, Ernst successfully completed his High School exam in 1913 and was rewarded with a motorcycle by his grateful father.

He was only 5 feet 3 inches tall and in 1914 was rejected by the military when he attempted to enlist at the outbreak of the war. Instead, he brought along his own motorcycle and joined a group of civilian volunteers who acted as couriers for the army.

When his contract expired, Udet was in Belgium and had only recently managed to have a flight with a nearby air unit. Udet was convinced of what he needed to do -join the Air Service. However it was not easy to get into the German Air Service at the time and it was likely to be a very long wait for him. Therefore he returned home, and rather quickly Udet convinced his father to finance lessons to become a private pilot. Udet began his studies in February 1915 and by April 1915 he had secured his pilot's license.

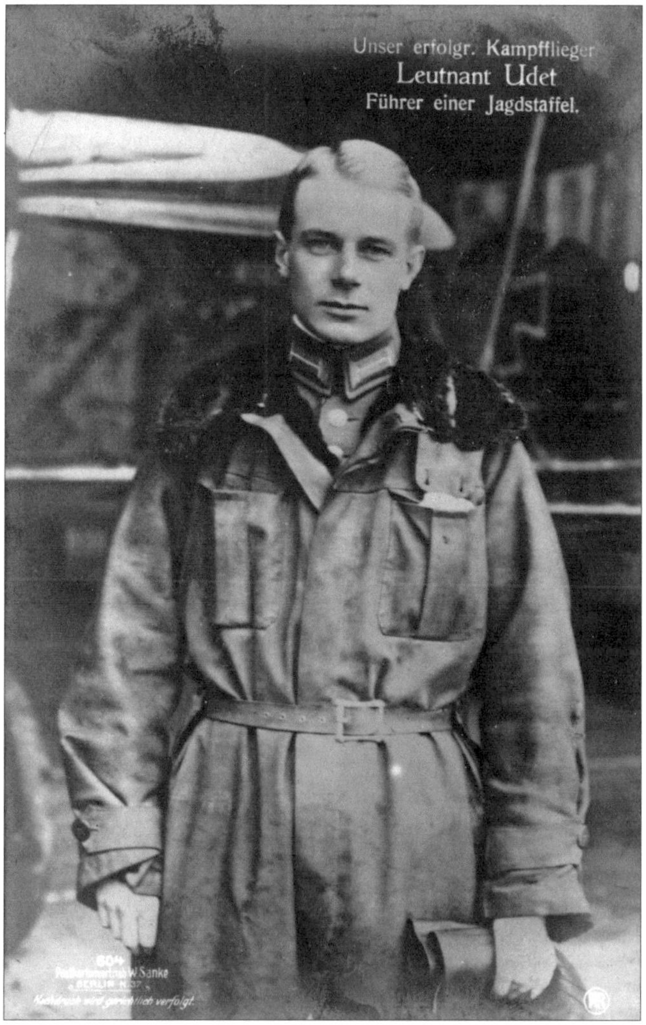

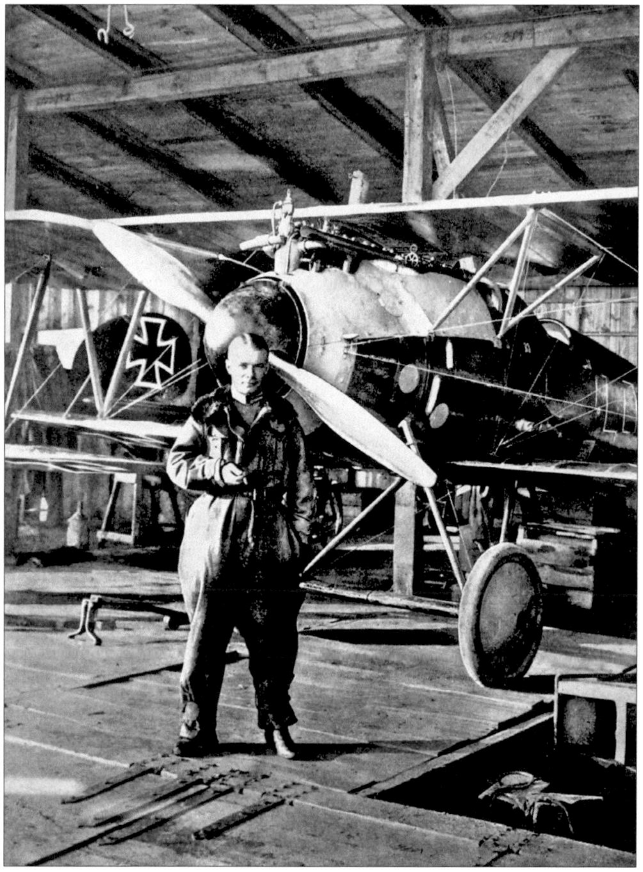

Above: Udet in front of his Albatros fighter , he is now the commander of Jasta 37.

Left: Udet strikes a determined pose for his first of several Sanke card images during the war.

Below: A good look at the Albatros fighter that Udet used as C.O. of Jasta 37. Note the very visible white chevron marking that indicated to his pilots that he was the C.O..

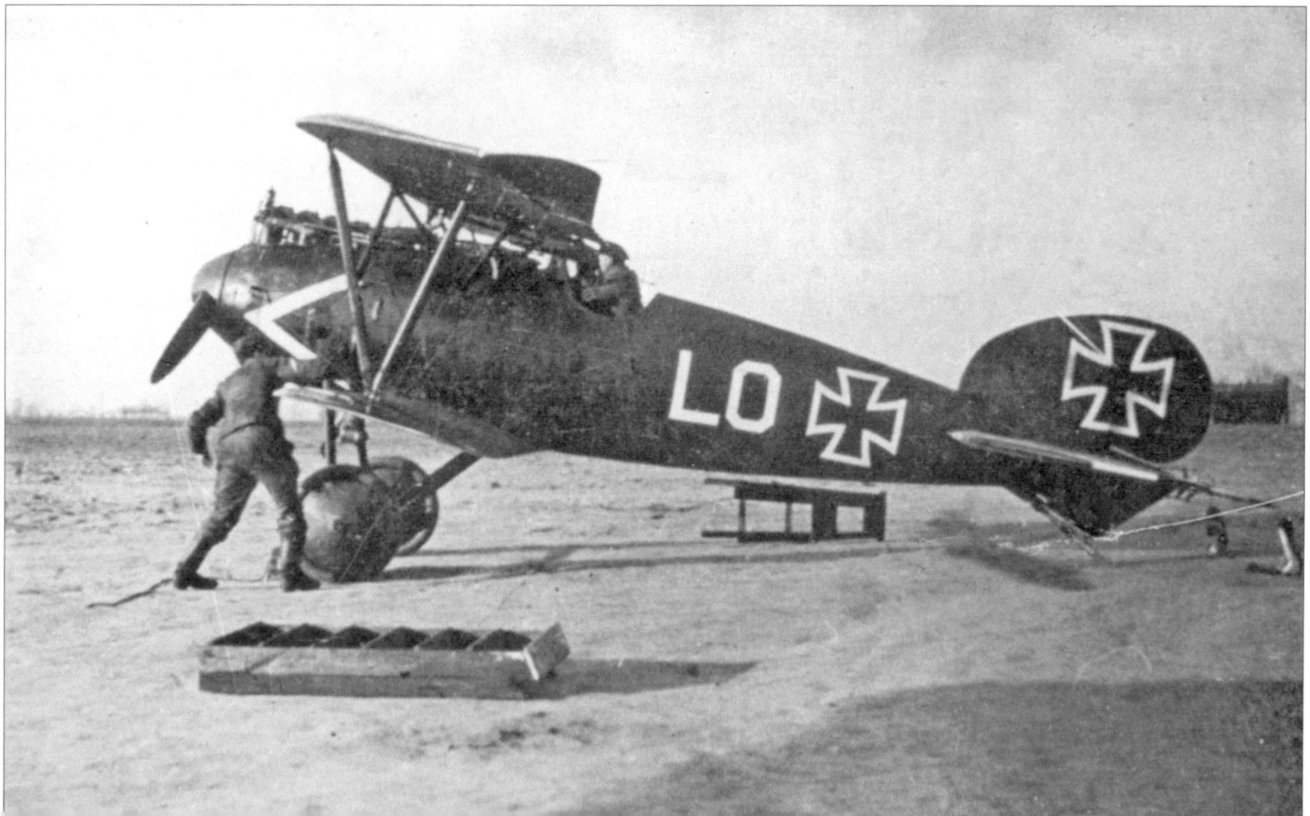

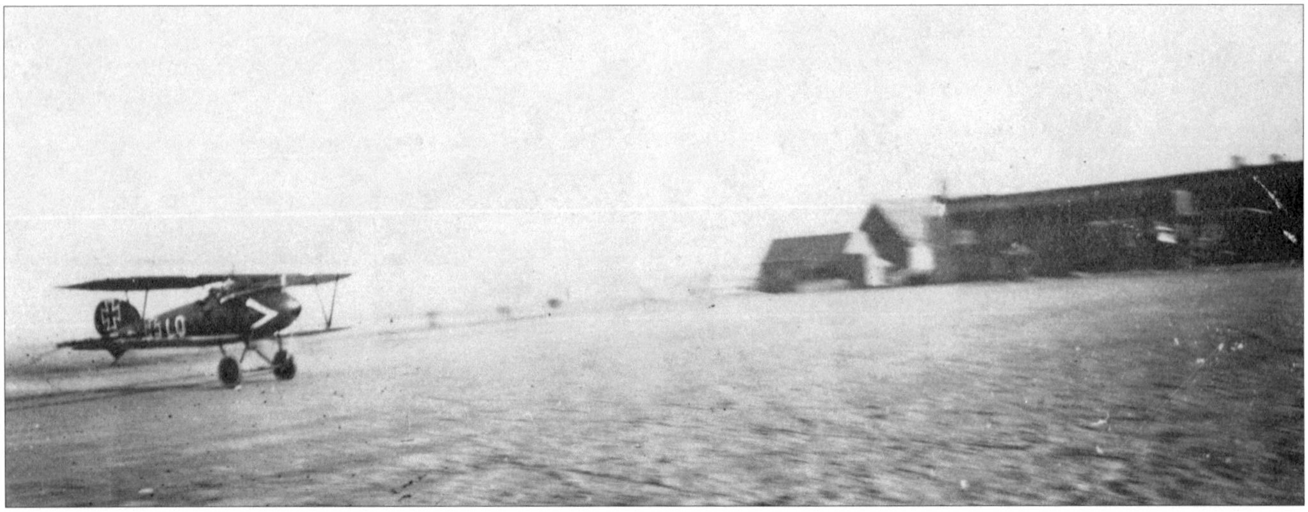

Above: Udet takes off in his Albatros, Jasta 37 had numbers on the front fuselage of each aircraft to better identify the pilots; Udet as commander marked his with a chevron, a practice that would continue in German fighter units in WW II.

Facing Page: Udet in his Albatros immediately after returning from a dogfight that did not go well for him; note the number of bullet holes. Renowned aviation historian Lance Bronnenkant believes that some of these bullet holes were manipulated into the photo. However, Udet does appear genuinely shaken.

Below: Udet back in the same cockpit; note that there are two bullet holes that have been repaired and marked with a British cockade – one is at the very rim of the cockpit on the left and one is in the fuselage towards the bottom right of the photo. Perhaps this was after the engagement with the RE8 that Udet wrote about.

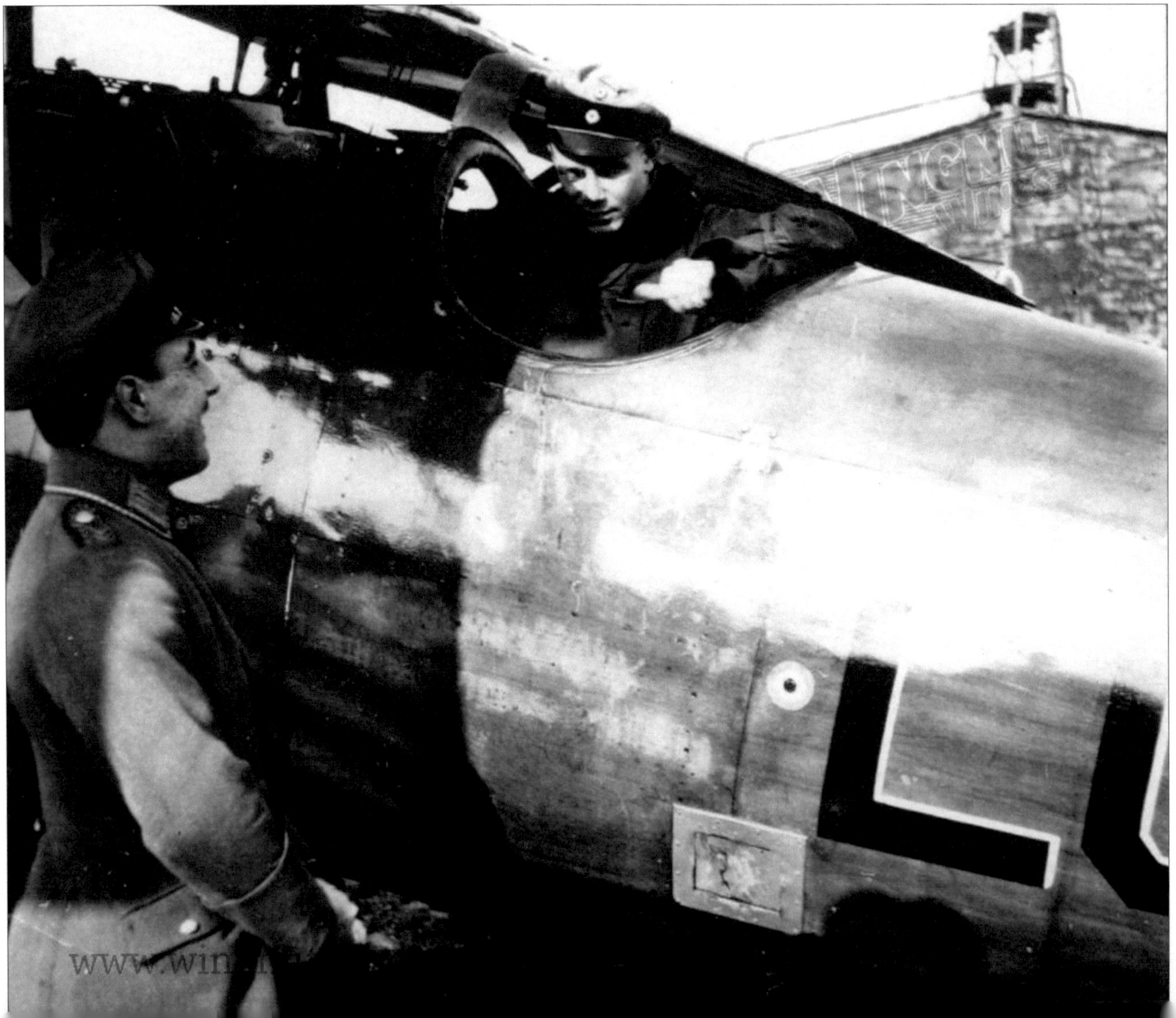

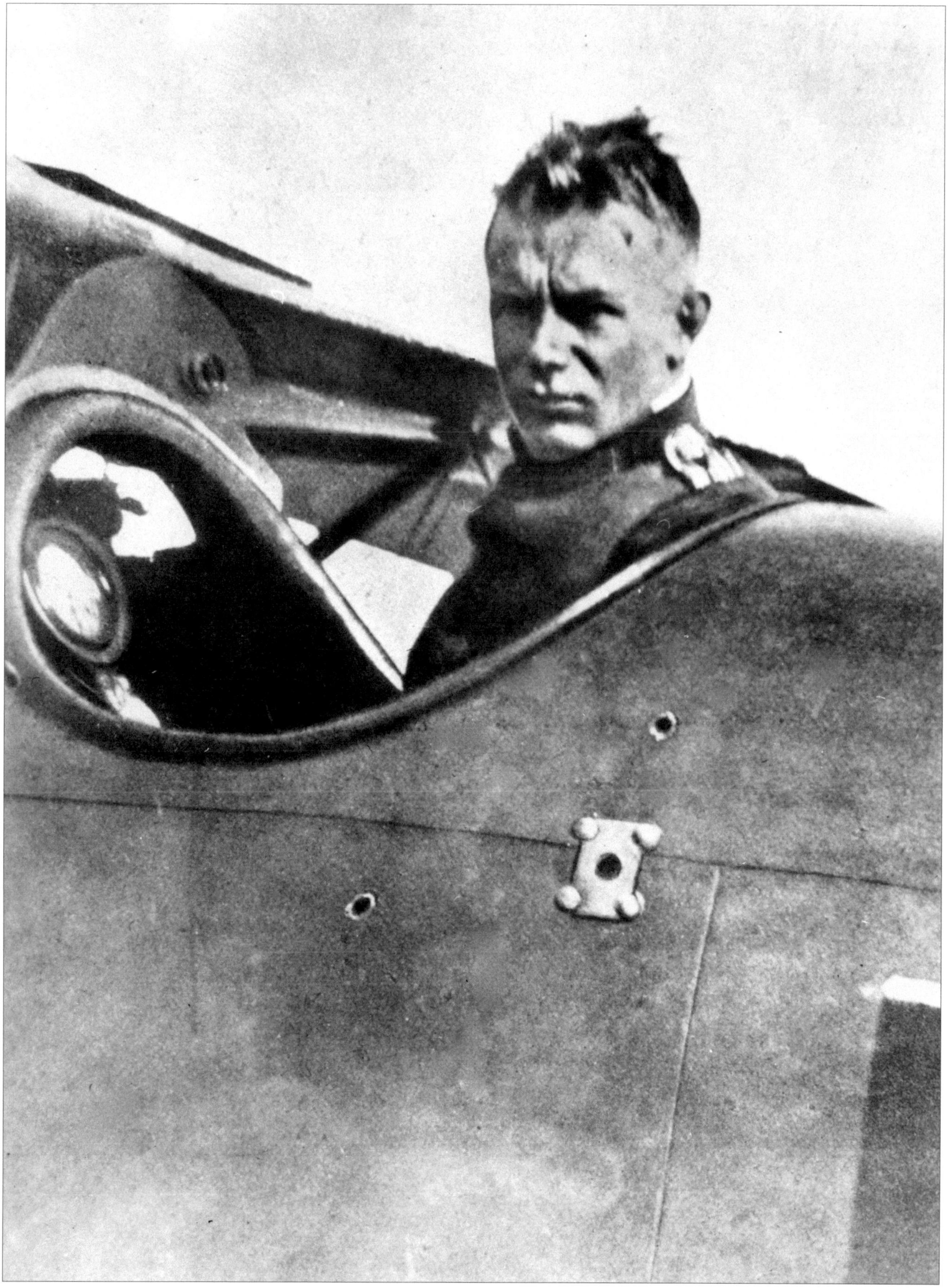

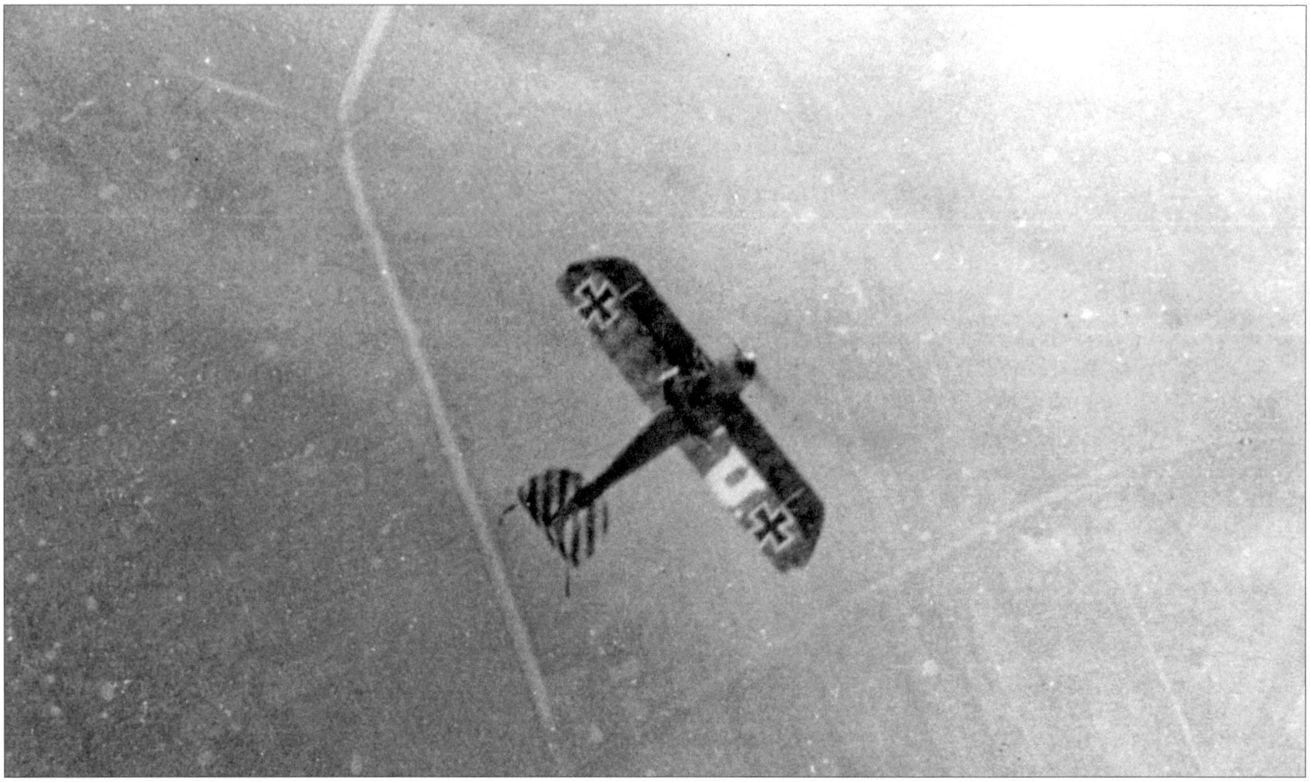

Above: Udet's Albatros fighter in Jasta 37 had a clear and distinct "U" on the lower wing to help ground observers know it was him in order to help gain confirmation of his victories in dogfights.

Facing Page: On the 29th June 1918, after being shot down and surviving by parachute, Udet had to make his way back through German lines during a gas attack. This incredible photo shows Udet shortly after he had returned to his airfield. Note the bandaged head and the gas mask that someone from the 16th Infantry Regiment had given him.

His first assignment was to AFA 206 and there he was partnered with Lt. Bruno Justinus as his observer. Already he had a desire to fly at the limits of his aircraft's capabilities. This impulse, which would stay with him forever, earned him some time in the stockade after a crash followed some reckless maneuvers immediately over the airfield that had been witnessed by his commander.

However, on the 14th September Udet's skills as a pilot were both evident and very much needed. Returning towards their base following a bombing mission to attack Belfort, a critical guy wire broke which forced Justinus to leave the cockpit and walk out onto the wing to help stabilize that portion of the upper wing of their Aviatik B. II.

Udet was able to pilot the damaged aircraft back to inside German territory where he managed a credible forced landing saving his aircraft (as well as Justinus and himself).

For this act Udet was promoted to Gefreiter and awarded the iron cross second class. More importantly, this coolness under duress showed that Udet had the "right stuff" and before long, in November of 1915 he was promoted further to Unteroffizier and transferred to KEK Habsheim where he would now fly the Fokker monoplane fighter.

His debut as a fighter pilot was almost his last flight as well. In December, Udet approached a French Caudron two-seater. As he closed the distance, Udet froze and did not open fire. His opponent had no such qualms and opened fire at close range. As a result of some excellent shooting, Udet's goggles were shot off and Udet barely managed to dive away to safety. He vowed that next time he would do better.

His next chance did not come until March 1916 when he was scrambled to intercept a flight of French bombers that had been detected on their way to bomb Mulhausen. When Udet found the enemy formation, he was able to count a total of 23 enemy aircraft of various types.

Despite the odds, Udet swallowed hard and dove into the middle of the formation and during the combat shot down a Maurice Farman for his first confirmed victory of the war.

Not too long after his victory, Udet's engine failed as he was flying back to his aerodrome with

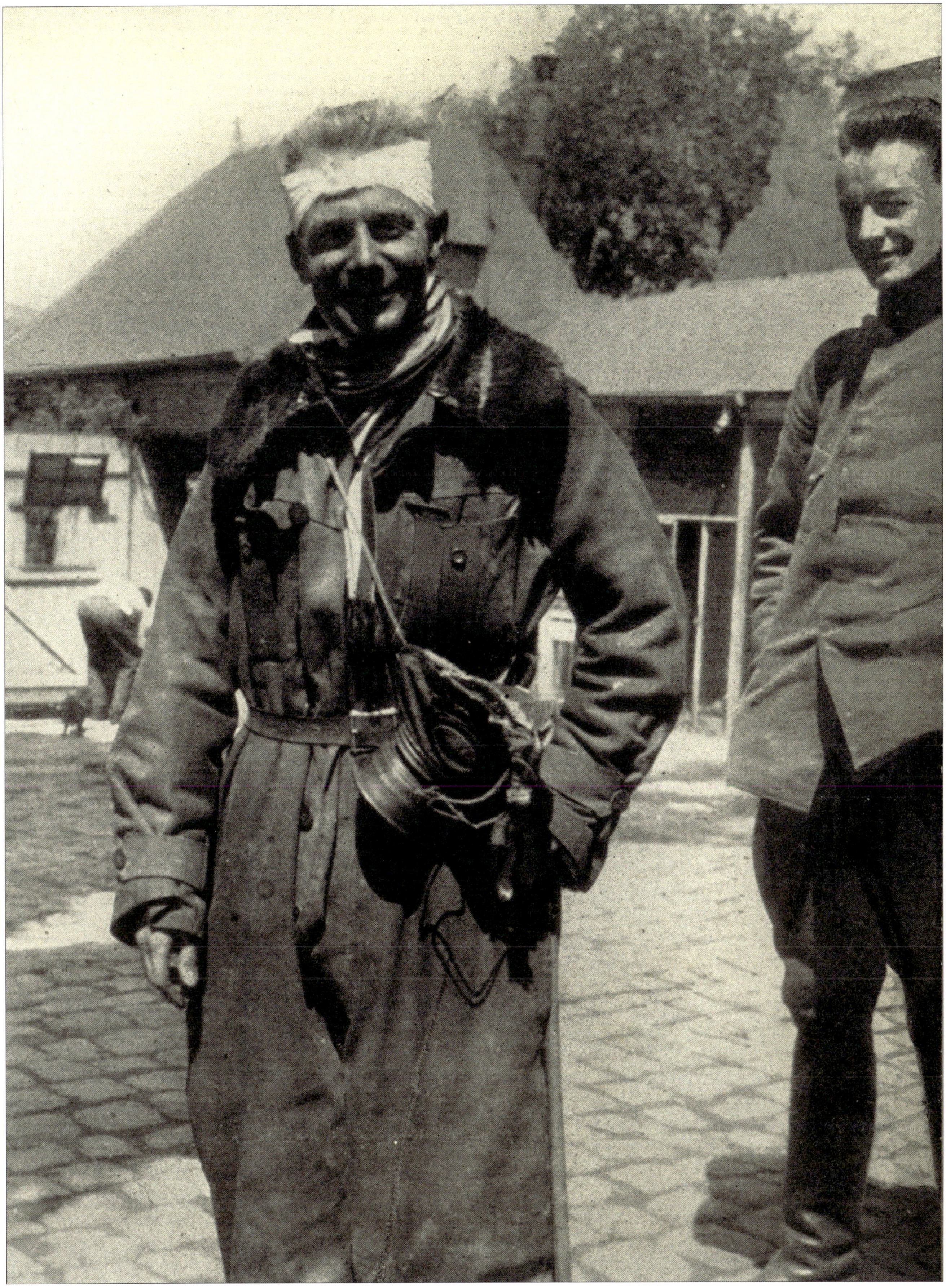

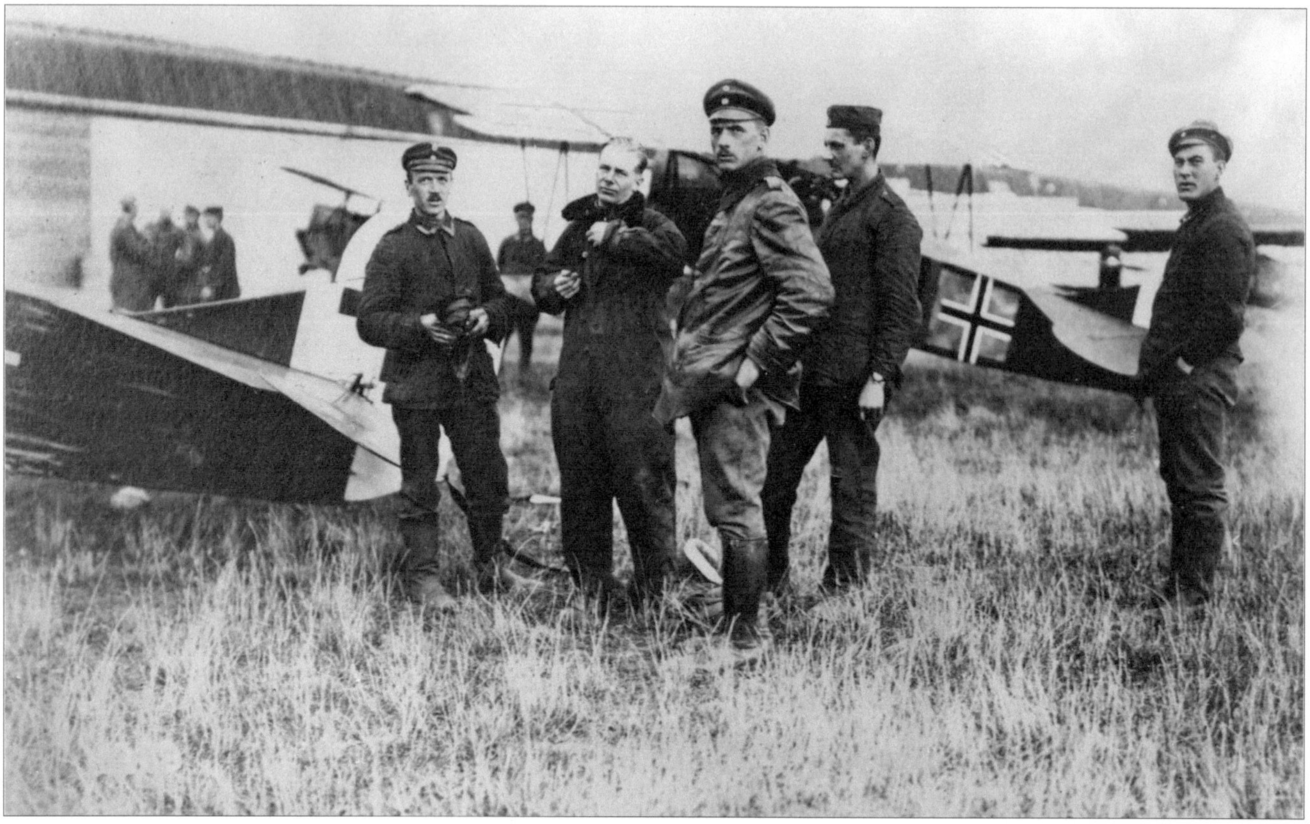

Above: Udet had already scored his last victories of the war by the time this photo was taken sometime in October 1918. Udet, along with his other comrades in JG 1, would fly on to the very end of the war.

Facing Page: A portrait that would hang in his study after the war shows the prominent PLM at his collar.

the result that he had to make a forced landing. As he lowered his aircraft to land, Udet noted that in the middle of the field he had chosen there was an obstruction that he had not seen:

"To make matters worse, in the middle of the field was a dam which was almost a meter high and with which I soon became acquainted….when his head cleared, discovered that he was trapped in the fuselage that had flipped over and dug itself into the ground. He too was upside down and his right knee had pushed through the side fabric, so he was unable to extricate himself...After about 15 minutes, all the while fearing that the airplane would catch on fire... he heard someone say 'I can't find him. He must be dead'...They all shouted to one another and a group of strong soldiers lifted the fuselage, freeing Udet." L. Bronnenkant

Shortly after this narrow escape, Udet was transferred to Jasta 15 and there he began to fly a Fokker D.II. He would score his second victory on 12th October 1916:

"About 20 kilometers behind our front, I observed a formation of seven Breguet bombers flying toward the Rhine, protected by two Nieuports. When they flew directly under me, I had an easy job of it...forced it to land after 350 shots. Since it landed safely and I wanted to prevent the airplane's destruction by its inhabitants, I landed next to it. As a result of a hit to one of my tires, I lightly overturned. It was a comical scene: the defeated safely landed and the victor with wheels up. The two Frenchmen climbed down from their framework and we shook hands with one another."

Udet would score his fourth confirmed victory on 20th February 1917 while flying a brand new Albatros D.III fighter against a French Nieuport fighter:

"He flew quite well, and demonstrated all kinds of silly tricks for us, such as loops, lateral somersaults, vertical zooms and the like...I managed to riddle his motor. His propeller stopped. I then flew close up to him and saw how the cheeky guy lifted his gloved hand and threatened me. I did the same thing...Over A., we were only 400 meter up and it was high time that the fellow came down... He set down safely on a small field between the two foremost lines. The plane had not finished rolling when I saw a guy jump out of it and hop in great bounds toward the enemy barbed wire."

The two recent quotes are from a book that

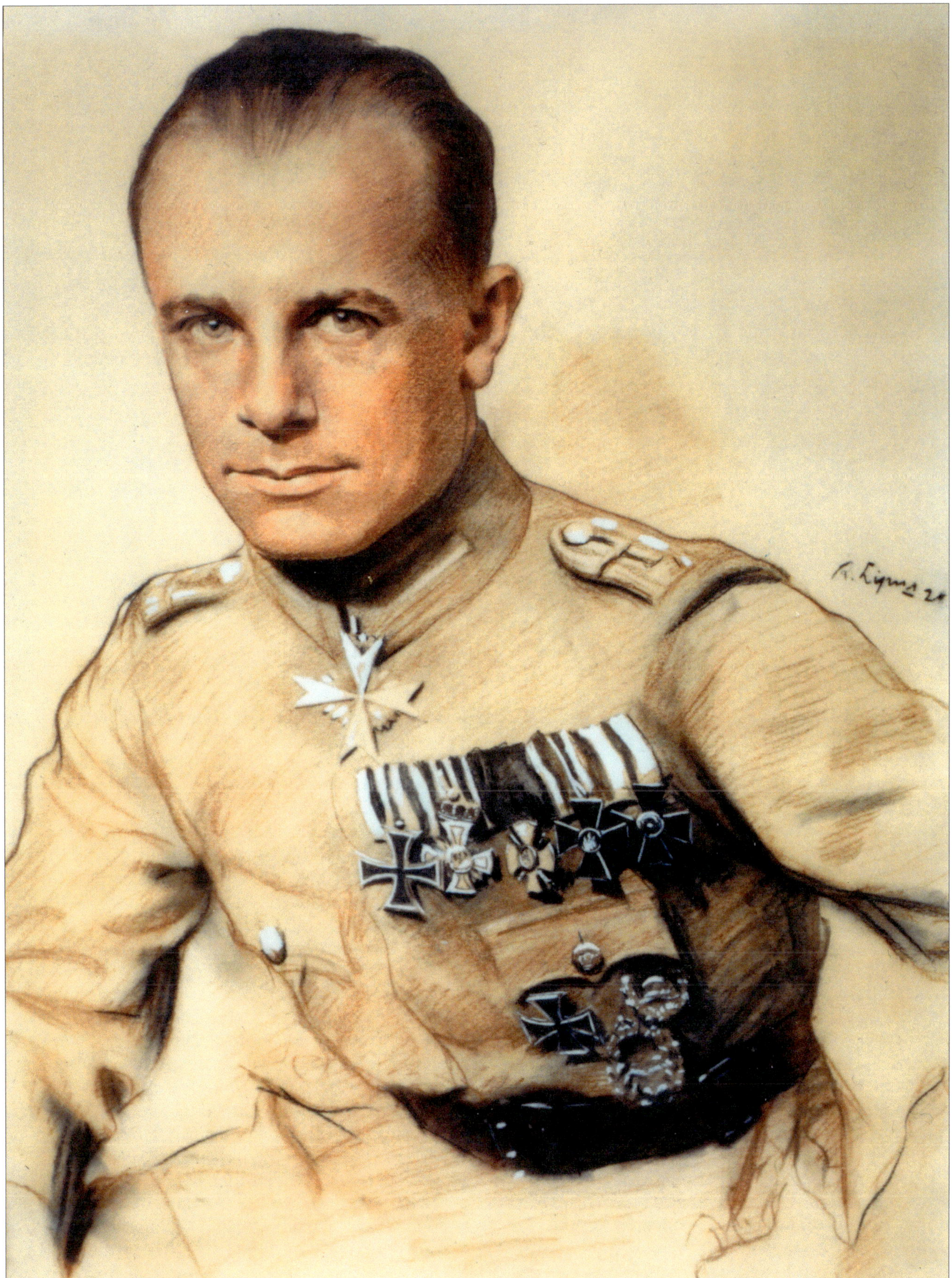

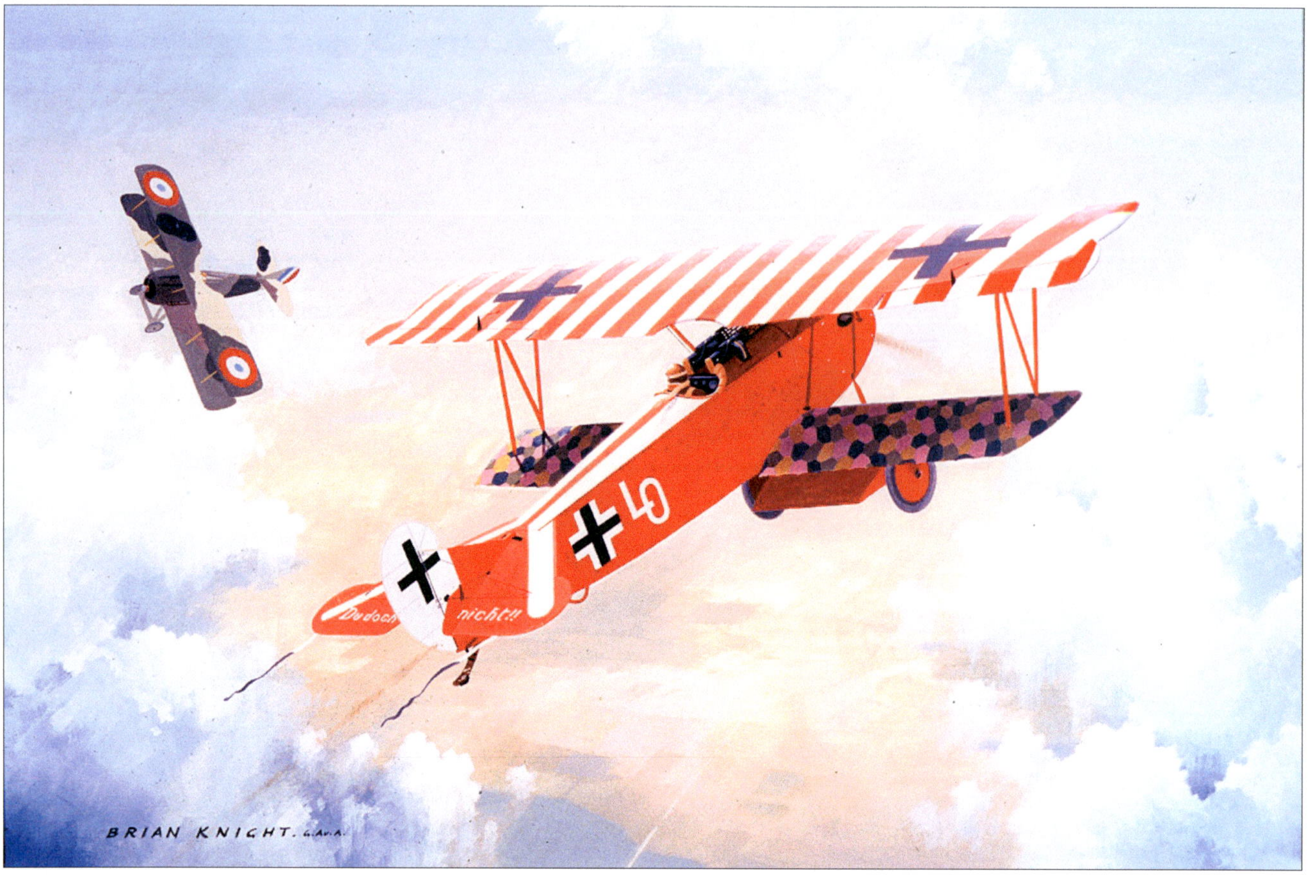

Above: One of the most famous and discussed aircraft of the First World War was Udet's Fokker D.VII fighter with the "du dock nicht" markings. For many years the overall color scheme of the aircraft was felt to be as rendered here. Brian Knight

Udet dictated just before the end of the war, Kreuz wider Kokarde and they help to illustrate how good a storyteller Udet was. The accuracy of this most recent tale was partially confirmed by aviation historian Jon Guttman when he interviewed the Nieuport pilot, Pierre de Casenova de Pradines of Esc.81, in 1978: "I had just landed in No Man's Land when I was off and running to get back to the French trenches, which seemed a long way off, sprinting from shell hole to shell hole." However, the Frenchman did not mention that he had waved his fist at Udet: which leads to the thought that Udet was not one to hesitate to add an embellishment of an already good story.

By May Udet had scored six victories and was recognized in Jasta 15 as a capable flight leader. When the unit's commander, Lt. Gontermann was sent home for four weeks leave after receiving the Pour Le Merite (PLM), Udet was chosen as temporary leader.

Udet was leading a five man patrol on the 25th May when a lone Spad attacked the formation from out ot the sun and in its swoop shot one of the Albatros fighters down:

"Udet had a sudden premonition of danger, looked to one side and saw Hanisch's Albatros in flames. Then, as he watched in horror, he saw Hanisch's right hand go up to his hemet- a last salute from a doomed comrade- before the flaming machine fell to oblivion. Below him, Udet spotted the lone Spad, and with hatred surging within him, dived as fast as possible to avenge his friend 'Puz'. As his Albatros reached its highest diving speed its wings began to shudder alarmingly, Udet was forced to abandon the dive before his aircraft shook itself into pieces." Armand van Ishoven

Some historians believe that the Spad was flown by Geroges Guynemer, who scored four victories that day. A few days after that combat, Udet's friend Muller was killed and later in the same day Udet's good friend Glinkermann was shot down and killed as well. A few days later another member of the staffel, Eichenhauer was killed in combat. As leader of the unit, the task of writing to the families of the fallen was Udet's unenviable one. Perhaps a bit shaken by the frequent letters he was forced to write to various comrade's families, he also found time to write a friend, Lt. Kurt Grashoff who was in command of Jasta 37: "I'm the last of Jagdstaffel 15, the last of those who once left Muelhausen to go to

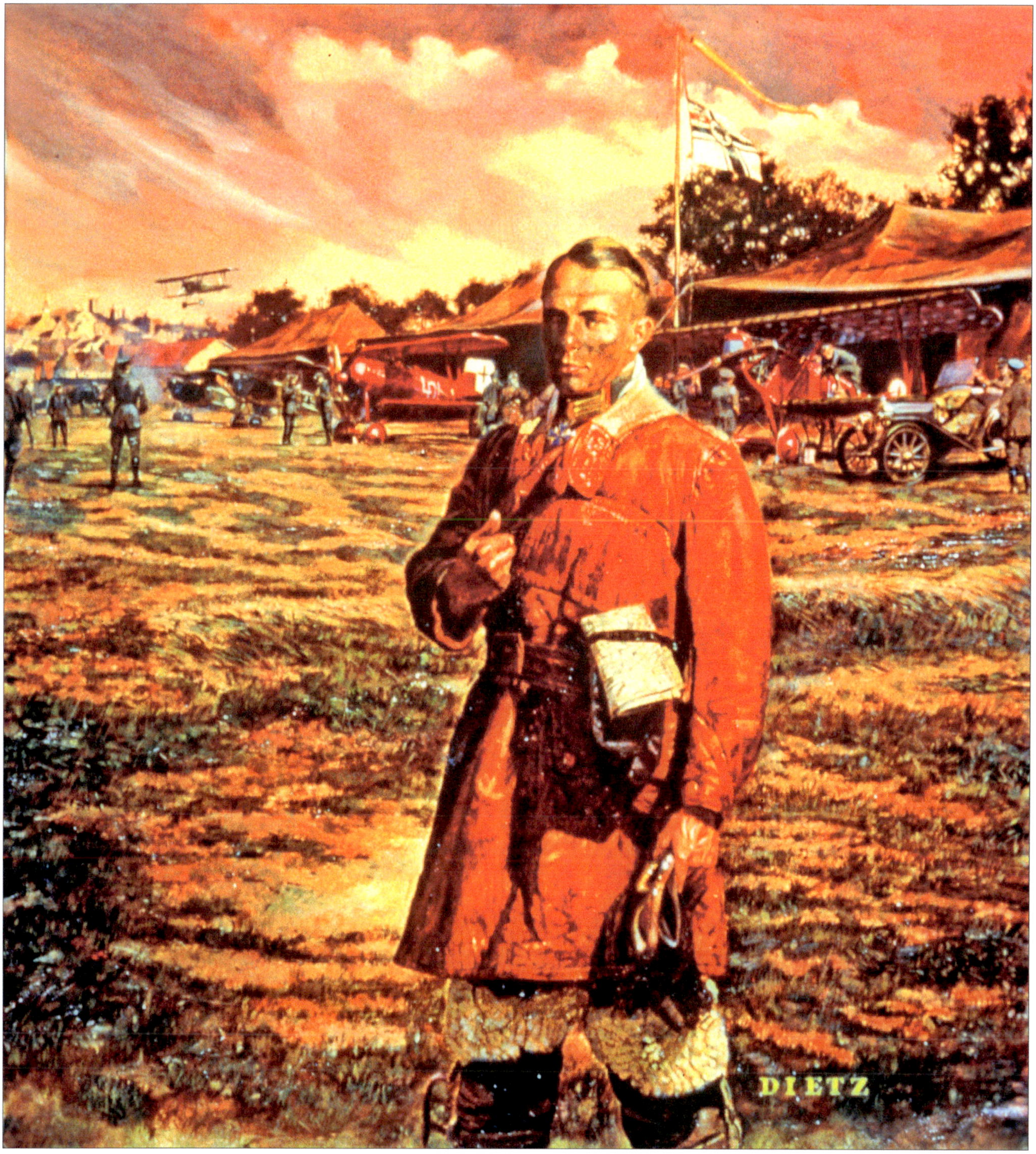

Above: A wonderful portrait of Ernst Udet. Note that he wears a captured British flight coat, a prize that was highly coveted by many German fighter pilots. In the background mechanics are working on his D. VII and there is a new SSW D.IIIa fighter also awaiting him, already marked in red and with LO! on the fuselage side. Jim Dietz

Champagne. I should like to move to another front, to come to you."

He would soon get this transfer to Jasta 37, but not before one of the more momentous episodes in Udet's career took place. Udet took off alone early one morning, with the hope of surprising and downing a French observation balloon across the lines. Udet detailed the action in his wartime biography Kreuz wider Kokarde:

"We go for each other at the same height,

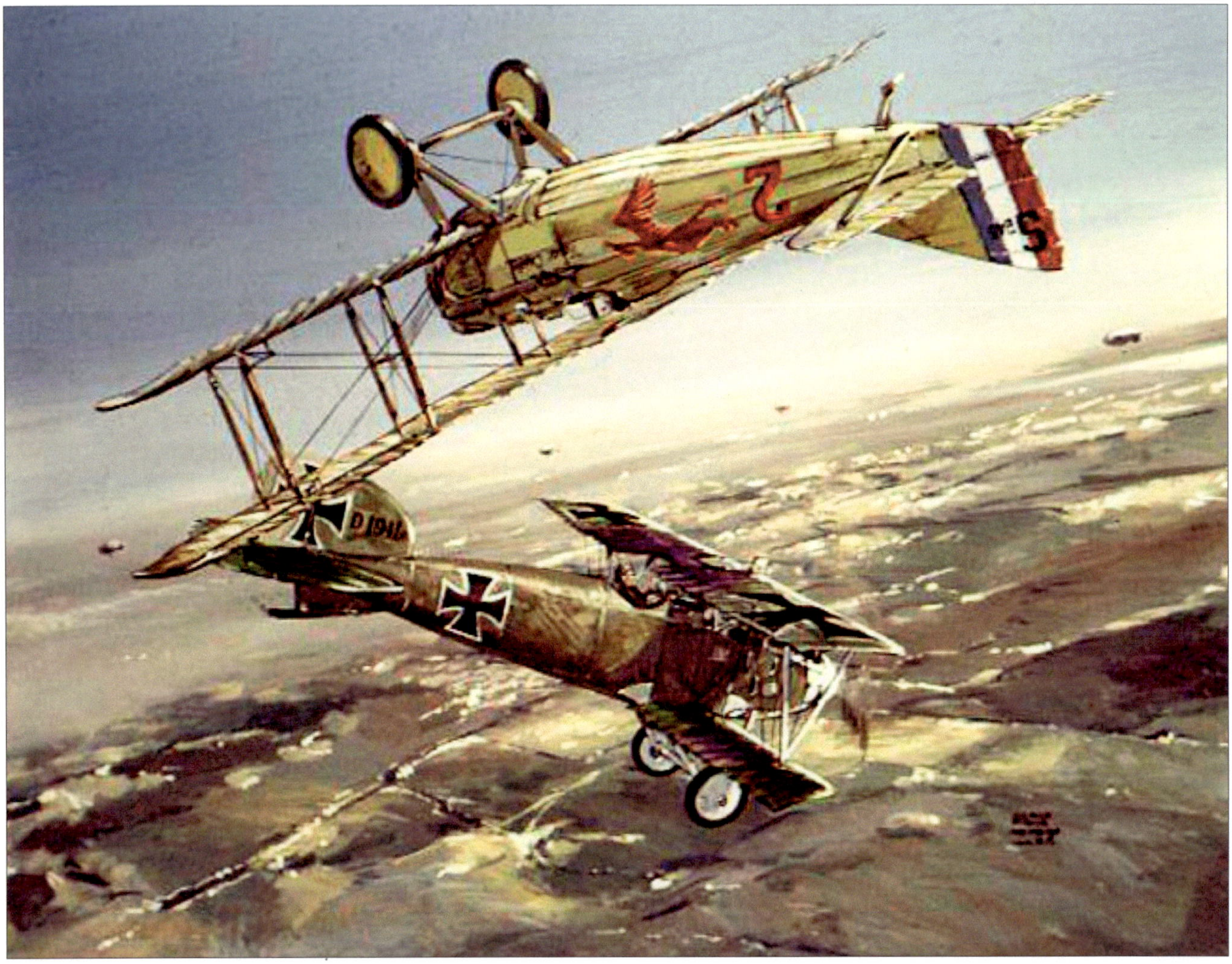

Above: The remarkable dogfight between Udet and Guynemer so painstakingly described by Udet, and never mentioned by Guynemer. Serge Stone

whizzing by each other at a hair's breadth. We bank into a left turn. The other's apparatus shines light brown in the sun. Then the circling begins. From below, it might appear as though two large birds of prey were courting one another, but up here it's a game of death...Sometimes we race so close together that I can clearly see a slender, pale face under the leather helmet. There is a word in black letters on the fuselage, between the wings. As he passes me for the fifth time, so close that his propeller wash shakes me back and forth, I can make it out- "Vieux" it says there- vieux- the old one. That's Guynemer's sign...I do a half loop in order to come down on him from above. He catches on at once and also starts to loop. I try to turn, and Guynemer follows me. Once, coming out of a turn, he latches onto me for a few seconds. Metallic hailstones settle through my right wing and ring out lightly as they strike the struts. I try everything I can, the tightest banks, turns, side slips , but he anticipates all my moves with lightning speed and immediately reacts to each one with lightning speed. Gradually, I realize he is superior to me. Not only is the machine over there better, but the man in it can do more than me. But I keep fighting. Another turn. For a moment he comes into my sights, I push the button on the stick...the machine gun is silent... a gun jam! With my left hand clutching the stick, my right attempts to reload. No use...the stoppage remains. For a moment I think of diving away. But that would be useless with such an opponent...he would immediately be on my neck and shoot me to pieces. We continue to twist and turn. Beautiful flying if the stakes weren't so high. I have never had such a tactically clever opponent... For eight minutes we were twisting around each other- It is the longest eight minutes of my life. Now, lying on his back, he races over me. I let go of the stick for a moment and hammer the machine gun with both fists- Guynemer observed this motion from above, he must have seen

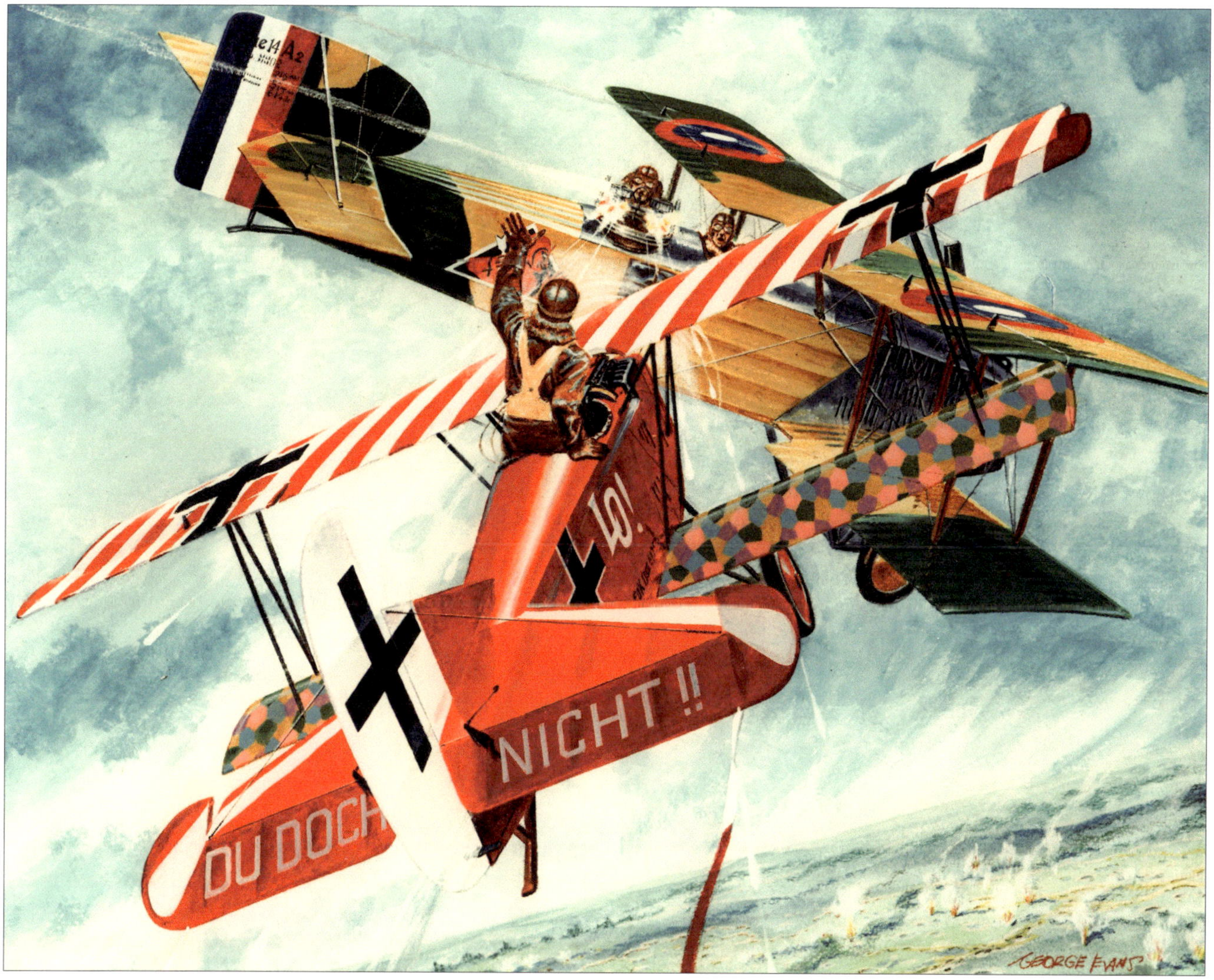

Above: Udet makes a near fatal mistake while attacking a Bregeut two-seater and is forced to parachute to save his life on the 29th June 1918. Subsequent research shows the wing stripes were black, not red. George Evans

it, and now he knows what is up with me. He knows that I'm his helpless prey. He sweeps over me again, almost on his back. That is when it happened: he sticks out his hand and waves to me, waves lightly, and dives down to the west, heading for his front. I fly home, dazed. There are people who say that Guynemer had a gun jam himself back then. Others claim he feared I might ram him in despair. But I don't believe them, I still believe to this day that a bit of chivalry from the past continued to survive."

Guynemer never mentioned such an encounter to anyone, only Udet spoke and wrote about the dogfight. We know Udet was a very good story-teller, but historians have no obvious reason to doubt his word either.

In August 1917 Udet was transferred to Jasta 37 and began to score against a new opponent, as he was now facing British squadrons. In September the squadron was moved to a new base at Wynghene, south of Bruges. Udet was billeted with a local family, whose son was actually flying in the Belgian Flying Corps!

On the evening of 28th September Udet displayed his skills as a fighter pilot:

"...as he [Udet] patrolled, five British Sopwith Camels in tight formation crossed his path. Maneuvering behind the arrow-head quintet, Udet cautiously crept up to the rear of the Camels. The British pilots flew on unsuspectingly. Eventually, Udet was literally in formation with them, at the rear of the Vee, with the tail Camels only 30 meters away on each side of him. Taking a deep breath, Udet slammed a short burst into the Camel to his left, which immediately erupted in flames. Switching quickly to the leading Camel, Udet fired again, sending it down to a forced-landing...The remaining trio finally woke to their danger and scattered wildly. The whole episode had lasted just

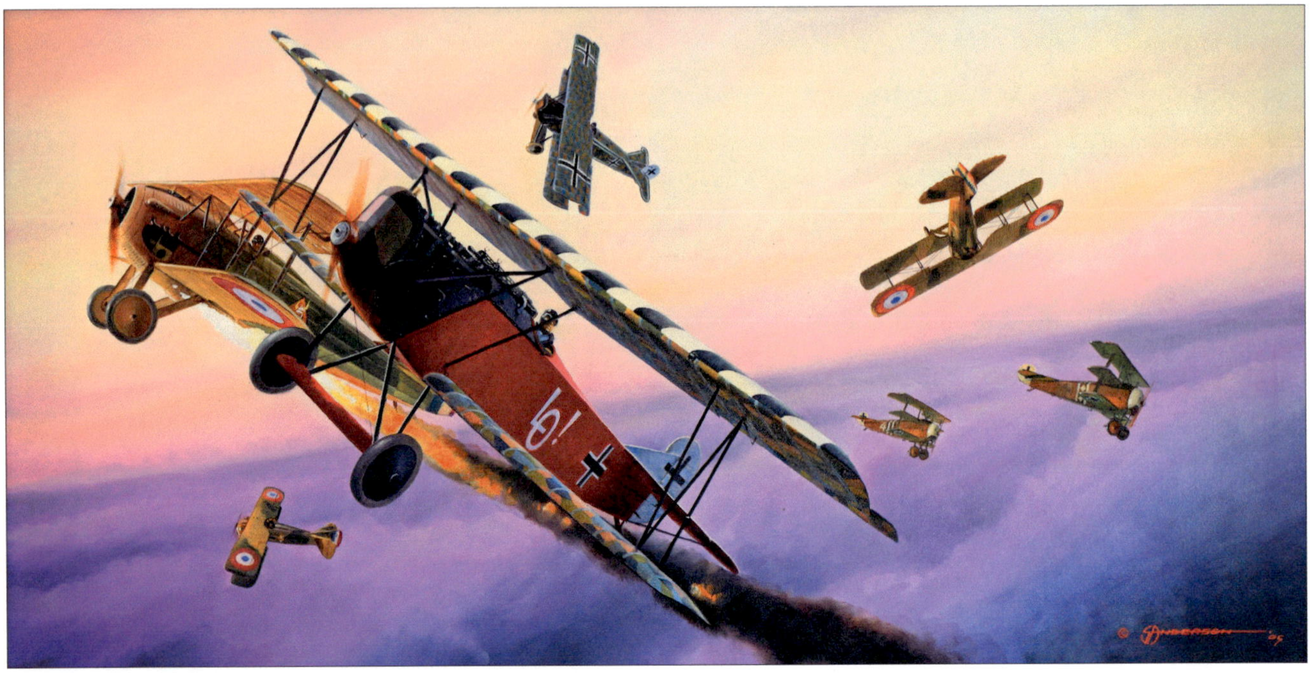

Above: Through the perseverance of many aviation historians, the colors of Udet's Fokker D.VII were almost certainly as represented here in combat with French Spads. Steve Anderson

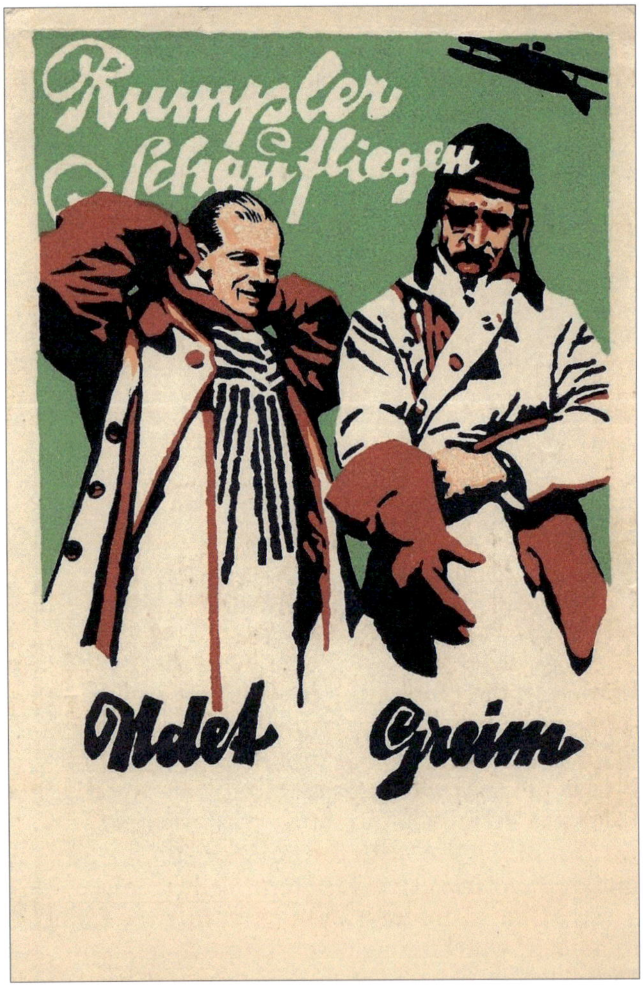

Above: A poster advertising one of the airshows that Udet and von Greim put on in 1919 to raise funds to help a German POW relief effort. Artist unknown.

20 seconds." A.van Ishoven

Udet also had his share of misfortune. During an attack on a British Re8 two-seater Udet swung in close, however the British crew must have been an experienced pair as the pilot made a sharp turn and at that very instant the observer-gunner opened fire. Udet was hit in his knee, and also gas was spraying about the cockpit from several holes in his fuel tank.

Udet switched off the engine and managed to coax his Albatros fighter back to a safe landing behind German lines. It would not be the last time that Udet was out-fought by a two-seater.

Overall though, Udet was hitting his stride and leading his Jasta 37 to success. In January 1918 he was sent along with other top profile fighter pilots from the front to Berlin to partake in the first aircraft competition.

These evaluation tests were composed of the various aircraft manufacturers bringing their newest designs to be flown by the combat pilots who could then offer their thoughts and their "votes" for the type that should go into production.

Hence, the pilots were treated quite royally by the various manufacturers with parties, dinners and escorts. Udet thoroughly enjoyed his visit and was likely rejuvenated upon his return to the front lines. In any case, he soon scored his 20th victory and thus was now to be awarded the PLM.

In March 1918 Jasta 37 was transferred to Le

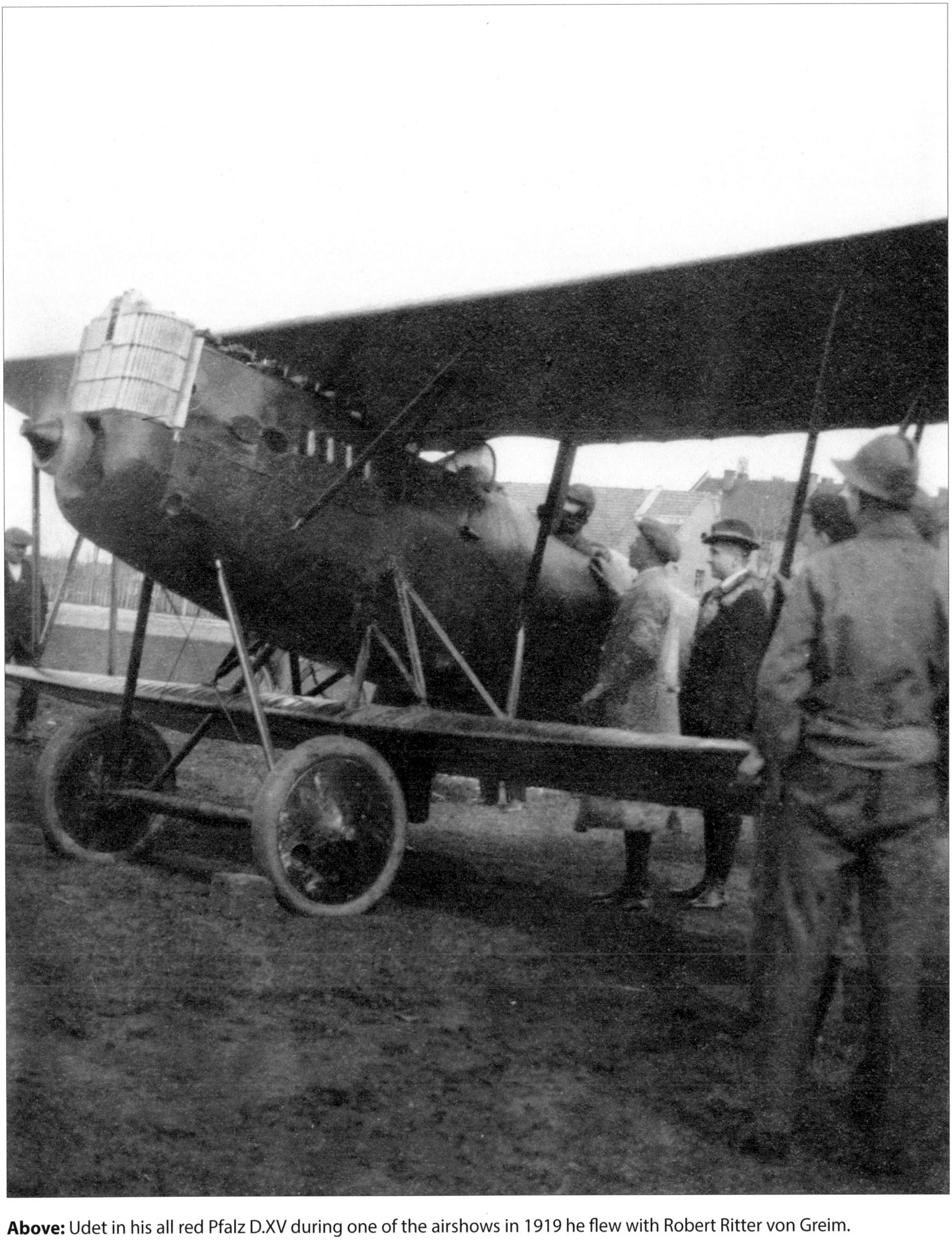

Above: Udet in his all red Pfalz D.XV during one of the airshows in 1919 he flew with Robert Ritter von Greim.

Cateau in preparation for the massive spring German offensive. Udet was helping to erect the canvas hangars during a heavy rainstorm when he felt a light touch upon his shoulder. He turned to see von Richthofen behind him, who proceeded to invite Udet to join his command. Without any hesitation, Udet accepted and quickly presented himself to JG I. Lothar von RIchthofen had been leading Jasta 11 before he was injured in a crash. Udet was offered temporary command of Jasta 11, a tremendous honor. JG I's adjutant, Karl Bodenschatz later wrote:

"In those days of heated battle a very cheerful young man joined the group, who had painted on his machine in delicate lettering the initials 'Lo': he had up to that point already shot down 21 enemy planes out of the sky and was greeted by Richthofen with great warmth and sheer pleasure, because he knew that this cheerful young man was a fighter pilot of the very first order."

Udet would continue his successes as he downed his 24th victory before he was ordered home by von Richthofen on medical leave. Udet had been suffering from severe ear pain, especially in the air. He was found to have a ruptured eardrum and a severe inner ear infection.

While at home on this medical leave Udet received a telegram which made the award of his PLM official as of 9 April 1918. His family was overjoyed with such an honor and had a special dinner celebration that evening. The next day, Udet set off for a well-known jeweller in Munich to acquire a PLM.

He was disappointed to be told that they had none of the awards in their stock. Udet was eager to be able to wear the medal not simply for his family, but to impress his girlfriend Eleonore "Lo" Zinke. As he walked along the boulevard he happened upon Kptlt. Ralph Wenninger, C.O. of UB-55, who was on leave in Munich following his award of the PLM on 30th March. Udet impulsively approached him and asked if he had a spare medal that Udet could use. Initially Wenninger was quite alarmed, but as Udet spoke of his situation Wenninger laughed and although he did not have a spare medal, he gave Udet the name of a jeweller in Berlin.

Udet sent a telegram and a few days later was proudly wearing the highest award for bravery that Prussia could award. Udet collected Lo for a stroll about town; with his pretty girlfriend on his arm, the couple strolled past the Residenz (the King of Bavaria's palace). As they approached the entrance, the sentry called out the guard to come to attention. Udet explained to Lo that the guard was to turn out in recognition of any PLM holder. Lo's response was that she wanted to walk past them again!

Udet would later write about that home leave as:"days made of blue silk. I have never had another spring like it."

Upon his return to JG I, Udet was transferred to command Jasta 4, another squadron in the newly named Richthofen Geschwader since the Red Baron had fallen while Udet was away. Udet resumed his scoring and soon began scoring even more rapidly than in the past with the arrival of the new Fokker D.VII fighter. On the 7th June, Udet scored an unusual victory when he shot down a Spad VII fighter of Esc. 86. Not that downing a Spad was unusual for him, but that particular fighter was piloted by Sgt. Kobayashi Shukunosuke, a Japanese citizen who had volunteered to fight in the French Air Service.

Between 27 May and 28 June Udet attended the second fighter aircraft competition in Berlin and once again likely thoroughly enjoyed all of the festivities. Upon his return to his command Udet found himself yet again being bested by a two-seater on the 29th June 1918:

"I took off at 7:15 a.m. with my Staffel on a hunting flight...I attacked an infantry-support plane flying at an altitude of 800 meters over the area which was under heavy French shell fire. Upon my first attack, the Breguet turned towards me and flew under me. As he did so, I observed that the observer was no longer standing in his machine-gun ring. I therefore assumed I had already hit the observer and, contrary to my usual habit, I attacked the enemy pilot from the flank. I suddenly noticed, however, that the French observer had reappeared from the body of the plane and, at that same moment, I took several hits, including one in the machine gun and another in the fuel tank. At the same time, the elevator and aileron cable must have been shot through, because my Fokker D.VII plunged out of control. I tried everything possible-sometimes with the throttle, sometimes with the rudder-to bring the plane under control, but in vain. At an altitude of about 500 meters, the machine went over on its nose and I could no longer bring it out of this position. It was high time to get out. I unbuckled myself, positioned myself up on the seat, and in the next moment, I was flung backwards by a tremendous draught of air. At the same time, I felt a violent jerk and realized that I had gotten hung up by my parachute on the leading edge of the rudder. Summoning up the last of my strength. I broke the tip off and I fell free behind the airplane, repeatedly tumbling head over heels.I had already assumed that the parachute had malfunctioned when I suddenly felt a slight braking and shortly after that hit the ground. The chute had opened after all at about 80

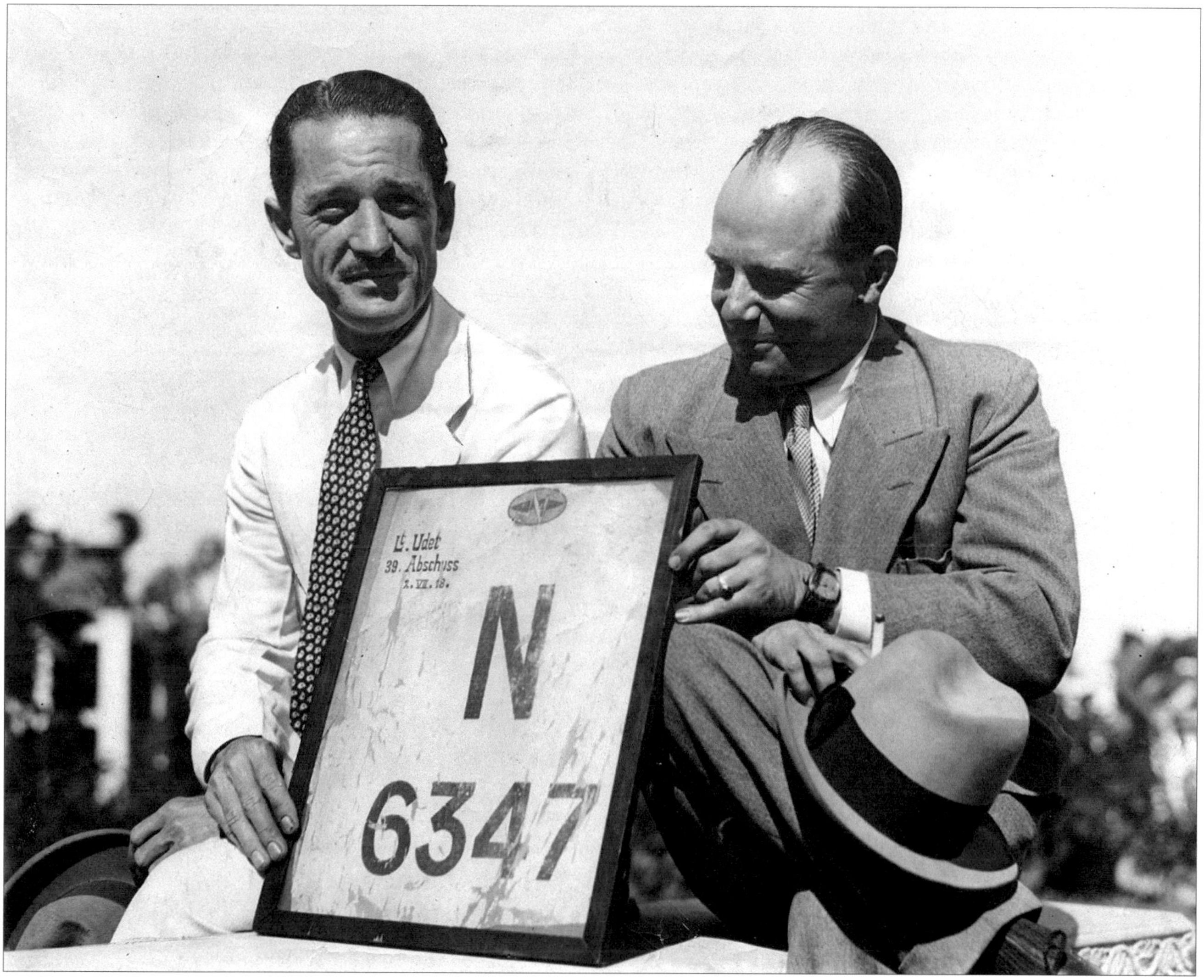

Above: Udet said: "I now give you what I took away" as he returned the tail number of his Nieuport 28 to its American pilot, Walter Wanamaker during the 1931 Cleveland Air Show: Wanamaker had been Udet's 39th victory when he was shot down on 2 July 1918. After the airshow, Udet accepted an offer by Wanamaker to come to his home and have dinner with his family.

meters. The landing was pretty hard and I sprained my left leg...The enemy fired on me with machine guns shortly before and after landing. I unbuckled myself from the parachute and ran in an easterly direction...I ran with all my might and luckily came to the edge of the ravine...where I was taken in by the 16th Infantry Regiment...I experienced heavy coughing and nausea, as I had covered the distance of roughly three kilometers without a gas mask... by that afternoon I was able to carry out a more auspicious hunting flight."

The commander of JG I, , who had been hand-picked by von Richthofen died while test flying a new fighter design in July. The pilots of JG I were in favor of giving the command to Ernst Udet. However, Udet was a reserve officer and it had been decided that the commander of the famed fighter group had to be a regular officer. Udet's new commander was a man named Hermann Goring.

Unlike his dogfight on 29th June, Udet was usually the victor and not the victim: with his Fokker D.VII he was now scoring victories on a daily basis. On the 16th August Udet managed to bring down Sous-Lt. Jean Eduard Cael in his Spad XIII from the famed escadrille Spa.3. After the war, the stork insignia from Cael's aircraft adorned one of the walls in Udet's apartment. In an interview after the war Cael gave a keen insight into Udet's character when he described a visit that Udet made to see Cael shortly after he had brought him down:

"Udet told me he was going to fly to Munich on leave in a Spad, which was his favorite aircraft in which he always travelled. At the same time he promised me that one of his pilots would drop a message over the lines to say that I was safely in captivity. As he was leaving he hesitated. It seemed to me that he was in two minds about offering me his hands. Finally he held it out, and I held mine towards him. We had no reason to withhold this sign of mutual esteem, as we had faced each other in battle with mutual respect."

Despite the successes, the overall war was turning against Germany and despite their losses Udet and his fellow fighter pilots continued to fly and fight. The war essentially ended for Udet on the 26th September when , after shooting down two DH. 9 bombers, he was once again downed by a two-seater:

"A third whips past me toward the west. The leader's streamer flutters on his tail. I go behind him. When he realizes that he is being followed, he turns around and flies at me. A salvo bursts out from over there; I feel burning pain in my left thigh, and gasoline comes spurting out of my holed tank like a shower. I cut the ignition and land..."

Udet would not fly another combat mission in the war. He found himself, Germany's leading ace of aces at 62 victories, a survivor of the war but uncertain to how he would handle civilian life. Udet was a congenial man, a fun-loving man who not only was the life of the party: he enjoyed the party immensely. He was gregarious and was a charming womanizer. He also dearly loved to fly and therefore set out to make his way in not only the Germany of the 1920's and 1930's, but also the wider-world.

Initially in the autumn of 1919 he and Robert Ritter von Greim staged several airshows in order to raise money for a relief fund helping German POWs. They used several surplus aircraft, but these were swiftly taken away by the Allied commission charged with ensuring that Germany would not have any front-line capable aircraft.

Heinz and Wilhelm Pohl were brothers from Munich who had moved to Milwaukee, Wisconsin following the war. They contacted Udet and convinced him that because of his fame as a fighter pilot, that an Udet aircraft could sell well in the USA. This was in 1921, when there were still severe restrictions on the aviation industry in Germany.

Those restrictions were loosened and as a result, the Udet Flugzeugbau (aircraft company) was formed in 1922. It produced over 10 designs of light, sport aircraft up until 1925 when the rights for those designs were sold to a well-known, fighter ace from WWI Paul Baumer. Baumer was a businessman whose company did well, Udet was a pilot who was only too happy to let someone else manage the hard work of running a company while he still received royalty checks as he went about the world.

He flew in various movies between 1929-1932. He was a star at many airshows and air races where the promoters eagerly sought out the charismatic and charming German ace. At the Cleveland Airshow in September 1931, Udet surprised an American pilot, Wanamaker with the fabric from his Nieuport aircraft that Udet had salvaged at the time and had now signed and framed for his old adversary.

His popularity did not go unnoticed and it was therefore not surprising when his old commander, Herman Goring approached Udet with an offer of joining the resurgent Luftwaffe. At first Udet hesitated, because it would also mean joining the Nazi party and Udet was decidedly apolitical. However, Goring enticed Udet by sending him back to tour the USA on an official basis with enough money to buy two Curtis Hawks for Udet's personal use.

Udet toured several American air bases and was able to see many of the front line aircraft the United States had at the time. He also bought the two aircraft and was overjoyed to fly them when he returned to Germany.

Unbeknownst to Udet, this was the beginning of the end for him. He had already shown a profound dis-interest in managing his own aircraft company, now Goring put him in charge of aircraft development for the entire Luftwaffe.

Udet enjoyed the idea of being involved with the development of the newest aircraft designs and he was thoroughly happy to fly them whenever he had the chance. However, the myriad of tasks associated with aircraft manufacturing depressed him. As stated, he was a man who loved a good party, and the attention of beautiful women: but he was not inclined to the intrigues and schemes that abounded in the upper echelons of the Nazi party heirarchy.

The renowned aviation historian Lance Bronnenkant quoted General Alfred Mahncke in his BLUE MAX AIRMEN: volume 15:

"...Udet was the typical, easy-going fighter ace-a reliable friend, sensitive and easily-influenced. He had lots of charm and possessed an artist's eye, and with pencil and brush produced exquisite, witty and humorous caricatures of people and situations-himself included. But first and foremost he was a wonderfully gifted pilot with a sixth sense, an aerobatic artist of the highest order. He was a Bohemian and enjoyed the company of artists, actors and painters, preferring their companionship to that of soldiers, and it must be said in his favor that he was reluctant to become a soldier again. But Goring overrode his reservations, appealed to his friendship and eventually made him into someone he never wanted to be."

Udet was not unhappy to be the unofficial ambassador of German aviation to the world before the start of WW II and he was certainly caught up in the excitement of being critical in the development of such new aircraft as the Bf109 and Focke- Wulf FW 190 fighters. However, he was much earlier in his recognition of the likely loss of the war than the other leaders of the German military.

He became depressed and began to drink heavily and also turned to prescription medications. He began to distrust his girlfriend, Inge Bleyle (Udet had divorced his beloved Lo in 1923 after only three years of marriage). Goring and others in the Luftwaffe used Udet as a scapegoat for the inability of the German aviation industry to meet the demands of the war.

On 16 November 1941, Udet shot himself. On the wall of his apartment he had written:" Iron one[Goring] you are responsible for my death".

Udet was given an enormous state funeral, the pageantry appropriate for such a hero of the German people. The official statement concerning the cause of Udet's death was that he had perished in a tragic flying accident. Goring himself gave the eulogy.

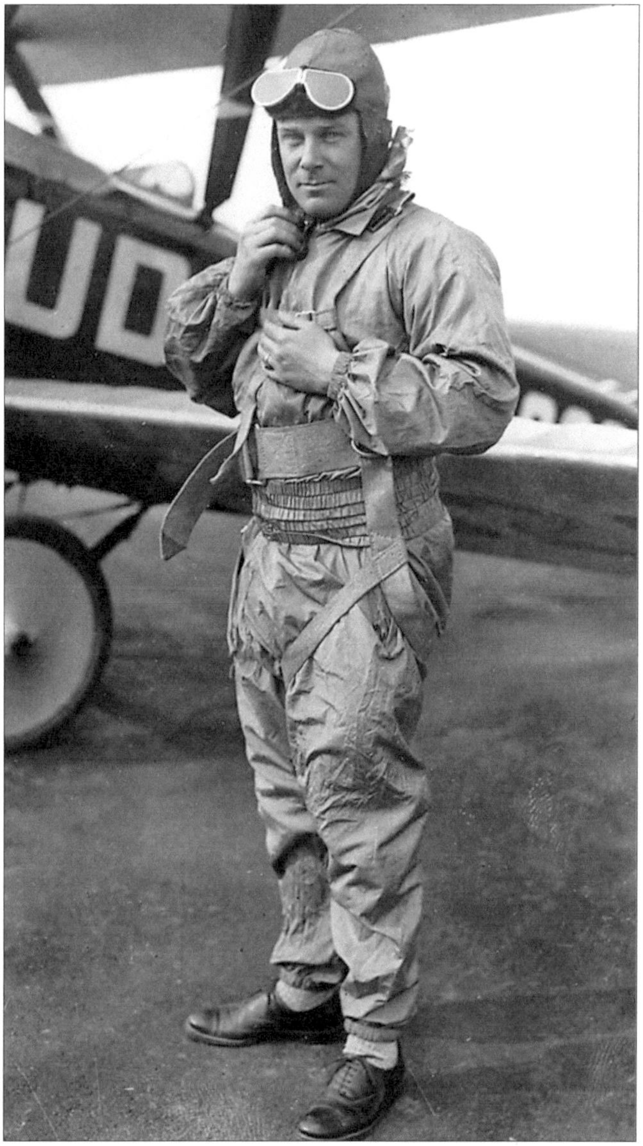

Right: Udet would be a well-known aviator throughout the world after the war. Here he stands before his famous Flamingo aircraft, in which he perfected his signature feat of picking up a handkerchief from off the ground by his aircraft's wingtip!

Below: This was the captured French Spad that Udet would fly back home on his leave. The words "GEUTE LEUTE" (good person) is on the top wing.

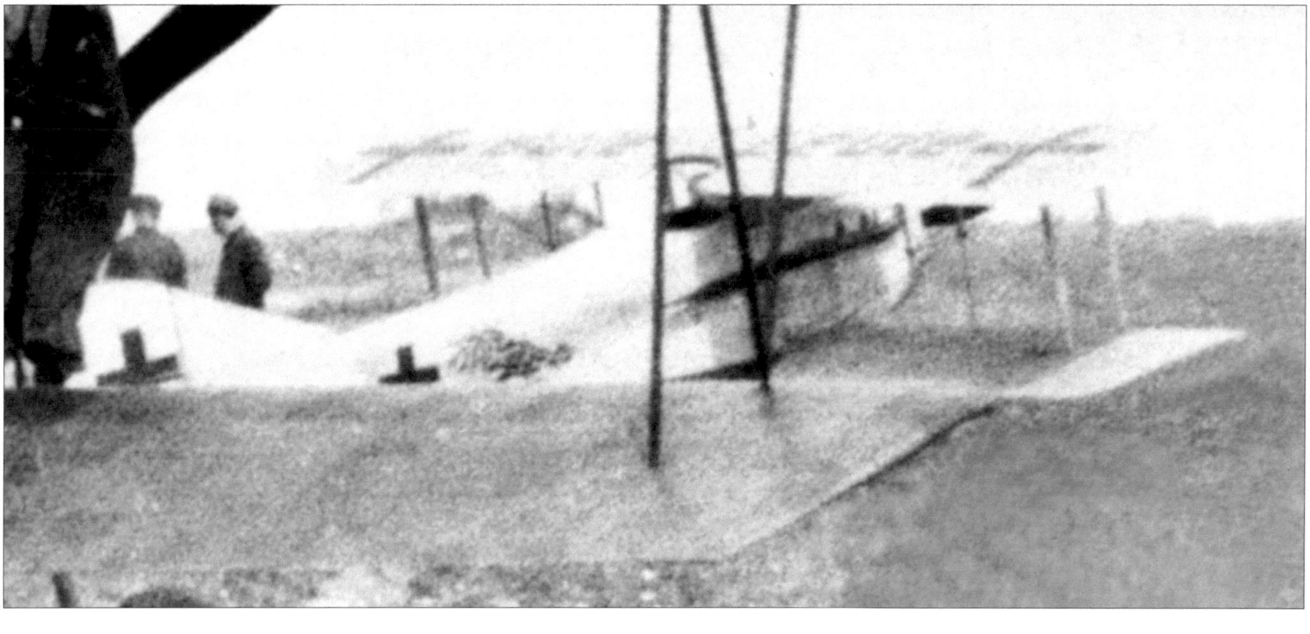

Home Field Advantage

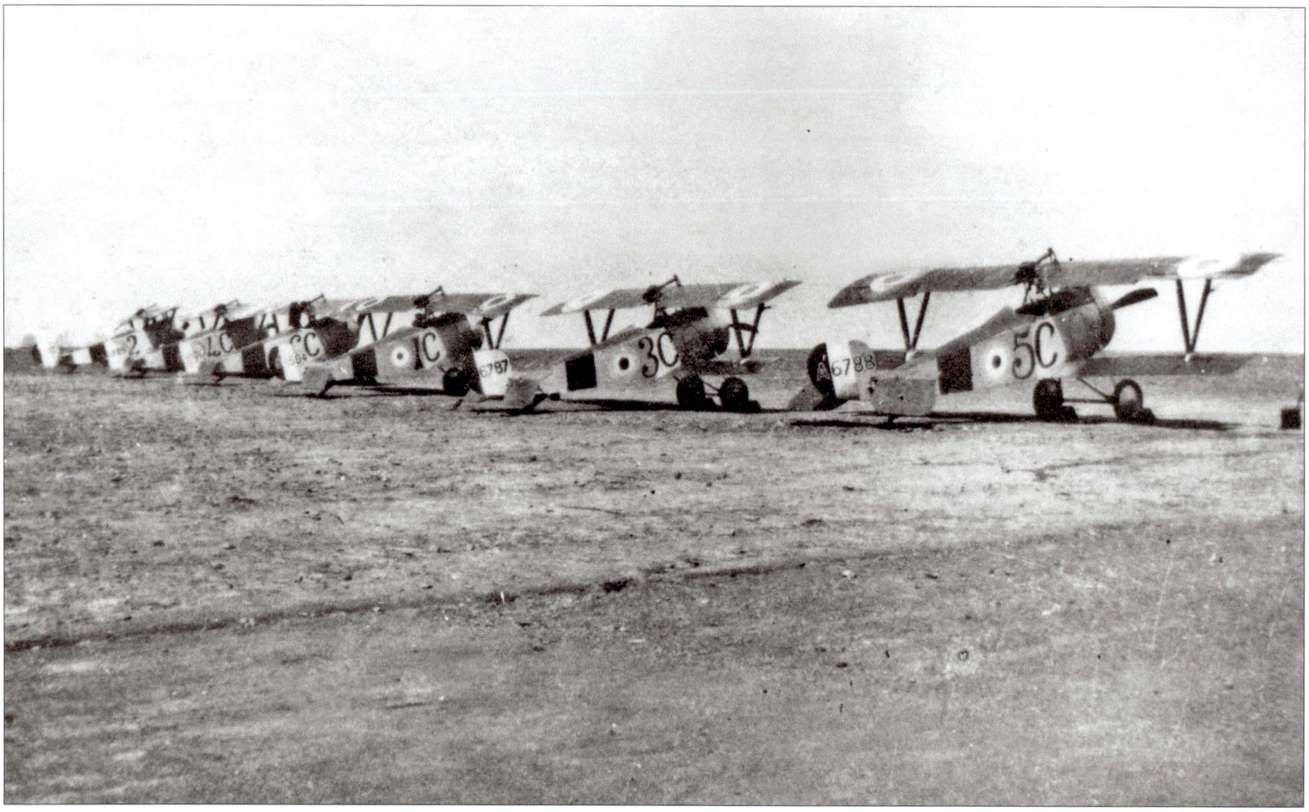

Above: A view of Nieuport fighers of No. 29 Sqdn. RFC gives an idea of how many WWI airfields were essentially open fields.

In the end, only a relatively small proportion of the belligerents in the First World War were aviators. And only those aviators would experience the war from their unique perspective of having the "luxuries" and conveniences of a "home field" that awaited their return from combat missions that were flown multiple times, day in day and out as weather permitted.

Many of the airfields of the First World War were converted farm fields, open space and a relatively level field were essentially the only required elements for an aerodrome. James Hamilton-Paterson wrote:

"The whole point of an airfield or aerodrome was that aircraft could land and take off in any direction according to the wind...All that was needed then was to stick up a pole with a wind sock on it- jocularly known in the RFC as the 'effel' from the initials of 'French letter'- and they were in business. It was from such lumpy fields, often dangerously ringed with trees, that aircraft from both sides took to the skies to do their armies' bidding."

As the Western front settled into its static nature of trench warfare, the airfields began to expand and tent hangars gave way to large wooden structures and nearby homes were requisitioned in the name of the airmen and ground crews.

Some airfields became established and names would be recognizable to all the members of the air service. As squadrons came and were transferred out, new squadrons would come in and begin their own service duty there.

The German fighter Jastas would be moved frequently, as they would be needed to establish air superiority over sections of the battlefront since the German Air Service did not have the overall numbers of aircraft or pilots to cover the entire Western Front. These units would fly to their new field, while the supporting elements would be transferred by rail or truck. Then canvas hangers would be set up and the unit would take up its duty at this new location. Thus, the famous flying circus was born.

The diaries of the pilots and observers are filled with discussions of their lives not only in the air and during combat, but they extensively write about their lives that centered in their squadrons, escadrilles and Jastas. Their existence was that of

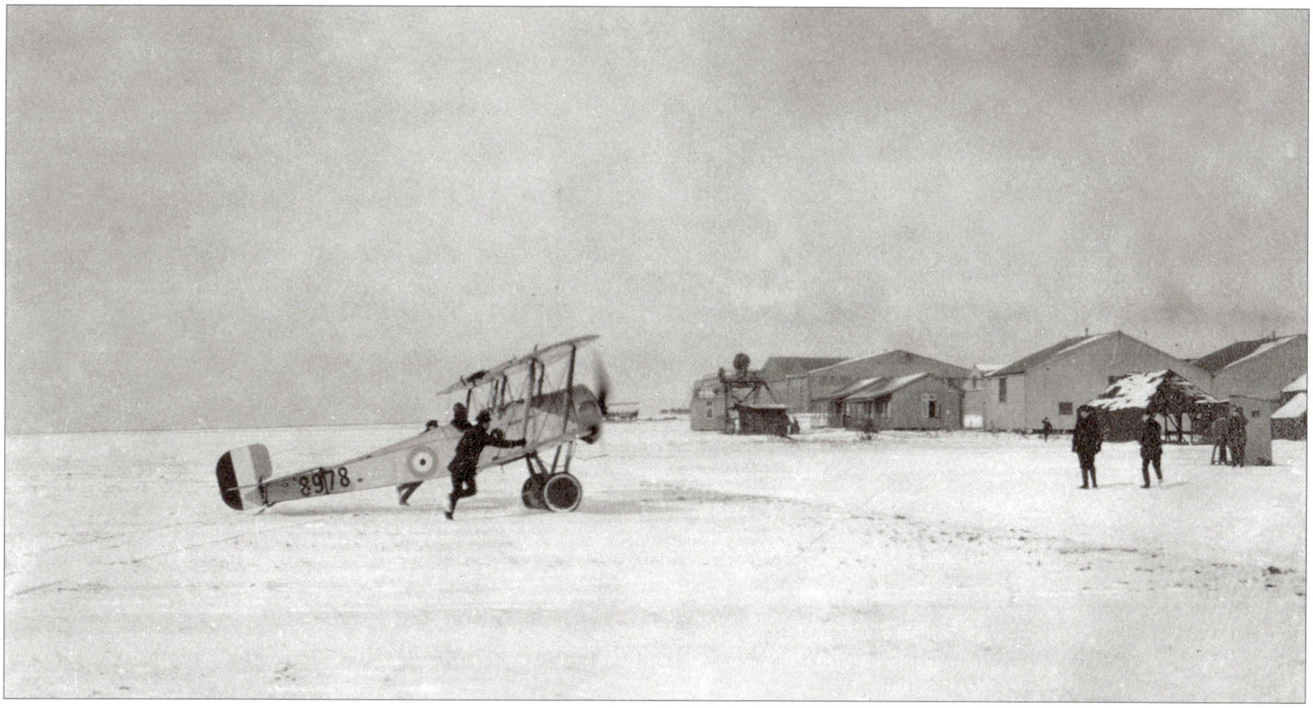

Above: Some of the airfields would be enlarged and the facilities made more permanent with many structures erected around the periphery of the landing field. This image also can give some idea of how open cockpit flying in winter must have been quite cold for the pilots.

Below: If you were lucky, your airfield was by a wonderful chateau and. for Allied airmen, a 48–72 hour pass meant a ride in the squadron's car to Paris. Jim Dietz

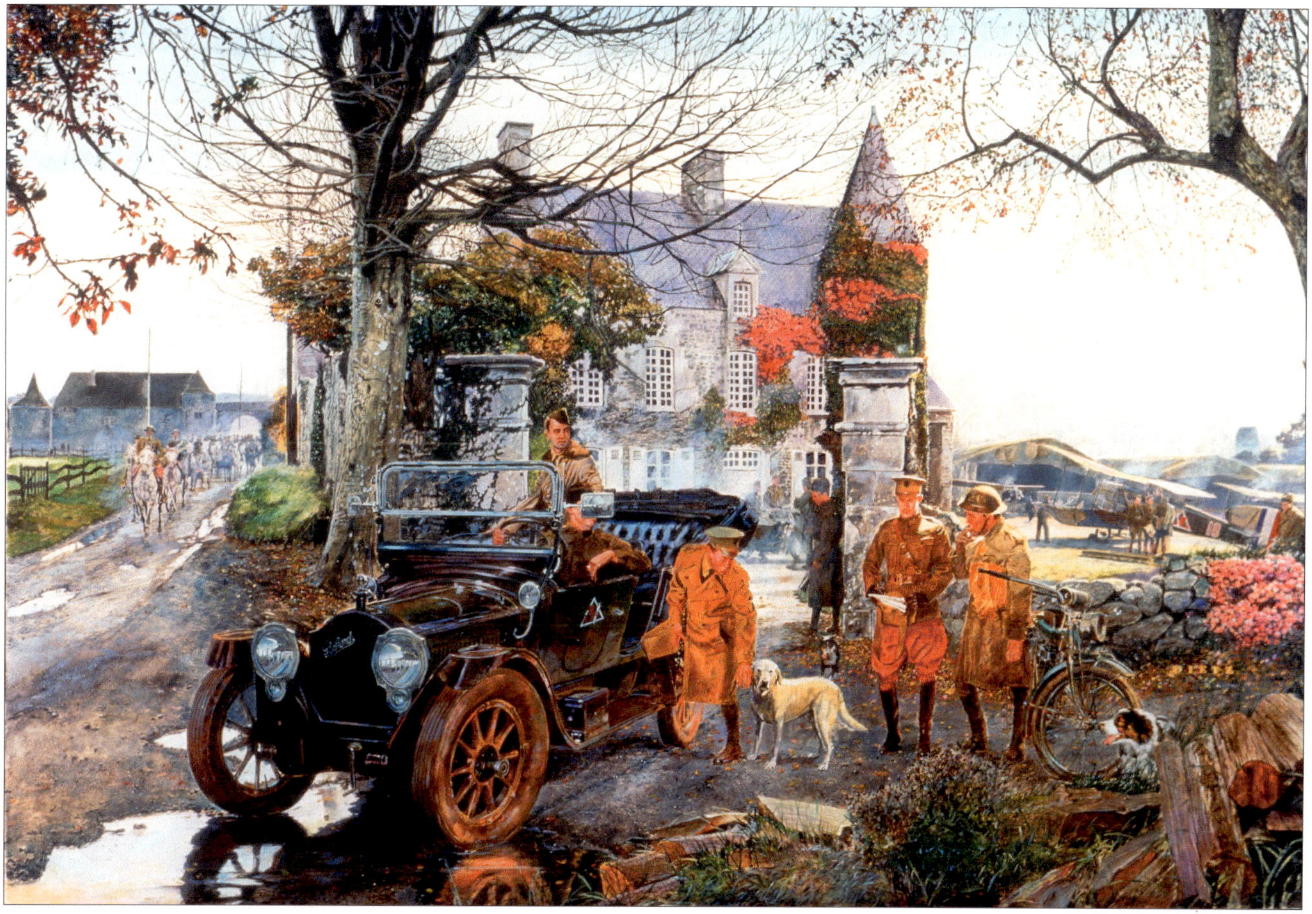

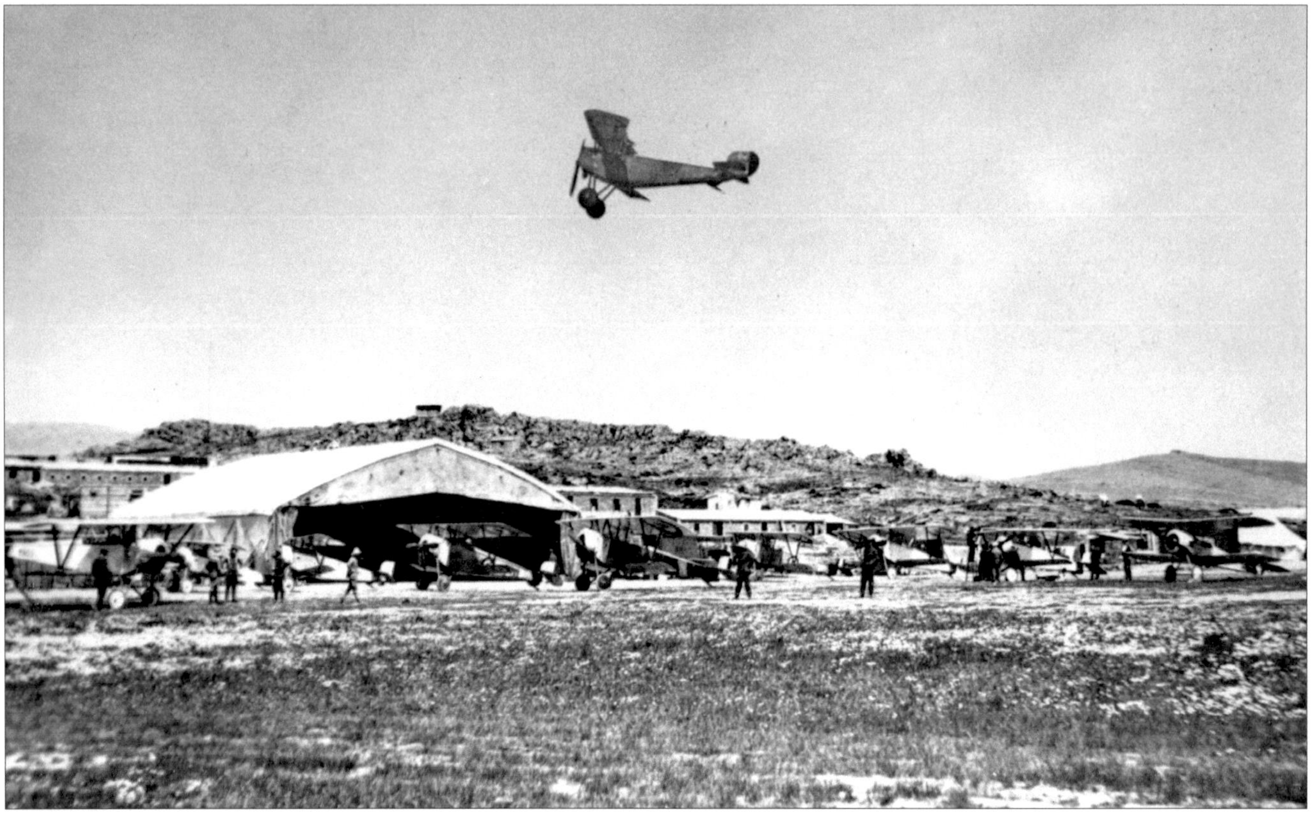

Above: No.2 wing RNAS at Imbros gives a good impression of how a First World War airfield must have appeared on any given day.

Below: A Fokker Triplane of Jasta Boelcke takes off from "its airfield near Douai", landmarks such as the large chimney in the background can aid aviation historians at properly placing photographs of aviators, or aircraft.

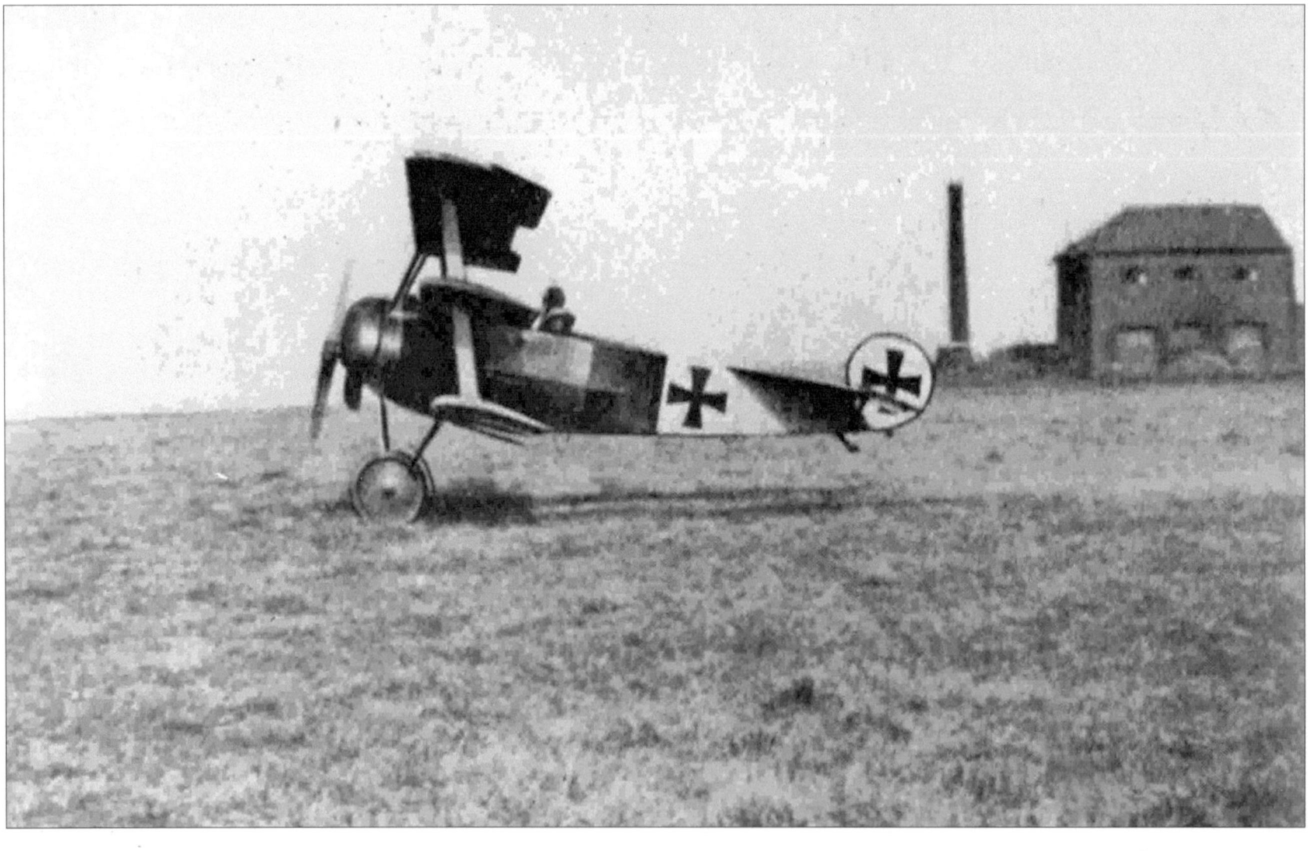

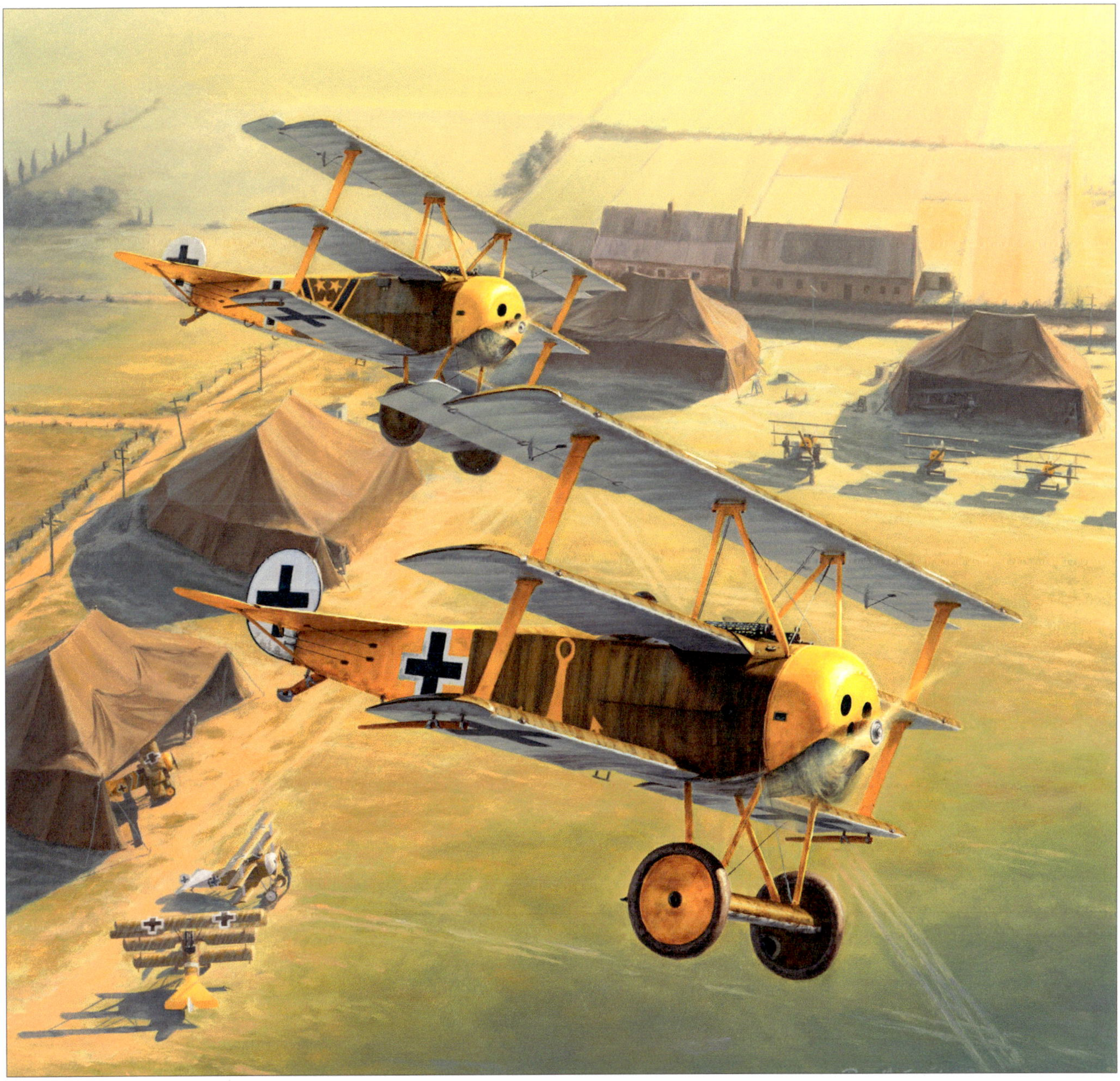

Above: Bringing to life the image of Jasta 12's airfield; however, in this case the Fokker Triplanes are part of Jasta 27, but the airfield and tent hangars would be remarkably similar up and down the Western Front Russell Smith

extreme contrasts. Unlike their comrades in the trenches, the aviators of World War I had mess halls with dinner in the evening, often a well-stocked officer's lounge with music via piano or gramophone. Their batman would attend to their "needs" and hot water was almost always available for washing or shaving.

The British especially seemed to relish the development of sports near their aerodromes by building tennis courts, football and rugby pitches nearby. During the warmer months, swimming in rivers, or nearby ponds was clearly a popular relief as many photos of pilots in their bathing trunks appear in memoirs.

The living quarters could vary from tents, Nissen huts, billets in local houses all the way to chateaus. The ability to act as if the aviators were part of the local, landed gentry (which of course many of the officer-aviators were) was especially true of the Allied pilots. Although German aviators were often grudgingly accepted by the occupied French locals. The British devised a new type of billets for their aviators on some of their airfields: Cecil Lewis wrote a description of the Nissen hut:

"Imagine to yourself a semicircular-roofed hut of corrugated iron. At either end a door for people to

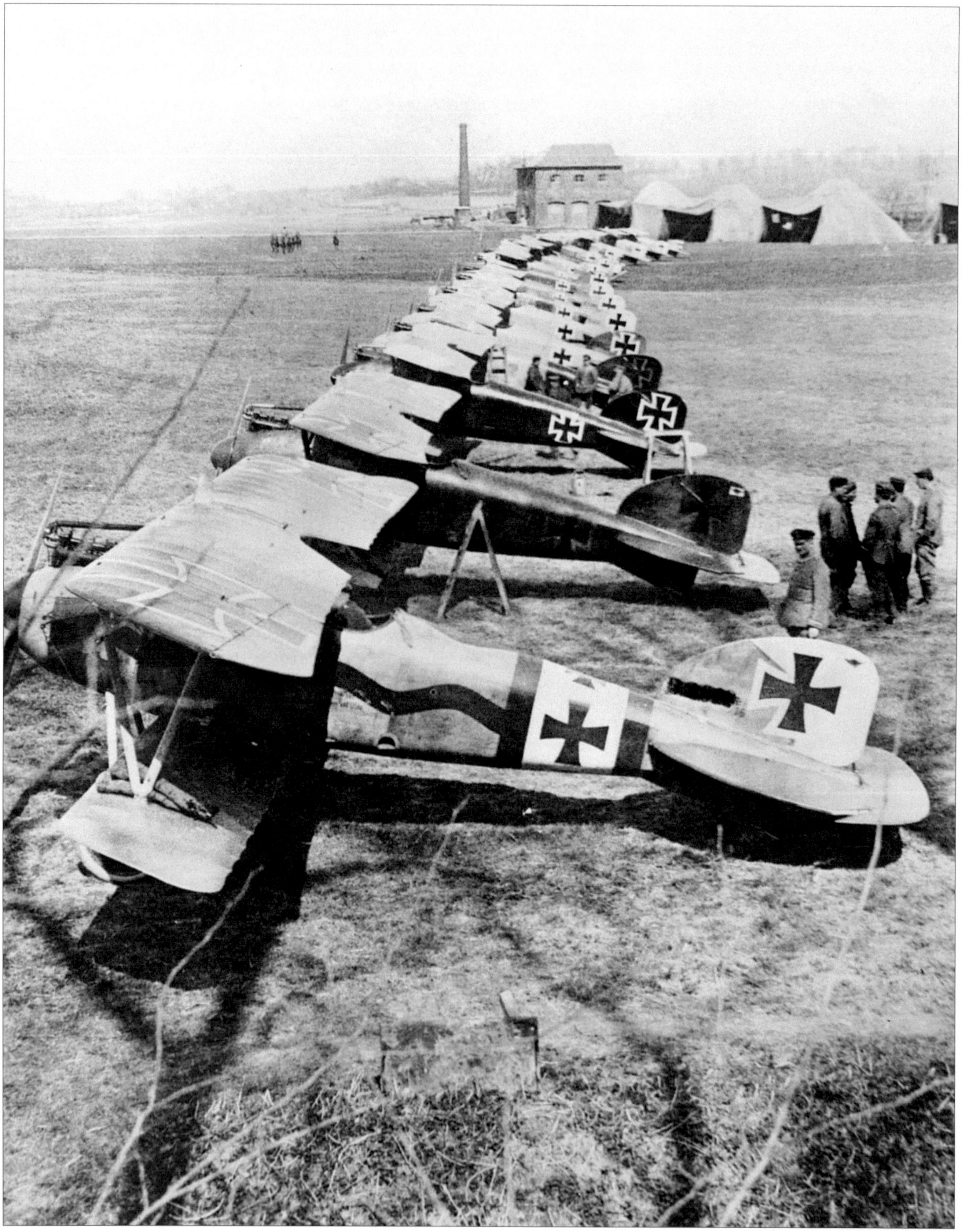

Above: Note the same structure in the background of this photo which shows a line up of Jasta 11 Albatros fighters at the same airfield which can be confirmed as LaBrayelle airfield that was 3 kilometers west of Douai. The airfield actually was from before the start of the war, the world's first aviation meet was held there and it was home for Breguet aviation.

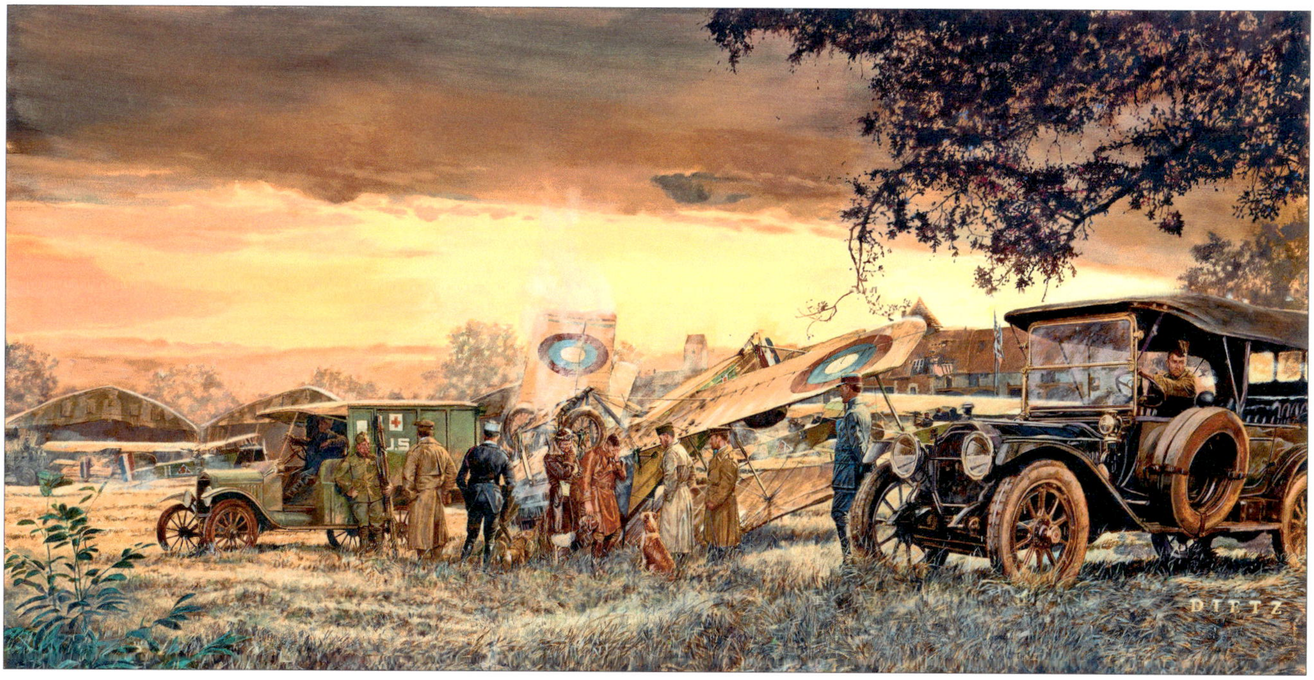

Above: One of the all too frequent scenes at a First World War airfield was that of a crashed aircraft, often one that was damaged and returning from its mission, at times with wounded comrades in the cockpit. Jim Dietz.

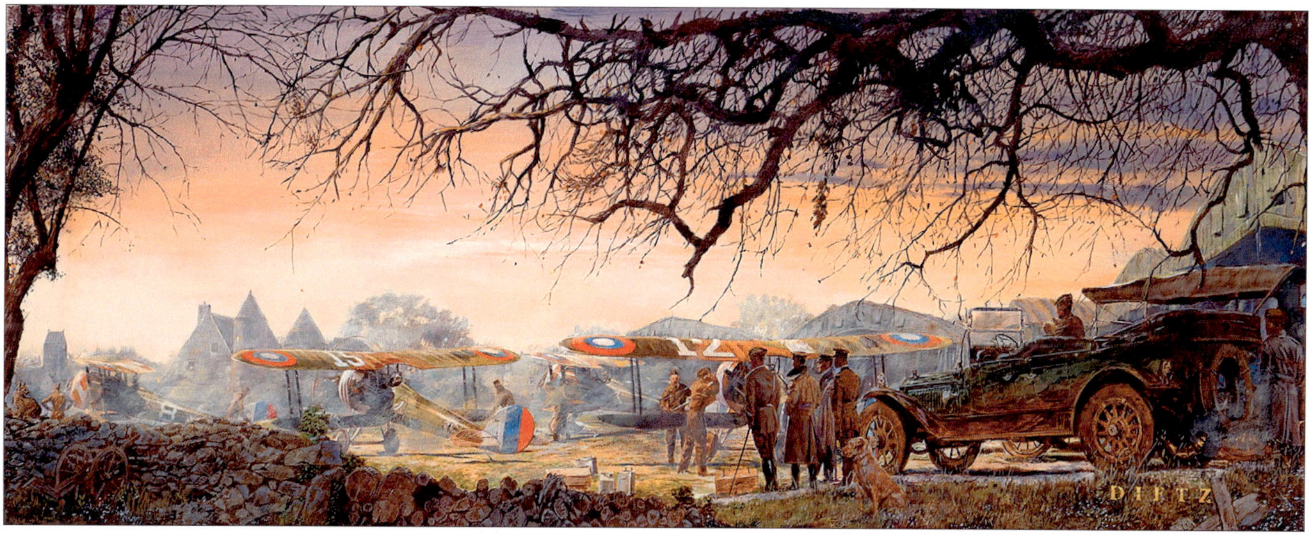

Above: At first light, the airfield comes alive and the pilots begin the first patrol of the day. Often, the aviators will have to fly away from what seems like a peaceful world and fly into the cauldron of war, hoping to safely return. Jim Dietz

enter by-also for draughts to enter by!- and on either side of these doors a window, making four in all, covered with canvas, none too airtight. In the centre of the hut a stove, made from an old oil drum, with a chimney going up straight through the roof. At the sides, tucked away under the curved roof, two little camp-beds, and about the rest of the hut towels on strings, tunics, overcoats hanging from hooks in the roof. In front of the stove there is a rough table-top-a piece of an old door- standing on two upturned petrol boxes. In the center of the table a solitary candle, which, as you can imagine, does not light the hut."

Albert Ball had a small garden that he tended, and because of his ace status was allowed his own hut at the edge of the airfield. Guynemer, when his unit was stationed close by, could often drive his personal car to visit his family. Von Richthofen had a red-colored two-seat Albatros aircraft for his own personal use.

Nungesser was given a roaming commission which allowed him to attach himself to any unit along the front lines, along with his several Nieuport fighters.

These freedoms allotted to the famous aces

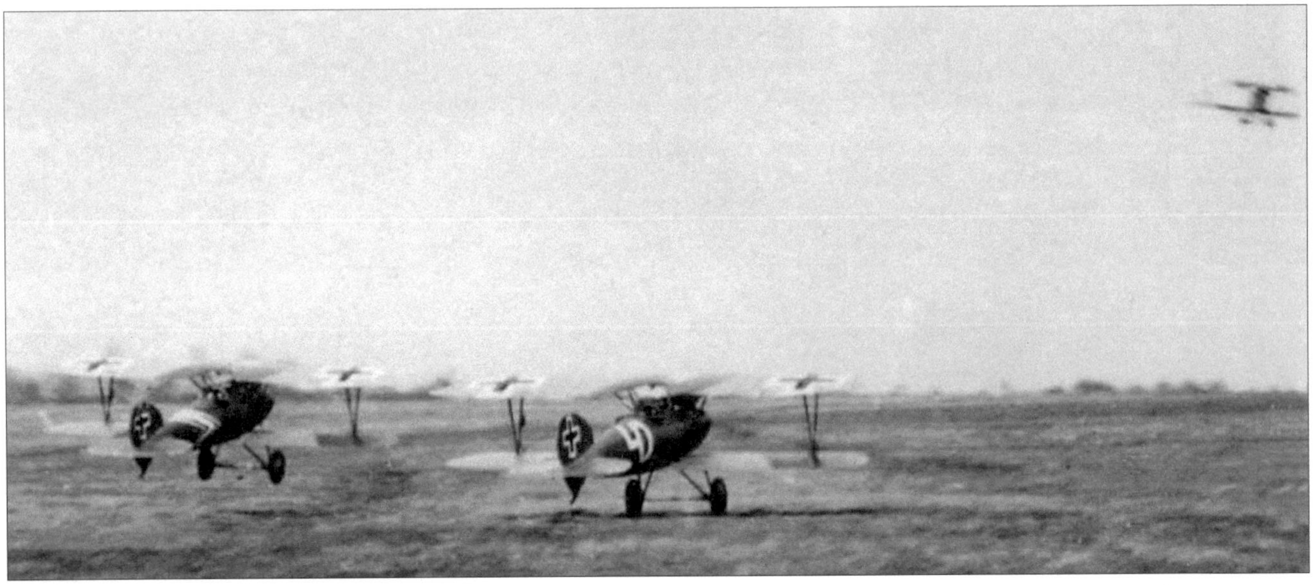

Above: The First World War airfield was generally a large, flat (as possible) field so that the aircraft could take off into the wind no matter which direction the wind was from.

Facing Page, Above: This well-known photo depicts Jasta 12 on 15 March 1918 and shows the perfect location of an airfield, large field near roads on which supplies, could be brought to the unit and a small village steps away to provide billets for the aviators. The Jasta is in the process of changing from Albatros fighters to the newly arrived Fokker Triplanes.

Facing Page, Below: One advantage for WWi aircraft was the ability to use any reasonably flat field; here RE8 observation aircraft are seen utilizing an improvised field during the retreat from the German Spring offensive in 1918. Fuel and parts, ammunition could be brought to them via trucks by the nearby road. Chances are very likely that by the next day, the aircraft will be at another field and will continue to change until the lines stabilize.

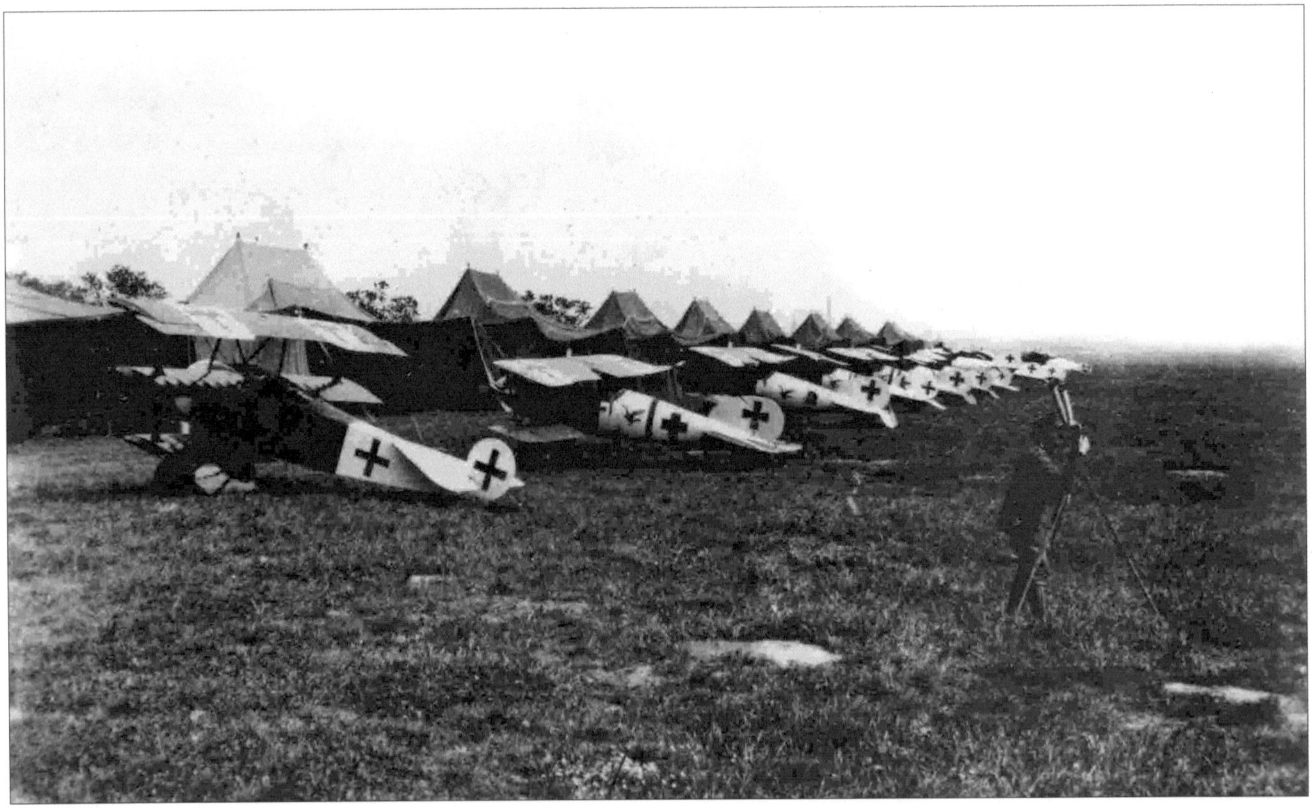

Above: A good example of a German airfield with tent-hangars, the person in the foreground is looking through high powered binoculars which were used to spot enemy aircraft and also sometimes to watch dogfights that happened within proximity to the airfield.

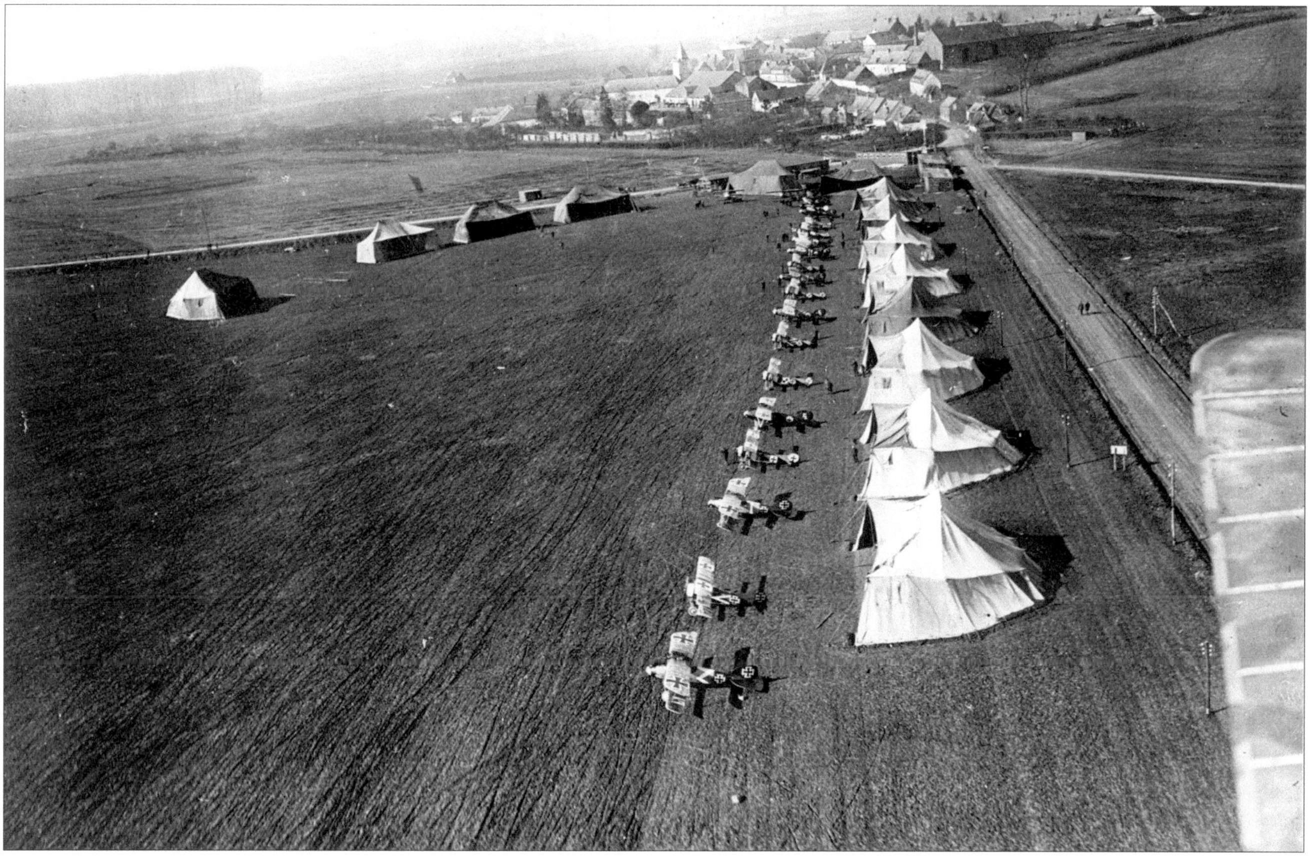

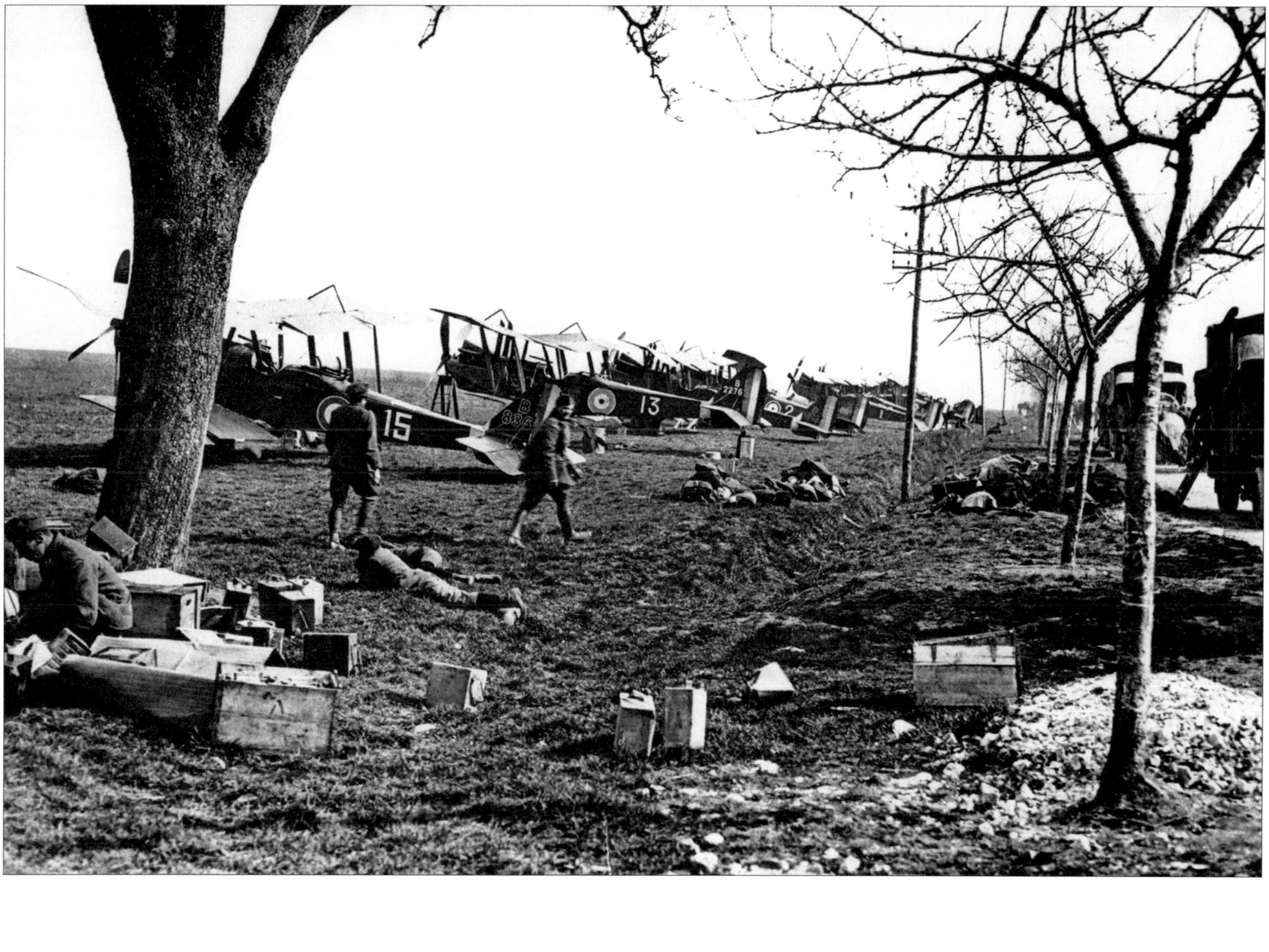
15
13

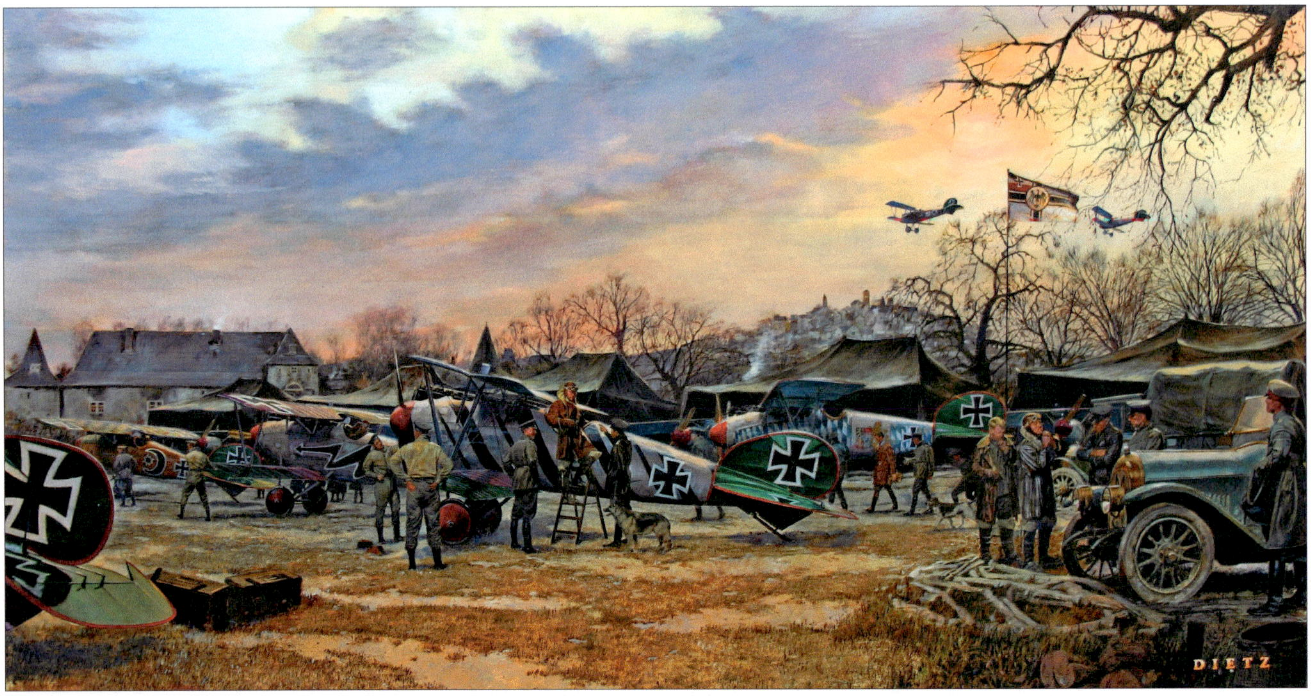

Above: On the other side of the Western Front, a German aerodrome readies itself for the beginning of the day; who knows what awaits these aviators? Jim Dietz

influenced how all of the aviators saw themselves. A relaxed atmosphere was apparent at the airbases on both sides of the Western Front. It was not unusual for Allied pilots to load themselves into a car after the day's missions and drive into a nearby town, or even to Paris if close enough and celebrate as best they could. Inevitably their mixed uniforms, with wings prominently displayed, would gather attention and the party would begin.

These portraits from the life of a First World War aviator are not wrong, however they only give a part of the story. Many pilots wrote of going into the front lines and to seek out help with confirmation for an aerial victory. Sometimes the aviator found himself trying to make his way back to a friendly trench after having been shot down. At times, a small group of airmen would visit ground units nearby in order to establish contact and foster good relations.

Ground troops inevitably looked upon the airmen as being crazy and often wrote in their own letters home how they could never for a moment conceive of going up in an aircraft.

Certainly, the average life-span of an aviator in the First World War which would usually be measured in mere weeks at the front lines would argue that the infantryman had a valid point to make.

Almost uniformly, for whatever reason the interaction, when aviators wrote of the trenches they did so in awe of how the infantry could stand such an existence. And the airmen would anxiously await the return to their airfields, back to their home field advantages.

However, such a privileged life did not come without some sense of guilt and exasperated feelings of the sharp contrasts between their lives at their air base versus the "moonscape" and desolate earth they saw from their cockpits as they flew over the Western Front.

One can frequently read about how empathetic airmen were for their comrades down below, as they flew above the poison gas, as they flew above the artillery shells, as they flew above the advancing infantry being decimated by machine gun fire.

Harold Buckley of the 95th USAS pursuit squadron wrote:

"The sight that met us when we reached the lines was enough to stop one's breath. From Verdun to the east as far as the eye could see, a broad belt of living fire cut the blackness below, a terrifying inferno which thrilled us with its magnitude and beauty, but sickened us with its ghastly portent. Like men of another world we watched it, fascinated by our soaring isolation yet trembling at our insignificance. It was a dream-a nightmare- a glimpse into Hell-or the morning of Judgment Day."

Of course, the airfields were not truly safe sanctuaries from the horrors of war. Airfields were bombed and strafed by the enemy. Perhaps even worse than those attacks, were the occasions when the aviators gathered on the landing ground

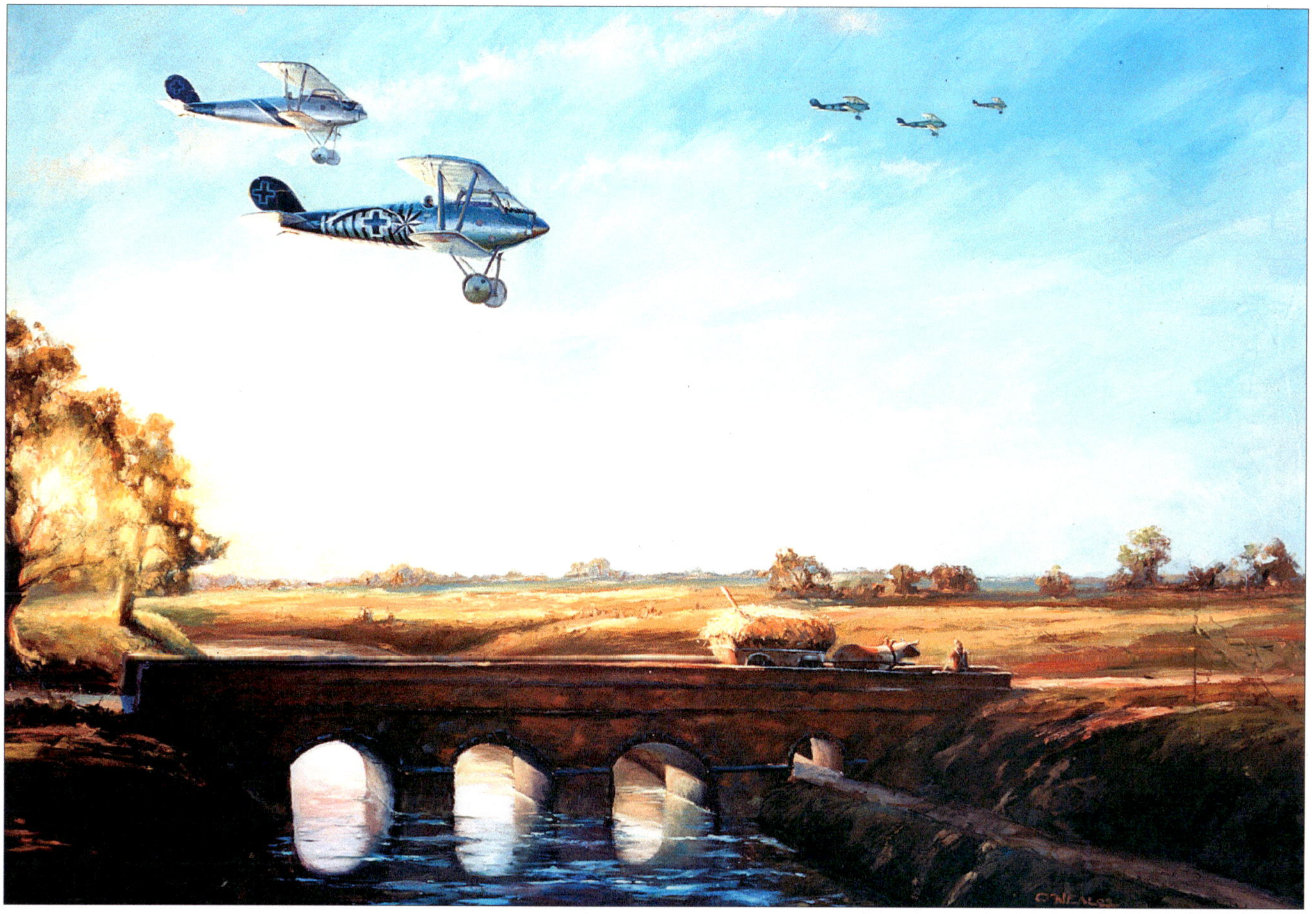

Above: A perfect image of the strangeness that enveloped the First World War aviator, who could see and be enveloped by bucolic, peaceful images when at or approaching his airfield and a way of military life that could be considered a home field advantage. Micheal O'Neal.

watching for a comrade that was not coming back. Or a comrade that nursed his damaged aircraft, perhaps having been wounded himself, trying to make it back to the safety of his airbase only to crash upon landing,

To watch a friend crash land, sometimes for the aircraft to catch on fire, knowing that his chance of survival was slim was a heavy burden that was not an infrequent sight during the career of a front line pilot.

It would be the incongruence of this life as an aviator in WWI that would catch so many off balance. One of the leading American aces of WWI, Bill Lambert (a Warbird who flew with the RFC/ RAF) wrote in his book COMBAT REPORT:

"Lunch finished, I was ready for a walk. A beautiful spring day- bright sun and not too hot. Apple trees in bud, I tried to promote a tour to Auxi but only Daley and Palmer were interested...On the land en route farmers were busy, the weather being made to order for them. In less than an hour we were entering Auxi...Everything was peaceful and quiet out there. One would never know that a war was going on except for an occasional low dull boom like thunder way off in the distance. The sun was bright and warm, the sky was blue, birds sang and rabbits scampered across the fields. Palmer, Daley and I relaxed deep in those soft cushions with our minds blank...We sat at a table covered with snow white linen and napkins-a rare item in those days. The crystal and silver looked as though it was the proprietor's own. On the table were bottles of white and of red wine..."

Gimpers

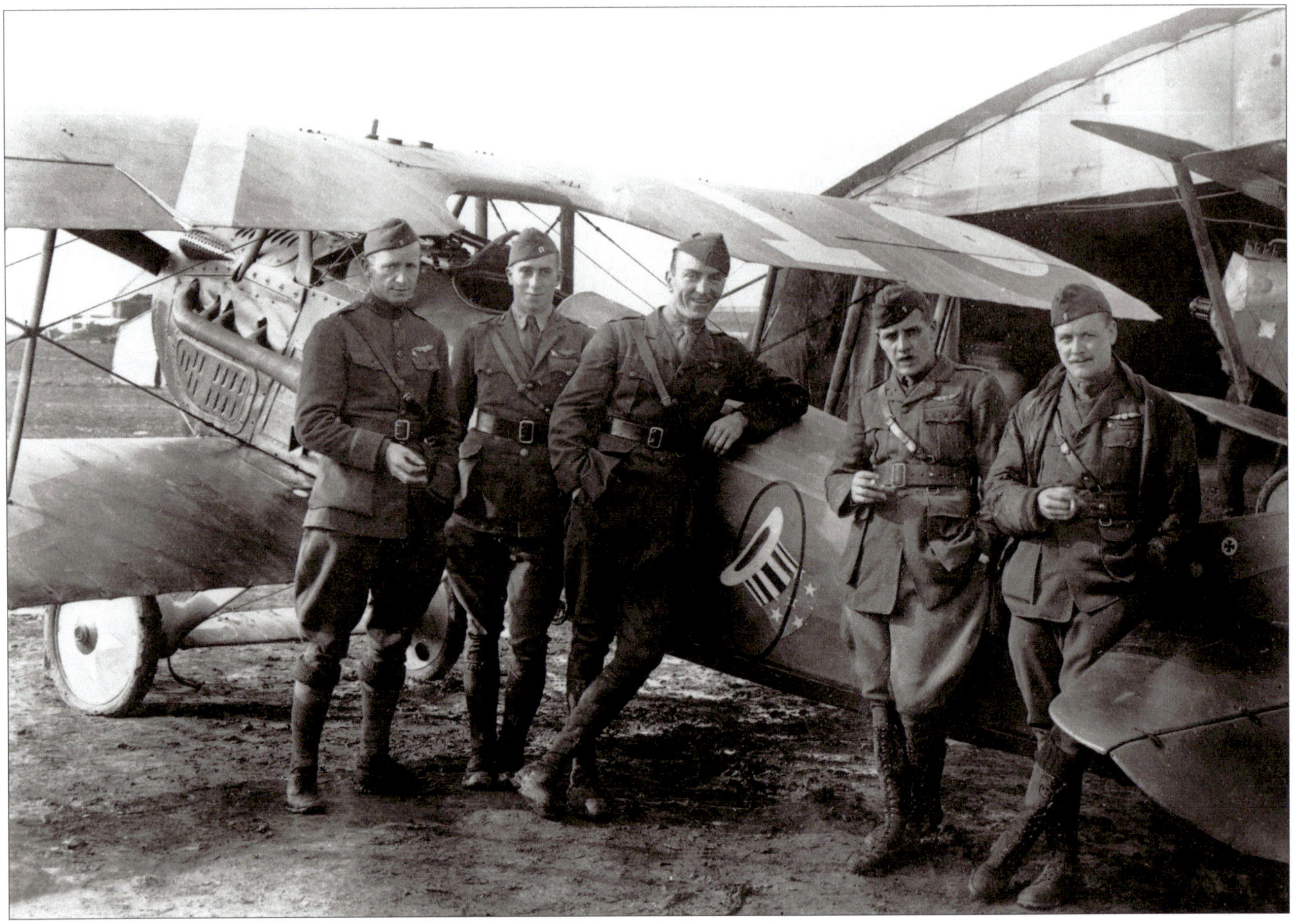

Above: 94th pursuit squadron "gimpers" in late October 1918, all were worthy of their squadron insignia seen of the Spad behind them, L to R: Joe Eastman, Jimmy Meissner, Eddie Rickenbacker, Reed Chambers, Thorne Taylor.

Although the United States Air Service (USAS) arrived only towards the end of the war, it is clear that the newly "minted" American aviators helped in a significant manner to shift the balance of air power to the Allies.

When the United States entered the First World War, its air service was woefully unprepared to participate in the air battles over the Western Front. It had very few pilots (26) and had no aircraft that could be considered as combat worthy. There of course had been some American pilots in the French Air Service (Lafayette Flying Corps) and in the Royal AIr Force (Warbirds) but the new USAS needed many more aviators and needed them in a hurry. A Yale professor famous for the discovery of Machu Pichu, Hiram Bingham III, was chosen to be in charge of recruitment and training for the new USAS.

Bingham had very entrenched concepts of who would be the ideal candidate to be an American aviator. In his wonderful book The Unsubstantial Air, Samuel Hynes writes:

"This ideal American pilot-wellborn, well educated, athletic, patriotic and honorable in all his doings-wasn't invented by Bingham; he was simply endorsing a type that already existed in upper and upper-middle-class American society. The ideal figure was there in the Yale men that Bingham taught...By accepting the assumption that such young men were the best possible material for combat pilots, the Service made a class distinction: flying- American combat flying- would be an occupation for gentlemen." (p.18–19)

It is therefore of interest to note that some of the most successful and famous American aces of the First World War did not remotely fit such a description. Aces such as Lufberry, Luke and Rickenbacker. In fact the American Ace of Aces, Edward Rickenbacker devised his own scheme to

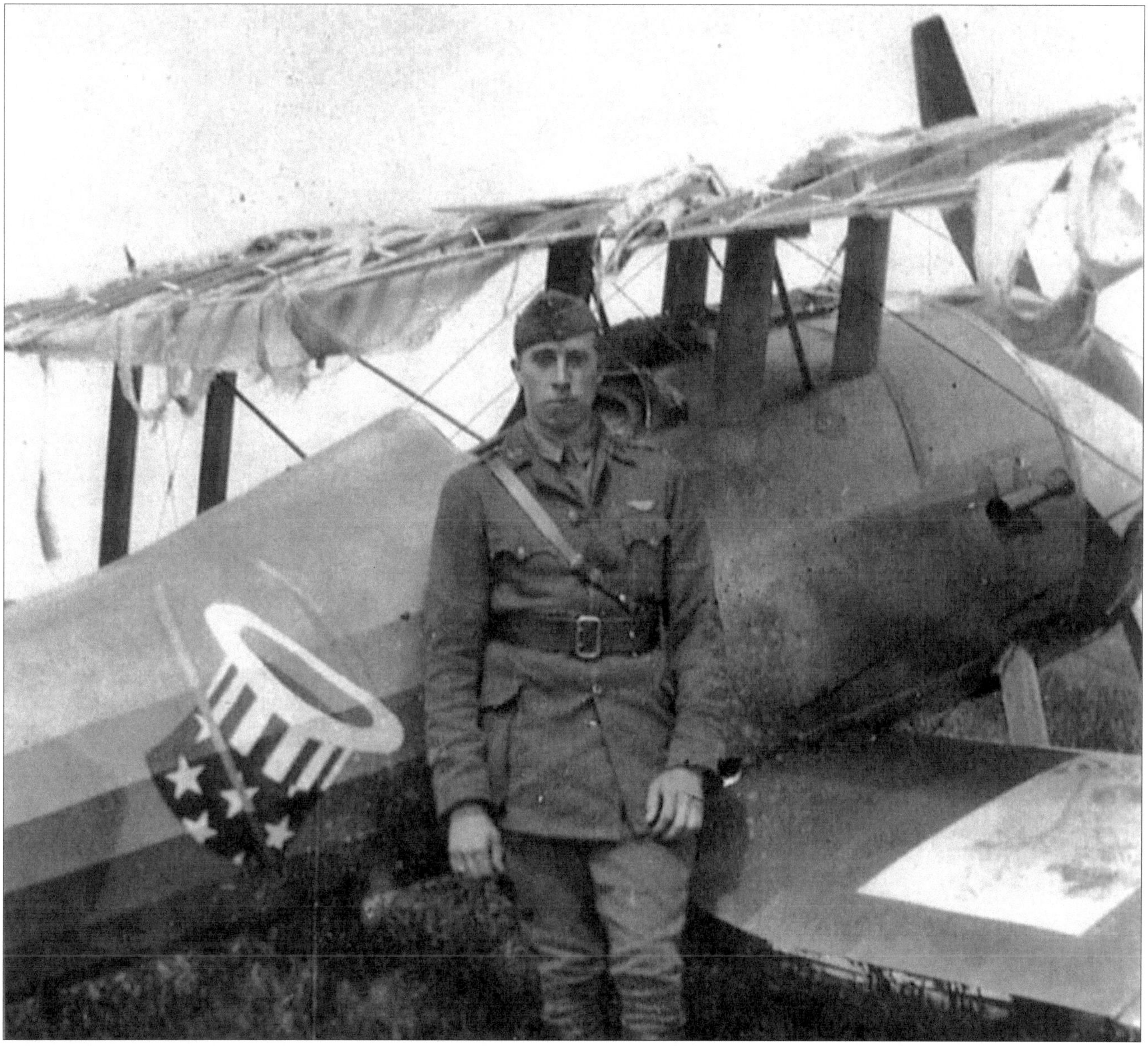

Above: Jimmy Meissner immediately after safely returning to his airfield following the loss of the canvas from his top wing.

rank combat pilots:

Egg: a newcomer, apparently not yet hatched as a combat pilot

Vulture: next step in the evolution of a combat pilot, someone who had several missions

Gopher: a pilot with enough experience to make a difference in combat

Gimper: the top accolade from Rickenbacker, "a bird who will stick by you through Anything...did everything just a little better than he has to."

Rickenbacker also had another ranking of a combat pilot, not a very favorable one. He listed a cuckoo bird as "an aviator who does all his fighting while none of the gimpers are around and then comes back and tells about it". In other words, an aviator who cannot be trusted in a real dogfight.

Rickenbacker would rise to be commander of the USAS 94th pursuit squadron and so had to make many judgements in regards to the actual capabilities of a front-line pilot and he would not tolerate any cuckoo pilots or pilots who were not clearly working their way towards being gimpers in the 94th. He would later write a very concise observation:

"It may seem strange that a man who had gone through flight training and had flown patrols in wartime would be leery of combat. Fighting, however, requires a different kind of bravery from that involved in flying. The combat pilot must be mentally prepared to shoot to kill and to be shot in return. Some pilots had sufficient motivation and bravery to learn to fly, but going out to kill or

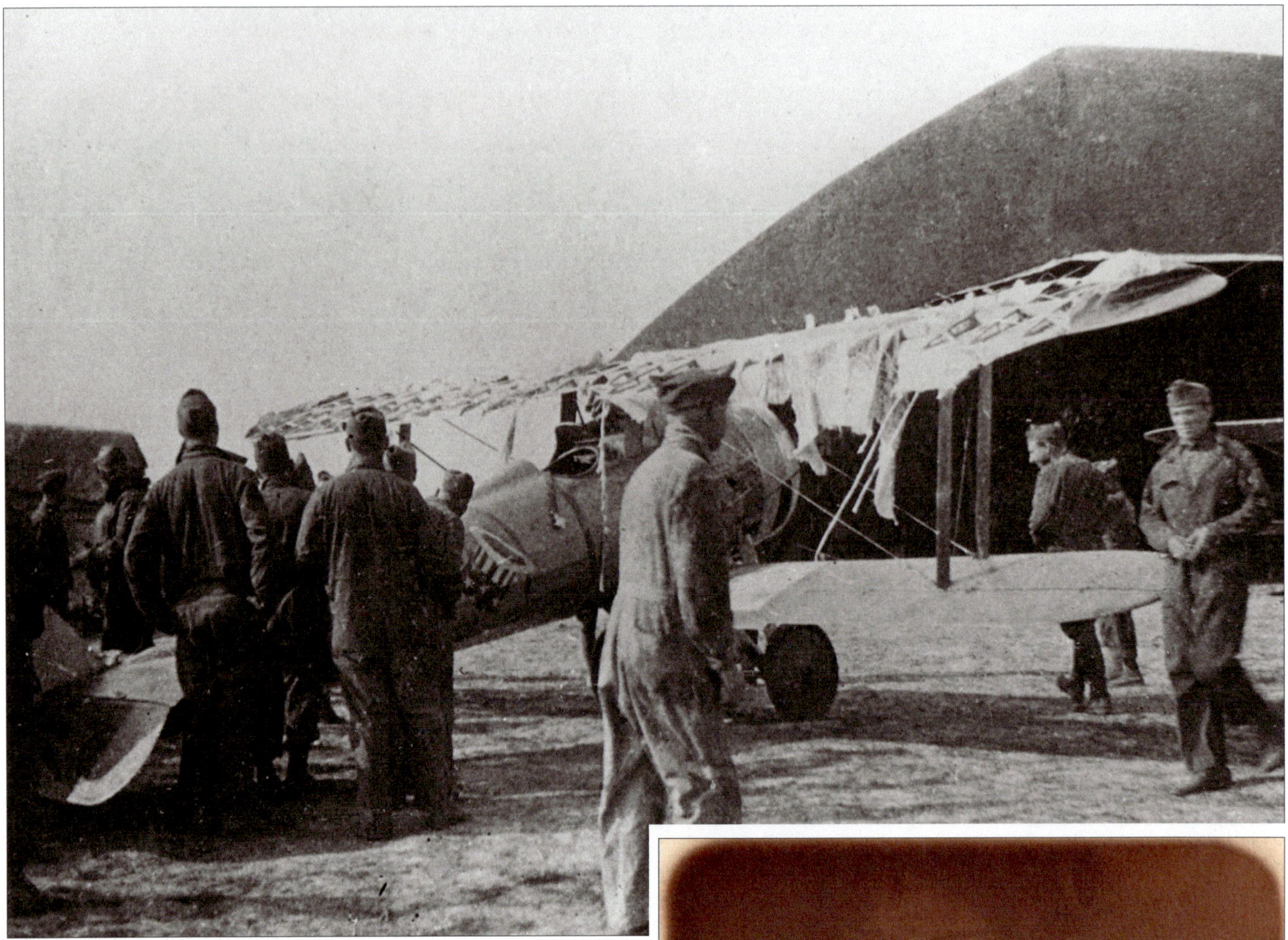

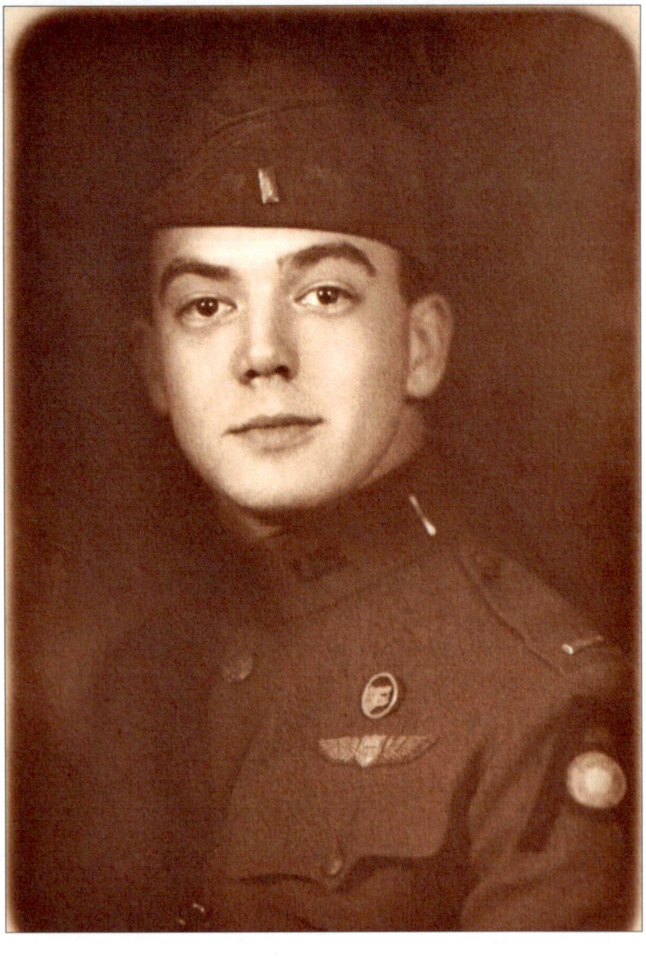

Above: Rickenbacker suffered the same thing while flying the squadron commander's Nieuport.

Right: A fine portrait of Charles Woolley.

be killed, over the lines required a different type of courage. Some did not have it."

The first two pursuit squadrons of the USAS to arrive on the Western Front for combat, the 94th and 95th, were both filled with "gimpers", anxious to get into the fighting. Because of the absence of American combat aircraft, the two squadrons were assigned the Nieuport 28. This aircraft had lost in competition to the Spad VII and Spad XIII aircraft which were being used in the French AIr Service. However, the production capabilities were already very good in the well-established Nieuport company and therefore, it was arranged that the new type 28 would go into production in order to fill the need of the American fighter squadrons.

At first the American pilots were ecstatic and were in a rush to engage the enemy in their fighter aircraft. However, several features of the Nieuport 28 soon became evident. The illustrious and well-respected fighter ace Raoul Lufberry was shot down in flames while engaging a German two-seater

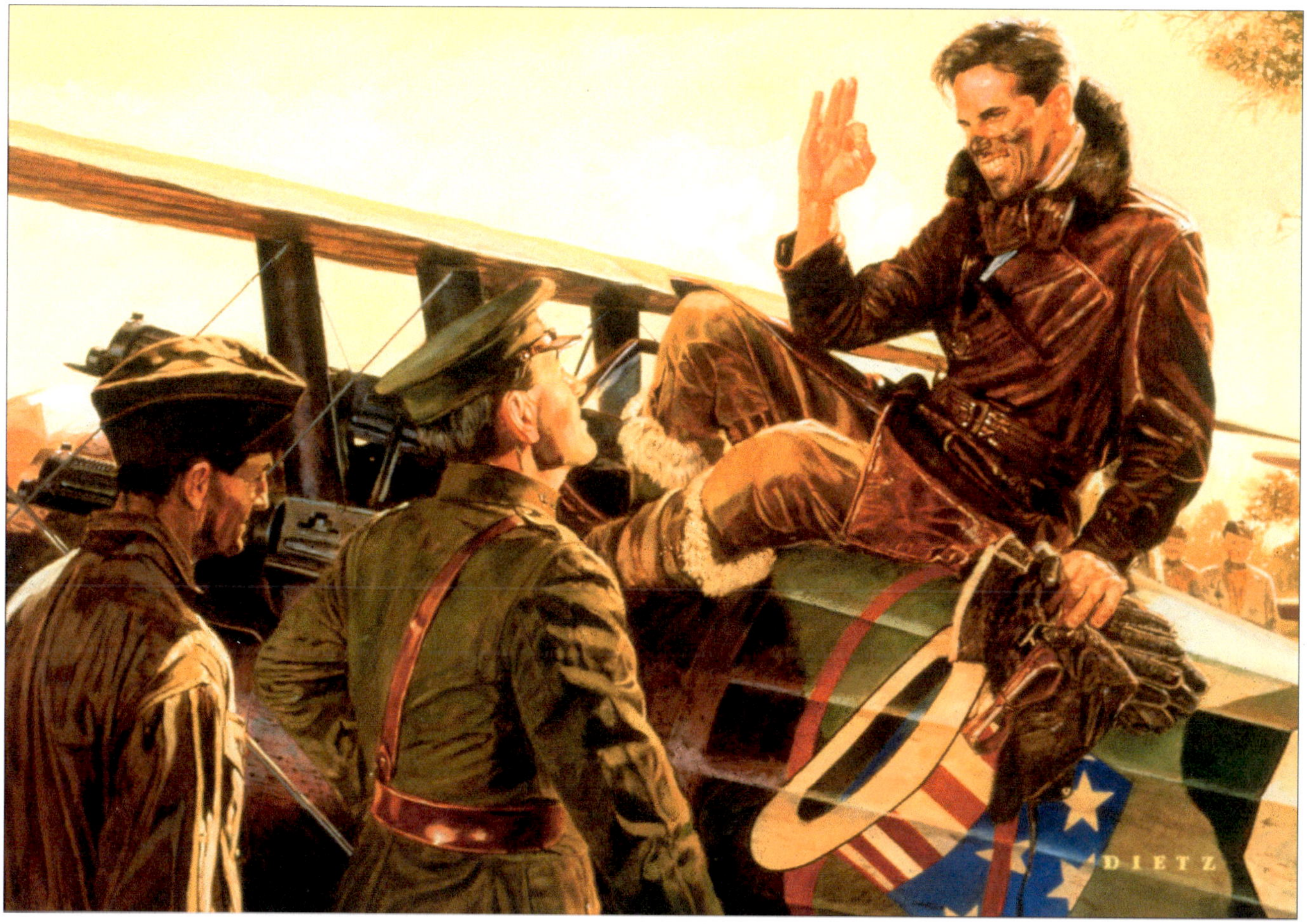

Above: Yale professor Hiram Bingham III was put in charge of recruitment for the USAS and he was in favor of finding young men who were of a certain character. Jim Dietz

aircraft within view of his fellow American pilots watching the dogfight from the ground. Analysis indicated that the reserve fuel tank was set on fire and was not self-sealing. The tank was also located on the bottom of the starboard side of the fuselage near the feet of the pilot. From that point onwards, most of the pilots did not fill the reserve tank on patrol, which limited their range.

Of more significance was the tendency for the upper wing fabric to shed during a steep dive or quick maneuver. The first pilot to experience the problem was Lt. Jimmy Meissner who later wrote about the experience:

"The [enemy] was going into a spiral when two tracer bullets went straight into its fuselage right back of the motor. Flames burst out from the cockpit instantly for I'd struck his tank, but to be sure the pilot was not alive and suffering, I kept up shooting. Diving with full motor in the excitement of the moment, I got so close that a lurch of the blazing enemy made me plunge almost vertically to avoid a crash. The strain was too great; with a crack, my top wing seemed to break loose and whip back overhead at the instant I shot under the Boche, so near I thought we had met. I wondered how soon the bottom wing would fold up and leave me in a wingless fuselage, pointed at Earth 15,000 feet below; what beastly luck to die just as my first victory had been won.

But suddenly, I realized the top wing wasn't all gone and the bottom wing hadn't folded up yet, the plane was still diving with the motor cut off, still answering to the controls as I gently rocked the stick and pushed on the rudder. Surprised at still being alive, I looked up to see what really had happened to the top wing. All the entering edge, back to the maine spar, had torn off, taking the linen completely off the upper surface from wing tip to wing tip so that only about two thirds of the covering on the lower surface was held against the ribs by wind pressure alone." Meissner quoted by Charles Woolley.

Other pilots also soon had similar events happen to them, several pilots not as lucky as Meissner. An experienced fighter pilot James Norman Hall had the same trouble, forced to fly straight and slow, he then

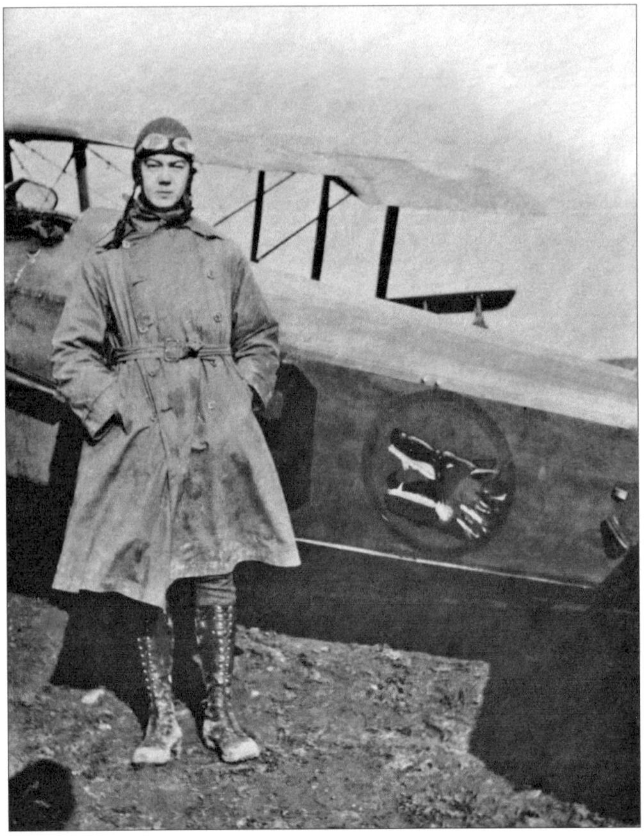

Left: A photo of Woolley next to his Spad fighter. The photo is thought to have been taken after he returned from the mission where he shot down two German Fokkers. After the war, Woolley would lend his flying coat to his friend, Amelia Earhart to use on her first transatlantic flight.

suffered a direct hit by anti-aircraft fire in his engine. Fortunately, the shell did not explode and Hall was able to manage a crash-landing behind German lines. Hall would survive the war and would find fame after co-authoring Mutiny on the Bounty.

Even the future ace of aces Eddie Rickenbacker would be forced to deal with losing his upper wing fabric. The cause was eventually found to be that the linen fabric was glued on to the upper wing's main spar which was not robust enough to handle the G-forces in combat.

American ground crew then devised a system to support the attachment by also nailing the fabric onto the wooden spars. The problem never did slow down the American fighter pilots as they continued in their duties without fail, proving themselves to be gimpers, now referred to as having "the right stuff".

Below: Rickenbacker was a well-known and successful race car driver before the war.

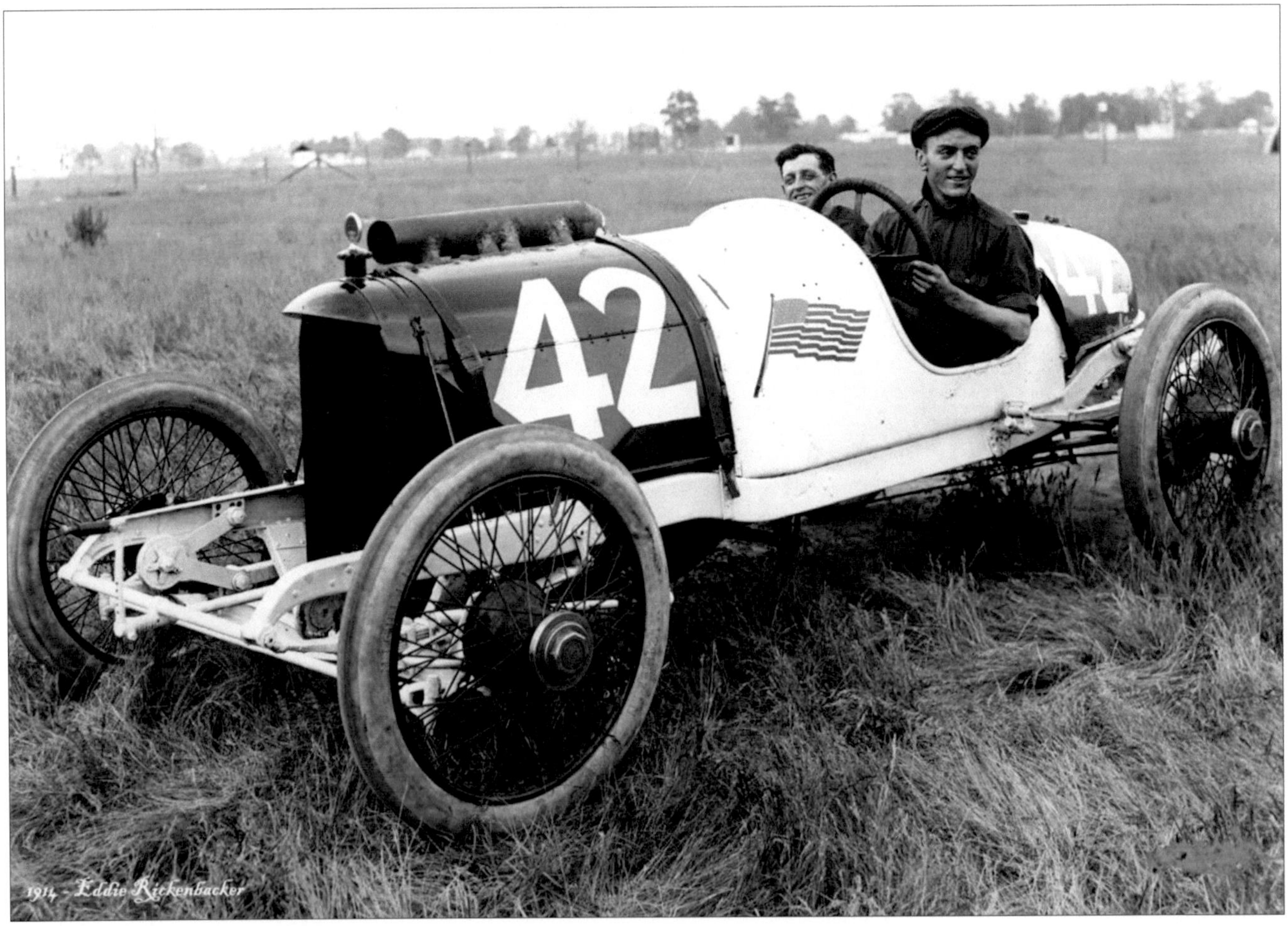

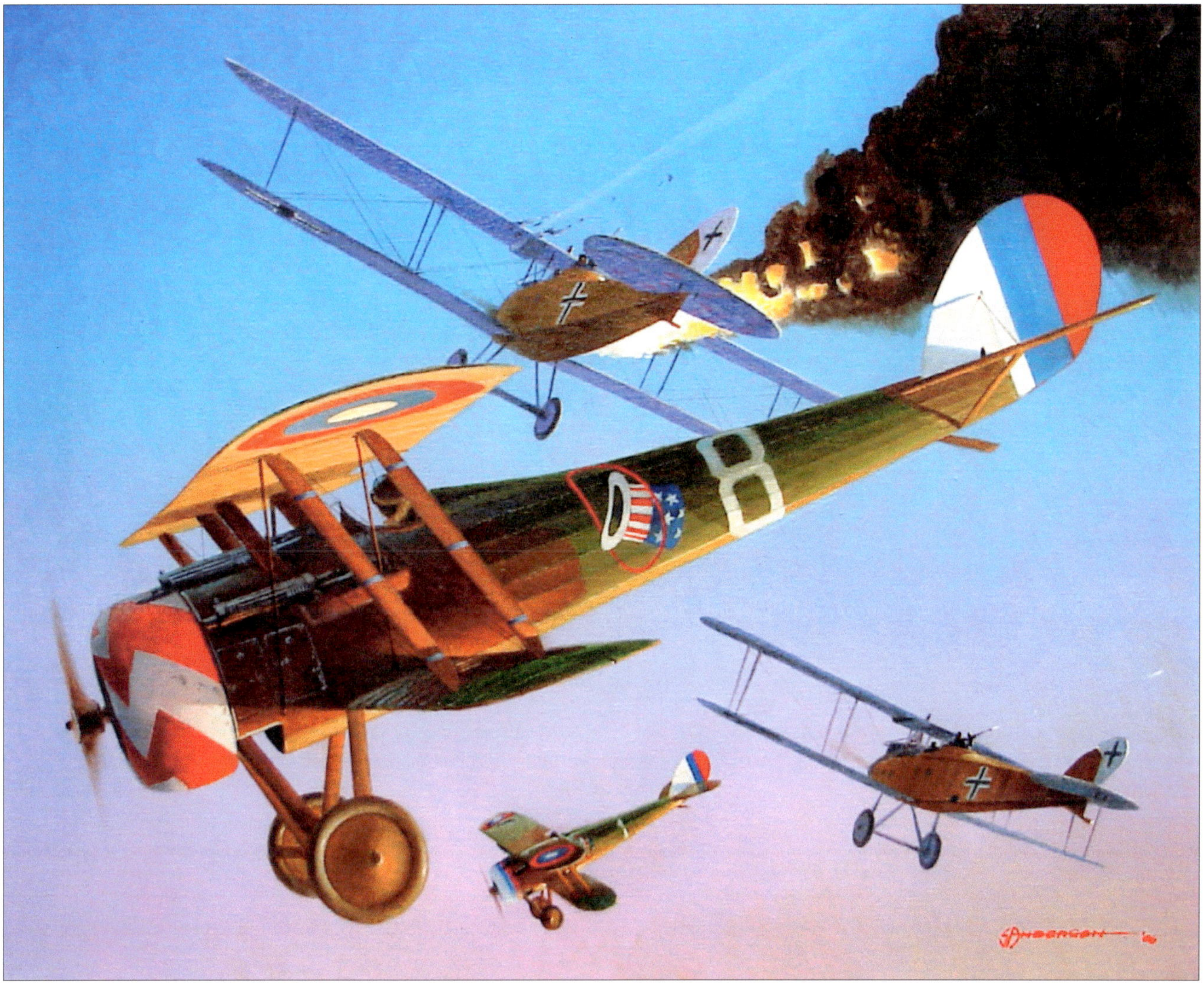

Above: Jimmy Meissner in #8 and Douglas Campbell in #1 share in the destruction of a German observation airplane on 5th June 1918. It would be Meissner's 3rd official victory and Campbell's 6th and last victory. Steve Anderson

Eventually the American front line fighter squadrons received new aircraft in the form of the Spad XIII. Even then, they noted problems with the aircraft as the Spad XIII proved to be a difficult fighter aircraft to keep on the flying line (mostly because of its engine).

During the battles of the late spring and summer of 1918, the Americans also faced another daunting challenge.Since most of the pilots were new to combat, they faced steep odds as they faced very experienced German fighter pilots who initially took heavy tolls of the Americans.

Eventually, enough American pilots survived to gain the experience and skill needed to survive. As these combat veterans then became flight commanders, they could look after the newer replacement pilots until they had several missions under their belts and the American Spad formations could begin to boast of having more gimpers than eggs: which began to inexorably turn the tables against their German foes.

A perfect example of the successful, experienced combat pilot that would help improve the USAS performance was Charles Woolley. Woolley had grown up in East Boston and left MIT early in order to join with fellow American volunteer ambulance drivers serving in SSU 9 in the spring of 1917, before the United States had entered the war.

With the declaration of war, Woolley eagerly and successfully sought the chance to enlist in the newly forming USAS. After completing his training, Woolley was one of the first group of American aviators to reach the from lines as a pursuit pilot in the 95th squadron. In fact, Woolley was in the first combat sortie over the western front flown by a USAS pursuit unit: he did so in a Nieuport 28 that

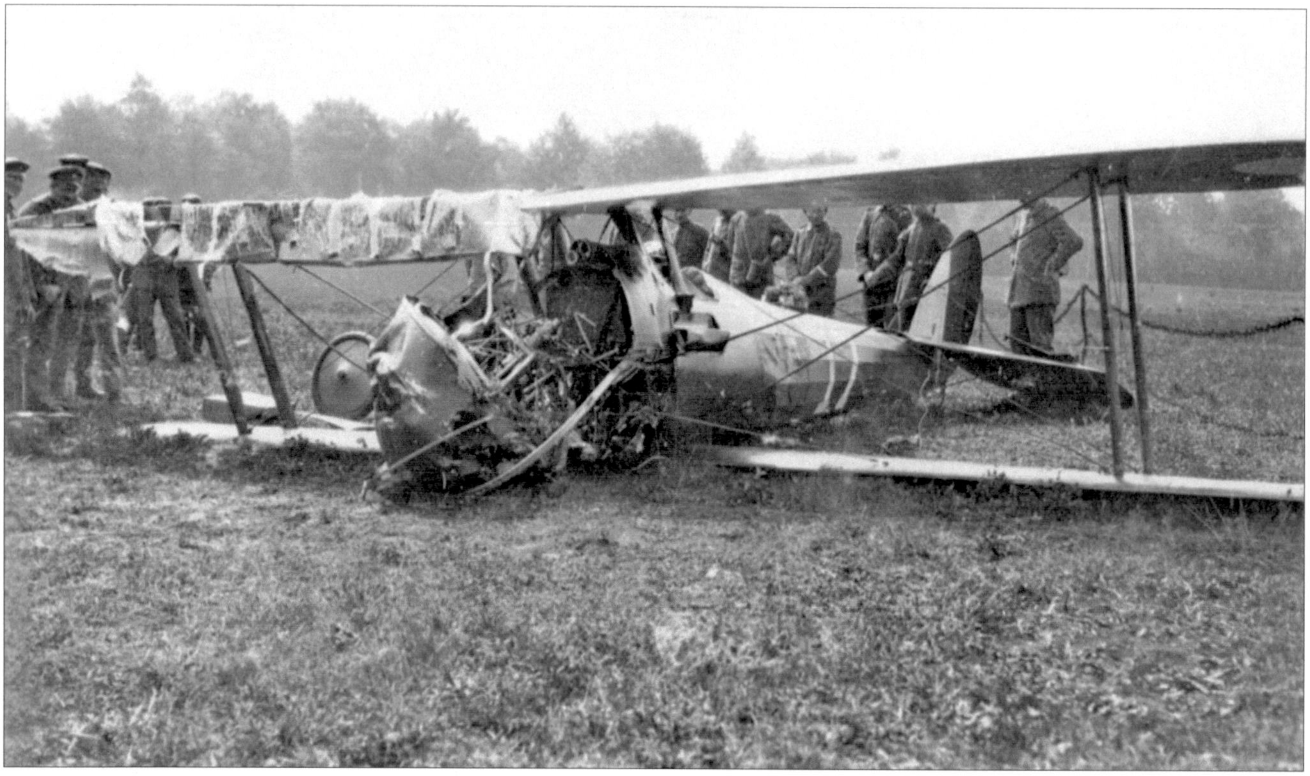

Above: James Norman Hall's downed Nieuport fighter; note the torn upper wing fabric and the severely damaged engine which suffered a direct hit by an anti-aircraft gun shell which fortunately did not explode. Note also the single Vickers machine gun.

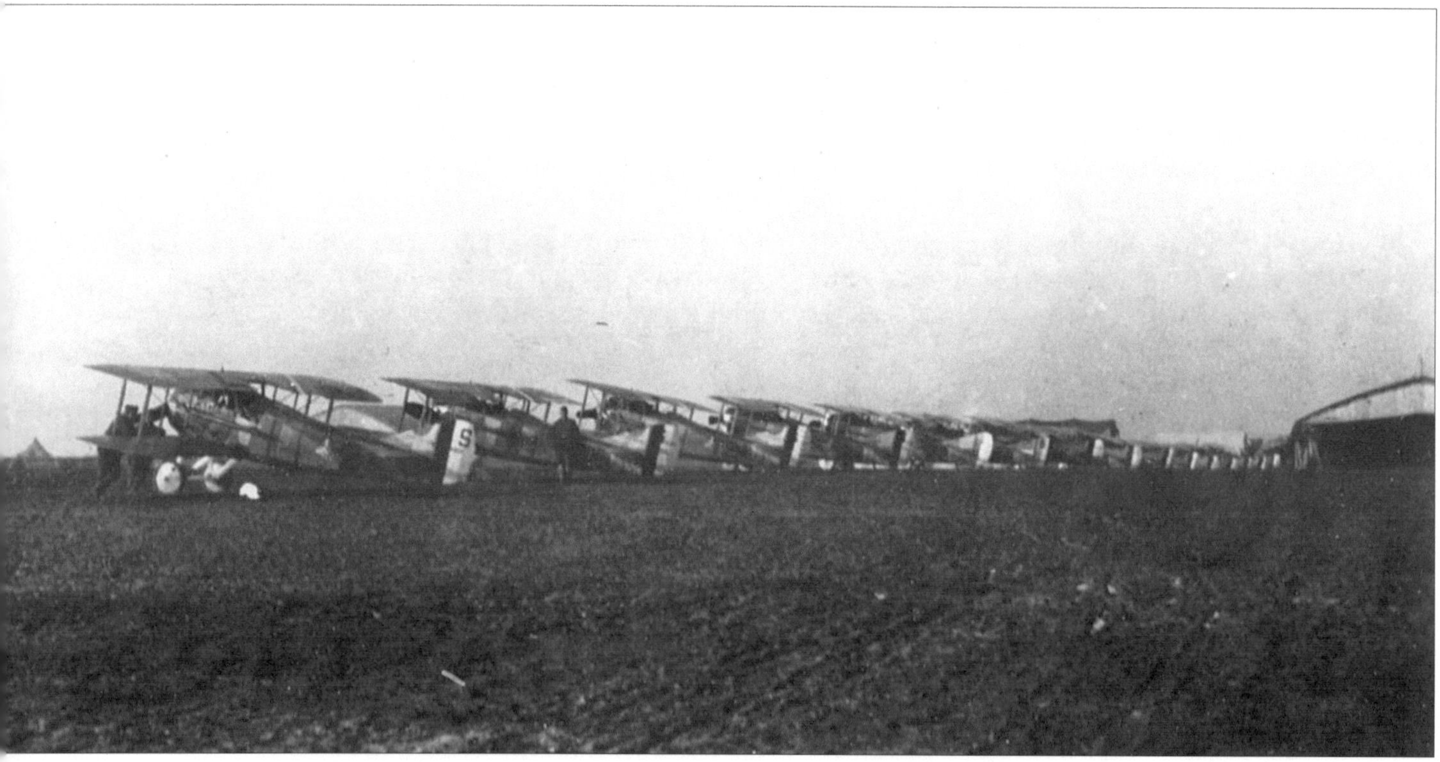

Above: Line up of USAS 94th squadron Spad fighters at Rembercourt airfield in 1918.

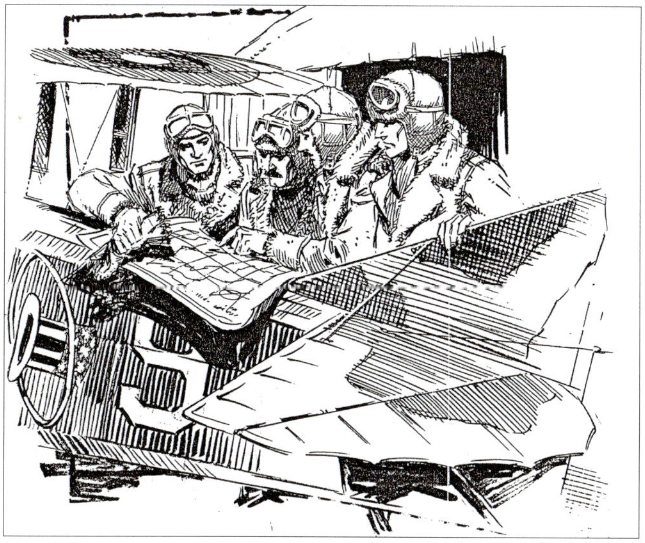

Above: The details of the flight plan are discussed just prior to take off on an offensive patrol by the 94th. George Evans

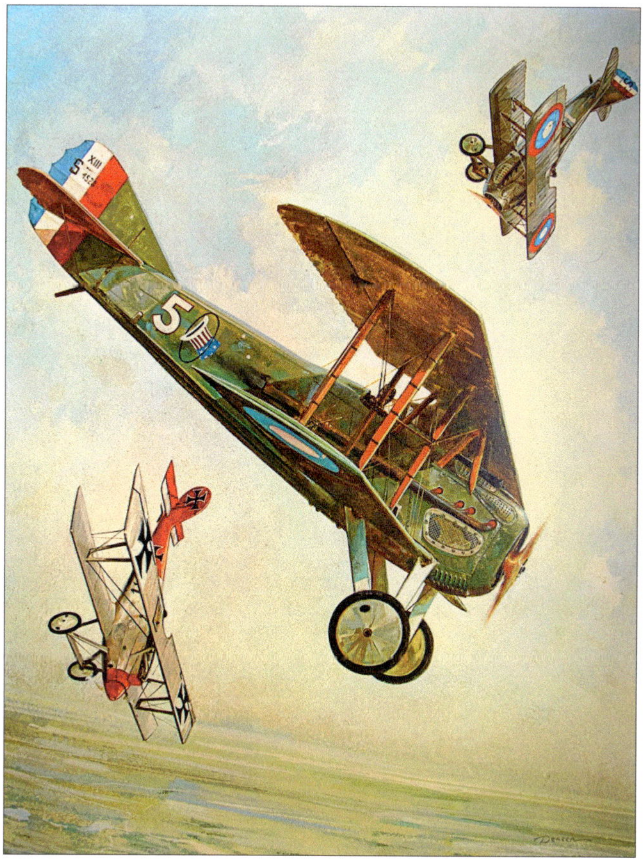

Right: A Gimper always stayed in the fight and protected the tail of his flight members. Here Rickenbacker chases a Pfalz away from another Spad of the 94th. J.B. Deneen

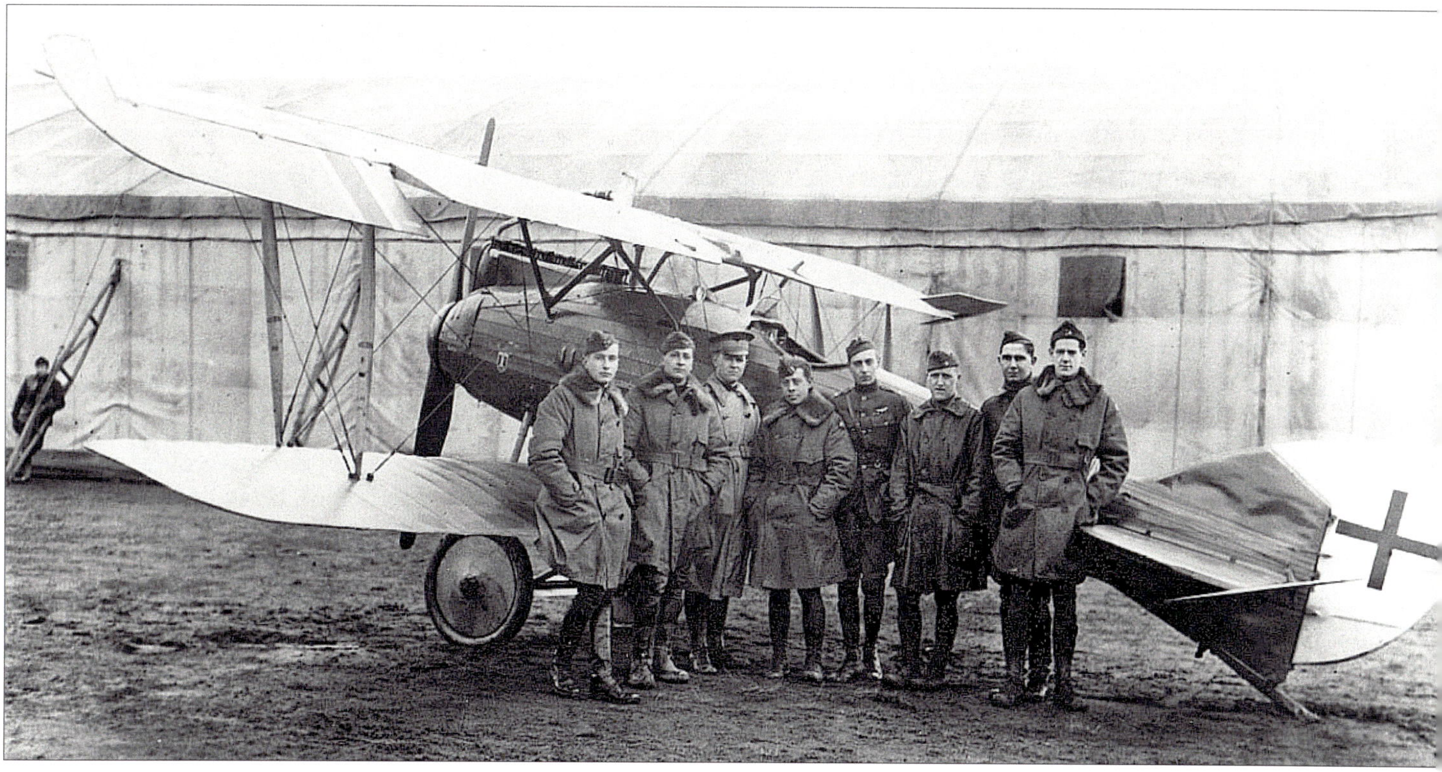

Above: A group of USAS personnel pose beside a captured German Roland D VIb fighter after the Armistice.

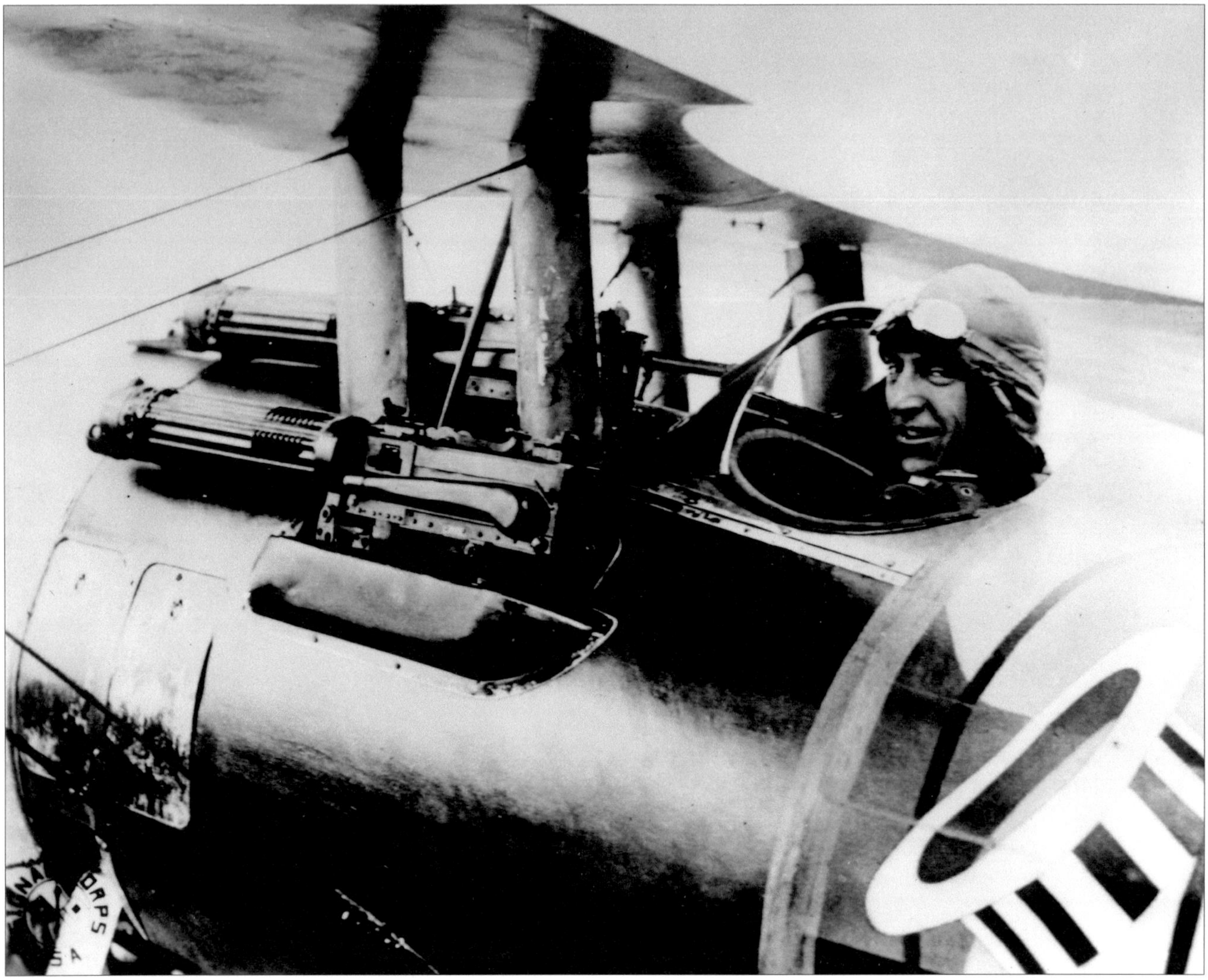

Above: Rickenbacker in the cockpit of his Nieuport 28.

was unarmed!

He would score an uncredited aerial victory and fly numerous missions as a member of the 95th: gaining that most valuable of traits a successful military pilot would need- experience. However, he was transferred from the front to recover from a medical condition for a short time.

During that time, the USAS began to increase the number of squadrons as more aircraft and more newly trained pilots were available. New pilots would not last long at the front if they did not have seasoned veterans to lead them during their very vulnerable initial combat flights. Therefore, instead of returning to his old unit, the 95th, Woolley was sent as a flight commander to a new fighter squadron, the 49th. In a perfect example of the way that a painting can be a form of visual history and bring to life an event described in words, Mike O'Neal the artist himself wrote the following to accompany his painting:

"23 October 1918:Four thousand feet above the French town of Doulcon, nearby the Meuse River, four American Spad fighter aircraft of the 49th Aero Squadron cruise through the forenoon clouds. The flight is lead by Lt. Charles Woolley. A veteran pilot who has yet to score his first confirmed victory.

At 11:45, Woolley spies a flight of German Fokkers below, unaware that the Americans are above awaiting their opportunity to attack. When the German flight crosses below, Woolley leads his flight down. A melee ensues. Chasing one of the Fokker fighters, Woolley watches his tracers pierce the engine and cockpit of his opponent. His guns jam. He veers away to clear them. The Fokker skids, then falls, trailing smoke. With no opportunity to see the results of the Fokker's tumble to earth, he turns to clear his guns before resuming the fight.

Now joined by another member of his flight the two dive wing to wing on a lone Fokker and fire in tandem. Woolley's wingman breaks off, but Woolley

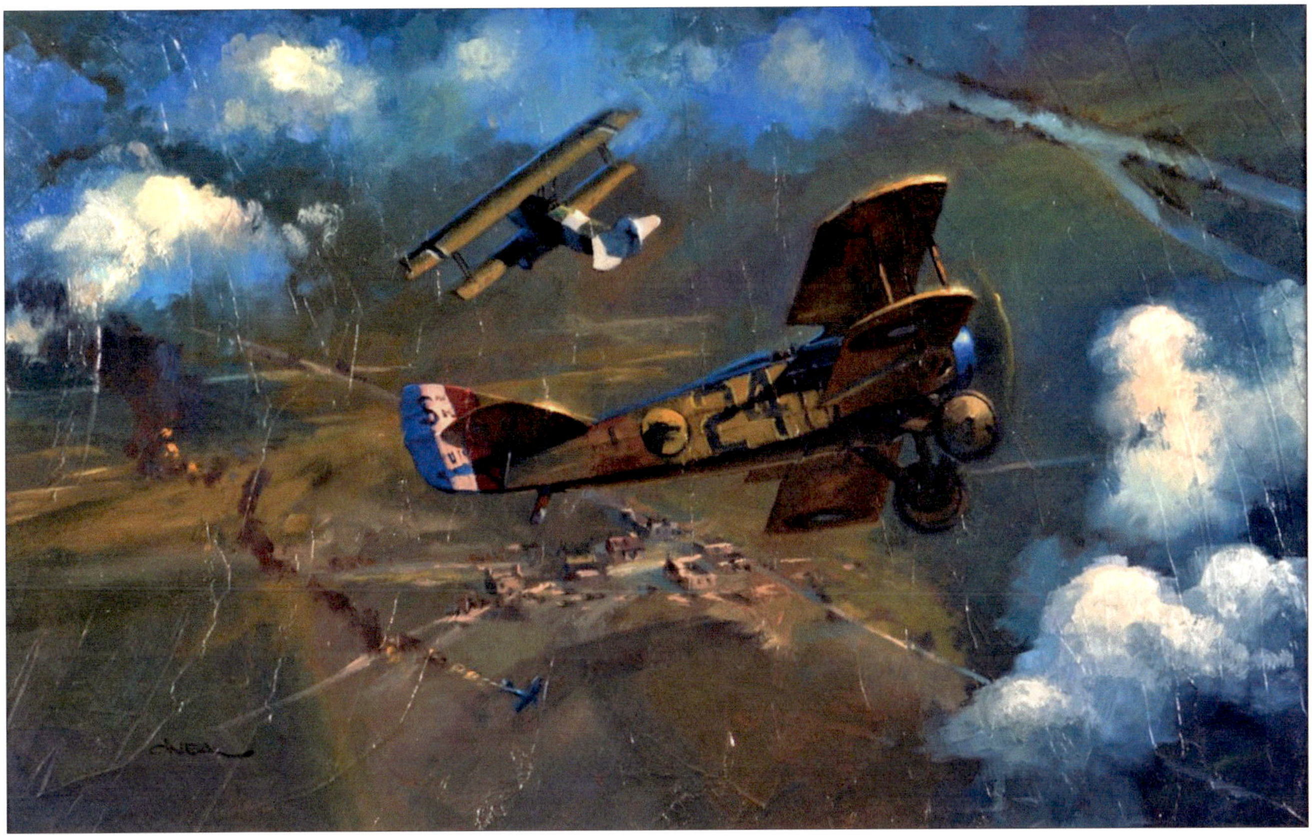

Above: Charles Woolley, flight commander in the 49th pursuit squadron USAS scores two victories in a single dogfight. Micheal O'Neal

presses his attack. The second Fokker droops into a spinning nose dive headed for the broken rooftops of Doulcon. The flight regroups, spies another eight Fokkers above them, hurrying to the fray and quickly retreat towards their own lines. Before the flight has returned to the aerodrome at Beltraine, forward observers would confirm the two downed Germans. They would be Woolley's only confirmed victories of the war.

"Echoes" [the painting's title] recreates the moment when Lt. Woolley has cleared his jammed guns to return to the fight. His first opponent falls away trailing smoke as Woolley rolls his Spad onto the tail of his next target. Doulcon lies below, one of the many towns demolished by the constant fighting during the Meuse-Argonne offensive. The Meuse River winds its way in the blue distance."

The American fighter pilots were beginning to turn the tables on their German foes and the most successful of the American fighter pilots was Edward "Rick" Rickenbacker. He was born on the 8th of October 1890 to Swiss immigrant parents in Columbus, Ohio. He was forced to leave school at the age of 12 when his father was murdered. Rickenbacker worked several jobs to bring money into his family to support his siblings and his mother. He eventually found his way into learning one of the newest trades in the world, that of an auto mechanic. From there in 1912 he began his extremely successful career as a racing car driver. In 1914, he would set the world speed record of 134 mph.

When the United States went to war, Rickenbacker was 27 y/o and was told that he was too old for the Air Service. Likely, as a 12 y/o school drop out, he also did not fit the "mold" as envisioned by Bingham as well. He joined the Army and once in France managed to be posted as the chauffeur to Col. William Mitchell who was in command of the USAS. Rickenbacker lied to Mitchell, telling him that he was in fact only 25 years old and pestered Mitchell until he secured a transfer to become a pilot. He finished his flight training in an astonishing 17 days!

Rickenbacker's wartime career would be well-documented in his book Fighting the Flying Circus. In fact, the 94th and 95th would face each of the German Jagdgeschwaders (I, II and III) at one time or another. The two pursuit squadrons maintained a rivalry throughout the war and were generally based in very close proximity and often on the same airfield. The 95th claimed the first to the front

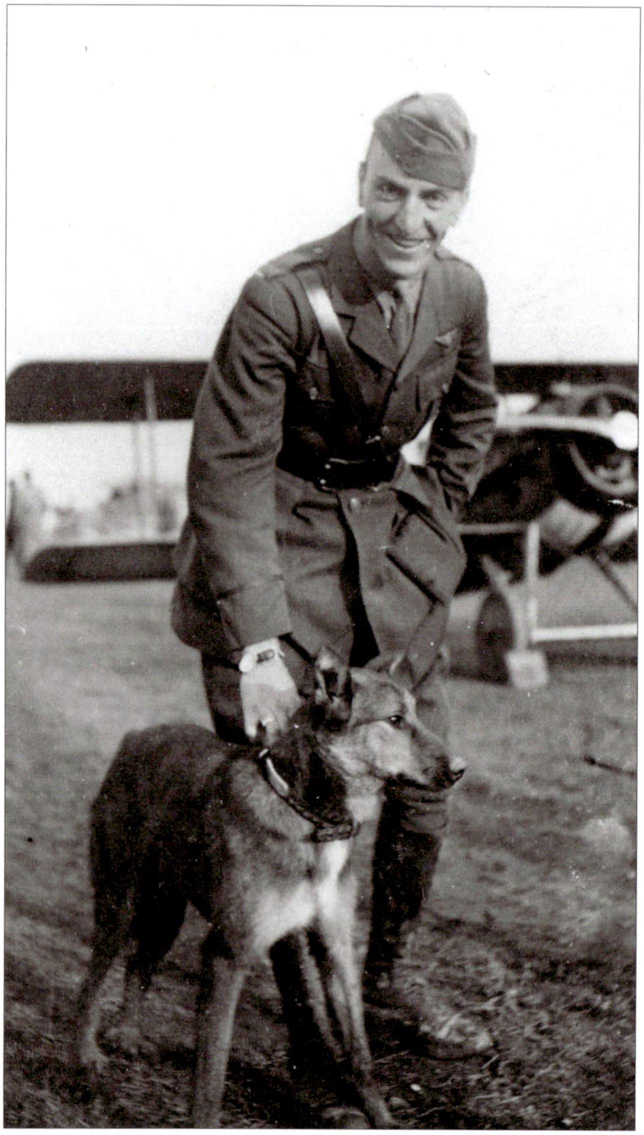

Above: A relaxed photo of Eddie with one of the squadron mascots.

honors and also the first patrol over the enemy in March, 1918. However, the 94th claimed the first German aircraft to be shot down and would end the war with a higher tally of victories.

As he was recovering from a severe ear infection in August, 1918 Rickenbacker was interviewed by writers from several American newspapers and it was in this moment that Rickenbacker's concept of gimpers was first published:

"A Gimper is a bird who would stick by you through anything. If you were up in the air and ran into a dozen enemies and were getting the worst of it, perhaps and the fellow with you stuck with you and gave it to them until they fled, you'd know he was a Gimper...Gimpers did not offer any one of a dozen excuses, such as motor trouble or jammed guns, to return home during a dangerous patrol… All Gimpers have to live up to the idea of the squadron emblem. Every man has a picture of a hat in a ring on his machine. That means he is ready to fight at any time, whether he wants to or not." composite of several newspaper articles

The commander of the 1st Pursuit Group, Major Hartney followed the victory tallies of each squadron in his command. He noted that the 94th was becoming the least successful of the squadrons and therefore placed Rickenbacker in command of the squadron when he returned to duty in late August, 1918.

"Rick" got down to business and it was largely because of his command that the 94th became the most successful American fighter unit of the war. However, it is interesting to note that the rivalry between the 95th and the 94th continued unabated until the very end of the war.

By the start of November, all of the American pilots could begin to sense that the end of the war was coming in the near future. And many of the pilots began to naturally re-calculate how aggressive to be so near to the end of the war to end all wars.

Rickenbacker's last victories were on October 30th and he went to Paris on leave during the first week of November. As a squadron the 94th managed only two victories in November. In contrast, the gimpers to be found in the 95th scored 13 victories between November 1st and the 11th!

In the final analysis, the American fighter pilots of the USAS managed to overcome serious complications from their aircraft , their initial lack of experience in order to eventually manage to obtain many triumphs over a very experienced and dangerous foe.

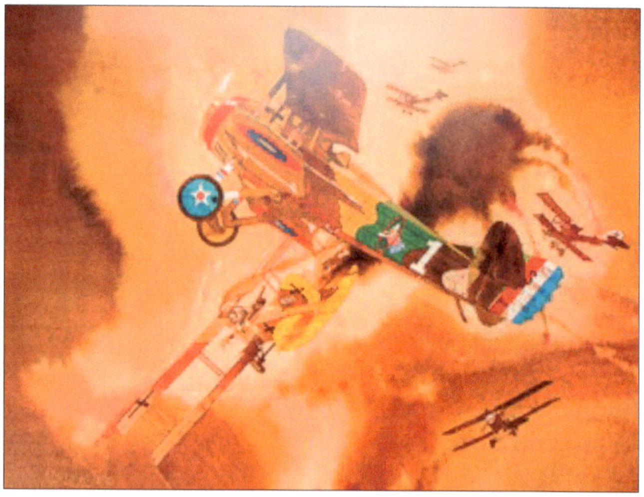

Above: Rickenbacker in his famous Number 1, downs a German two-seater and now looks to escape as the escorting German fighters belatedly come to the fight. Bill Marsalko

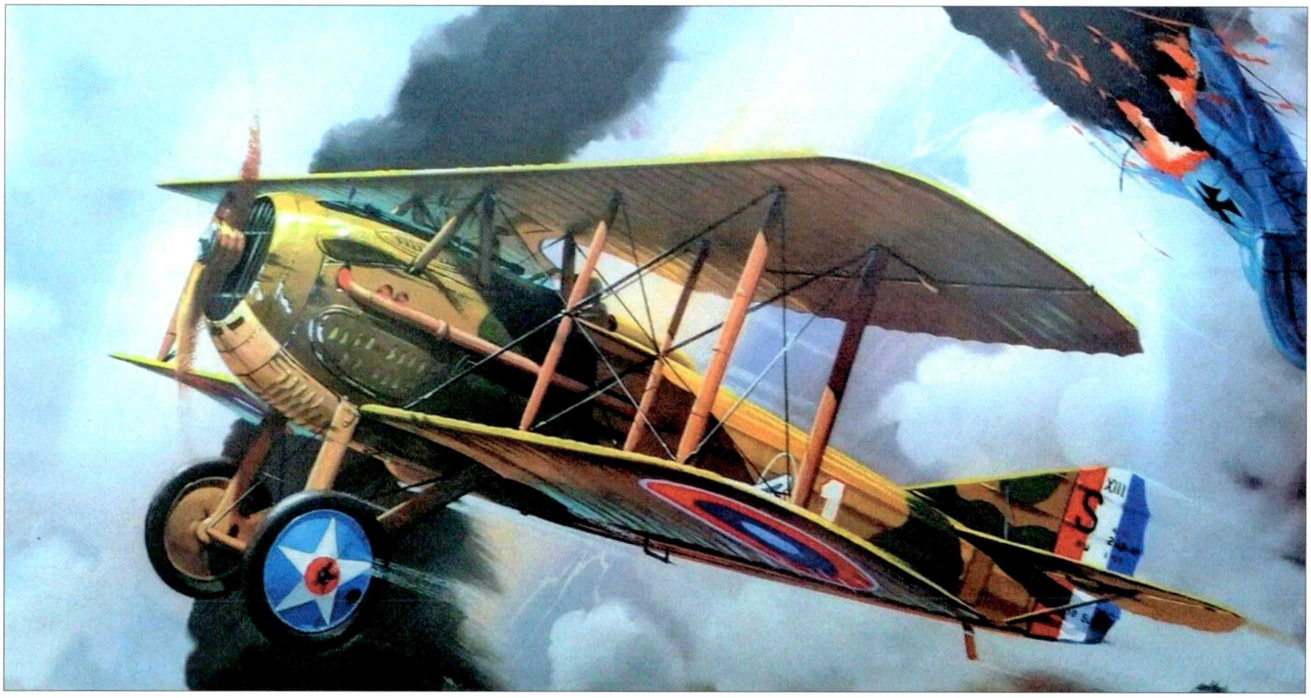

Above: Rickenbacker's famous Spad fighter known as "old number 1". Jack Leynwood

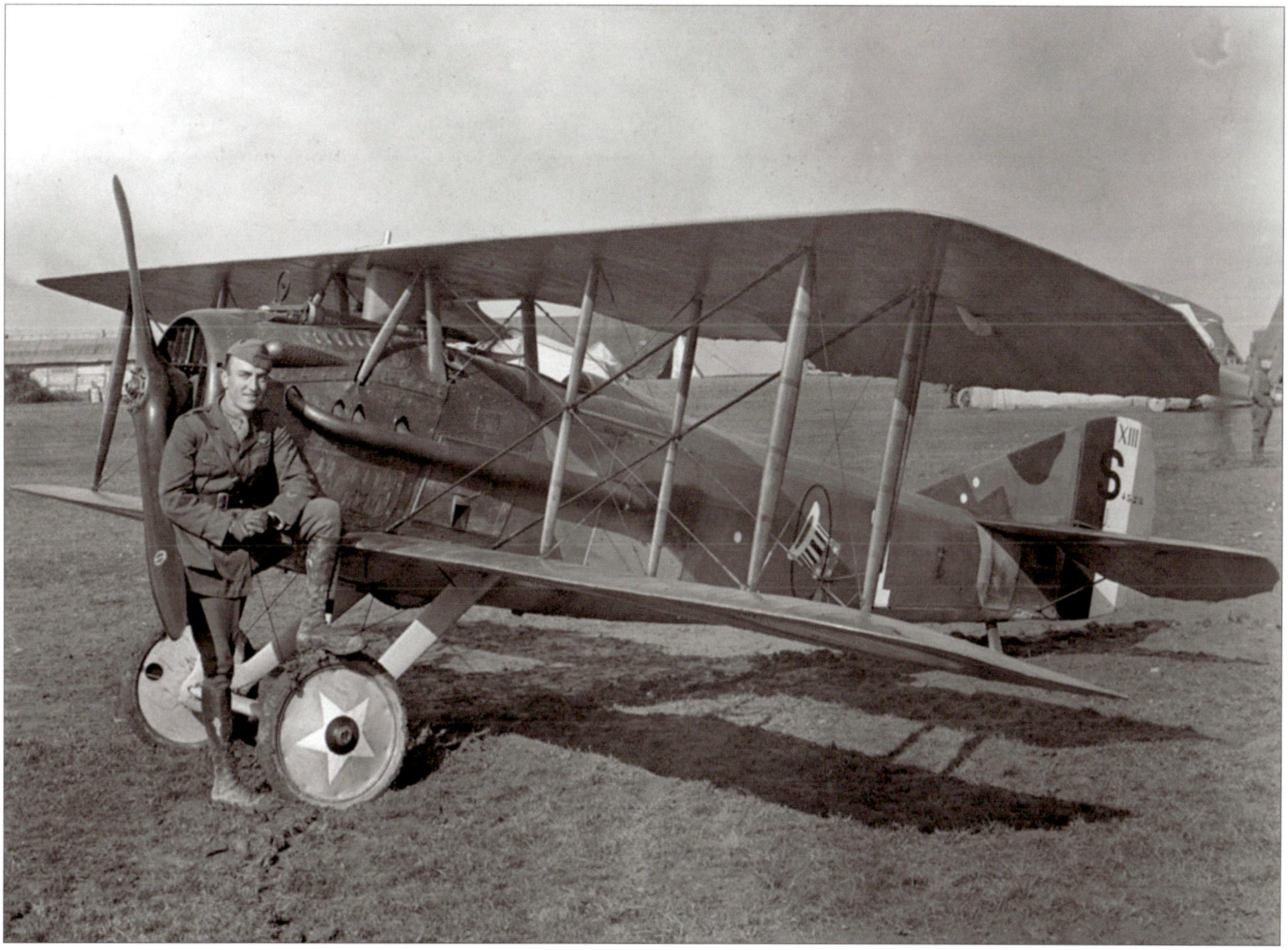

Above: A staged photo of Rickenbacker with "old number 1", his Spad XIII fighter.

Eagle of Trieste

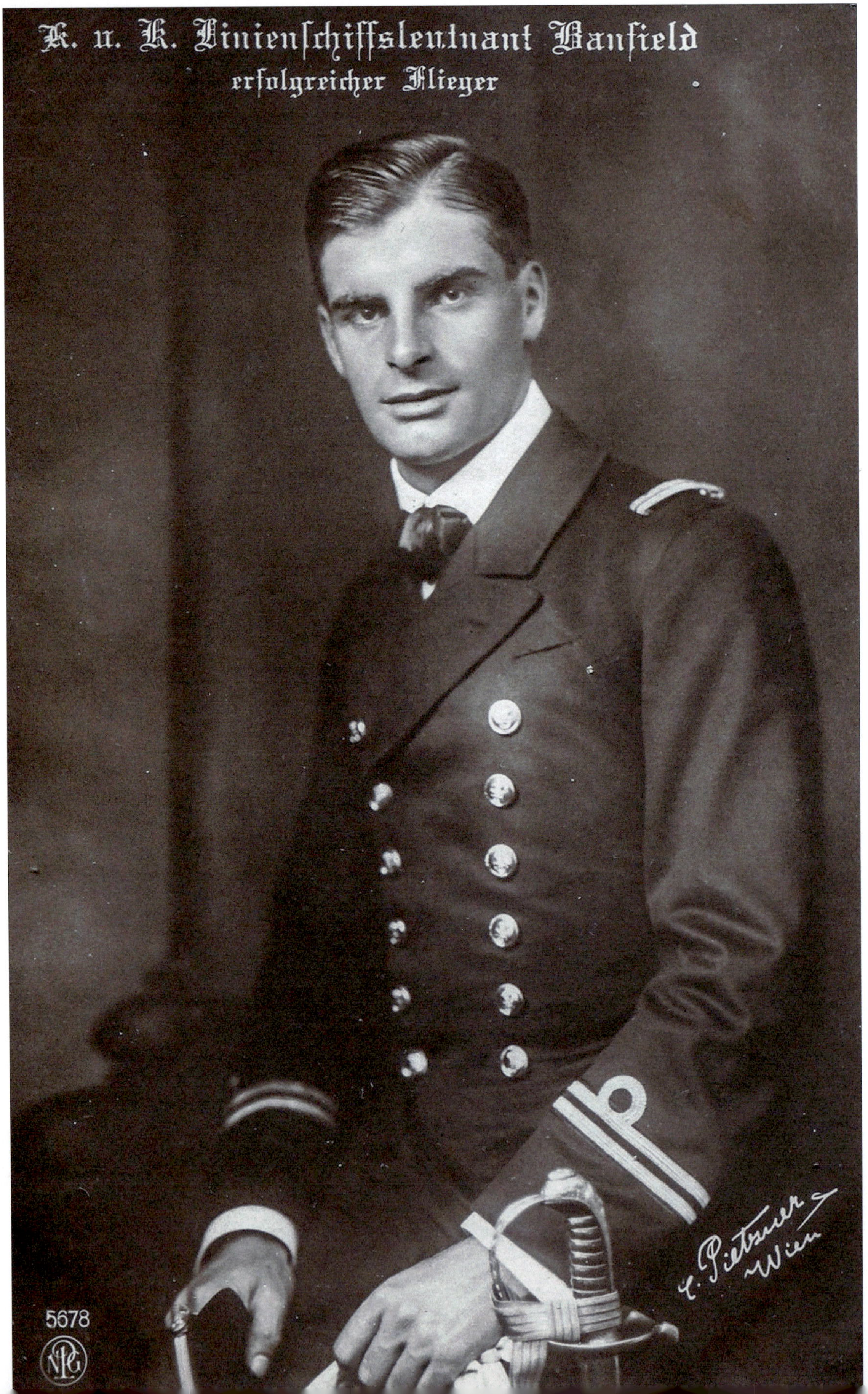

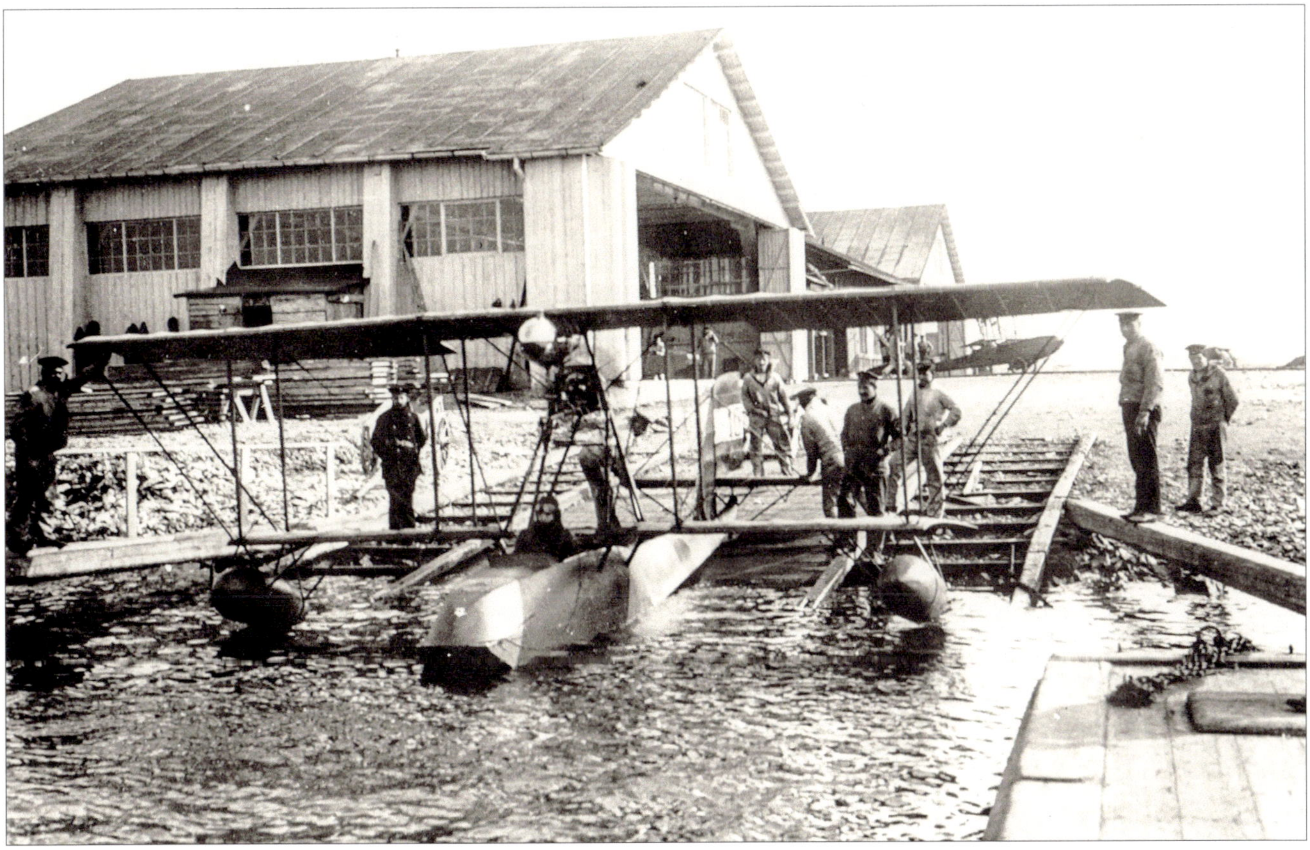

Above: Banfield taking off in a Donnet-Levesque flying boat from the naval seaplane base on Santa Catarina Island before the war.

Left: Gottfried van Banfield, the Eagle of Trieste.

"A charming, gentle man whose generosity towards English,
Austrians, and Italians alike was legendary, he always
remained modest in talking about his adventurous youth. It
was left to others to tell of his countless acts of chivalry,..."

The TImes of London

Geoffrey Thomas Banfield was born in February 1890 in Castlenuovo in the Austro-Hungarian empire. He was the fifth child and third son of Richard Matis Banfield, who was a career officer in the k.u.k. Kriegsmarine (the Austro-Hungarian navy). At birth, Geoffrey was a citizen of the United Kingdom since his father was actually Irish. In his lifetime, Geoffrey / Gottfried/ Goffredo would become a citizen of the Austro-Hungarian empire, re-claim his citizenship as a member of the United Kingdom and then finally become an Italian citizen.

He was raised along the Adriatic Sea and it is there that he spent almost his entire life. In order to attend the naval academy at Fiume he became a citizen of the Austro-Hungarian empire. Sadly, his father died several years before Gottfried graduated in June 1909.

Young Banfield was not destined to follow the typical career of a naval officer. His growing interest in aviation was fueled by his adventurous inclination to join this newest modern aspect of military capability as well as his interest in his brother Karl's success as a pilot in pre-war competitions. His personal interests dovetailed nicely with the aspirations of the commander of the fleet, Admiral Montecuccoli, who wanted the k.u.k. to develop its capabilities in modern naval warfare: principally with aircraft and submarines.

The aviation section of the k.u.k. had only begun in 1910. Gottfried went to Weiner-Neustadt to train on land-based aircraft and in 1912 earned his international pilot's license. He was chosen to stay there and immediately continue his training so that he could earn his military pilot's badge (#67)- which took him only an additional ten days!

In the couple of years since it was started, there was a newly-developed instructional course for naval pilots and Banfield went to the large naval base at

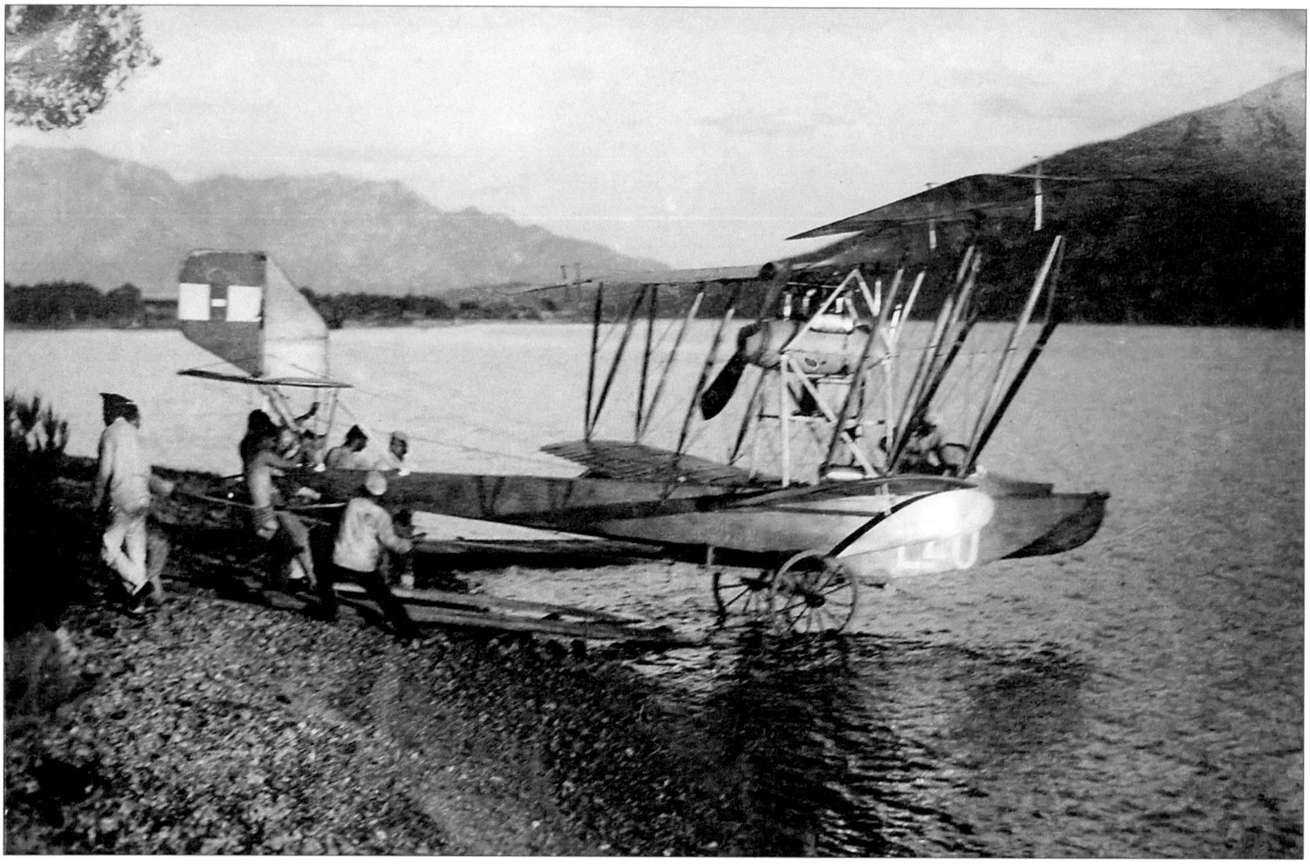

Above: Launching a M-type Lohner seaplane at the beginning of the war; at first the Austro-Hungarian naval aviators had improvised canvas shelters for their aircraft in the midst of an olive grove at Kumbor.

Below: The base quickly grew, here a flight of three M-types are prepared for a mission. Possibly to bomb the Allied base at Antivari, which was a critical supply point for their forces in Montenegro.

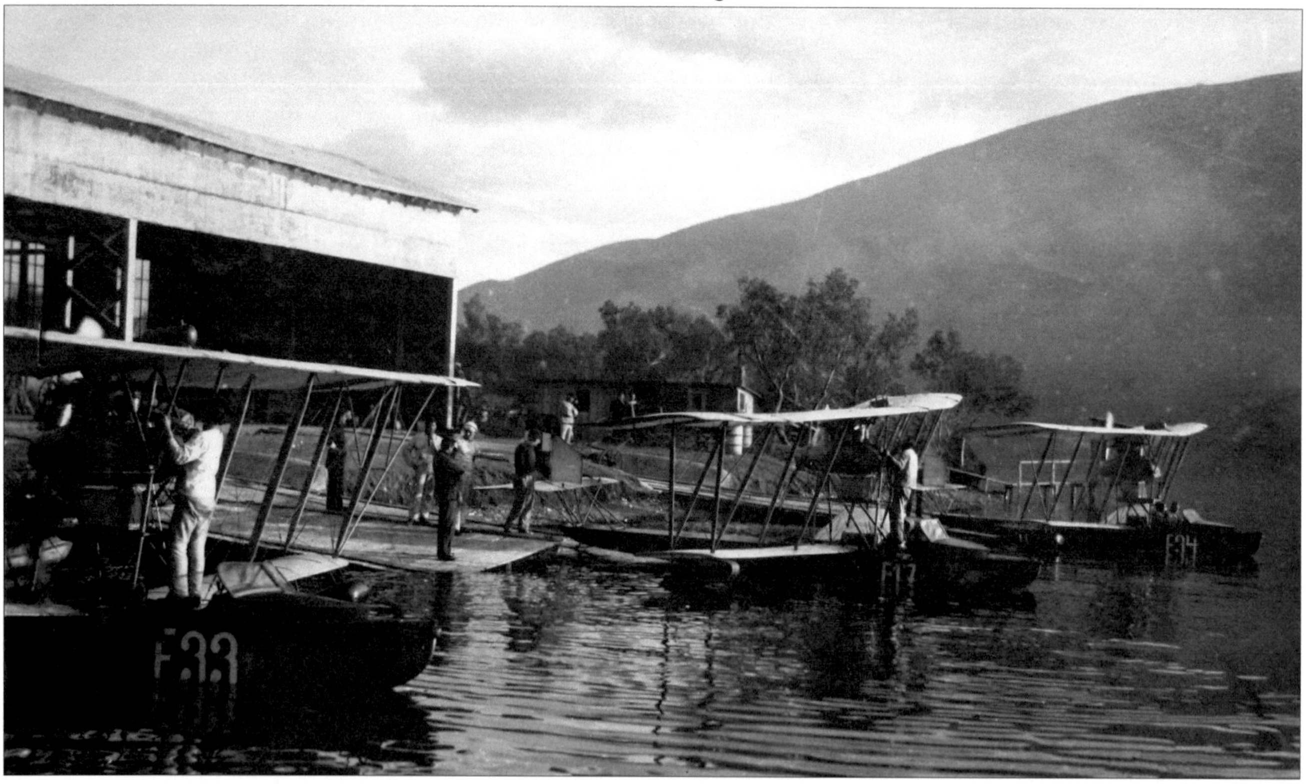

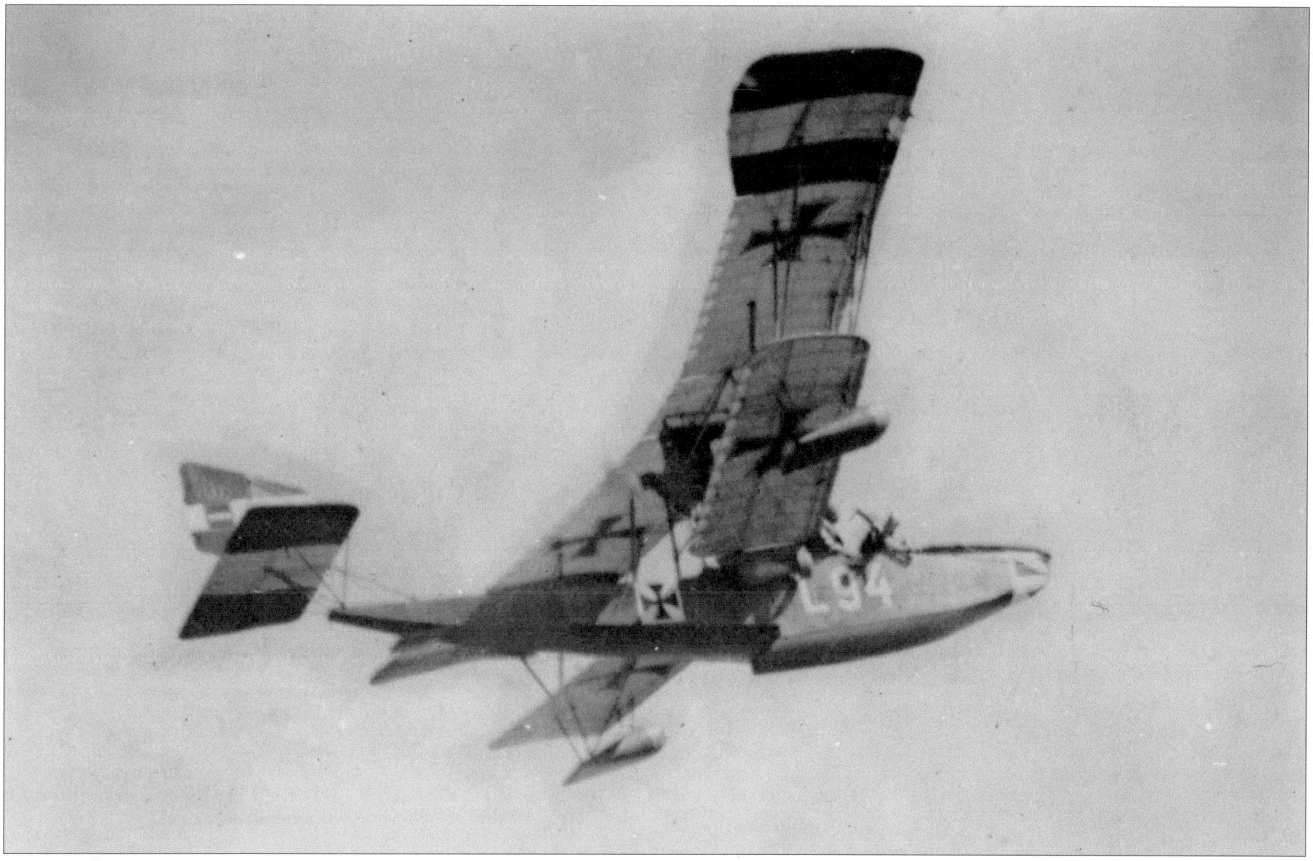

Above: A Lohner T-type, numbered L.94 in flight. Note the observer standing and the heavy infantry machine gun mounted by him. This type would be the backbone of the k.u.k. flying service in the war.

Pola to earn his seeflugzeugfuhrerdiploma (naval aviators badge): Banfield's award was #4.

At this time, despite several efforts, no successful seaplane design had been achieved by any of the Austro-Hungarian aviation firms.

Relatively soon after earning his naval aviation badge, Banfield and another qualified seaplane pilot, Lschlt. Wosecek, were sent to Paris on a mission to test fly and then chose a seaplane type for the k.u.k. There they received additional training from the renowned and world-famous chief pilot of the Donnet-Levesque aircraft company, Jean Louis Conneau.

The Donnet-Levesque was technically a flying boat design and it was selected for the k.u.k.

They flew the aircraft from the Seine River, in the Argenteuil section of Paris. There were significant risks involved in this location as noted by Banfield as he described one flight to the recognized expert of Austro-Hungarian aviation in the First World War, Dr. Martin O'Connor:

"As you know, the testing of the flying boats was done right on the Seine River. I had misgivings about this because of the many bridges that cross the river. Sure enough, on one flight my lift-off was not good. As I skimmed along just above the water, a bridge came into view. I applied full power, but with the closeness of the bridge I realized that continued application of full power would allow me to rise high enough to smash squarely into the bridge! Therefore, I throttled back and dipped down enough so that I flew safely under the bridge and soon landed. I thought little of it until I was shown the next day's paper with huge headlines 'Austrian Dare-Devil Flies under Seine Bridge!"

A new seaplane base was built on the island of Santa Catarina in the bay where the largest k.u.k. base was located at Pola. From here, Banfield and his compatriots flew the newly purchased Donnet-Levesque flying boats and taught more pilots so that by the end of 1913 there were a total of 14 aviators who had qualified for their seaplane rating. Banfield would also fly missions between Pola and Trieste and Fiume in order to familiarize himself with the aircraft and to practice reconnaissance missions.

It was at this time that Banfield suffered a tragic accident. Archduke Franz Ferdinand was aboard his yacht, the Lacroma, steaming towards Pola on an inspection visit. Banfield and Wosecek each took off in a seaplane with the intention of flying above the

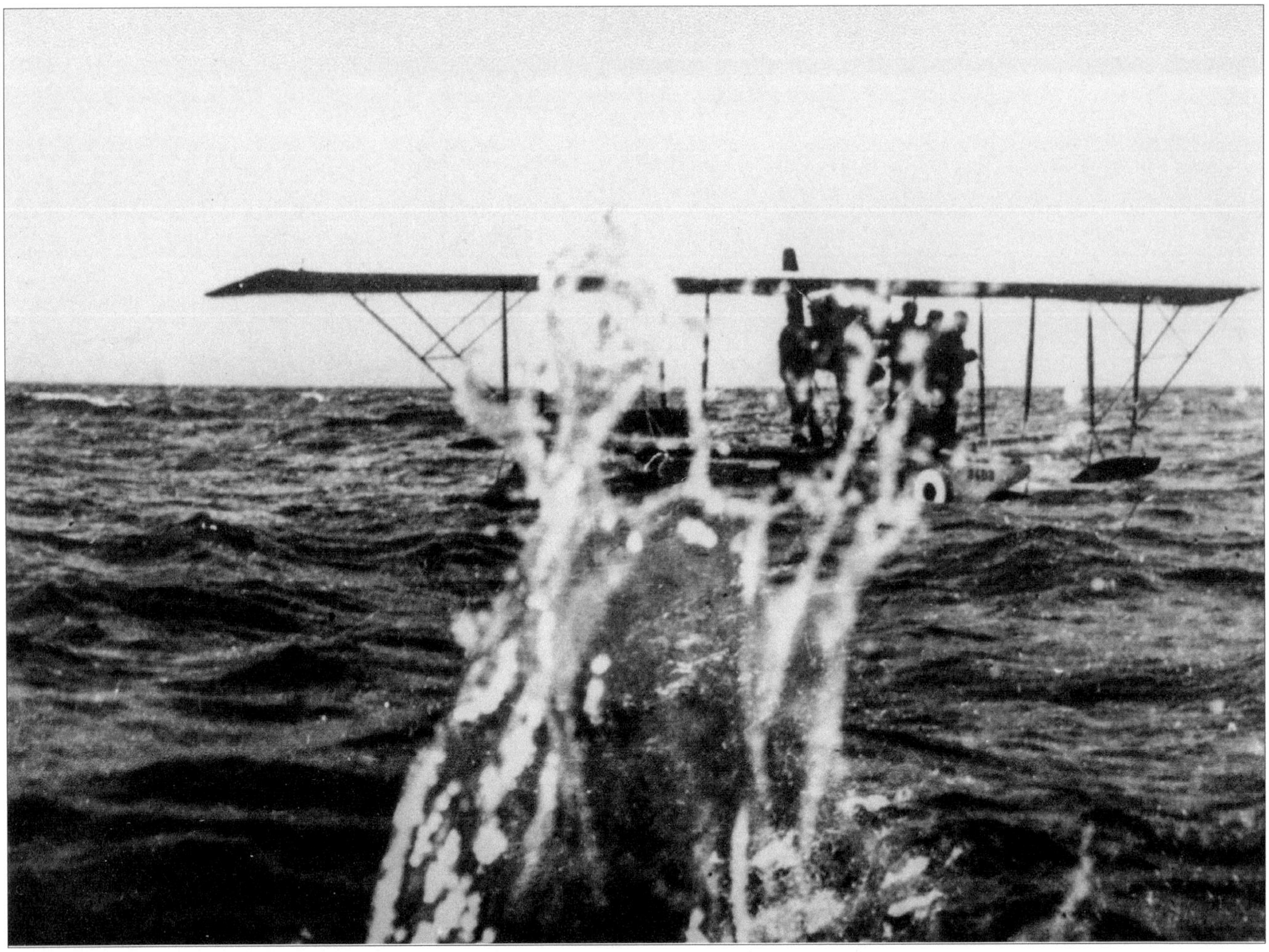

Above: Throughout the naval air war over the Adriatic in WWI both sides of the conflict readily considered each other as marooned mariners if they were forced down onto the sea, Typically, most would try to help rescue the stranded aviators by either landing nearby themselves or summoning a small patrol boat to the downed aircraft. In this extraordinary photo taken by a crew member of the Austrian torpedo boat Tb75, both Italian and Austrian aviators are seen standing on an Italian L.3 seaplane. The Italian crew had gallantly landed to help rescue the Austrian naval aviators after their own seaplane had capsized and sank after being forced down in combat. The Italians damaged their aircraft such that it could not take off again, and so the would-be rescuers became POWs.

incoming Archduke to show what the fledgling naval aviation unit could do.

After their overflight, on the way back to base Banfield's engine failed. He was forced to ditch his aircraft into the open water. However, with no power at low altitude and the wind behind him, Banfield bounced badly on the choppy water and several supporting spars in the fuselage snapped:

"[They] snapped below my feet and smashed into my right leg...I managed to keep myself afloat for about twenty minutes thanks to my life jacket, when I spotted the Tb45, which had accompanied the Lacroma, coming towards me and I was quickly hoisted aboard via a hammock. The Archduke himself later came on board to make sure of my condition. A formal communication would later state that his Imperial Highness consoled the injured pilot."

It is likely that the Archduke intervened on Banfield's behalf because Banfield would be treated by the best surgeons of the k.u.k. Instead of having an amputation which would have ended his career, he had the newest orthopedic procedure which involved placement of a steel plate and screws to fixate the broken bones into proper re-alignment.

Still, Banfield was removed from an active duty post for the duration of his recovery.

Uncertain about his ability to return to military flying, he accepted a position with the Lohner Co. to test fly their newest seaplane design, the type M. Banfield would recover by the fall of 1914, just in time to join two other pilots, Maglic and Woseceki

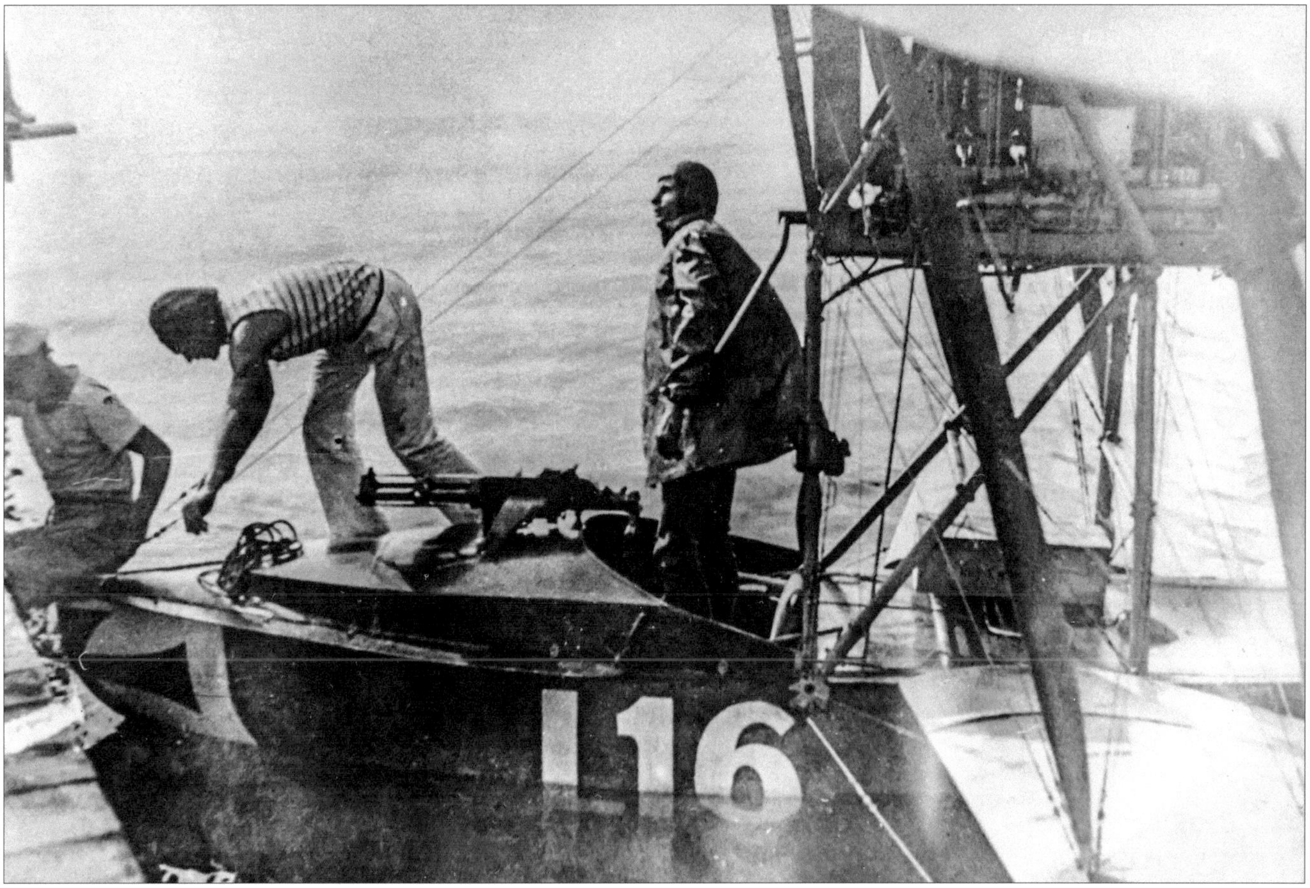

Above: Banfield in one of his two unique aircraft, the Lohner L-16. Note the machine gun that could fire at a very rapid rate, no need for synchronization in this pusher aircraft. Banfield would begin to earn his fame as his successes mounted with this speedy fighter.

(again) as they went to war- flying the Lohner type M seaplane.

Their first posting was to Kumbor, where they had to erect tent hangars amongst olive trees. Their mission was to fly reconnaissance missions over the adjacent Montenegrin border as well as bomb the port of Antivari which was the key supply port for the Allies as they tried to develop a Southern front supporting Montenegrin and Serbian forces.

Through May 1915, Banfield and his comrades flew multiple missions, including missions to bomb the transport ships noted approaching or in Antivari harbor, as well as to try to destroy the single rail line that led out of the city.

Banfield had been recognized as a talented pilot and during this time he also would be brought back to Pola to help test fly the next Lohner seaplane design, the type T(for some reason these aircraft were numbered using an L designation). On 19 April Banfield suffered another crash following damage sustained in a combat mission. During the open water landing the L5 was destroyed, but Banfield escaped with only minor injuries.

On 23rd May 1915 Italy declared war on Austro-Hungary. In anticipation of this event, Banfield had already been re-assigned to the seaplane base at Pola. In fact, in Lohner L-47 Banfield had flown an extensive reconnaissance flight from Pola to Ancona to Venice and back on 20th May. Then, on the 24th May Banfield joined five other Austrian seaplanes in a bombing mission over the shipyard at Ancona. On the 27th May, Lschlt. Wosecek was forced down by an engine malfunction in L-40 while near Venice. He was captured, but more importantly the Italians had captured his Lohner seaplane intact.

The aircraft was given to the Macchi Company located in Varese and was told to reverse engineer the aircraft and begin to produce the seaplane. The Macchi company was able to build its first copy in only 40 days and by the end of 1916 had built over 200 machines!

In addition, their aircraft would actually have better performance and a more reliable power plant.

However, in the summer of 1915 the Italian navy's aviation capabilities were still quite minimal.

Therefore, Banfield's principle opponents of that

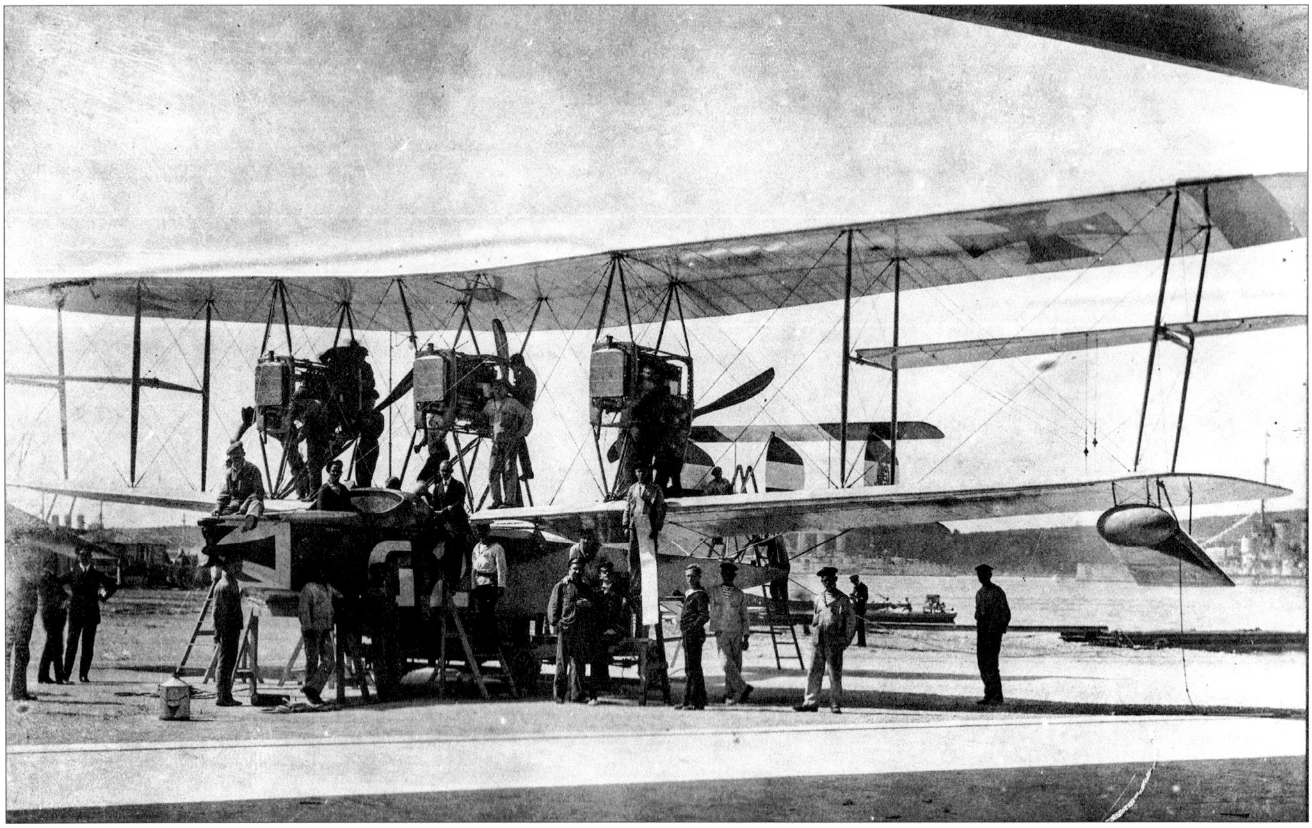

Above: The enormous size of the Mickl designed G seaplane bomber. This type would never enter front line service; it was intended to be a large bomber with sizable payloads capable of striking targets throughout Northern Italy. It was left to the Austro-Hungarian naval air units to provide the strategic bombing of targets in Italy during the war.

Facing Page, Above: Banfield taxies in his new Hansa-Brandenburg fighter. Note the characteristic star-strutter design similar to a land-version design.

Facing Page, Below: Sketch of Banfield using his newest fighter, now designated A.12, in downing an Italian Caproni bomber. George Evans

summer would be the French aviators of Centre Aeromaritime (CAM-Venice). Amongst the pilots there would be Banfield's old tutor, Jean Conneau.

Both sides in the conflict realized that the Northern Adriatic was going to prove to be crucial and so Banfield was transferred to the naval base at Trieste in order to develop and command the seeflugstation (SFS) there:

"Trieste was a strategically-placed base, only 18 miles from the important Italian base at Grado and less than 70 miles due east of Venice. Trieste also was quite close to the rivers along which the front lines of the ground fighting tended to be oriented, lying 10 miles from the mouth of the Isonzo River. The result of this was the involvement of the aircraft and fliers of the Trieste Air Station in fighting of high intensity and almost incredible variety- involving land planes, flying boats, long range bombers, all manner of warships, and all types of various missions flown in support of the infantry. Trieste was an extremely exciting, interesting and dangerous place to be." Dr. Martin O'Connor

Banfield later wrote that he and his fellow aviators at Trieste SFS called their base "the station of death". Banfield would command this base from its beginning in February, 1916 until the end of the war. During the course of the war, the naval aviators at Trieste SFS suffered a 55% mortality rate.

The Italian submarine RN Jalea was due back from its operational patrol by the middle of August 1917. Therefore, the Allies sent out aircraft to find it. In fact, the submarine had already struck a mine in shallow waters in the Gulf of Trieste and sunk with the loss of all but one of its crew.

On the 27th August, while on a reconnaissance flight Banfield noted the submarine and not knowing that it had already been sunk in the shallow waters, he dropped four bombs onto the "submerged" submarine.

An Italian crew in a float plane had also spotted

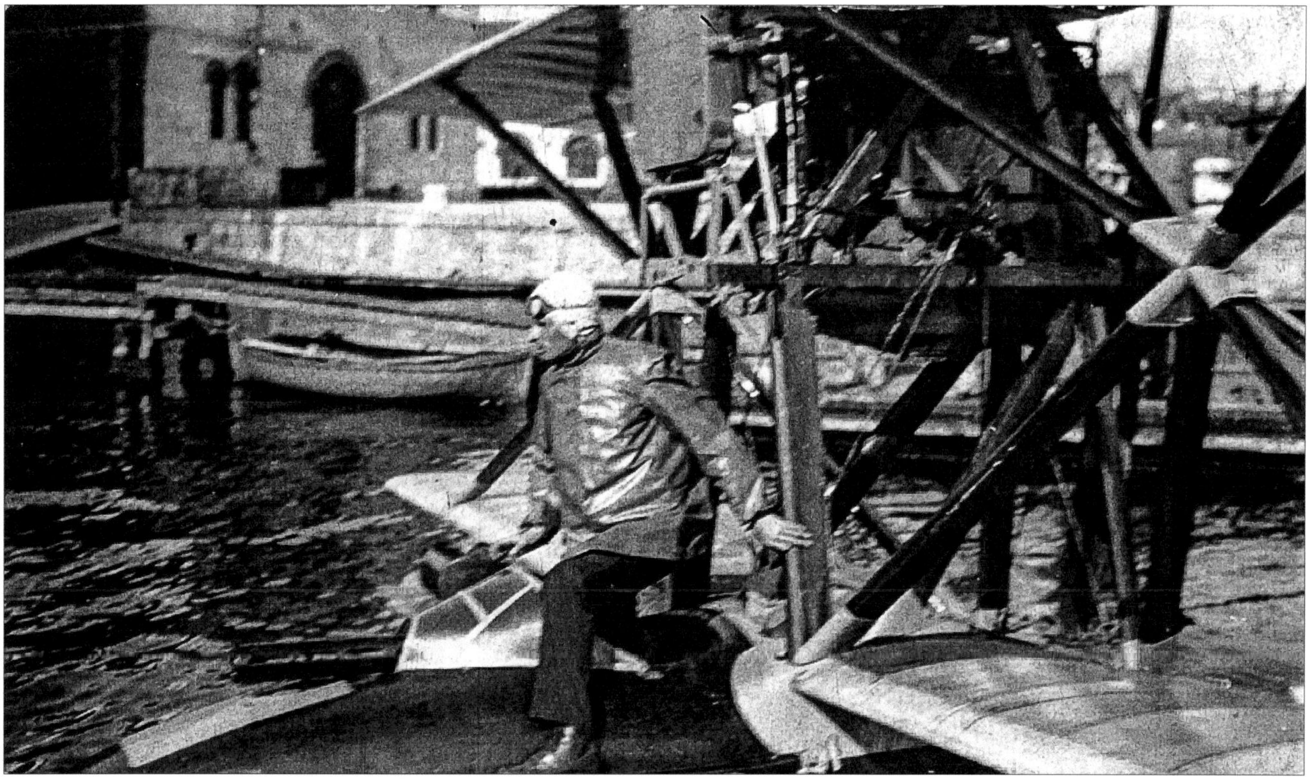

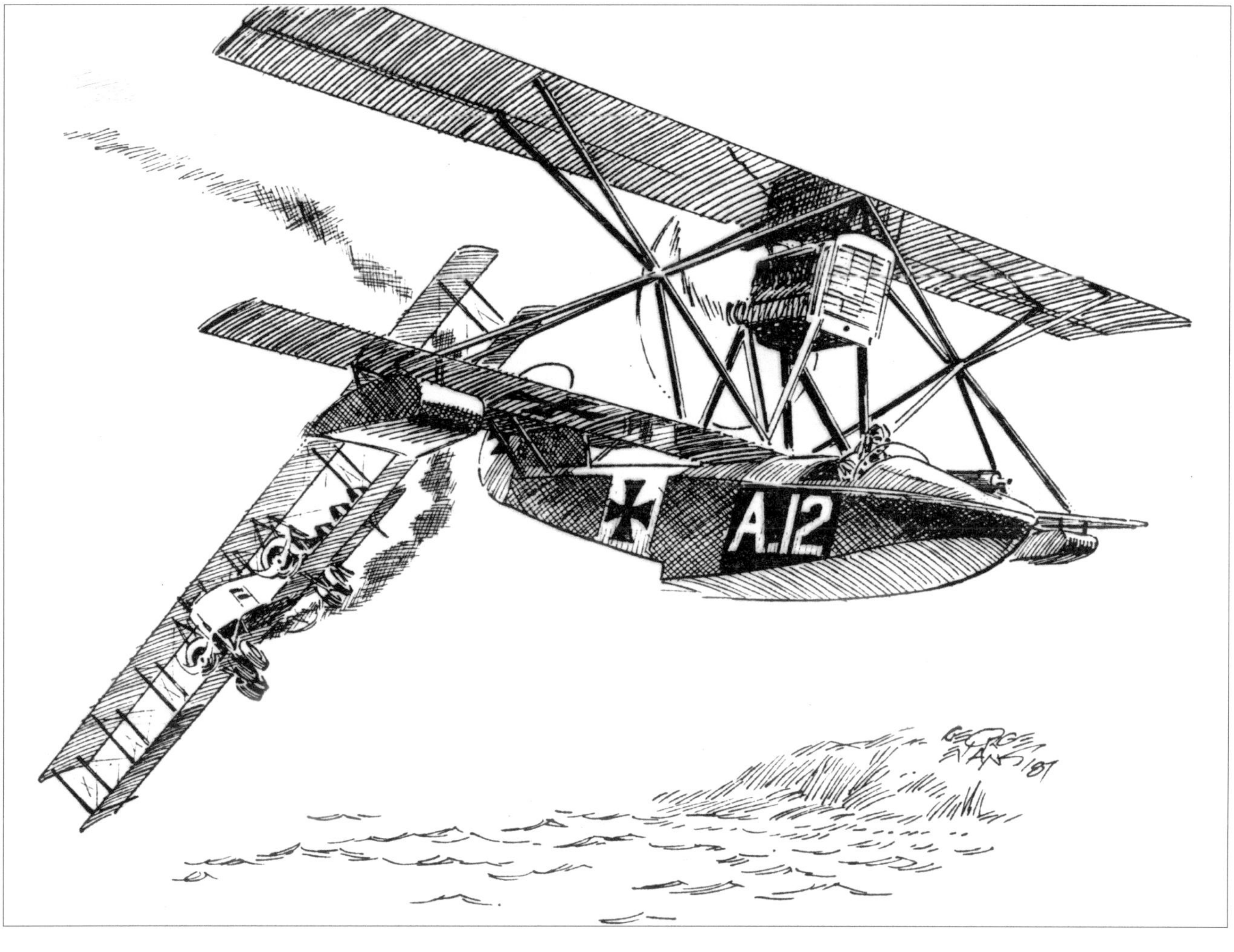
A.12

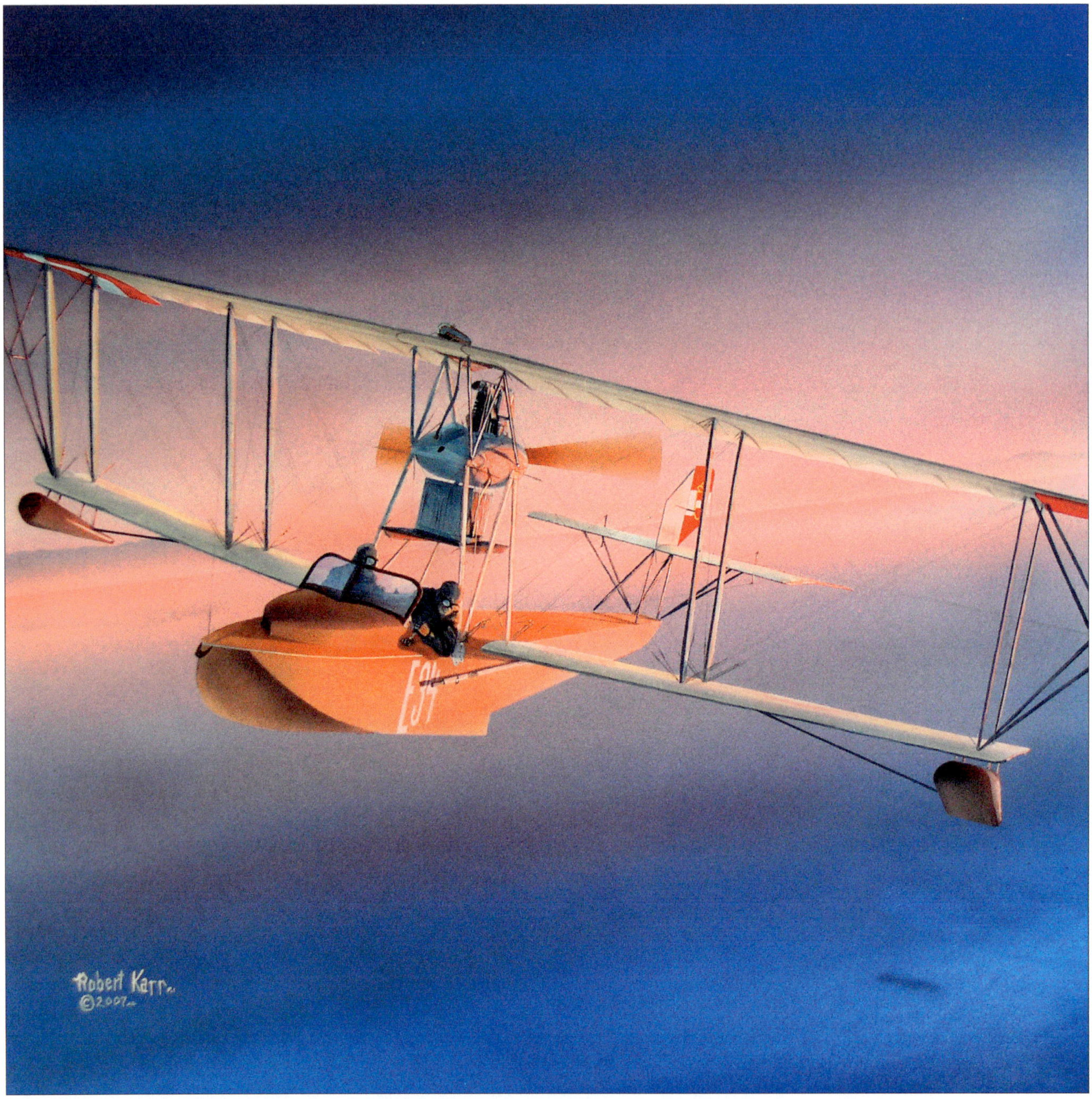

Above: At the outbreak of the war, the naval aviators based at Kumbor were kept very busy on a multitude of mission types – here anti-submarine patrol. Robert Karr

Facing Page, Below: A pleasing photo of a relaxed Banfield and friend.

the submarine and so the Allies planned a sortie on the following day in order to investigate. The French aviators of CAM-Venice flew in their two-seat FBA Type C flying boats:

1) LV Roulier piloted #10 with Lt. Romeyer as his gunner

2) LV Conneau piloted #2 with the Italian commander of the seaplane base at Grado as his gunner, tenente di vascello Giambernardini

As he was about to fly over the sunken submarine, Conneau received a message from a MAS boat below: "a seaplane released a bomb and is attacking the MAS boats." This was the Lohner L-46 piloted by Banfield with Strobl as his gunner.

Banfield dove upon the French FBA piloted by his old instructor Conneau. Meanwhile Conneau dove towards the nearby base of Grado. Roulier was in the process of taking off to engage the Lohner seaplane.

Banfield now saw Conneau drop down to land

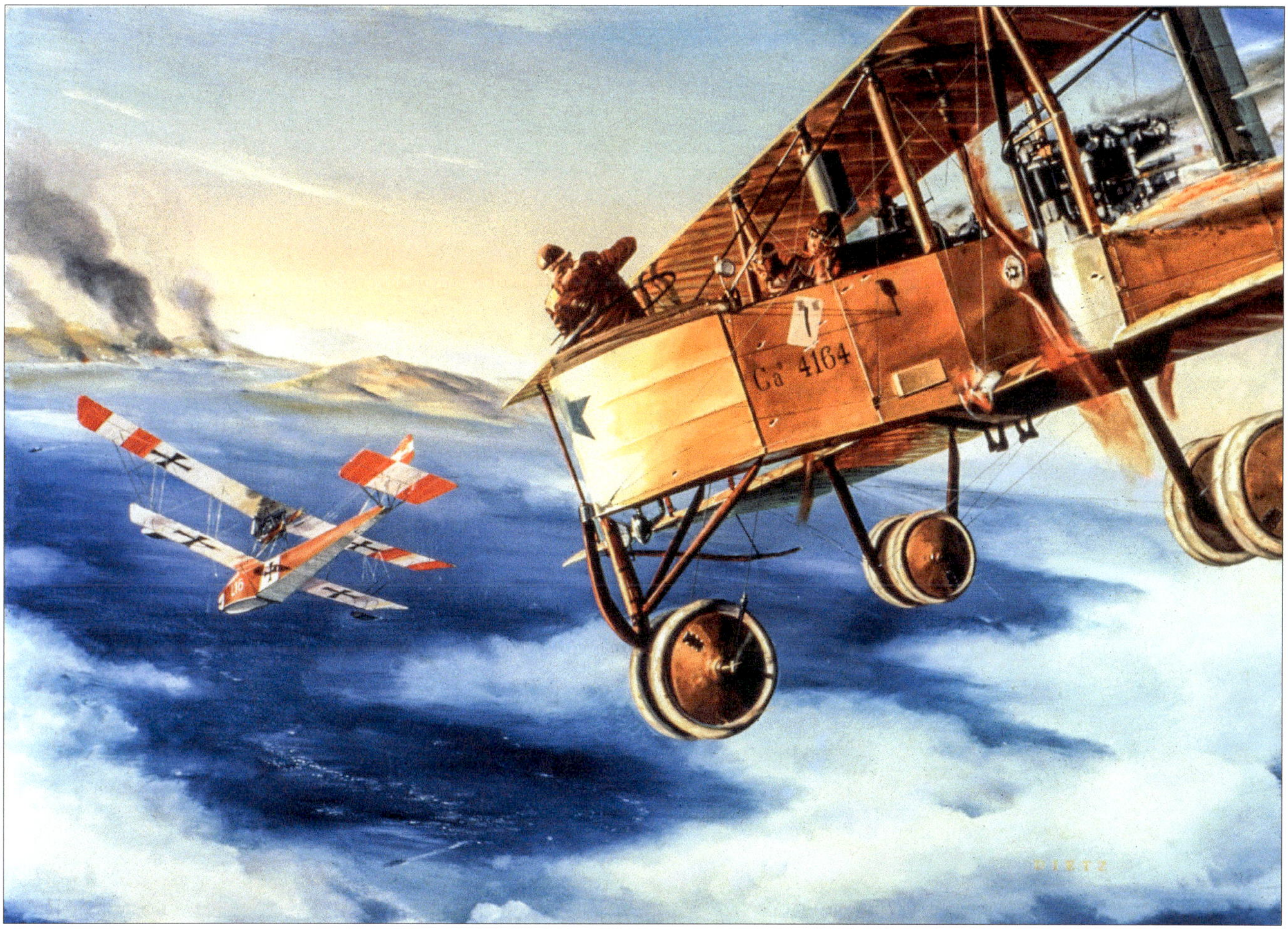

Above: Once they entered service, the Italian Caproni bomber units would attack the Austrian naval bases at Trieste and Pola on a regular basis. Here, Banfield earns one of his several victories over a Caproni as he speeds away in his iconic L-16 fighter. Jim Dietz

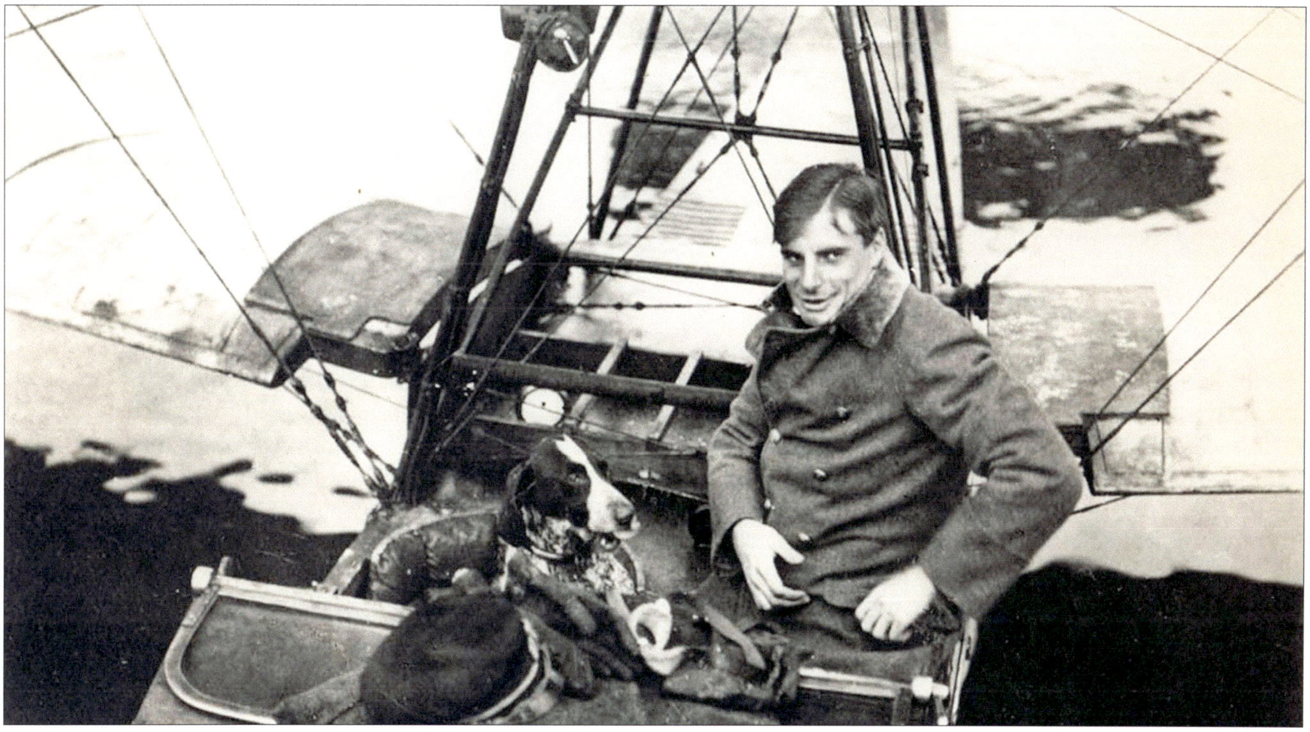

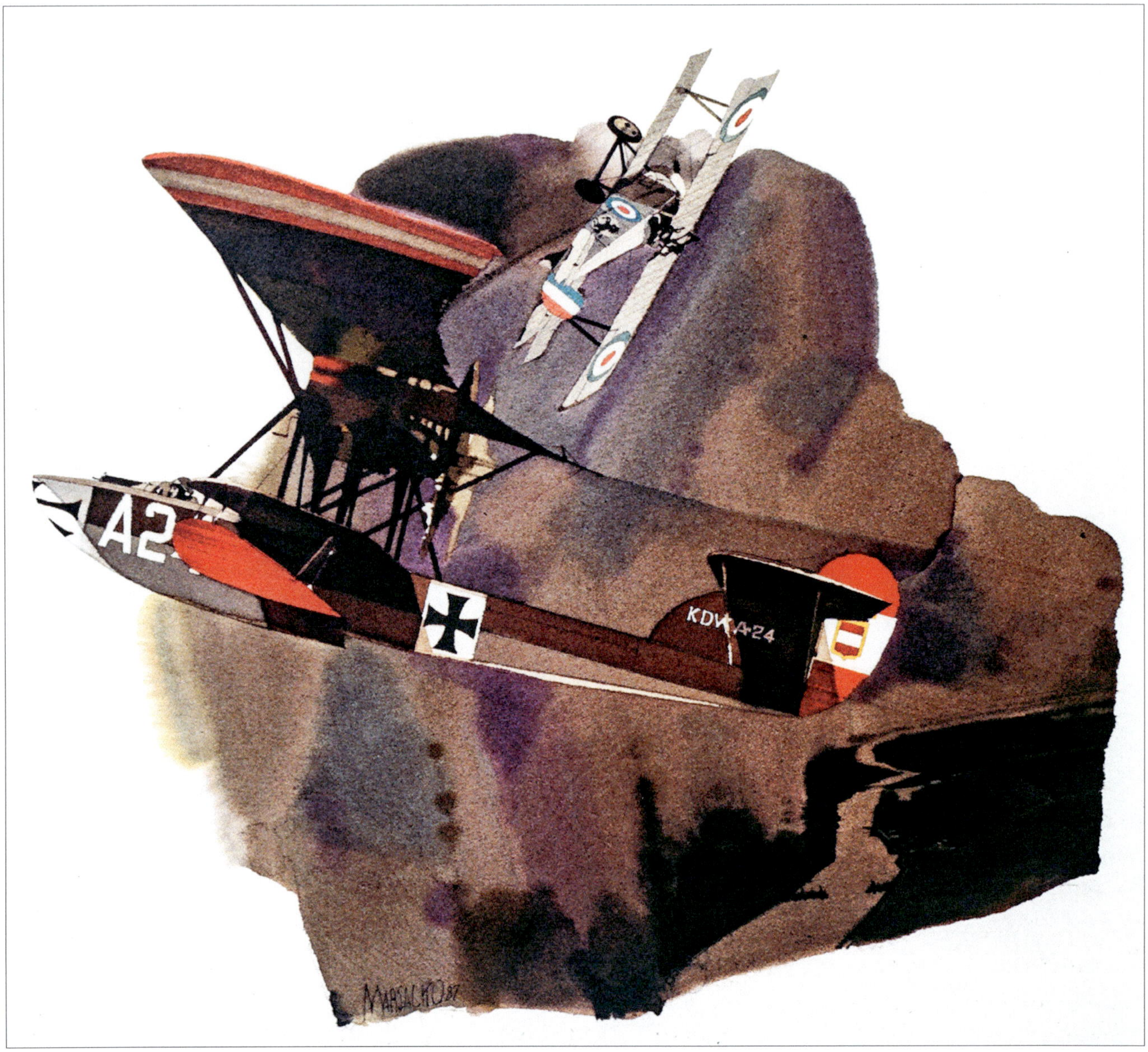

Above: A New Year's day dogfight that never happened. Similar to Udet's story of having to fight Guynemer, Banfield told of fighting the leading Italian ace of the war Baracca to a stalemate. However, like Guynemer, Baracca never mentioned the combat and records indicate that he flew missions that were a distance away from where he might have crossed paths with Banfield on that day. William Marsalko

on the water, and thought that several bursts from Strobl had damaged the aircraft. Banfield now quickly turned his attention towards Roulier as he still had the advantage of altitude. He forced Roulier to turn around and head back towards Grado. Again, Banfield thought he had damaged/shot down an enemy aircraft. In fact, Strobl's fire had been accurate and struck FBA#10 six times.

Despite this damage, Roulier was merely trying to gain the protective cover of some anti-aircraft guns at the Grado base.

Meanwhile, Conneau had landed and Giambernardini disembarked from the aircraft. It had only been his second flight and he was violently airsick, unable to man the machine gun. This was the reason Conneau had dropped out of the fight. Meanwhile, Conneau had grabbed a French seaman named Rouanet as his gunner and took off once again.

After taking off, Conneau went straight after Banfield to cover the retreat of Roulier. Banfield turned his attention back towards his old teacher and quickly forced him to dive away towards Grado's protective anti-aircraft guns. Conneau used that

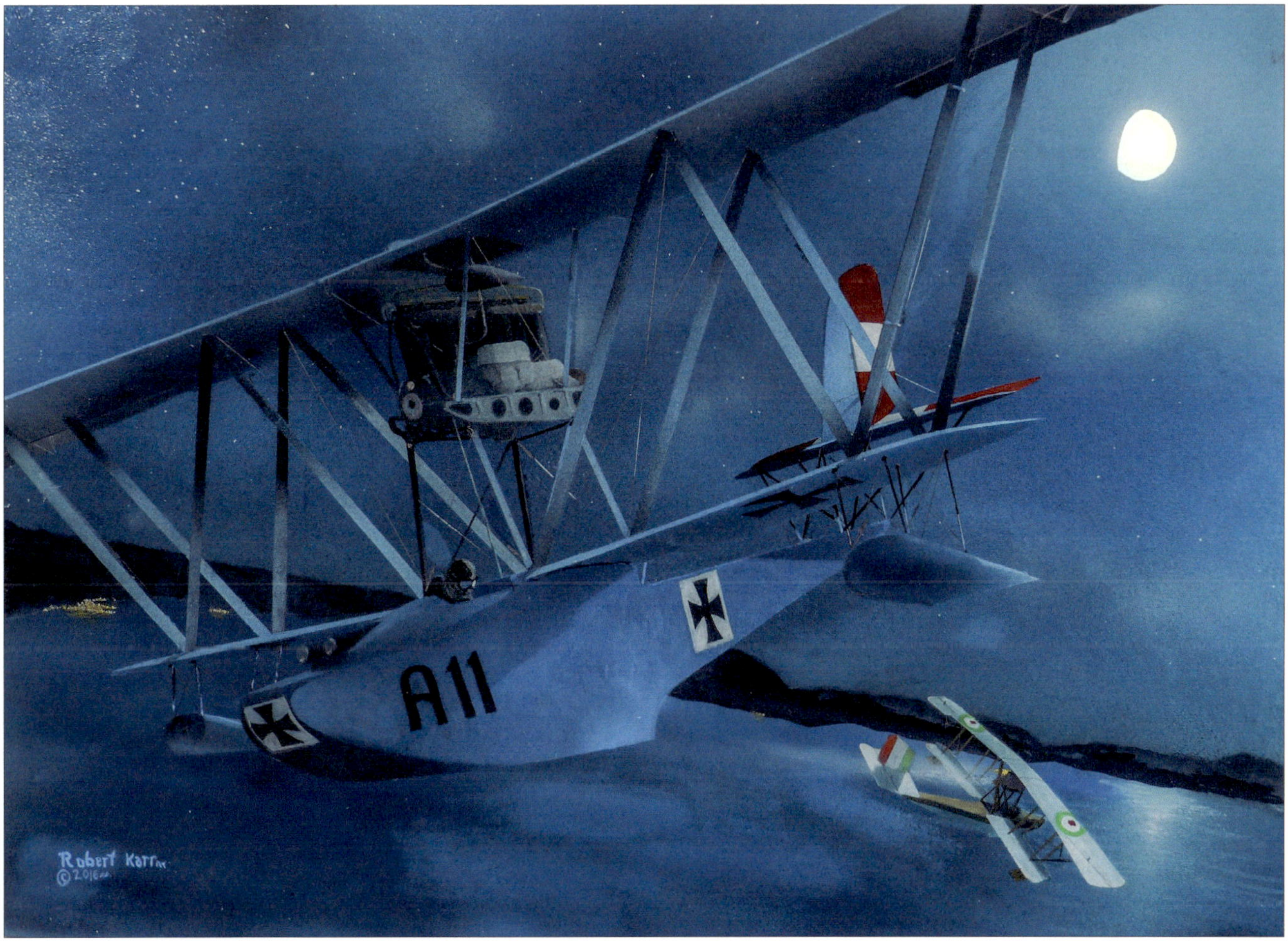

Above: Banfield flying his Blue wonder scores the first nocturnal victory for Austria on the night of 31 May 1917. Robert Karr

protective fire in order to spiral upwards to gain the same altitude as Banfield.

To Banfield and Strobl, they thought they had engaged and damaged, possibly downed three different French seaplanes!

By this time Banfield was low on fuel, and Strobl had fired all of his machine gun ammunition. As they headed towards Trieste, Conneau now attempted to engage with his new gunner firing short bursts from his Chauchat machine gun. Banfield and Strobl replied with a pistol and a carbine, but they continued on back to Trieste in order to make their claim of three victories. He would not receive any confirmed victories for this action, Banfield's most dramatic air combat to date.

One of the distinctions that Banfield had when compared to other successful fighter pilots of the first world war, was that he still continued to fly bombing and reconnaissance flights throughout the war. At this point in the war, one of the main tasks given to the naval pilots of Trieste was to fly overland and bomb critical targets around Venice. Venice was a major supply point for the Italian military forces during the first world war because of its location near the front lines. During the evening of the 22nd May 1916 Banfield led eight other seaplanes on a bombing mission against the train depot at Portogruaro. He later wrote of the mission:

"On the night of 22 May 1916 we attacked the train station at Portogruaro. During the return flight one of our aircraft, Lohner-120 [Soupper/Rappensberger] was damaged and gave the prescribed red flare signal while making a forced landing in the Bay of Lignano. I decided to land my airplane in the vicinity of the damaged aircraft, if possible. Despite the moonlight, it was difficult to find the airplane. At first I was deceived by the light of an enemy motorboat, but after further searching I found the red emergency signal of the downed airplane two miles further away. According to the prescribed routine, I ordered the other aircraft to be punctured and set afire, with the crew to swim to my aircraft after carrying out the order. After the damaged aircraft went up in flames, some suspenseful moments began. In picking up the other crew from the water, the engine of my aircraft quit and we began drifting

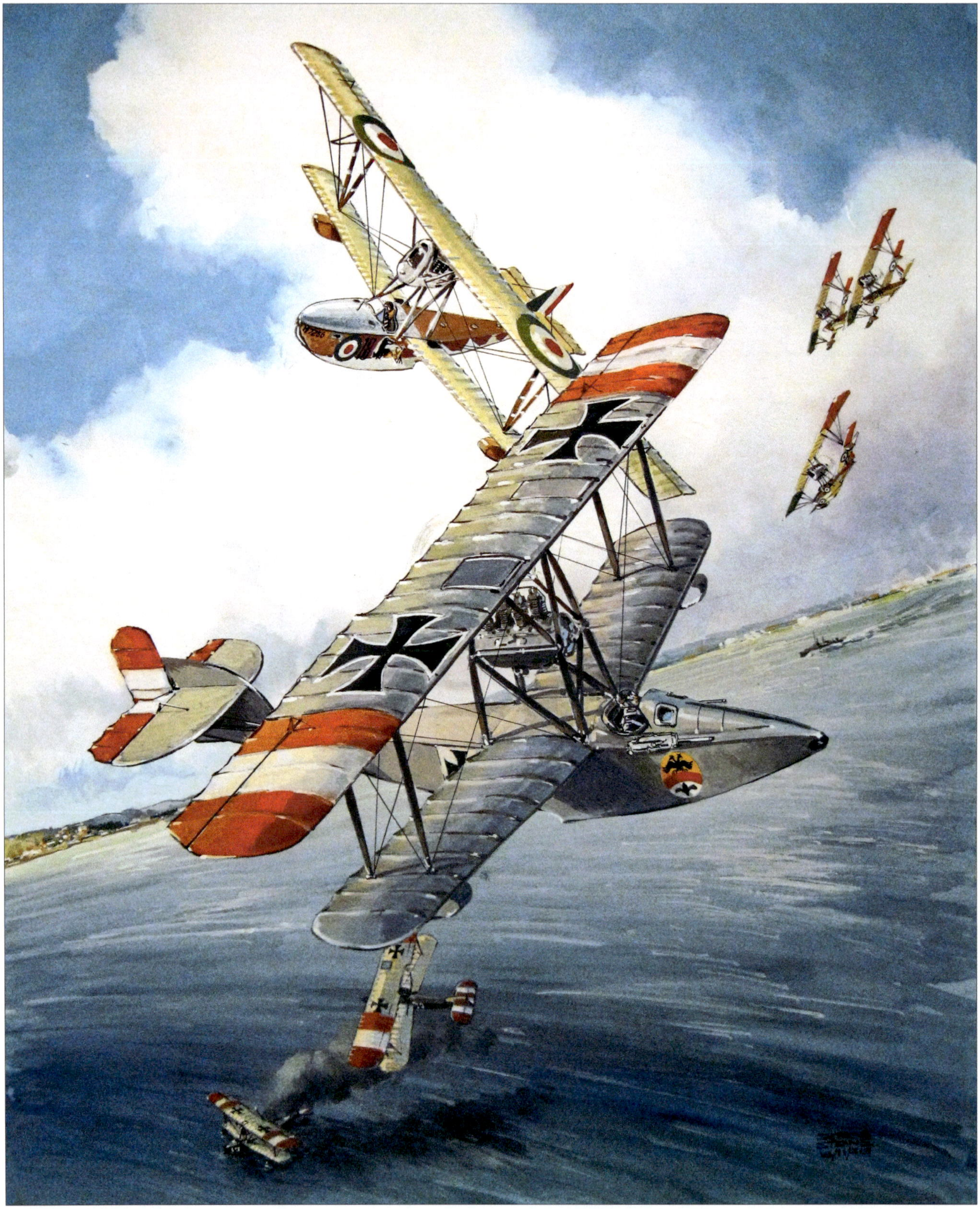

Above: The intense dogfight of 4 May 1918 illustrated just how deadly the Italian naval fighters were. The flight from Trieste, in their new W.18 fighters were led by Banfield. Both Niedermayer (seen on the water already) and Boros (seen circling above Niedermayer and about to be shot down himself) were lost and captured by the Italians. Banfield had his aircraft (in the immediate foreground with the special marking of the crest of Trieste) had his plane shot up and so damaged he had to force land and be rescued by a torpedo boat. Only one member of the flight, Schneider made it back unscathed. Serge Stone

toward the burning wreckage of the other plane. Fortunately, after a few minutes my engine started up and I headed across the water, followed by two enemy motorboats. Rather strong ground swell prevailed and made it hard for me to reach the speed necessary to get my airplane off the water, but after some time I finally did it. At about 3:00 A.M. I landed with the rescued crew at my seaplane station in Trieste. Not until then did it become evident that the bottom of my aircraft had been damaged during the difficult takeoff. This rescue of an aircrew forced down was later repeated many times by other seaplane pilots."

As Banfield mentioned, the naval aviators of both sides during the conflict looked upon each other as stranded mariners in need of rescue and there were many stories of fliers landing upon open water to help their comrades or their enemies or just a moment ago.

Another aspect of Banfield's first world war fighter pilot career was that he was the only pilot to have two, separate and unique aircraft made just for him. He received the first of these two aircraft in March of 1916. The Lohner M-type Nordenflug had been built with a special 140 hp Heiro engine before the war to participate in an aviation meet. This single example was now fitted with a Schwarzlose MG 07/12 and put at Banfield's disposal as a single-seater fighter.

The aircraft was designated L-16 and was kept ready for flight at any time. On the 23rd June 1916 Banfield took off to intercept a French FBA type C (#12) which had been spotted patrolling near to Trieste.

In his first firing pass, some of his bullets killed the gunner, E. Grammaticopoulo. The French pilot was now essentially "unarmed". Banfield's L-16 was much faster than the FBA and he used the speed to gain altitude and then dove again on a firing pass that forced the pilot, EV Victor Vaugeois, to land on open water about four kilometers away from Trieste. Once on the water, Vaugeois used his engine to begin to taxi towards the Allied base at Grado. Banfield dove upon him several times and finally disabled his engine.

Now dead in the water, Vaugeois proceeded to man the machine gun and fire at Banfield! Soon, an Austrian motor boat arrived to capture the Frenchman and tow his seaplane back to Trieste. Banfield wrote:

"The pilot, while having been forced to ditch the aircraft, had continued to shoot at me which required me to defend myself by shooting at him, but I did not kill him. Only after the pilot was wounded was he convinced that further resistance would have been pointless and he surrendered. He was picked up by one of our boats which towed his seaplane to Trieste. We afforded military honors to this brave pilot, a naval officer himself. Having had medical attention we procured clothes for him and invited him to dinner. In a short time he was more like a guest than a prisoner, which was not very pleasing to the area commander Vice-Admiral Koudelka."

Several days later, Banfield flew over the CAM-Venice base and dropped a note that Vaugeois had written: "Met Austrian fighter type Etrich[sic]-shot down by a very chivalrous Austrian master pilot. I am healthy."

As mentioned in the chapter BIG BIRDS, the Italians mounted an effective strategic bombing campaign against Trieste utilizing their Caproni bombers. On the 1st August 1916, Banfield flew his L-16 fighter to down his first Caproni bomber: nearly half of Banfield's confirmed victories in the war would be Caproni bombers indicating how often the Italians launched bombing sorties against Trieste.

In the summer of 1916, the French naval aviators received two of the latest FBA Type H flying boats with a 180hp Hispano-Suiza engine. One of the pilots stationed at CAM-Venice wrote to his brother:

"The Austrians have impressive results with their fighter hydroplanes. One of them, named Banfield, shot down three planes. We have now two Hispano engines for six pilots. This is a ridiculous number as our six pilot's flight should have at least twelve machines. Our Hispanos will be ready this week. They are fast, but climb slowly and don't turn very fast. Aerial combat seems to be problematic with this kind of plane. In front of an agile fighter, what would I do in this plane? I fear to think about it."

The pilot who wrote this letter , EV Roulier, was to tragically be forced to try his best in just that scenario on the 15ht August 1916. He was part of a formation of Allied aircraft in a sweep of the Gulf of Trieste near enough to Trieste to bring Banfield, in his L-16, into the air.

Banfield attacked two of the aircraft and in his first attack Banfield badly damaged one of the French FBA flying boats and forced it to land upon the open water. His attack on the second aircraft, which at that moment was engaged in a bombing run, caught the crew completely by surprise. The FBA was shot about, and its pilot, EV Roulier, fell out of the aircraft at about 30 meters height.

The heavily damaged aircraft and its pilot's body were brought to the base at Trieste.

The Hispano-Suiza engine from this aircraft was salvaged and given for study to Ferdinand Porsche, who was designing aircraft engines for Austro-

Daimler company at the time.

The well -known expert of First World War aviation, historian David Mechin shared this footnote to the events of 15th August 1916 with this author:

"This story ends after the war. I [DM] made contact with Jean Roulier's nephew, a catholic priest. He told me that a few years after the war a gentleman came to the house of Roulier's mother(his grandmother) living in Paris. The old lady does not want to be bothered and orders her servant to tell this gentleman that she does not want to see anybody. The servant comes back to Madame Roulier:'He insists. He has a German accent... 'Madame Roulier finally receives him. This was Baron Gottfried Banfield, who came to tell her how sorry he was for the death of his valorous opponent Jean Roulier." [DM correspondence]

The two confirmed victories of 15th August was the only time Banfield scored two victories in a single day.

Because of his successes, and likely because he was generally held to be a very good, natural pilot Banfield was often given the task to test fly the newest seaplane designs. In one circumstance, Banfield was test flying an enormously large design, the Oeffag-Mickl G-1, with six engines. The aircraft was to be a bomber that the Austrian navy could use to respond for the need of strategic bombing. Banfield was not impressed with the aircraft and after only a couple of flights made some suggestions for improvement. In November 1915, Banfield was joined by his closest friend, Lschlt. Ockermuller, in a test flight of the revised aircraft. Tragically, the aircraft crashed from a low height upon an attempted take-off and Ockermuller was killed in the crash. In the fall of 1916 Banfield was sent two new seaplanes to trial as a test pilot. One was the Oeffag A.11 designed by Josef Mickl and the other was the Hansa-Brandenburg type CC fighter designed by Ernst Heinkel. Banfield was very enthusiastic about the Hansa-Brandenburg fighter and because of his approval the type went into production. However, with time, the A.11 design would become Banfield's favorite aircraft of the war and similar to his Lohner L-16, the A.11 would be the only aircraft of its design ever built.

For the fall of 1916 though, Banfield enthusiastically embraced the Hansa-Brandenburg fighter. The new, diminutive fighter had a 185hp Austro-Daimler engine that gave it a top speed of 160km/ hr. Banfield kept the prototype for his personal use and the aircraft was re-designated the A.12. Banfield later wrote about a particularly memorable event in this aircraft:

"January 1, 1917 was a beautiful day and I decided to take a pleasure flight in my A.12 flying boat. About 30 minutes later while flying along enjoying myself, I suddenly realized that I was being attacked by an Italian Nieuport, I was really angry that this pilot was ruining my lovely pleasure flight. I turned to meet his attack with the thought that I would teach this pilot not to ruin my good time! It was obvious right away that he was extraordinarily skillful. We went around and around and around for almost an hour, with neither able to attain any advantage. At the end of this time, almost by unstated understanding, we broke off the fight. As the Italian headed west and I headed southeast, we passed quite close to one another. Almost at the same time I noticed the black horse painted on the aircraft and the fact that he was waving a salute to me! I gladly returned the salute, feeling honored to have fought him. I knew it was Baracca [Italy's ace of aces]."

A marvelous story, and one which fits so perfectly into the type of stories being told after the war by surviving pilots. Sadly however, the event could not have happened and it was merely a good tale, told in the spirit of camaraderie that Banfield felt towards his fellow aviators from both his side and the Allied side of the war.

A journalist of the time described Banfield in this manner:

"Everything about him led to the extraordinary impetus with which he defended the Adriatic city of Trieste...His slender body, thin, in good shape and dominated by a head that resembles that of a hawk more than any other I have seen...A long nose, thin, finely carved and under thick eyebrows were dark, gray eyes that knew how to penetrate and appear cold-a man accustomed to pursue with perseverance and also stubbornness to his own ends..."

The port of Trieste had historically been Italian, but had been gained by Austro-Hungary following a successful conflict with Italy in the late 1800's. Therefore, the population of the city considered themselves to be Italians, under the control of a foreign government. However, this emotion did not cloud the reality that the populace of Trieste faced during the war.

The fact was that Italian Caproni bombers were frequently attacking the port and wayward bombs were killing civilians.

Those same people gathered a collection of money to purchase a special silver wreath and present it to Banfield, who they viewed as their protector. They also began to refer to him as the Eagle of Trieste.

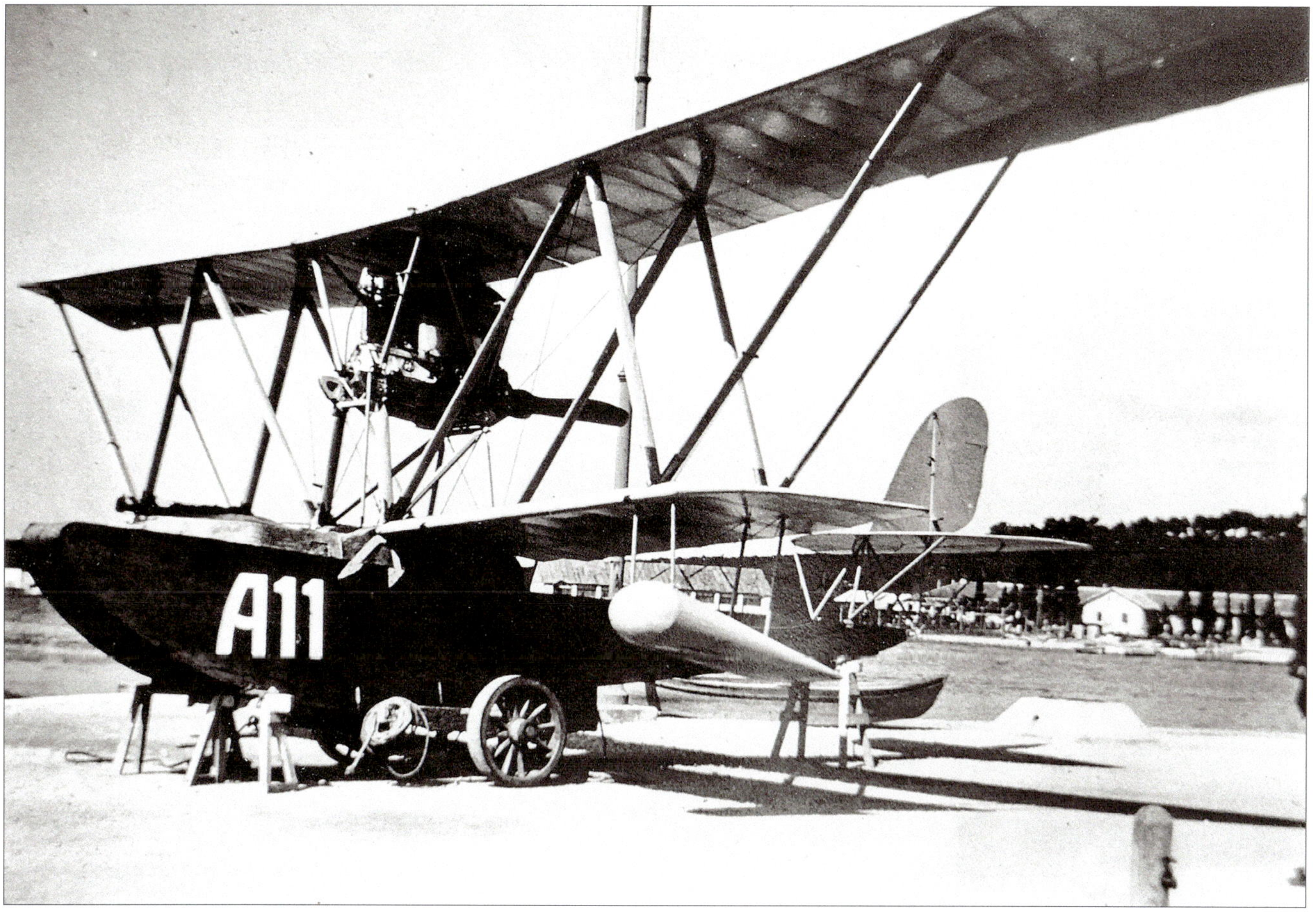

Above: The Blue Wonder, Banfield's second one-of-a-kind fighter, the Mickl A.11. He would comment that this was his favorite aircraft of the war.

Below: The superlative Macchi M.5 seaplane fighter aircraft. Under the command of Pierozzi the Italian naval fighter pilots would make good use of this fighter and make life very dangerous for their Austrian opponents.

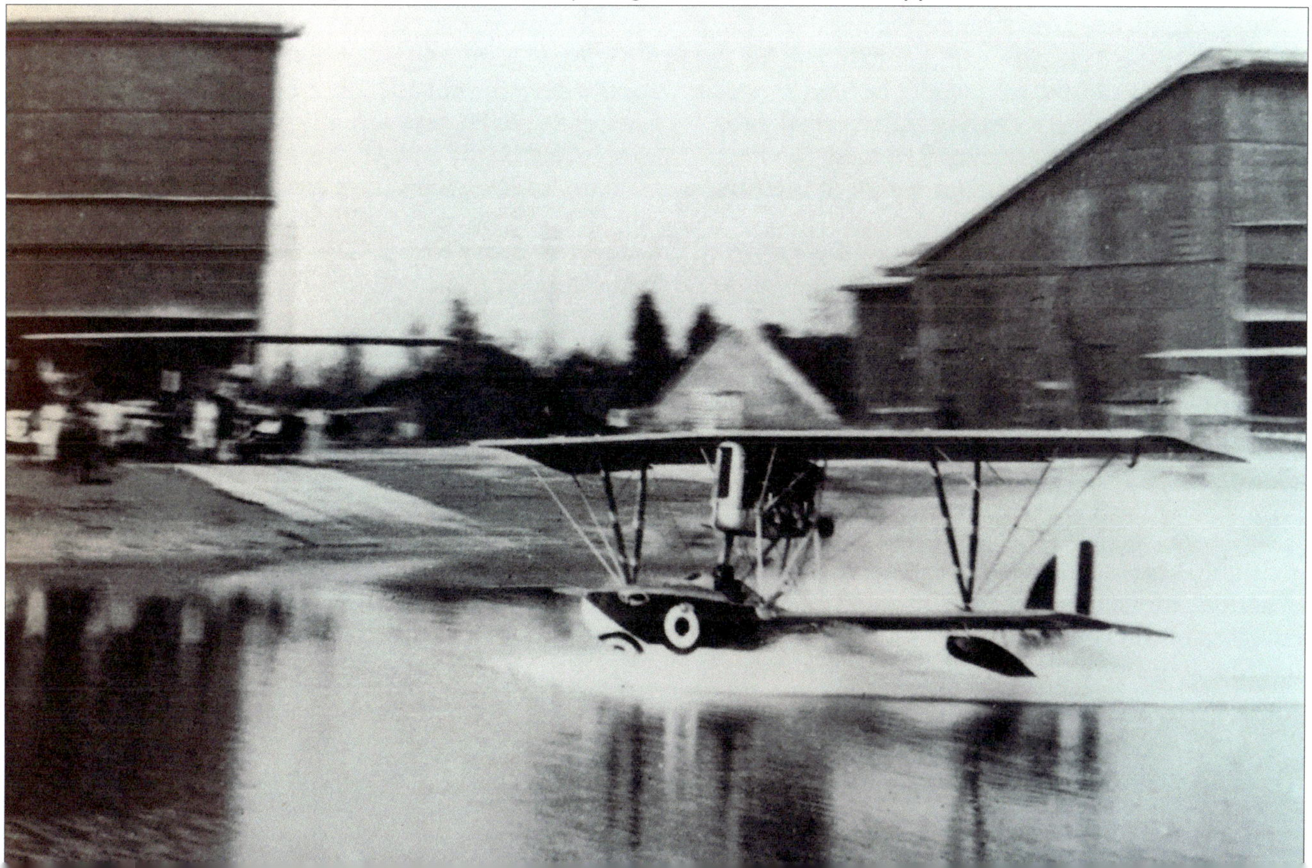

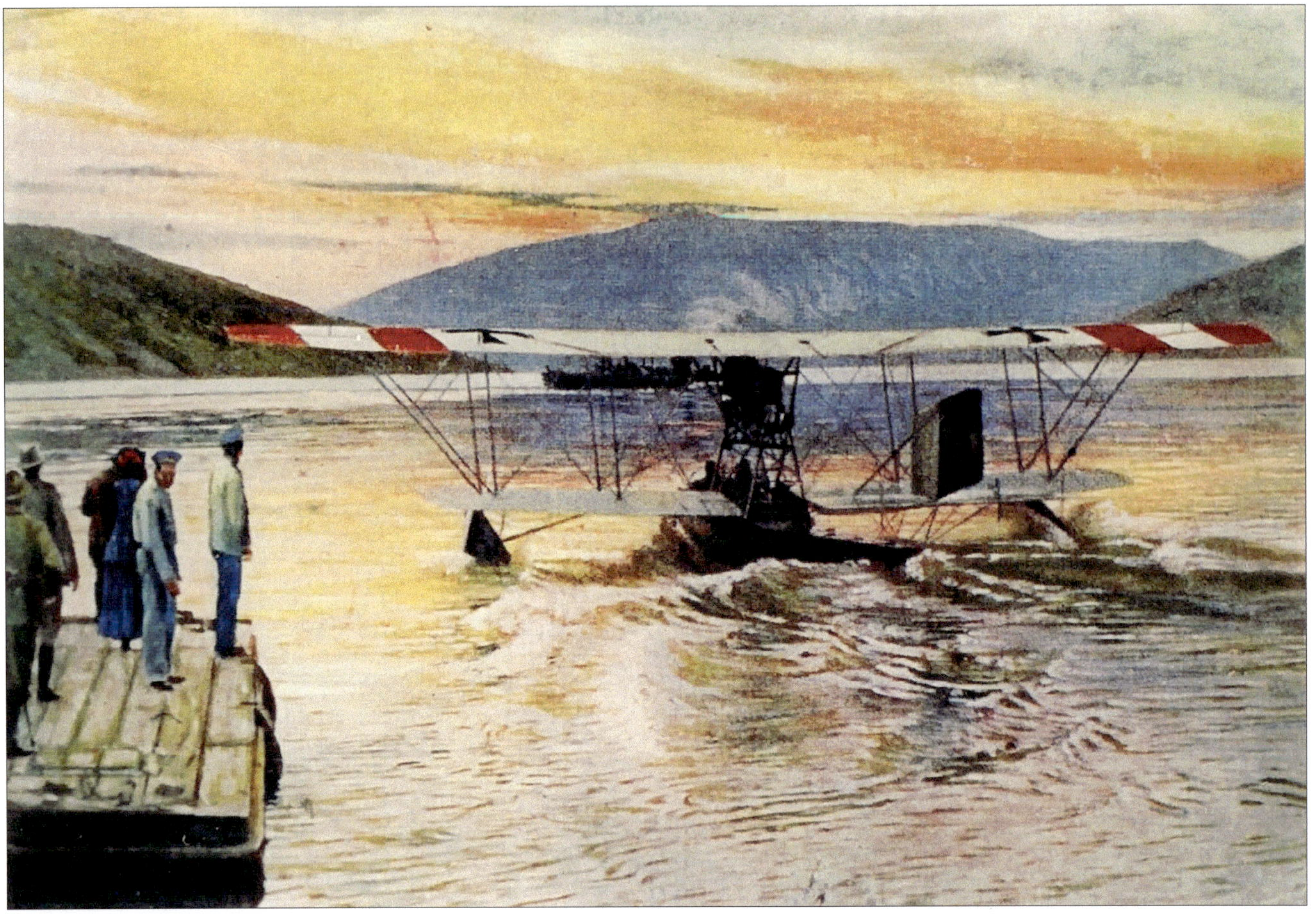

Above: The naval air war above the Adriatic Sea in the First World War was especially unique in the fact that it was fought primarily by seaplanes. A contemporary artwork by Harry Heusser

Facing Page: The last official knight of the Holy Roman Empire, Gottfried van Banfield.

Because most of the Caproni bombing missions were now being flown at night, Banfield had his A.11 fighter painted all blue so it would be hard to see at night and/or over the water. He called this fighter his "Blue Wonder" and managed to score the first nocturnal victory of the war for a pilot of the Austro-Hungarian empire.

The highest and rarest award that the Austro-Hungarian empire could award to a member of its military was the Knight's Cross of the Military Order of Maria Theresa. Since, the Austro-Hungarian empire was considered to be the last remnant of the Holy Roman empire itself this award would indeed be quite special.

An oddity for the award was that in order to receive the medal, one had to request it and since it was not considered the thing a gentleman would do, there was a quandary about the correct course of action. For example, the leading ace of Austro-Hungary in the First World War, Godwin Brumowski felt that if his heroic actions did not justify his superiors to award the medal to him, then he would not request the KCMOMT.

It had been suggested to Banfield that he should apply even in mid-1916, but by late 1917 he still had not done so. He was actually formally ordered to apply and he did so by simply forwarding the combat reports of several of his missions.

Therefore, on the 30th birthday of the new emperor Kaiser Karl I, 17th August 1917, Banfield presented himself to Karl's Villa Wartholz to receive this award, which also elevated the recipient into the nobility. Banfield would be the only aviator in the First World War to receive this prestigious award. He would also be the last surviving recipient, which would technically make him the last knight of the Holy Roman empire.

The naval aviators at Trieste received a new fighter in the spring of 1918. The Hansa-Brandenburg W.18 was an improved design to the earlier seaplane. Unfortunately for the Austrians, they were losing the battle for air supremacy over the Northern Adriatic Sea to the Italians as a rapid rate. The Allies were able to produce more aircraft, were able to have

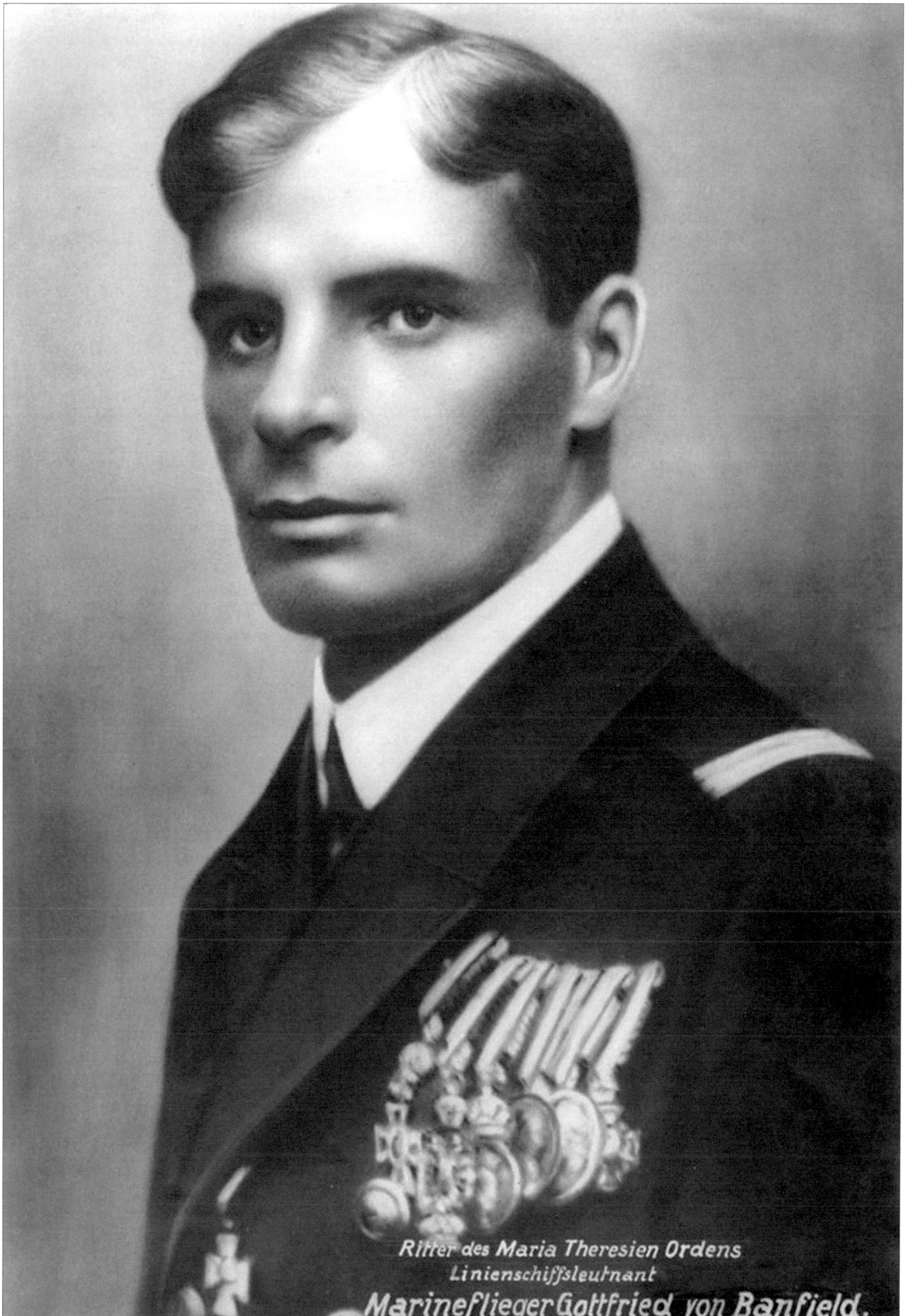
Ritter des Maria Theresien Ordens
Linienschiffsleutnant
Marineflieger Gottfried von Banfield.

more capable and well-trained pilots and in addition, had the superlative Macchi M.5 seaplane which would prove to be the best seaplane fighter of the war.

By this time in the war, the French naval unit at Venice had been disbanded, it was no longer needed because of the great strides forward that Italian naval aviation had made. Two seaplane fighter squadrons were now stationed at Venice under the command of Orazio Pierozzi who:"...proved to be a good leader, an able trainer...He was a zealous advocate of formation flying and teamwork over individualism."

Part of why Banfield and his comrades referred to SFS Trieste as the "station of death" was certainly due to Pierozzi and his fellow aces Umberto Calvello and Federico Martinengo.

On the 4th May Pierozzi led a flight of five other Macchi M.5 fighters on an escort mission for a Macchi L.3 reconnaissance seaplane.

They were engaged by Banfield and three other pilots(Niedermayer,Boros and Schneider) in their new Hansa-Brandenburg W.18 fighters. In the ensuing melee, Niedermayer was shot down and forced to land on the open water. Sometime during the dogfight, Boros chose to fly low over his comrade to check on his situation and Boros was promptly shot down as well.

Pierozzi himself was likely to have been the pilot that damaged Banfield's fighter to the extent that Banfield had to dive away from the fight and land near the entrance to the harbor of Trieste, forcing an Austrian motorboat to come out and tow him back to the safety of his base.

Pierozzi would lead a flight of Macchi seaplanes to land at Trieste after the armistice. Awaiting him was the only commander the SFS-Trieste had ever known, Lscht. Banfield. The Italian fighter pilot stated what an honor it was to meet the Eagle of Trieste and that he was sorry that the two had not met earlier(although they likely had).

Whereupon a steadfast Banfield replied it was not a cause for regret: 'If we had met earlier," he said,"one of us would be unlikely to be present today!"

Despite having been treated as a hero by the civilian population of Trieste during the war, after the armistice Trieste became part of Italy again and Banfield was promptly thrown in jail!

Just days before, he had become officially engaged to the "most beautiful girl" in Trieste, the Countess Maria Tripcovich. Her father was one of the wealthiest men in the city and owner of the most successful shipyard in this port city. Likely through his influence, Banfield was released. Banfield sought out his uncle in England and actually became a citizen of the United Kingdom once again. He and Maria married in England in 1920.

Before long, the couple returned to Trieste where Banfield became an Italian citizen and joined his father-in-law's shipping firm. He began to specialize in maritime salvage and during the Second World War assisted the Italian navy in recovering several vessels, including the RN Conte di Cavour, a battleship that was significantly damaged during the Battle of Taranto.

At the end of WWII, as the Germans were retreating from the Allied advance in Italy and the Adriatic, Banfield was able to convince the German officers in charge of Trieste not to destroy the harbor and thus allowed the city to escape most of the devastation wrought upon other cities and ports in Italy.

The grateful citizens now referred to Banfield as Il Nostre Barone, our Baron. His maritime salvage company was utilized to help clear the Suez Canal of 113 vessels in 1956. For this work, Banfield would be made an officer of the Legion of Honor and be given the title of honorary French consul to Trieste by Charles de Gaulle himself. He remained a popular figure in Trieste for the remainder of his life, he cherished sailing the Adriatic Sea in his yacht, Veruda. During his funeral service following his death at age of 96 on 23rd September 1986, it was stated that "an epoch has come to an end."

Biff

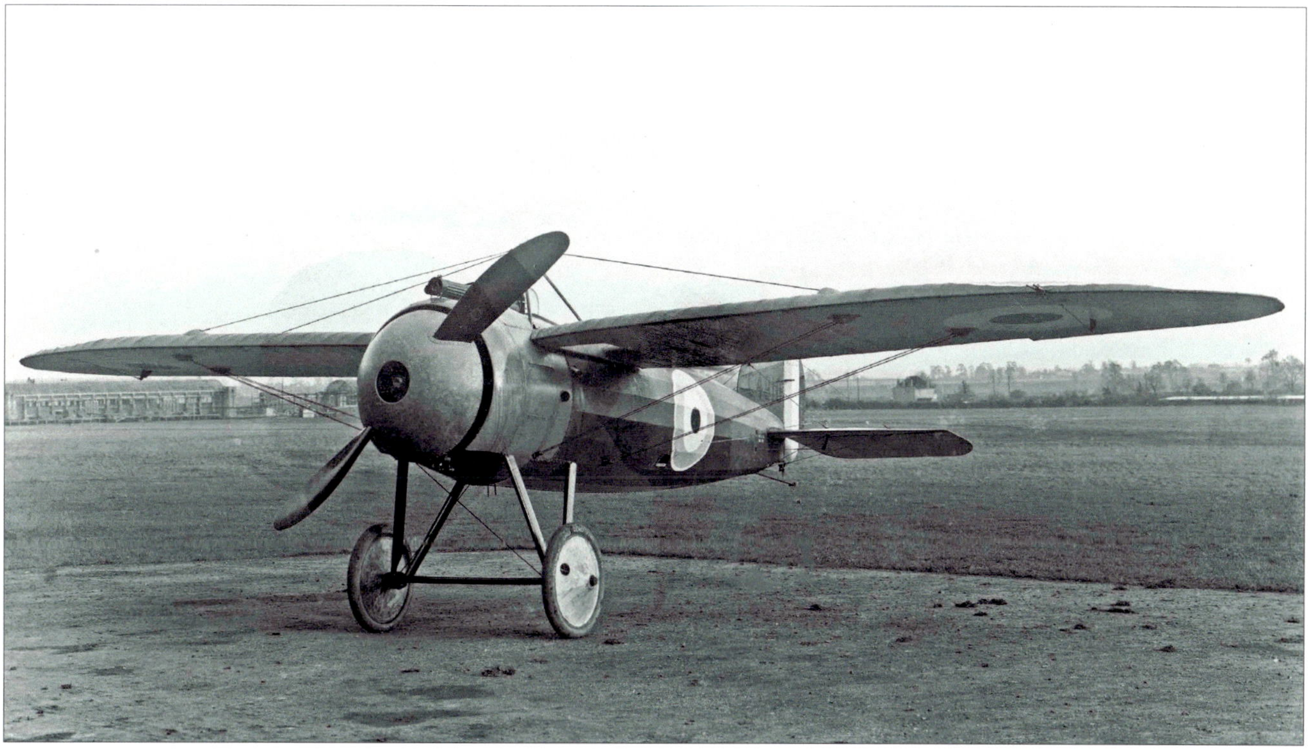

Above: Frank Bamwell's most distinctive design was a monoplane fighter. The Bristol M1.C was a very fast aircraft, and had some design elements well ahead of its time; at this point the biplane was considered a better design configuration.

One of the most successful fighter aircraft designs of the war, the Bristol F2A and B variants had a disastrous beginning to its frontline career. The Bristol Aeroplane Company was originally called the British and Colonial Airplane Co. and was owned by Sir George White.

One of its chief designers was Capt. Frank Barnwell who joined the company in 1911. Barnwell had already had success with his Bristol Scout and his Bristol M-1 monoplane; which was a very good design for its day that was completely undervalued and unappreciated by the Royal Flying Corps which had a true bias against monoplane aircraft.

The initial two-seater design by Barnwell was designated the R.2A since it was intended as a reconnaissance aircraft and a replacement for the Be2 series. The prototype flew in 1916 and its performance was considered average. However, the newly built Rolls-Royce 190hp Falcon engine was fitted to the R.2A and the combination of superb power plant with phenomenal aircraft design so transformed the design's performance that the aircraft was now re-named the F.2A (F for fighter).

Very few F.2A aircraft were built(50 total) until the F2.B variation entered production with the difference between the two designs primarily involving tweaking of the Rolls-Royce Falcon engine into a slightly lighter, more powerful variant.

The Bristol F2.B developed two nicknames during its career. It was called the Biff during the war (for uncertain reasons) and it was called the Brisfit after the war apparently for Bristol Misfit, because the type served for so long and had so many modifications as it soldiered on into the 1930s.

It had a ceiling of 20,000 feet and an endurance of 3 hours. At 10,000 feet it could go at a respectable 113mph and at sea level the Biff had a top speed of 125mph.

An unusual feature had the Bristol Fighter equipped with ailerons on both wings which helped give it dramatically improved maneuverability. The fuselage was positioned in the mid-gap position between the wings which effectively lowered the top wing in relation to the pilot. This design gave the pilot much better visibility above the top wing and was cherished by the Bristol Fighter crews.

The pilot had a synchronized Vickers machine gun with over 900 rounds to fire forward and most of the observers had twin Lewis-guns and seven drums of ammunition. However, what made these crews

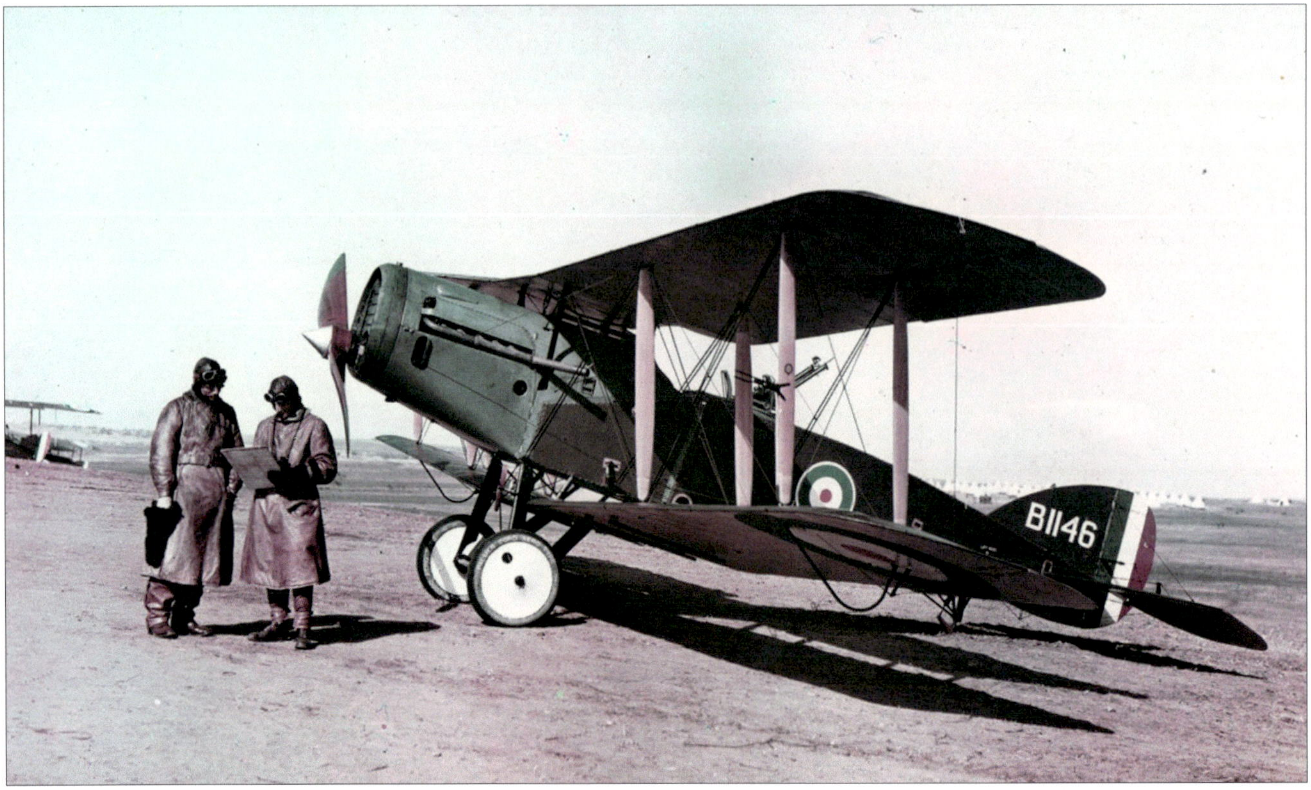

Above: An original color photo of a Bristol Fighter crew in 1917.

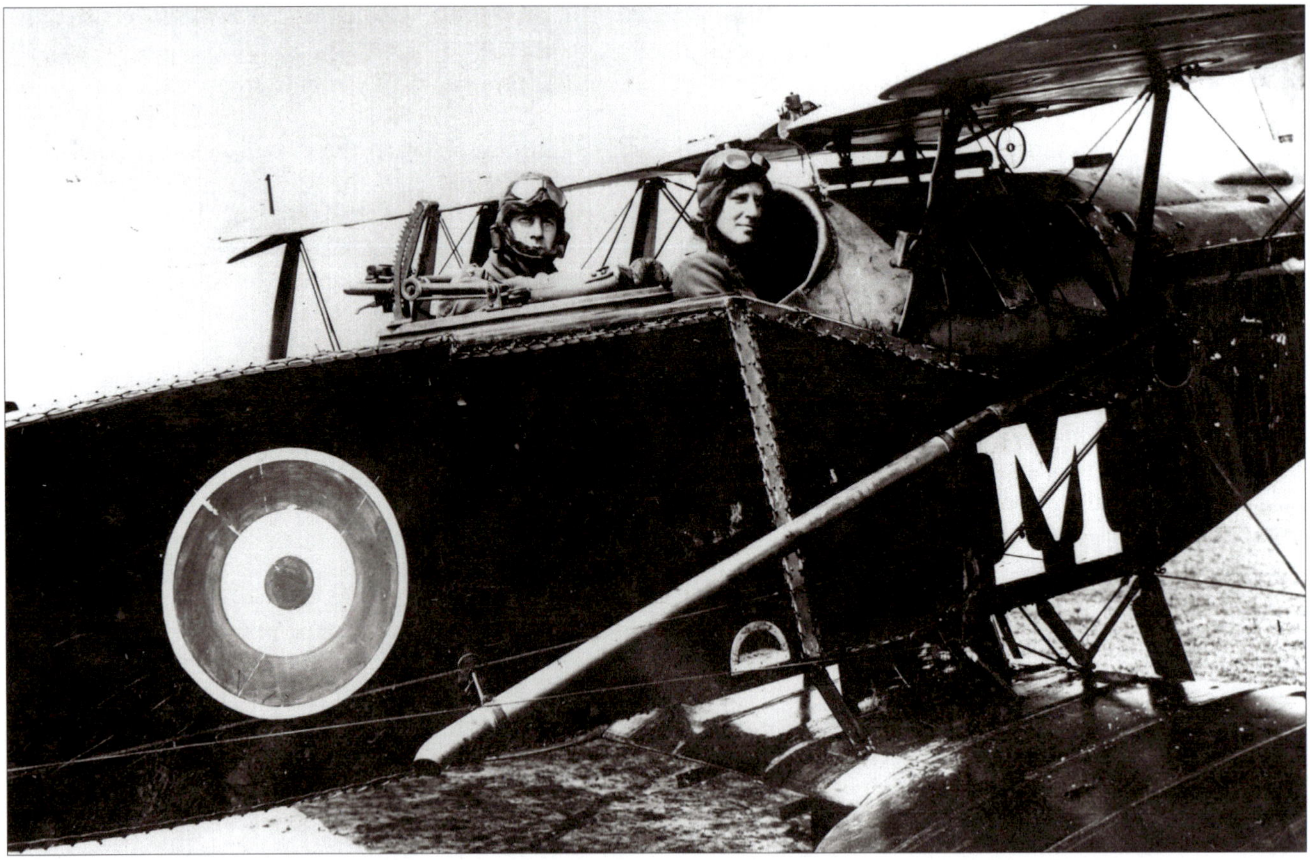

Above: One of the key design features was the proximity of the two crew members which greatly aided their cooperation during combat.

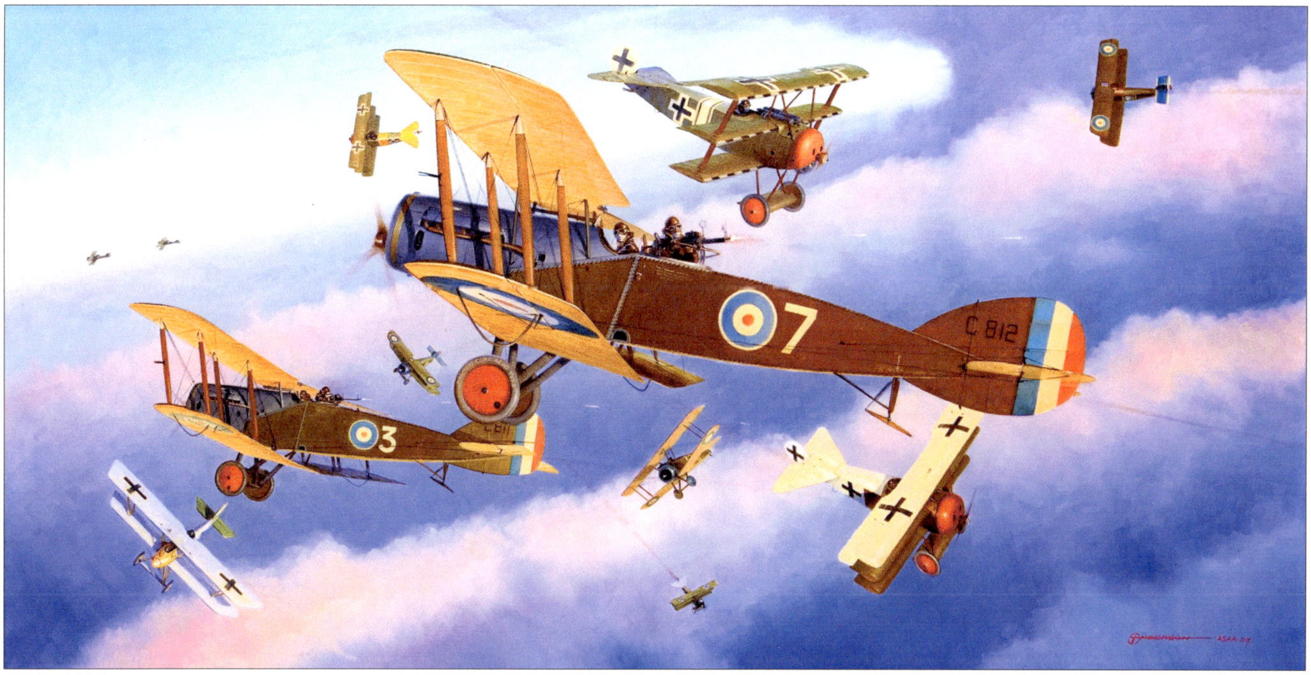

Above: The Biff was introduced on the Western Front during the Battle of Arras, and it was in the thick of the air battle from the very start. Steve Anderson

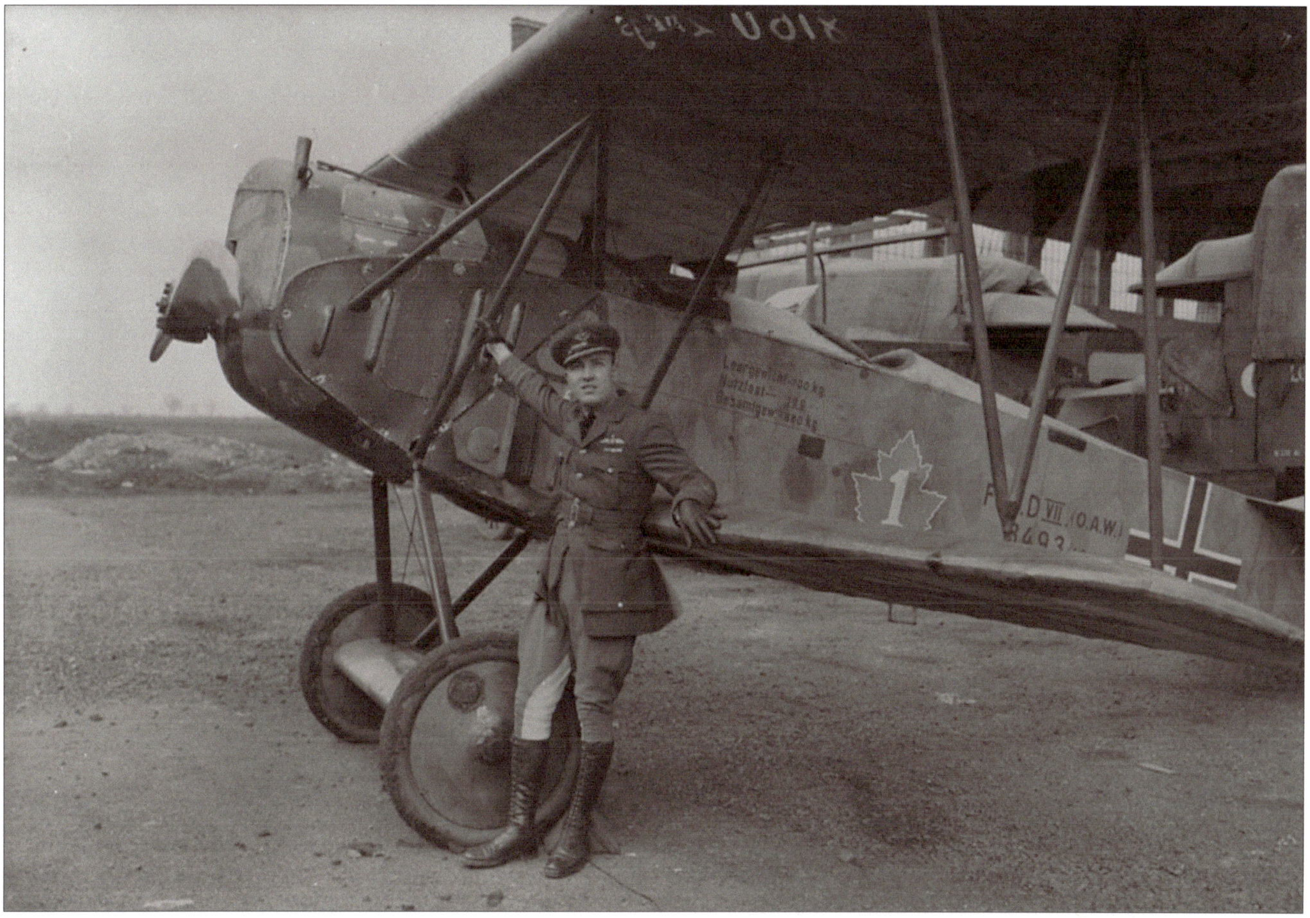

Above: A Canadian, Major A.E. McKeever and his gunners would shoot down 31 victories in six months to help make him the top scoring Bristol Fighter ace of the war He is seen in this photo immediately after the war when he was C.O. of No. 1 squadron of the newly formed Canadian Air Force. He is leaning on a Fokker D.VII, note the maple leaf /1 insignia on the fuselage.

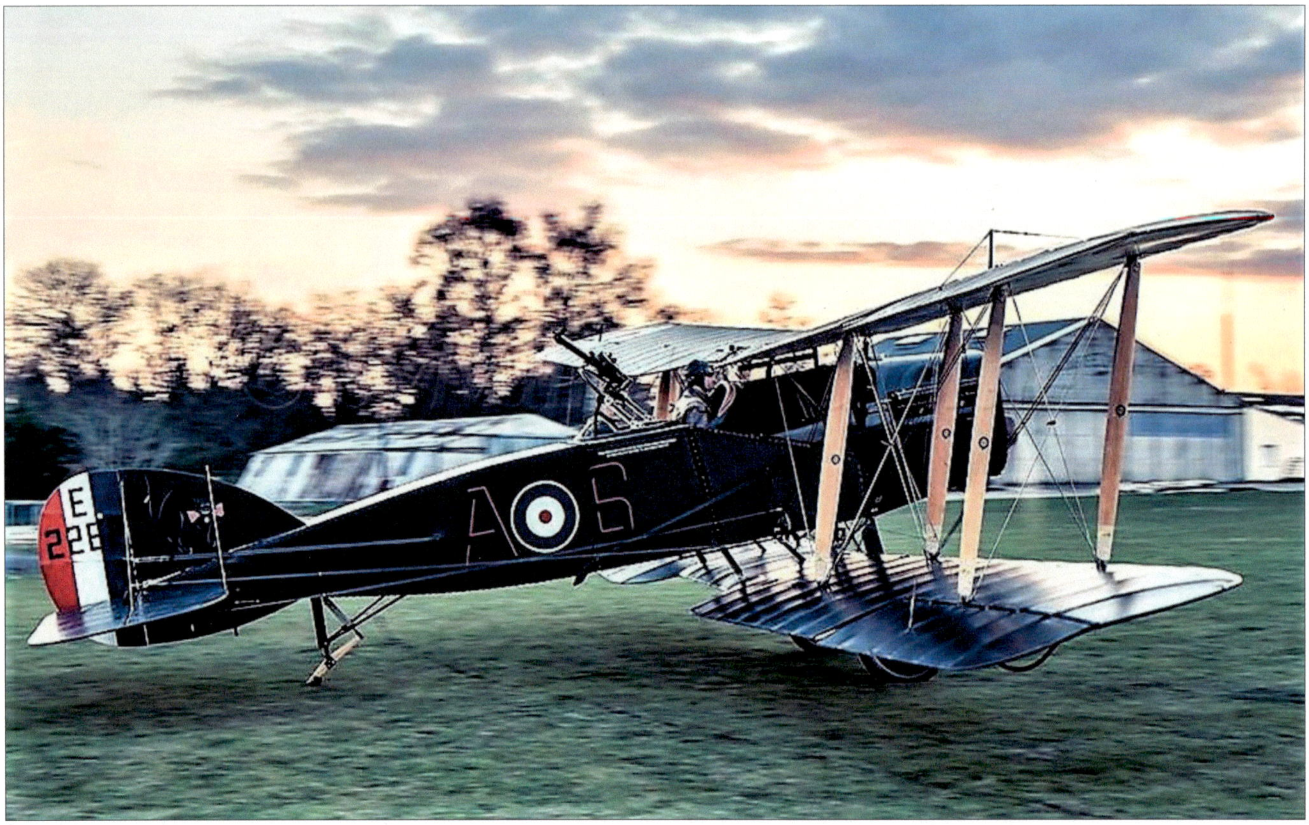

Above: The Biff also served an important role as a home defense fighter, including as a night fighter as seen in this modern reproduction of the type.

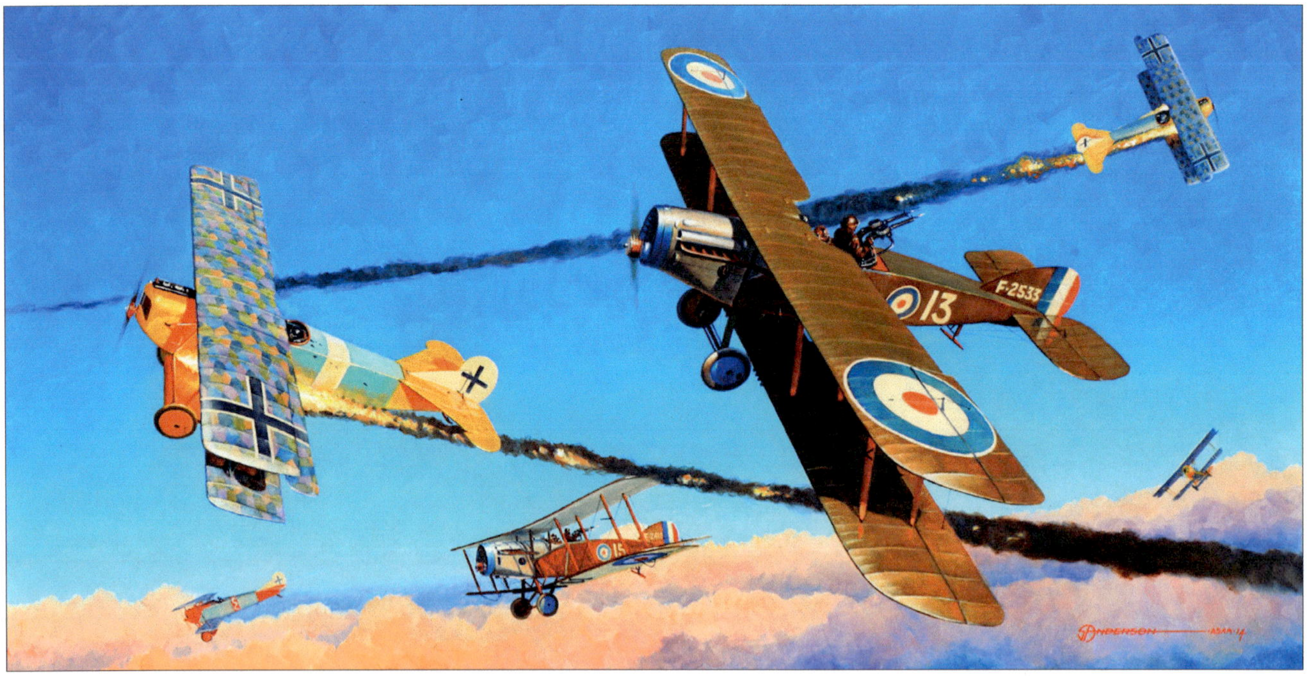

Above: Capt. Lale (pilot) and 2Lt. Edwards (gunner) demonstrated the lethality of the Bristol Fighter when they simultaneously downed two German Fokker D.VII fighters from the front and the back! Steve Anderson

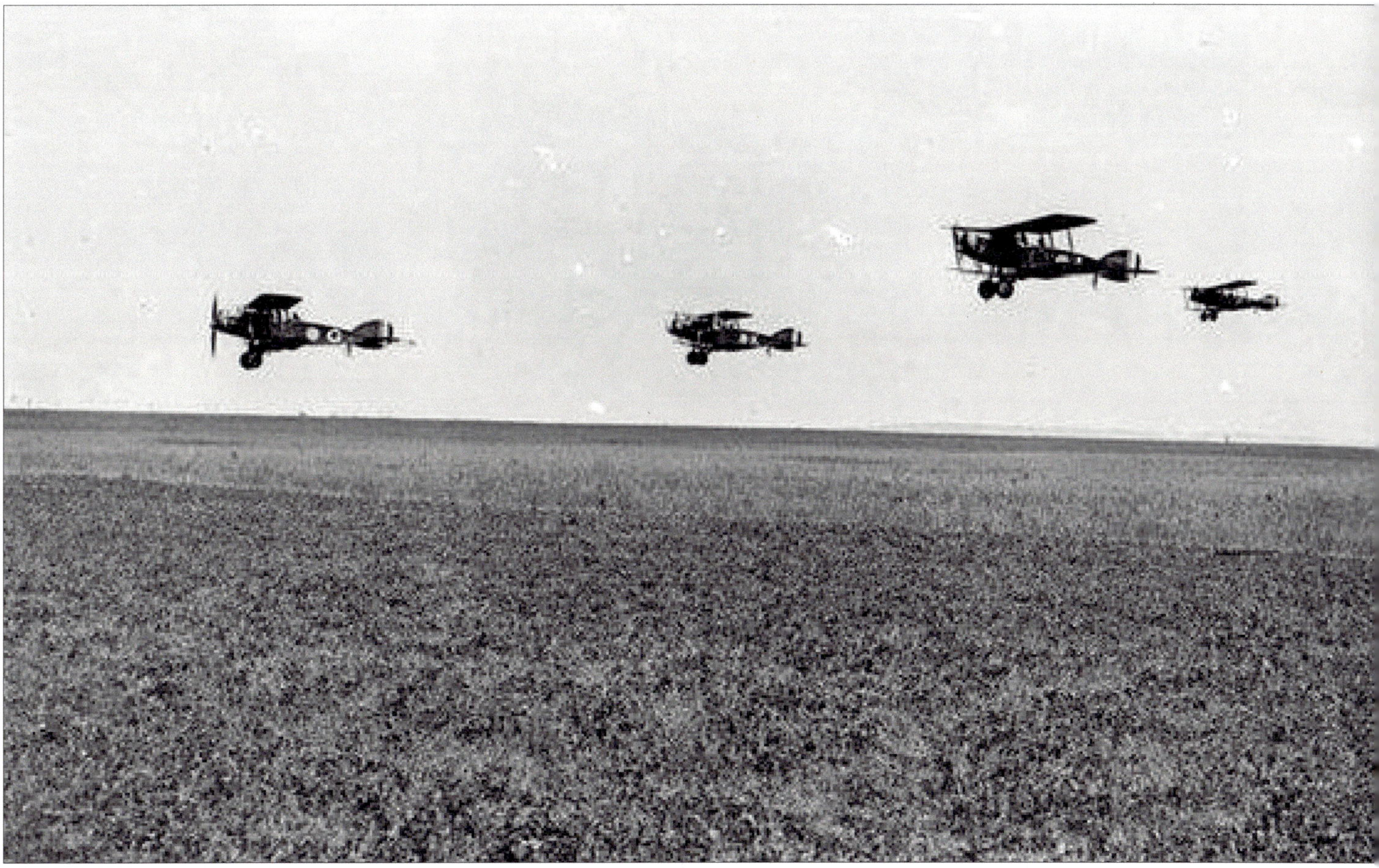

Above: A flight of Bristol Fighters takes off. Once flown in combat as if it was a single-seater fighter with a "sting" in its tail, it proved to be a very successful fighter aircraft.

Below: A modern Biff in New Zealand shows what a German fighter pilot might have seen during a combat in the skies over the Western Front.

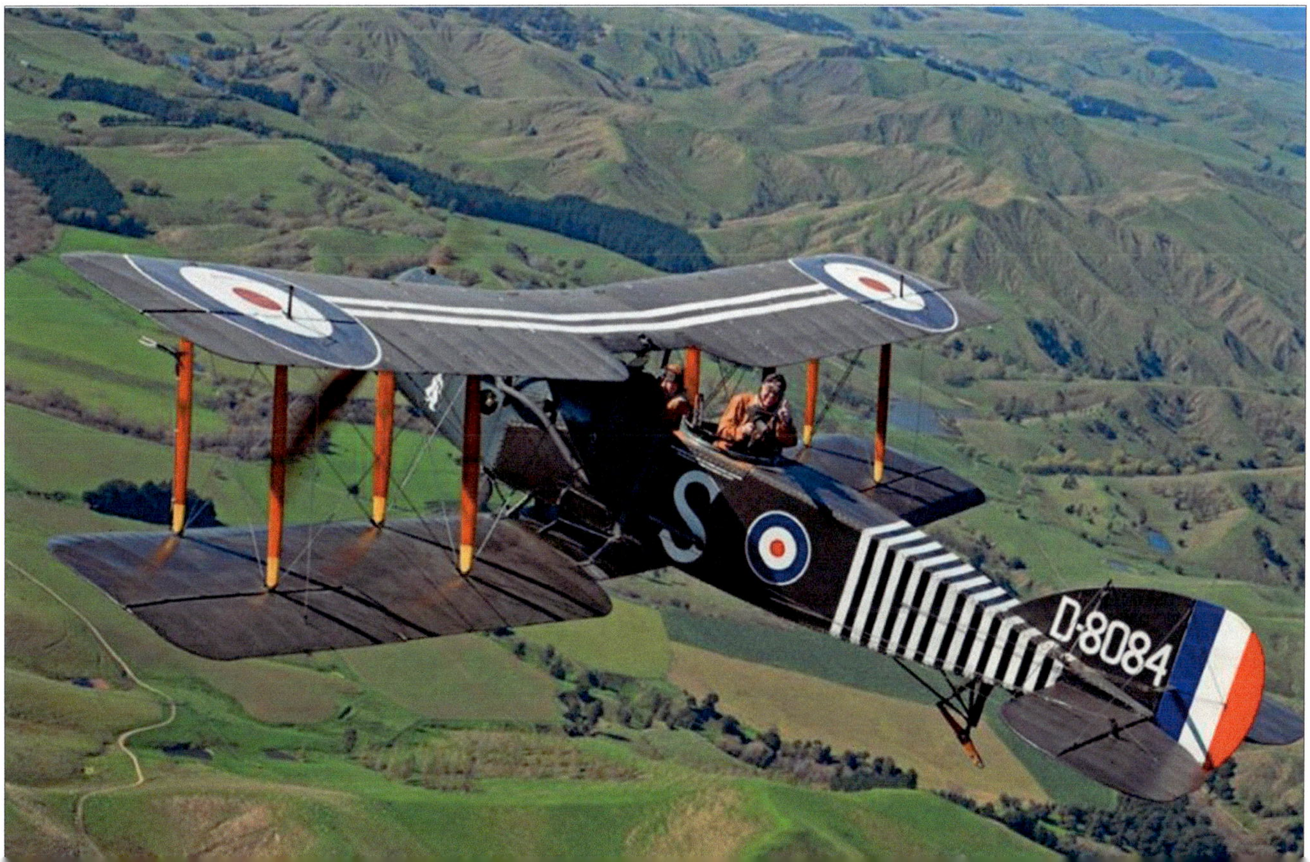

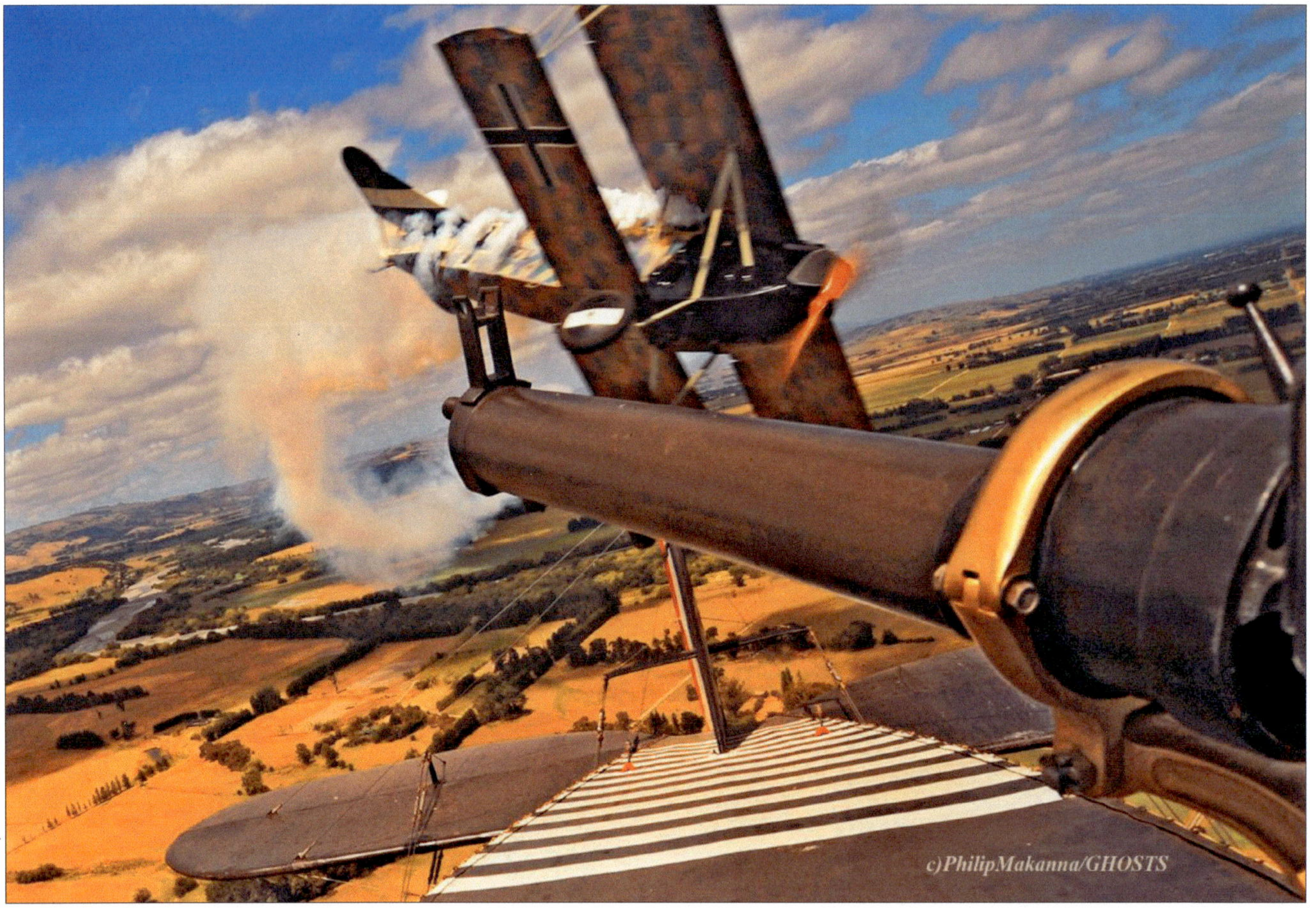

Above: A German Fokker D.VII finds out the dangers of ignoring the capability of the gunner of a Biff. Photo by Philip Makanna

especially effective was the tandem location which gave the two aviators close proximity to each other and allowed them to communicate much better during the flight and while in a dogfight.

For some reason, most Allied two-seater designs of the First World War placed the pilot and observer far apart and therefore severely hampered their ability to communicate with each other (later designs such as the Re8 were better in that regard).

The Royal Flying Corps was desperate for improved aircraft types in the Spring of 1917. As the British Spring Offensive was about to begin, an offensive which would become known as the Battle of Arras, there were high hopes as the Bristol F.2A began its operational service in March, 1917.

The first squadron to be equipped was no. 48 and the squadron was given one of the best known pilots in the RFC. Its lead flight commander was Capt. William Leefe Robinson, V.C. who was famous for destroying a German zeppelin. However, Robinson had never before served on the Western Front and therefore had very little experience in leading a flight into combat against front line German units. He also had very little experience in piloting the two-seater Bristol fighter before heading into battle in the aircraft.

Robinson led a patrol of six Bristol Fighters over the Western Front for the first time on 5th April 1917.His offensive patrol had the misfortune to be intercepted by Manfred von Richthofen leading a flight of his soon to be famous Jasta 11 comrades.

In the ensuing combat, the Bristol fighters of no.48 squadron entered a traditional two-seater formation of closing together and trying to protect each other via a concentrated fire from the six observers. The flight was devastated with four of the six aircraft shot down and one of the two that returned to base was severely damaged. Robinson was the pilot in one of the crews that had been shot down and he was captured. The Red Baron scored his 35th and 36th victories in the fight.

Hardly an auspicious start. Soon though, the Bristol Fighter crews began to fly their aircraft as if it were a single seat fighter and they began to achieve significant successes. Instead of approaching combat in a defensive attitude, they began to convert their Biffs into the successful raptor-like fighter that would lead to its impressive performance.

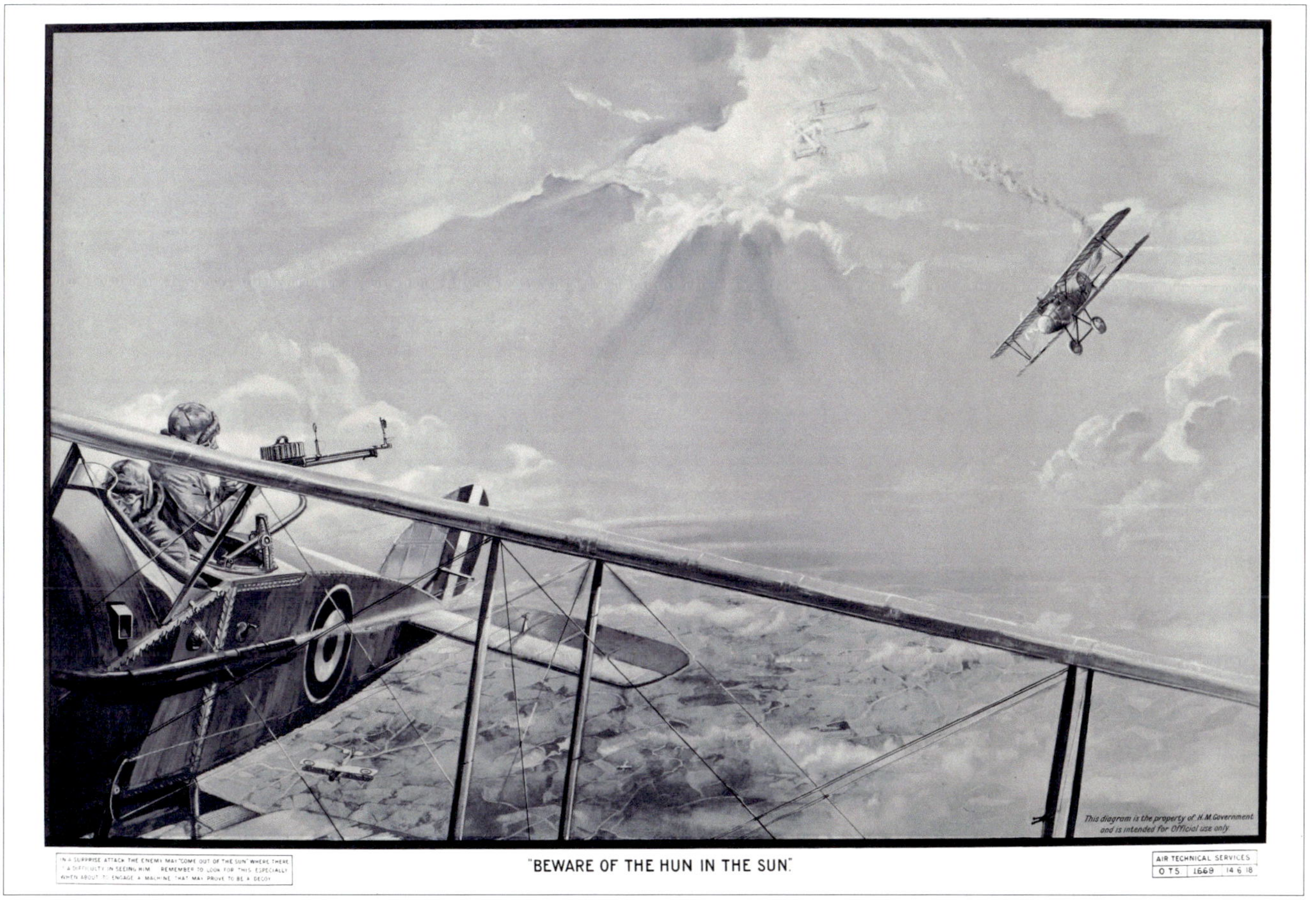

Above: A training poster shows the need for vigilance as the gunner of this Biff looks towards the Albatros which as a decoy will quickly dive away and allow the Fokker Triplane to drive home the attack from "out of the sun". Unknown

Below: An original color photograph depicts two Biffs in no.1 squadron AFC, with this new and capable aircraft, the Australian and British airmen would quickly gain air superiority over Palestine and the Middle East.

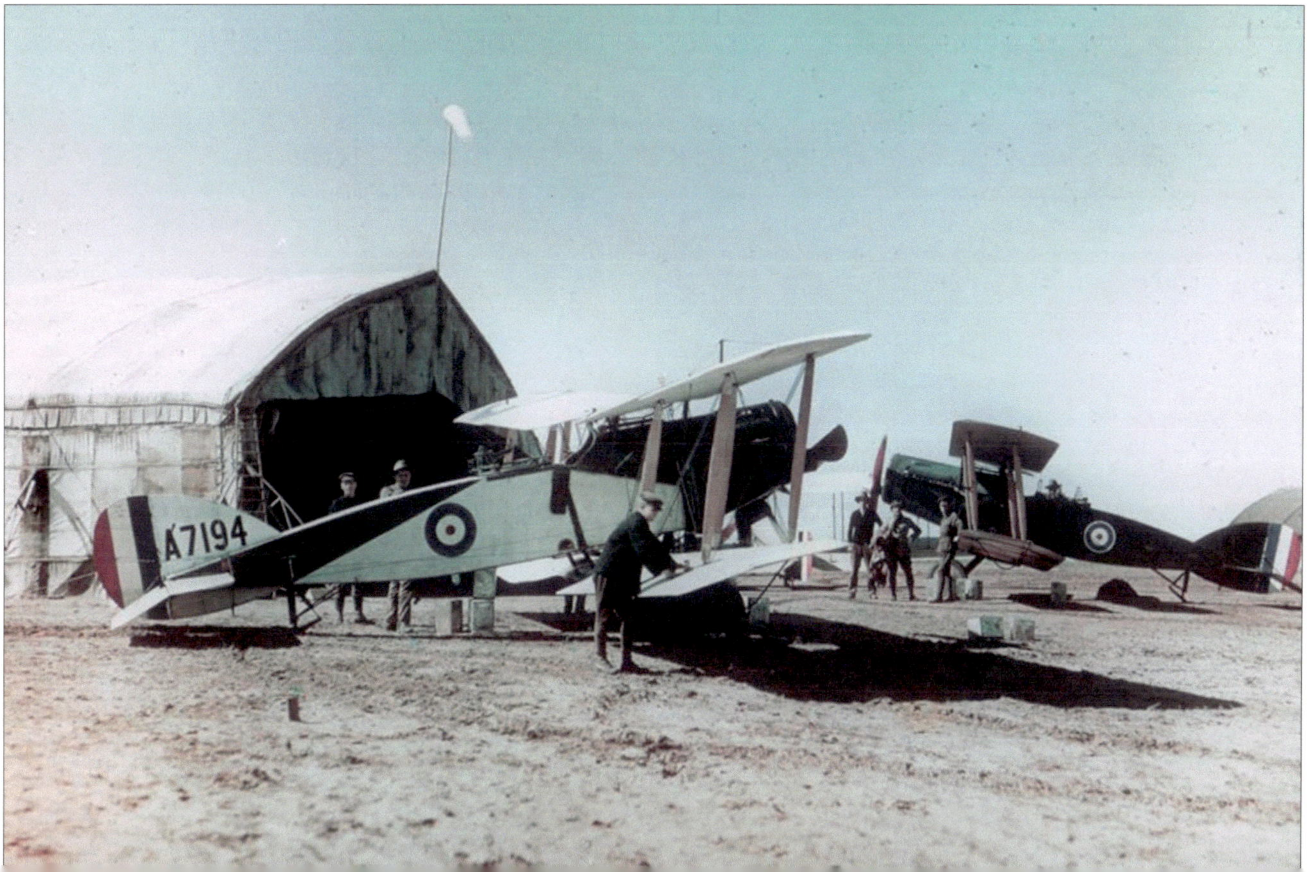

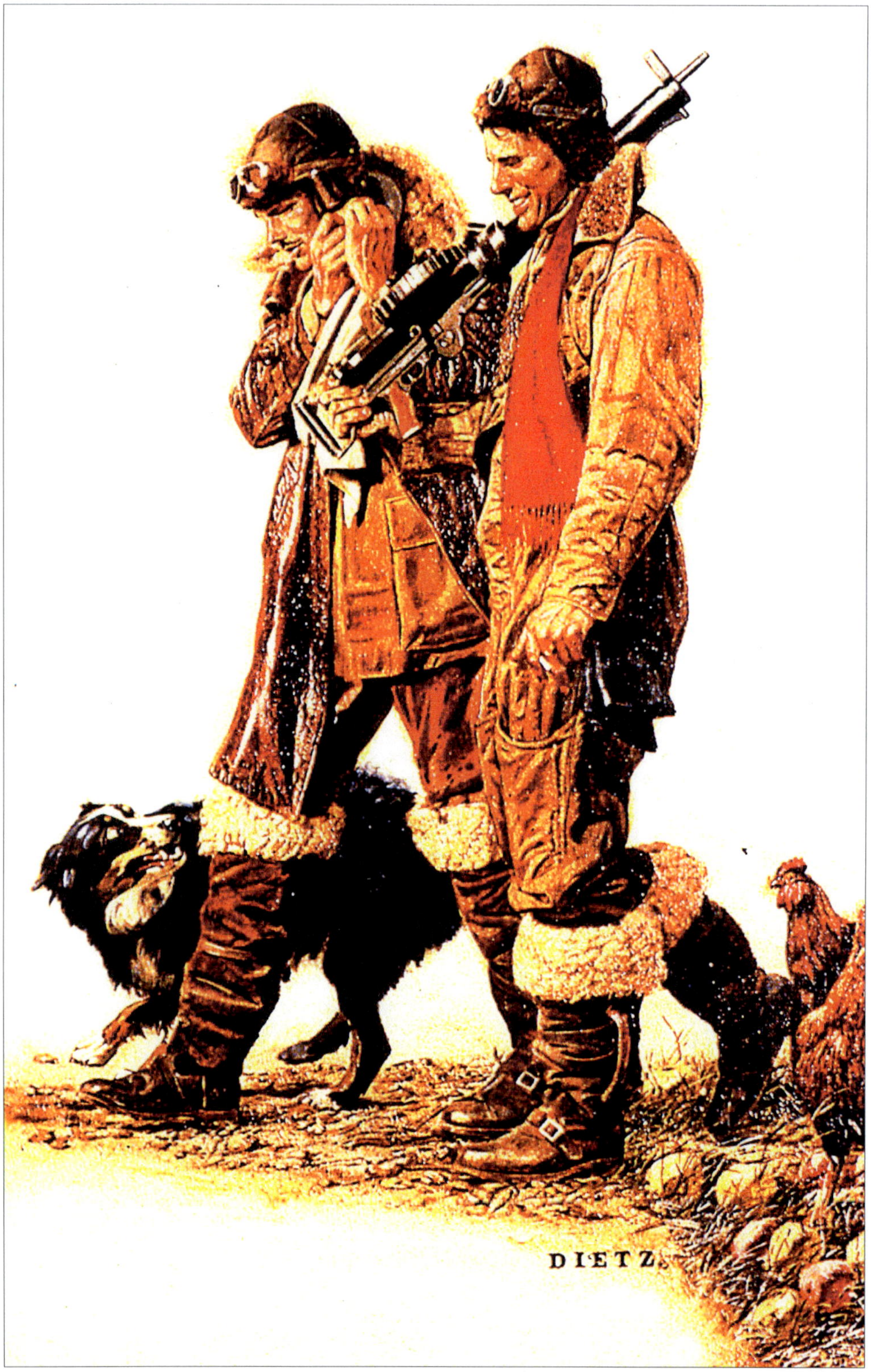

Left: The team of a Bristol Fighter make their way to their aircraft, during the air combat of WWI the observer-gunners of Biff fighters faced a disproportionate death and injury rate. Jim Dietz

Canadian pilot Capt. Alfred Atkey along with his British observer Lt. Charles Gass(No. 22 sqdn) scored 13 victories in the month of May 1917, while the entire squadron was given credit for 29 total victories.

Another Canadian pilot, Lt. Andrew McKeever along with his British observer Lt. Leslie Powell(No.11 sqdn.) achieved 31 victories in the six months between June and November of 1917.

The change in tactics led to the Bristol Fighter to being a tremendously successful fighter aircraft and these successes were quickly noticed by those in command. Eventually there would be 13 squadrons equipped with the Bristol F2B fighter on the Western

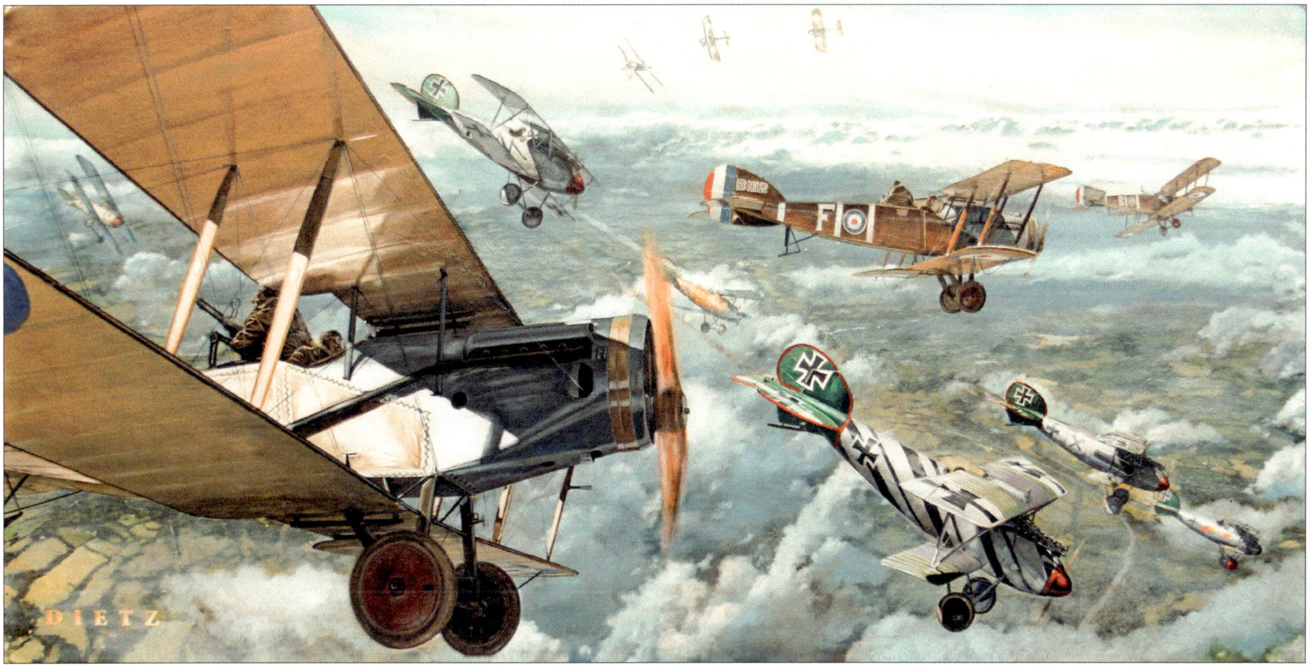

Above: The crews of the Bristol Fighters were fearsome fighters during the dogfights of WWI. Jim Dietz.

Below: Once the Bristol Fighter was introduced to the Middle East theatre, it became the dominant aircraft and lead to complete dominance in the air by the British forces there. Seweryn Fleischer

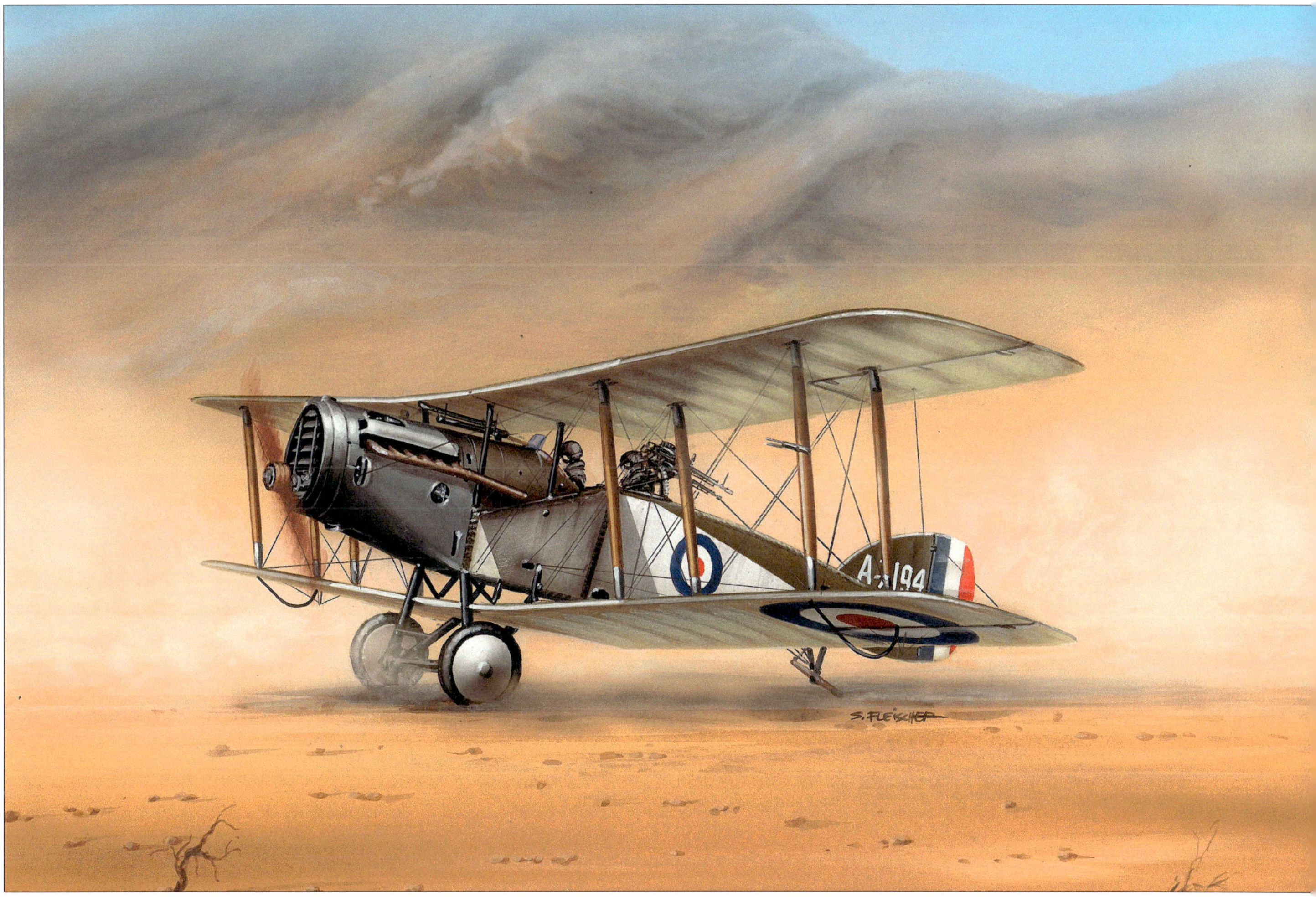

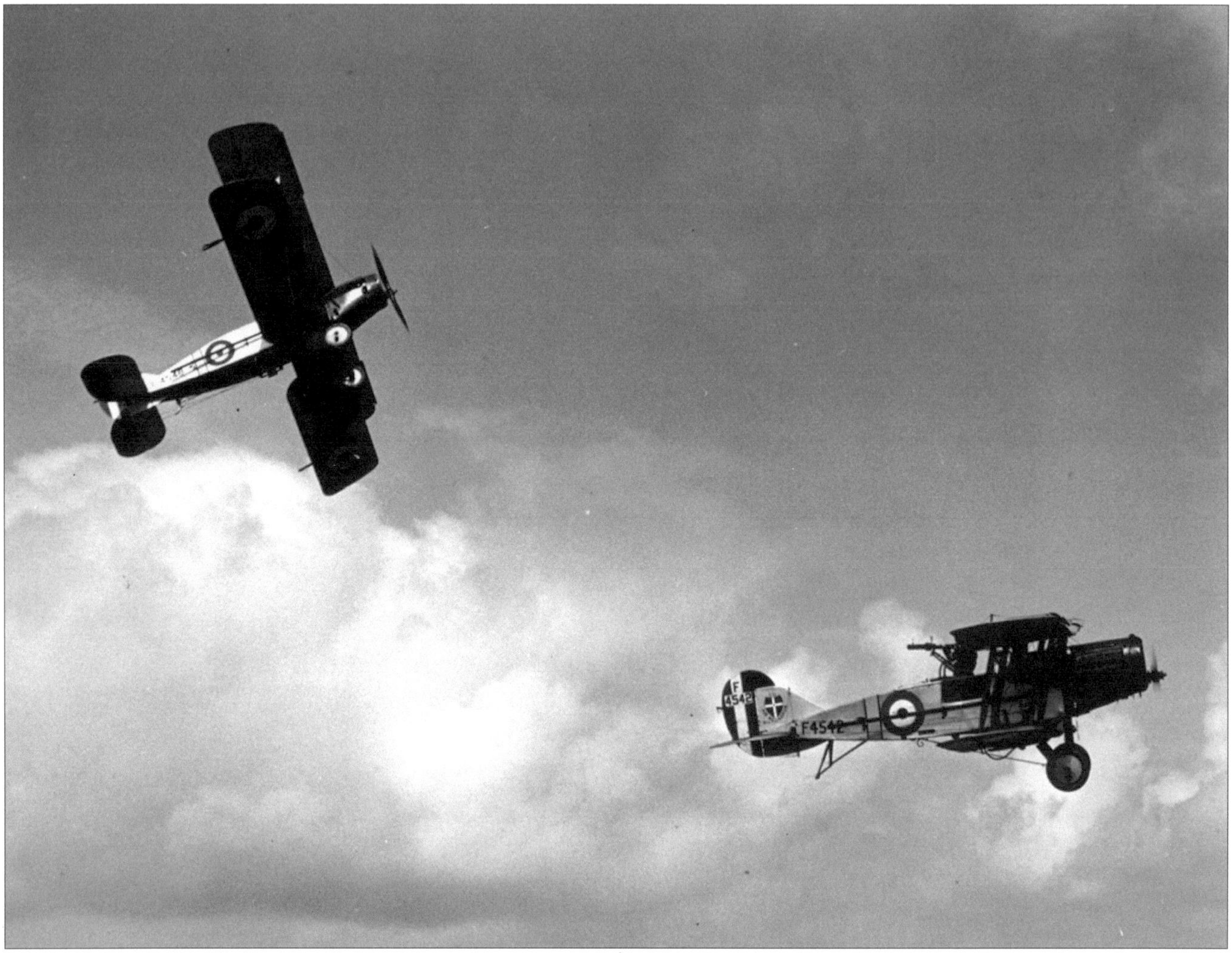

Above: The Bristol Fighter soldiered on after the war for another couple of decades. It acquired its other nickname, the Brisfit, after the war: apparently because of the several modifications and sometimes parts from one used for another type so it was a bit of a "misfit". Here are two Brisfits of a university squadron engaging in practice flight: thus the type contributed to the protection of England in WWII as well, by helping to train the fighter pilots of Fighter Command.

Front and the type saw service in Italy as well as a Home Defense night fighter.

One example of its potent lethality can be found in the combat report of Capt. H.P. Lale and 2/Lt. H.L. Edwards of No. 20 squadron:

" Capt. Lale put 50 rounds into it[Fokker DVII] and sent it down in flames. The observer was in the meantime firing with double Lewis guns at another E.A. about 40 yards away on the left. This E.A. first spun, then appeared to gain control, but a few seconds later burst into flames."

Two German Fokker DVII fighters shot down simultaneously, no other Allied fighter could do that! The Bristol Fighter could thus strike forward and rearward at the same instant. This feature allowed the pilot to fly as if he was a fighter pilot and maneuver in the dogfight like a single seat fighter that had the advantage of a potential lethal sting in its tail as well. However, this advantage came with a price.

In his book *The Clouds Remember*, Leonard Bridgman wrote the following about the F2B:

"Many stories are told round the Bristol FIghter, but the thing which ought to be remembered about them above all else was the heroism of the observers. Machines came back from patrol again and again with their observers dead in the rear cockpit, and the casualties among observers were exceptionally high. The reason is to be found in the tactics used. As has been explained, the pilots fought their machines as single-seaters and left the rear defense to the observer. Acts of prodigious courage were entailed in this defence work, for the observer, with no engine to protect him, had often to stand up in his narrow cockpit and face the fire of diving fighters and, at the same time, to attempt to bring effective fire to bear upon them. At such moments the observer's twin Lewis guns must have seemed a poor counter to the fixed guns of the German machines. Yet, in spite of the difficulties and dangers of the observer's task, in spite of the heavy casualties, there was never a lack of volunteers for work with the Bristol fighters."

In his outstanding book, Aces Falling, Peter Hart quotes two separate aviators who describe actual events to support Bridgman's statements:

" On the way back with the sun more or less behind us I heard a machine gun firing followed by an agonized yell. My observer was a fairly new boy and had not yet learned to look for hostile aircraft diving from the sun. I also had not been warned how difficult it was to spot aircraft diving from the sun. I turned my head to see how Lamb was. He was badly wounded and was obviously in great pain." 2/Lt. F. Ransley, no.48 sqdn

"Engaged eight Pfalz scouts-I got one in flames who was on a Bristol's tail and did not see me coming...Then, when the formation was a bit scattered, ten Fokker Biplanes came out of the sun astern...Jones tapped me on the top of my head, our "extreme emergency" signal and I kicked full left rudder, stick hard forward to the left and ducked.. I felt a blow on my left elbow as if I had been hit by a hammer, the aileron controls went out of action... and a bullet went through the windscreen 6 inches in front of where my nose had been a moment before...Having found the aeroplane would fly, [I] pointed her for the lines. I looked round to see Jones sitting quietly on his little seat, his head resting on the butt of his gun. When I put my hand on his shoulder I knew he was dead." Lt. T. Traill, no. 20 sqdn.

It seems that not only did the observer/ gunners have a high casualty rate - as they certainly paid that higher price, it appears their courage helped to lower the casualty rate amongst the pilots of the Bristol fighter.

The Bristol fighter would be in production from 1917 until 1926. A total of over 4,400 aircraft were produced (3,100 during the First World War). As previously mentioned, the F2B and variants would serve on the front lines of the British Empire until 1928. It was a robust, dependable and adaptable aircraft capable of performing the many different missions that faced the RAF in the Middle East and the Northwest frontier of India for example.

Several Mk. IV variants would continue to fly into 1931 as trainers with the Oxford and Cambridge University Training Squadrons, likely to have trained some of the pilots that would serve in fighter command during the Battle of Britain.

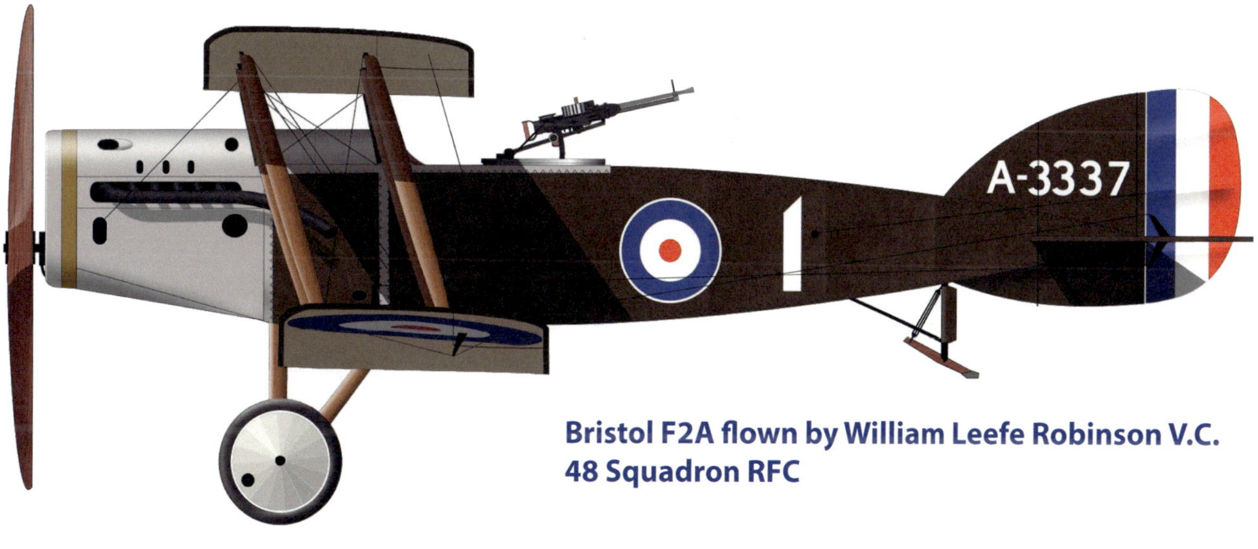

Bristol F2A flown by William Leefe Robinson V.C.
48 Squadron RFC

Fabulous Fokker

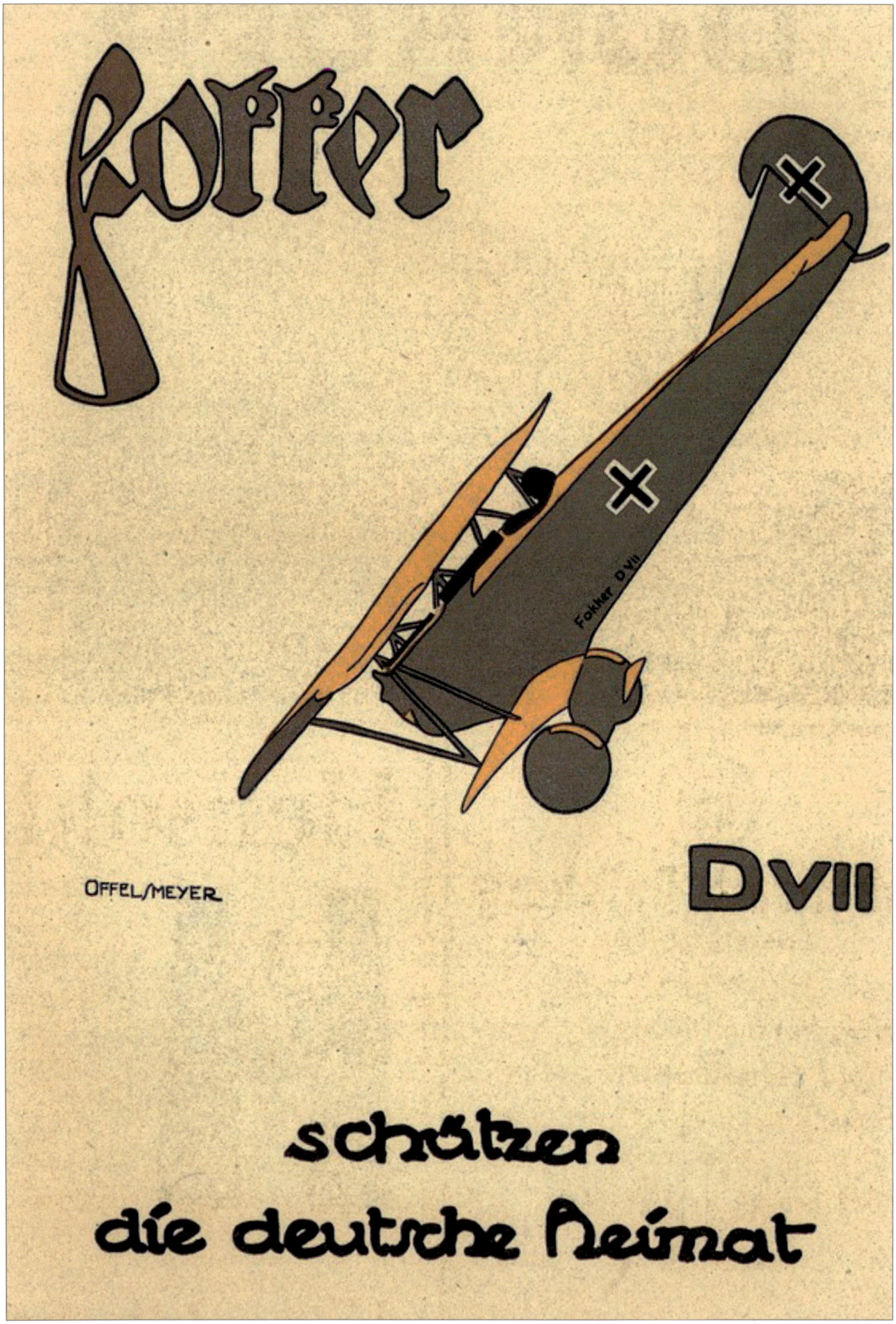

Left: A clear winner of the First Fighter Competition of 1918 and requested by some of the most experienced fighter pilots Germany had, the Fokker D. VII was definitely viewed as defending the German homeland.

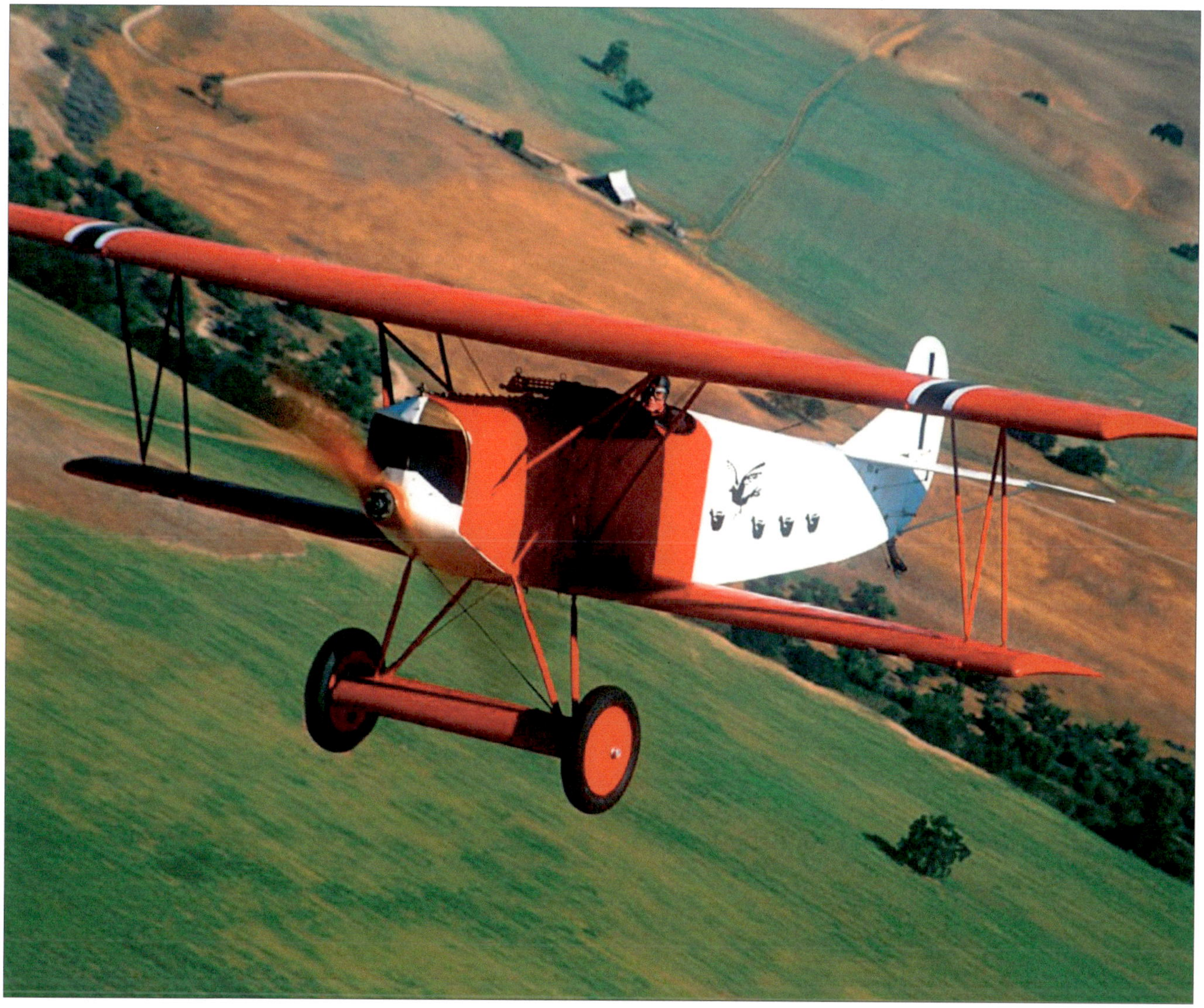

Above: This photo of a replica Fokker D.VII in flight demonstrates the clean lines, note no rigging to produce enormous drag in flight and the type's thick, effective cantilever upper wing. Philip Makanna

"The Fokker was a thoroughbred that answered to the slightest movement of the hand and could almost guess the rider's will in advance." Rudolph Stark

"Powerful, maneuverable, rugged, easy to fly, it was a fighter pilot's dream." Jeff Ethell

There were 31 aircraft from several aircraft manufacturers poised to be tested in January of 1918 at the airfield in Johannisthal. The goal of these flight trials was to find the next front-line fighter aircraft for the German Air Service. Naturally the award of this production contract would mean a large profit to whichever aircraft producer was deemed the winner.

This would be the first large scale evaluation to include some of the most successful fighter pilots that Germany had. Naturally, these pilots would be very eager to participate since they knew that within a few months, they would be at the controls of the chosen fighter in combat against the latest Allied designs.

Indeed, these kanone's opinions would be carefully weighed and a great deal of attention paid to what the pilots wanted by the German staff responsible for making the final choice of the next generation of German fighter aircraft.

Standing at the head of this line of distinguished fighter pilots was Rittmeister Manfred von Richthofen himself. Historians have credited the Red Baron with evaluating the initial offering from Tony Fokker, the Fokker V11 and making several negative observations regarding its handling qualities that led to Fokker modifying the design (adding a fixed fin to the tail of the aircraft and lengthening the aircraft

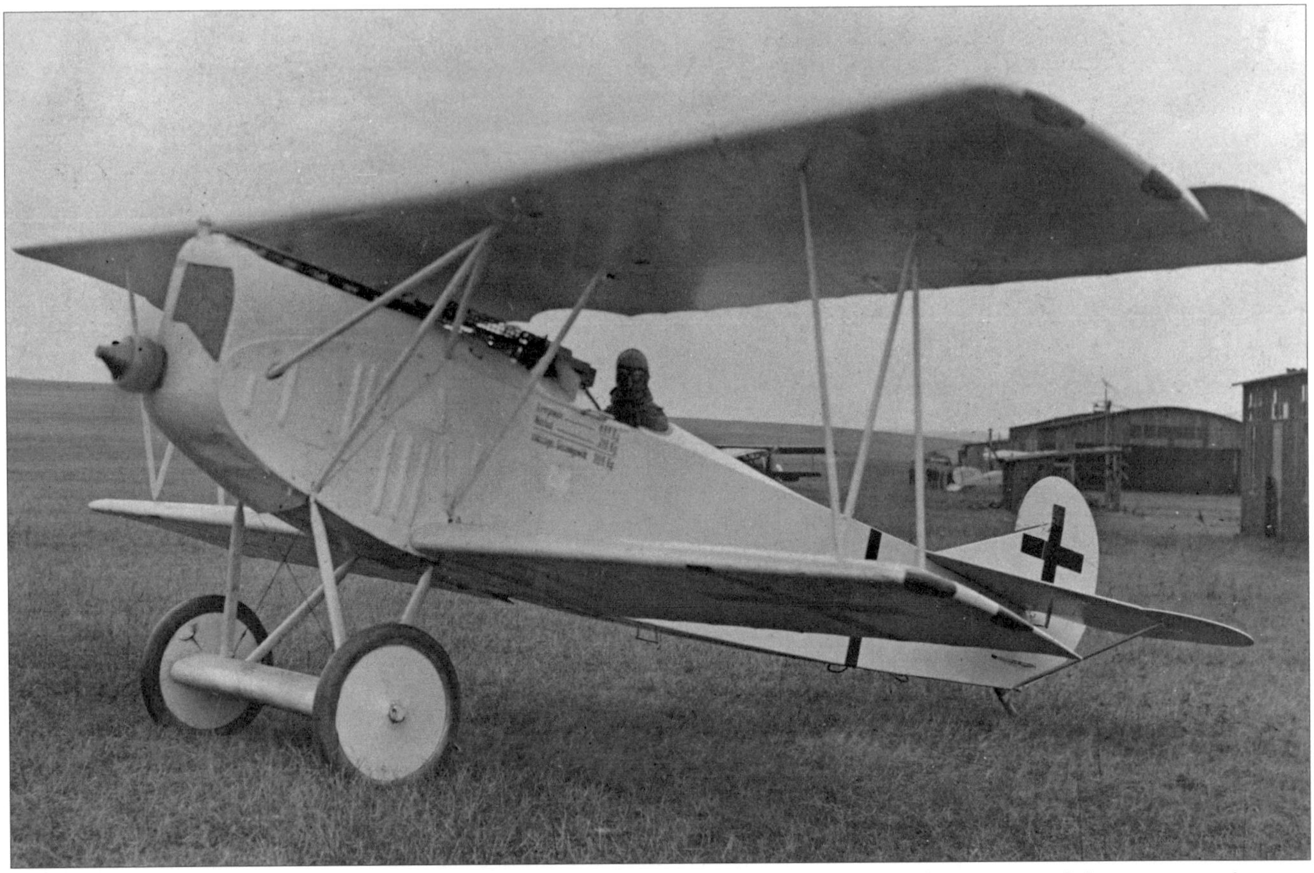

Above: Ever the shrewd businessman, Anthony Fokker realized that courting of the leading German fighter aces was key to success, since their evaluations of aircraft could lead to large production orders from the German military. In 1918, he sent an all-white Fokker D.VIIF to the commander of JG-1, the Richthofen Geschwader.

Below: Here is a photo of that commander, Herman Goring standing beside his especially decorated Fokker D.VIIF fighter. In his had he holds the Geschwaderstuck, indicative of his being the C.O.

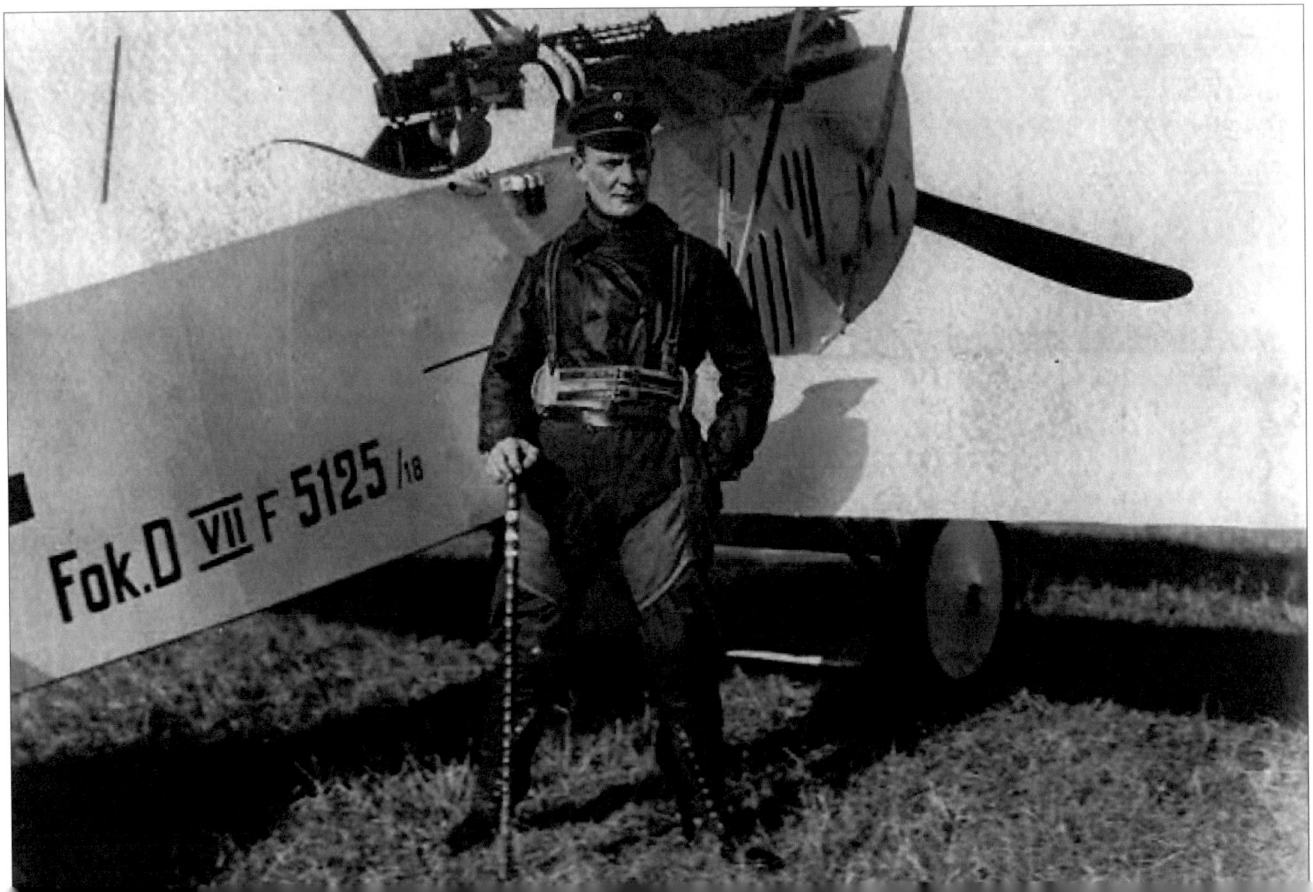

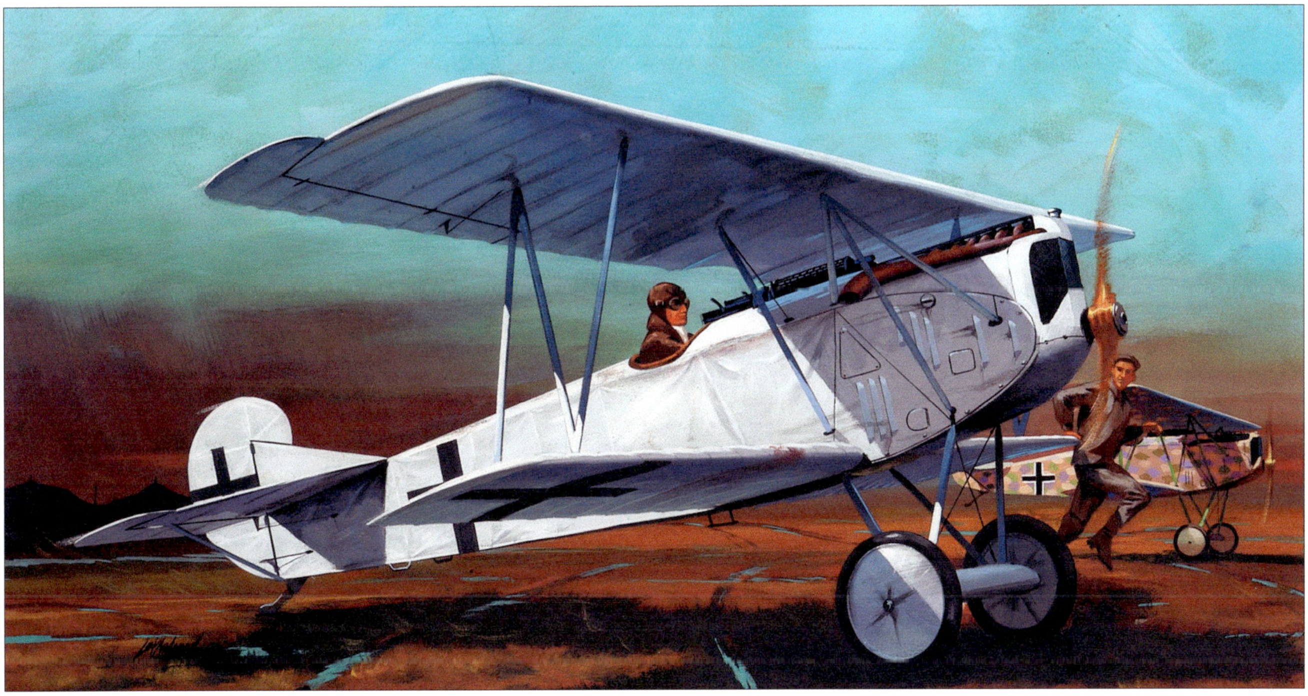

Above: The title of this piece by Jack Leynwood is "Black knave in a white Fokker", the image is of Goring's special Fokker D.VIIF. Jack Leynwood

Below: Although the Fokker was said to make an average pilot a good pilot and a good pilot an ace, the odds of survival towards the end of the war for the more inexperienced German pilots became much worse. J.B. Deneen.

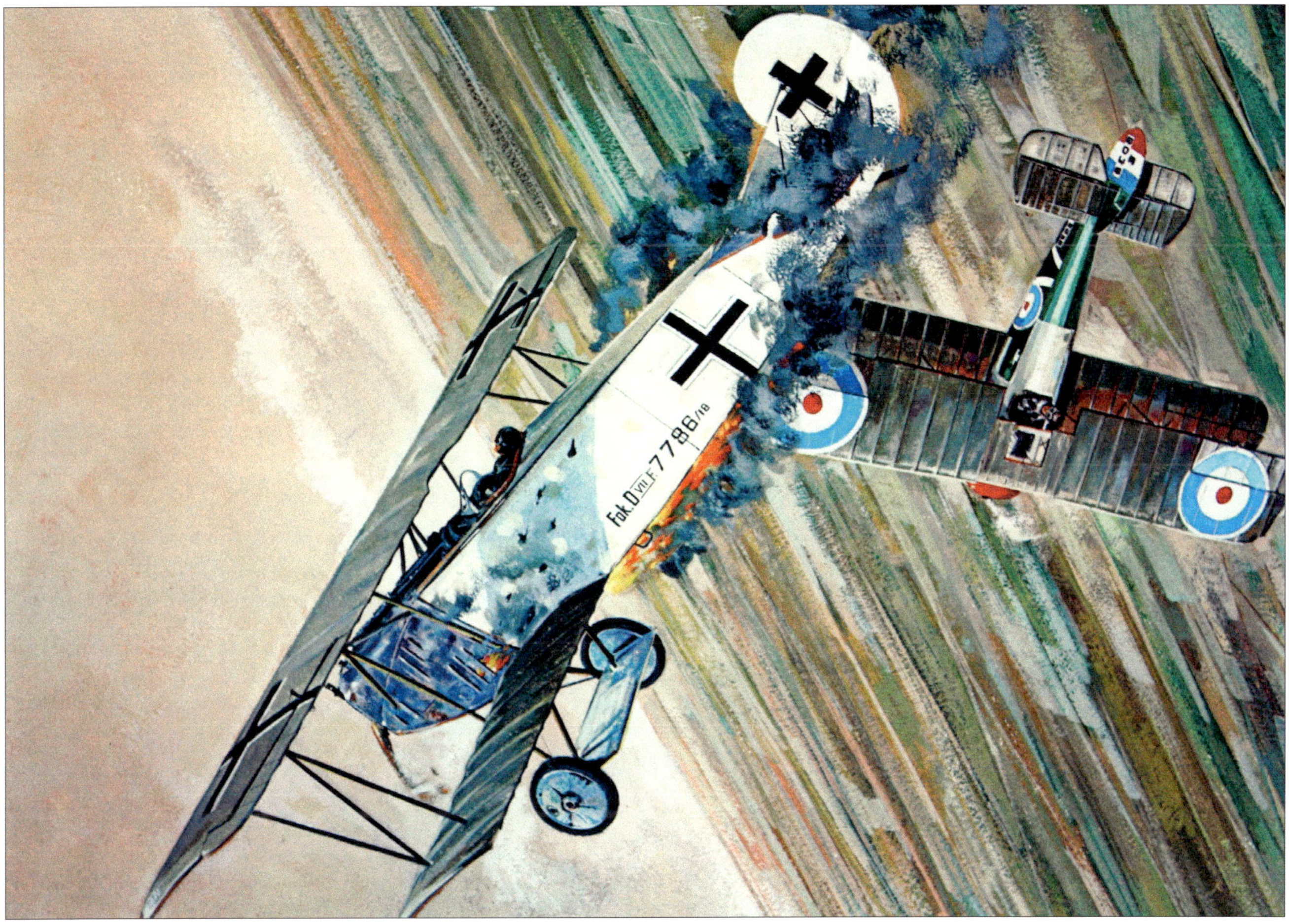

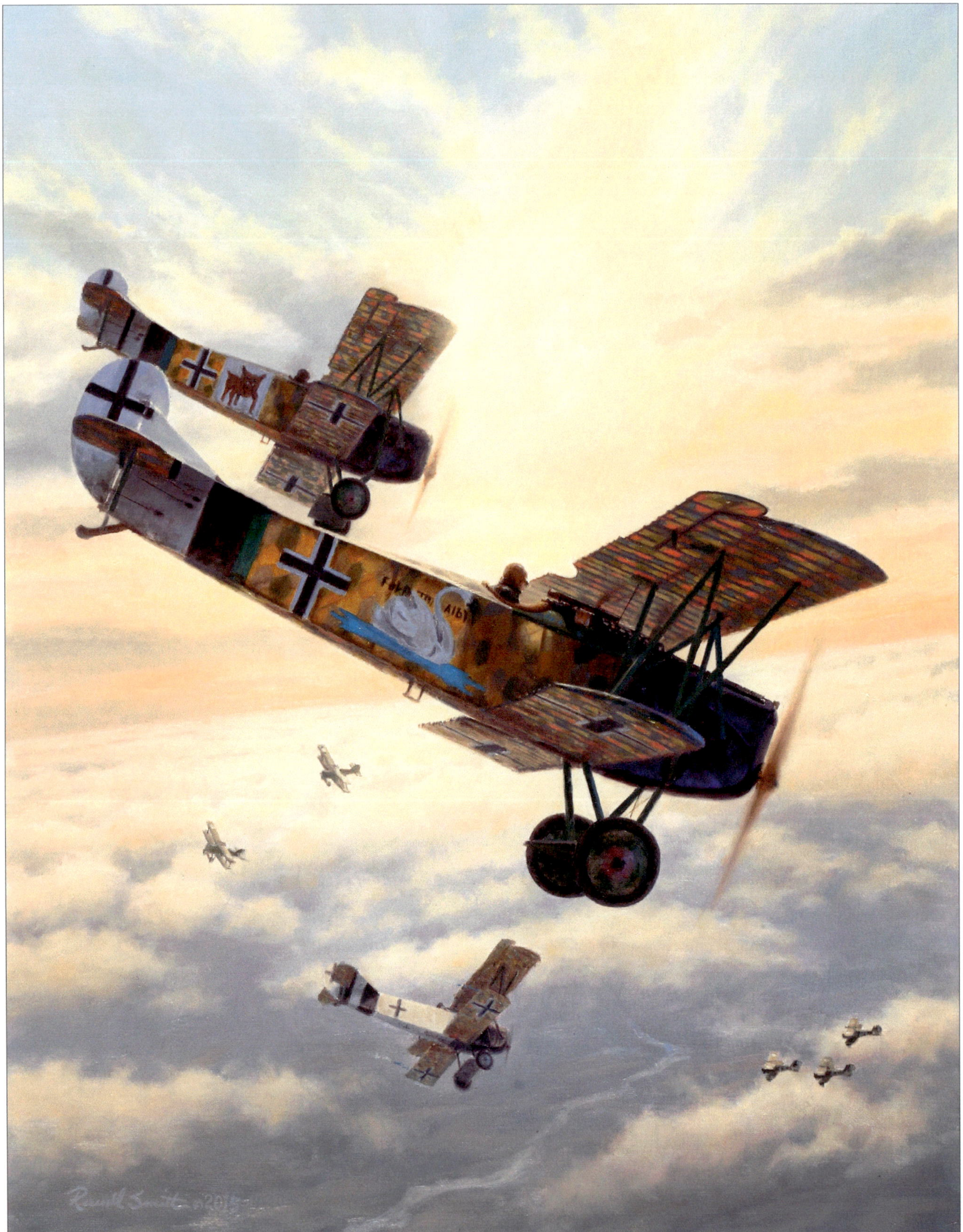

Above: The superior capabilities of the Fokker D. VII allowed the German fighter units to fight until the very end of the war, despite facing ever increasing numbers of Allied aircraft. Here two Jagdfliegers of Jasta23b dive to the attack. Russell Smith

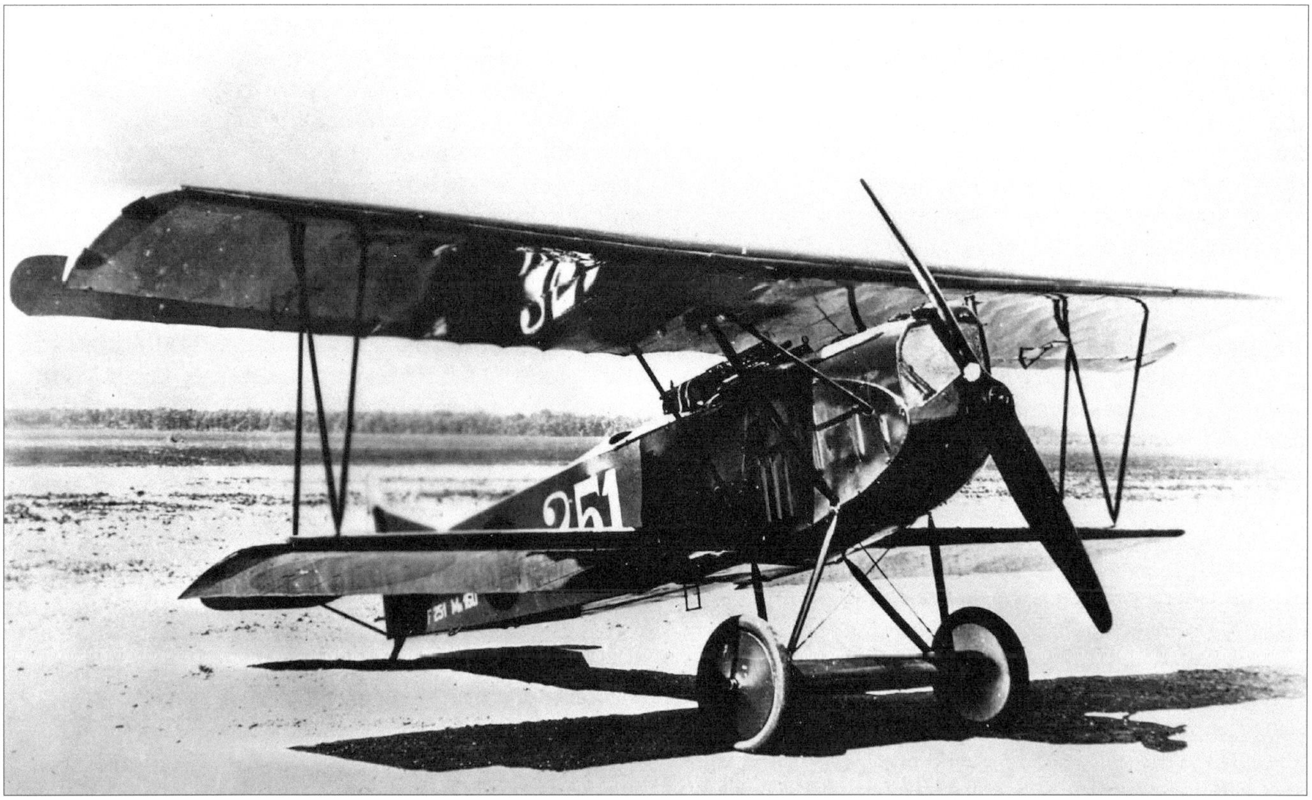

Above: The Fokker D.VII design would continue to be in the front line units of many nations for years after the end of the war. Here is an example of a Dutch Fokker D.VII.

to improve stability) which would ultimately end in the Fokker D.VII.

Fokker modified the aircraft right there at Johannisthal during the trials. From that point forward, the aircraft's performance "spoke" for itself. It would be the unanimous choice by all the gathered fighter pilots at the trials. From then on, von RIchthofen frequently wrote inquiries to the German staff about when he should expect the arrival of the Fokker D.VII fighters for his pilots.

Unfortunately for him, he would be shot down and killed in his Fokker Triplane before having a chance to fly the D.VII in combat. However, it would not be wrong to state that von Richthofen had used his compelling influence to posthumously provide his comrades- his jagdflieger- with a superior weapon with which they could once again fight to control the skies over the Western Front.

"Fokker is still the same old Fokker, for every new machine he brings out is a considerable improvement on the last...we practice our new birds instead. The more we get to know them, the better we are pleased with them. Whatever one tries- steeply banked turns, dives, loops- they always answer the controls beautifully ;one understands the machine at once. They give one a magnificent feeling of safety." Rudolf Stark

Anthony Fokker(see Flying Dutchman) had an understanding of the need to change and experiment in the new field of aeronautical engineering. Many of the factors that affect drag, lift, power to weight ratios and other fundamental principles of aircraft design were just beginning to be understood. Fokker was always willing to try something new, something different: even it was the opposite of what everyone else thought was best.

For example, while most fighter aircraft of World War I featured a wooden airframe, Fokker used welded steel-tubes for construction. This method allowed for a faster production schedule and a lighter, but stronger airframe as well.

Fokker also did not mind incorporating another person's ideas into his own aircraft. Therefore when Fokker had the chance to study the groundbreaking cantilever wings designed by Hugo Junkers, he modified the concept into using thin sheets of plywood instead of the costly, complex corrugated metal of Junker's designs.

Most airplanes at this time used very thin airfoils, which necessitated the addition of a box-like construction of struts and cables for bracing in order to give the wings strength and stability. Many aircraft designers thought a "thick wing" would slow the aircraft down because of producing drag.

By using a thicker, cantilevered wing with internal box-spar construction, internally braced for a very rugged strength. This design also eliminated the need for the multiple struts and numerous cables which were actually the major cause of drag-induced slowing of an aircraft's potential speed.

Having tested his new wing on several experimental aircraft, Fokker knew he had a winning design in the V11. The aircraft was coupled with the reliable and well-loved inline six-cylinder Mercedes engine of 160 hp. The resultant Fokker D.VII design proved to have a remarkably docile stall, and was very reluctant to enter into a spin. It offered a high degree of maneuverability, and the pilot could dive the D.VII without risk of structural failure (a flaw in the Albatros fighters that was well-remembered by the German pilots). These features led to the observation that the Fokker D. VII could make a mediocre pilot into a good one, and a good pilot into an ace.

The German Air Service was desperate to replace its aging Albatros and Pfalz fighters. A production contract was given to Fokker, but the number of aircraft required forced Fokker's rival, Albatros, to immediately begin license production of Fokker's newest design. For every aircraft produced by Albatros, it had to pay Fokker a royalty fee that undoubtedly pleased the Dutchman very much.

In fact, of the estimated 3,300 Fokker D.VIIs produced before the end of the war, Albatros would produce more than the Fokker Works itself. At its peak production rate, the Fokker factory was producing eight machines a day.

The Fokker D.VII entered frontline service in April 1918 with the first examples going to Jasta 10, of the Richthofen Geschwader. By May 1918 the D.VII was able to equip the entire jagdstaffel. As was sadly so common with a Fokker aircraft design, there were some initial quality issues but these were quickly remedied.

An excellent fighter was soon made even better when the first product of the new Bavarian Motor Werke was incorporated as a replacement for the Mercedes engine. The new variation was designated the Fokker D.VII F (for the engine's designer Max Friz). The over-compressed 185hp BMW IIIa engine gave the D.VII increased speed, climb and ceiling.

"Lately the Geschwader has succeeded in regaining complete domination of the sky. The enemy is much more on the defensive...The superiority of the Fokker D.VII with the BMW IIIa motor has increased quickly." Herman Goring

Its advanced design and the successes achieved by the pilots who fought in the D.VIIF made it the aircraft selected by most aviation historians as the outstanding fighter aircraft of the First World War. Such was its dominance that the Fokker D.VII was the only aircraft type that was specifically mentioned in the Armistice agreement: all examples of the aircraft were to be immediately turned over to the victorious Allies. In fact, the Fokker D.VII would ultimately fly in the Belgian, Dutch, American, Swiss, Canadian and Polish air forces after the end of the war.

As late as 1929, the Swiss Air Force was still ordering new Fokker D.VII aircraft.

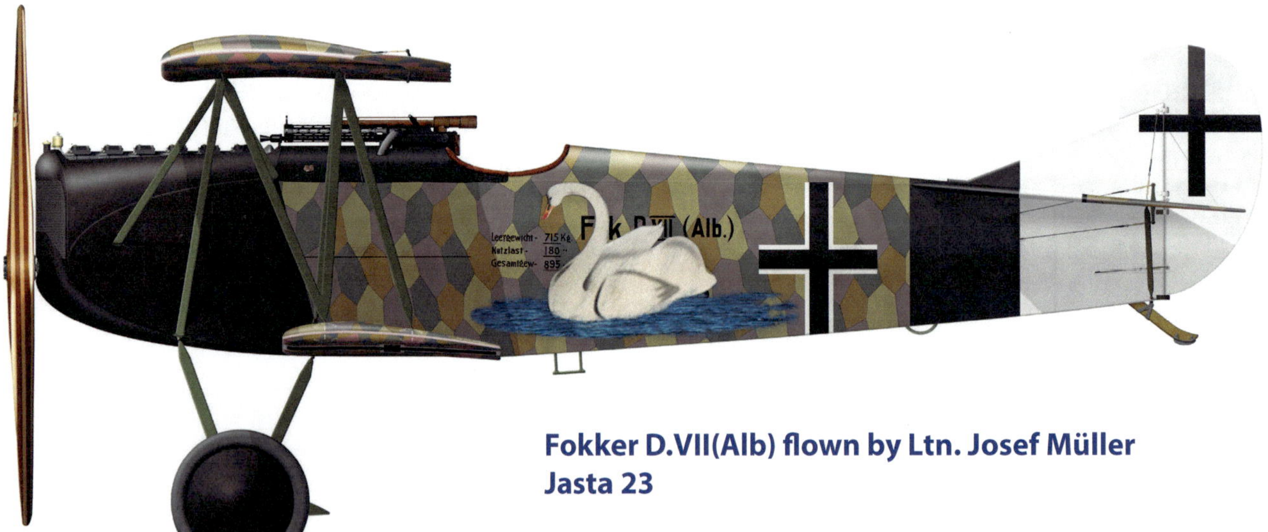

Fokker D.VII(Alb) flown by Ltn. Josef Müller
Jasta 23

Aviator Artists

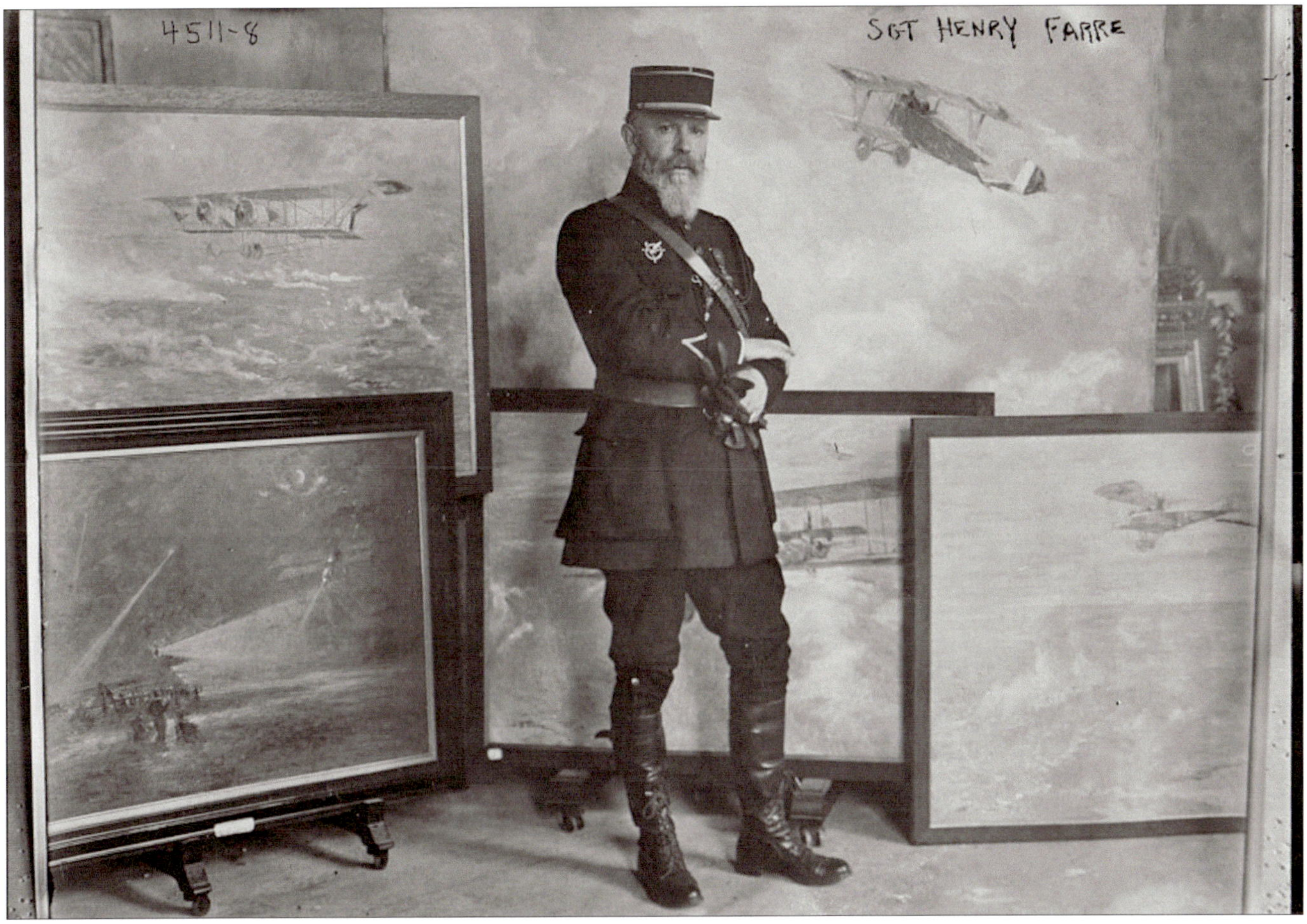

Above: Henry Farre poses during the exhibition of his artwork in Chicago. The show's success led to a nationwide tour to help raise money for widows and orphans of Air Service members.

The emergence of powered and controlled flight at the beginning of the 20th-century quickly captured the imagination of many people, including artists. In Robert Wohl's wonderful book, A Passion for Wings, the concept of man entering into the skies above quickly opened the realm of new possibilities in art, ideas that were also beginning to be explored and gain acceptance in the early years of the new century.

The pilots who would become heroes for their exploits were quickly seen as special: gallant and brave. The romance of flight now began to border upon the almost mythical, as the earth bound tried to imagine what those few individuals who actually cavorted amongst the clouds saw and felt.

With the onset of the First World War, although some of the art continued with the latest trends, such as the futurists, much of the art was now to be considered as necessary to promote nationalist and loyal sentiments within the populations of the belligerent nations. Therefore, the art had to be more based on classical and realistic styles of painting.

A near perfect example of this idea can be found in the most well-known aviator artist of the war, the Frenchman Henry Farre. He was born in Foix, France in 1871. After attending the prestigious Ecole des Beaux-Arts, Farre moved to Buenos Aires where he established a very successful career as a portrait artist.

However, with the onset of the war Farre immediately returned to serve his native land. Over 40 years of age, but fervently desirous of being in a front line assignment, he became an observer/bombardier and would fly in combat missions throughout the war. Farre was quoted in an article in the New York Times in March, 1918:

"Sometimes we hover over the battle as long as half an hour. Shells burst around us. Other airmen plunge to the ground. But we escape. Then my pilot whirls around and we fly back to the rear. We

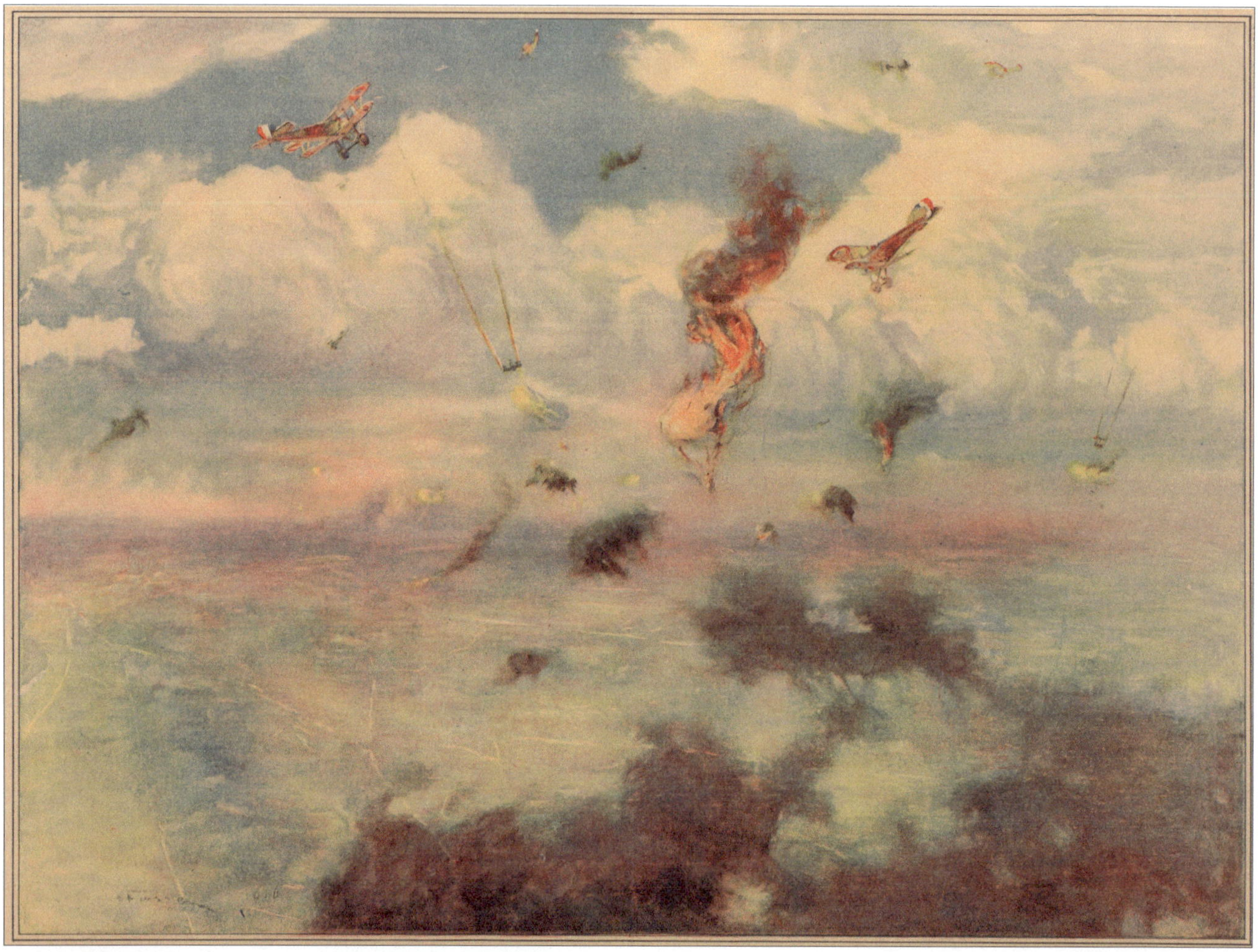

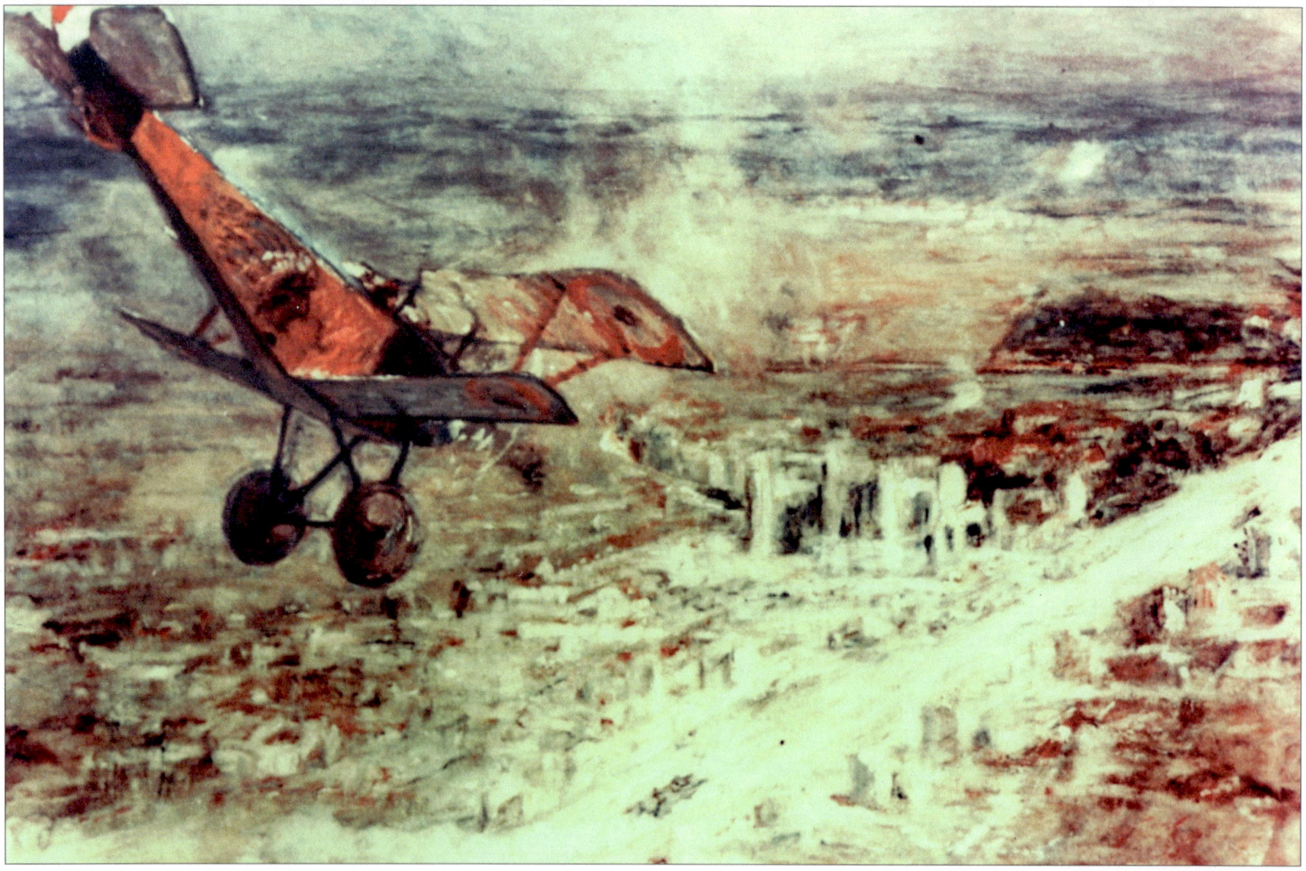

Right: Photo of Farre from a contemporary newspaper after he had received the Croix de Guerre.

Facing Page, Above: On the 22nd May 1916, French Nieuport fighters utilized the Le Prieur "aerial torpedo" system for the first time. Six of eight targeted German observation balloons were successfully destroyed H. Farre

Facing Page, Below: A Nieuport of N.124, the famed Lafayette Escadrille dives to strafe enemy positions in a town H. Farre.

Below: A Voison, similar to the one in the prior photo, returns from a mission while being shot at by anti-aircraft guns. H. Farre

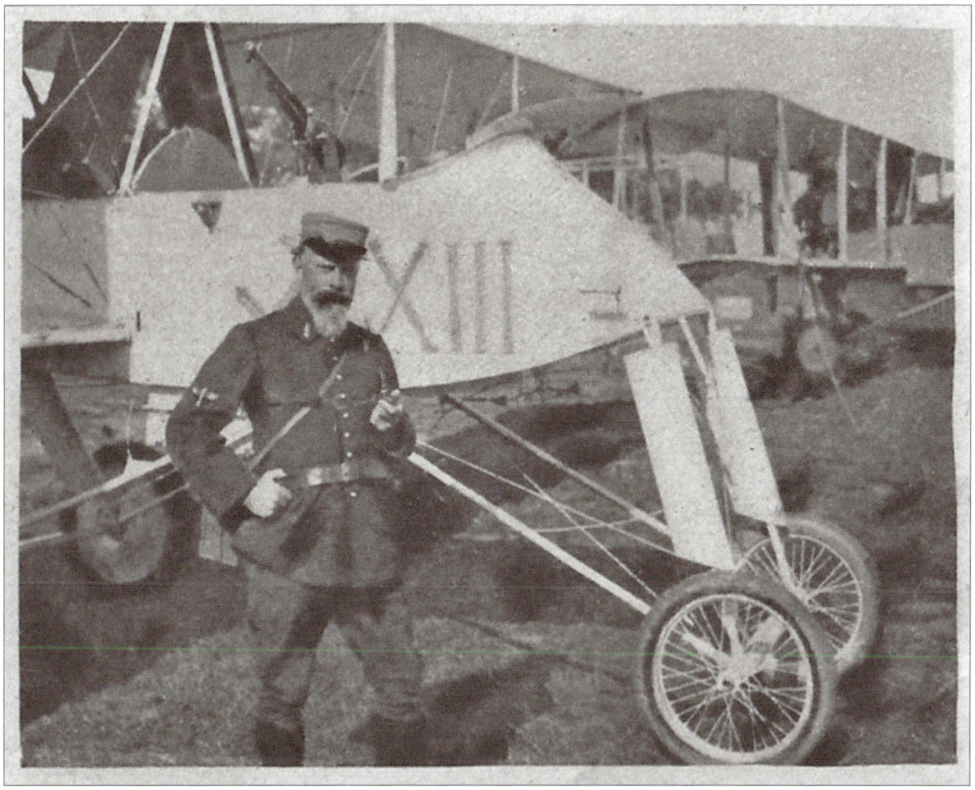

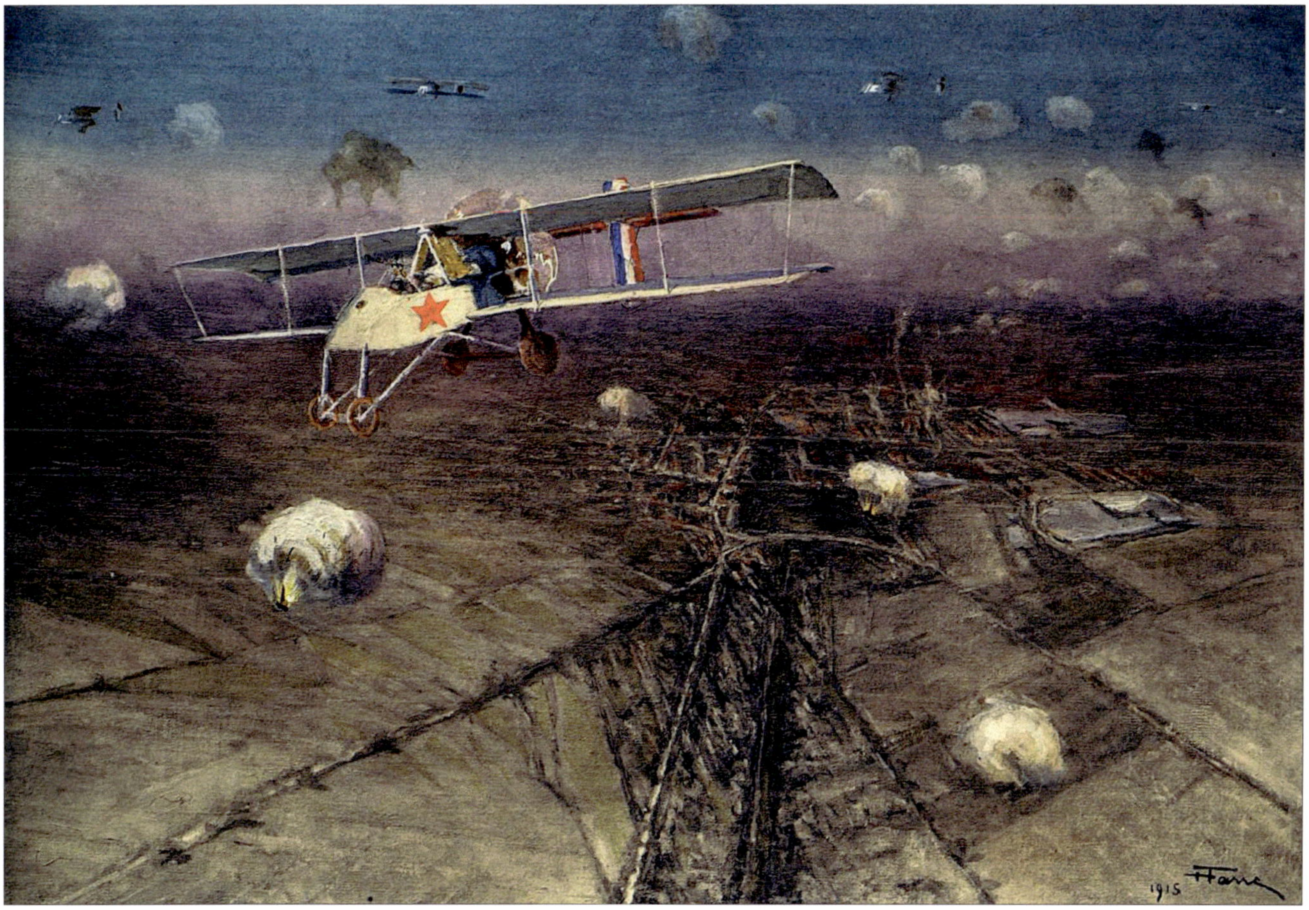

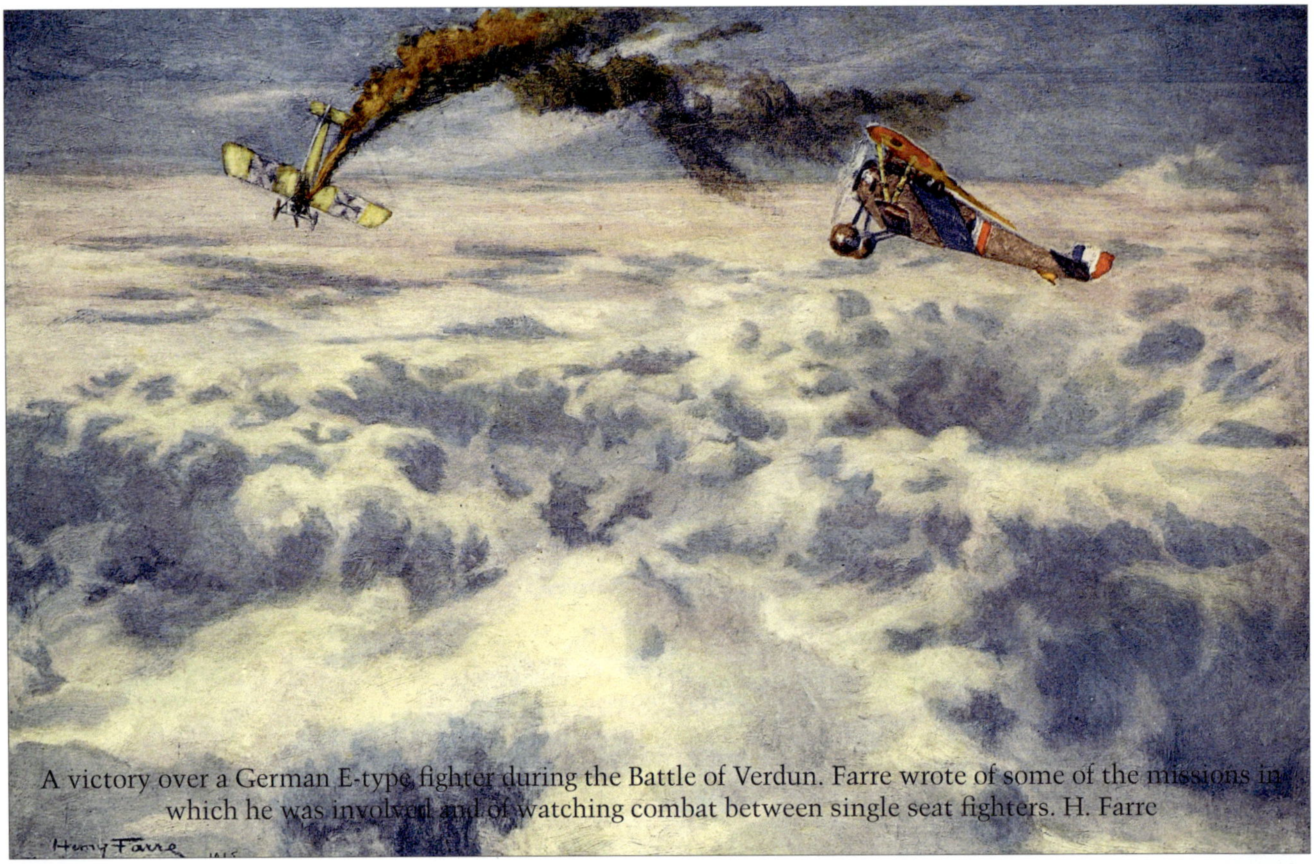

A victory over a German E-type fighter during the Battle of Verdun. Farre wrote of some of the missions in which he was involved and of watching combat between single seat fighters. H. Farre

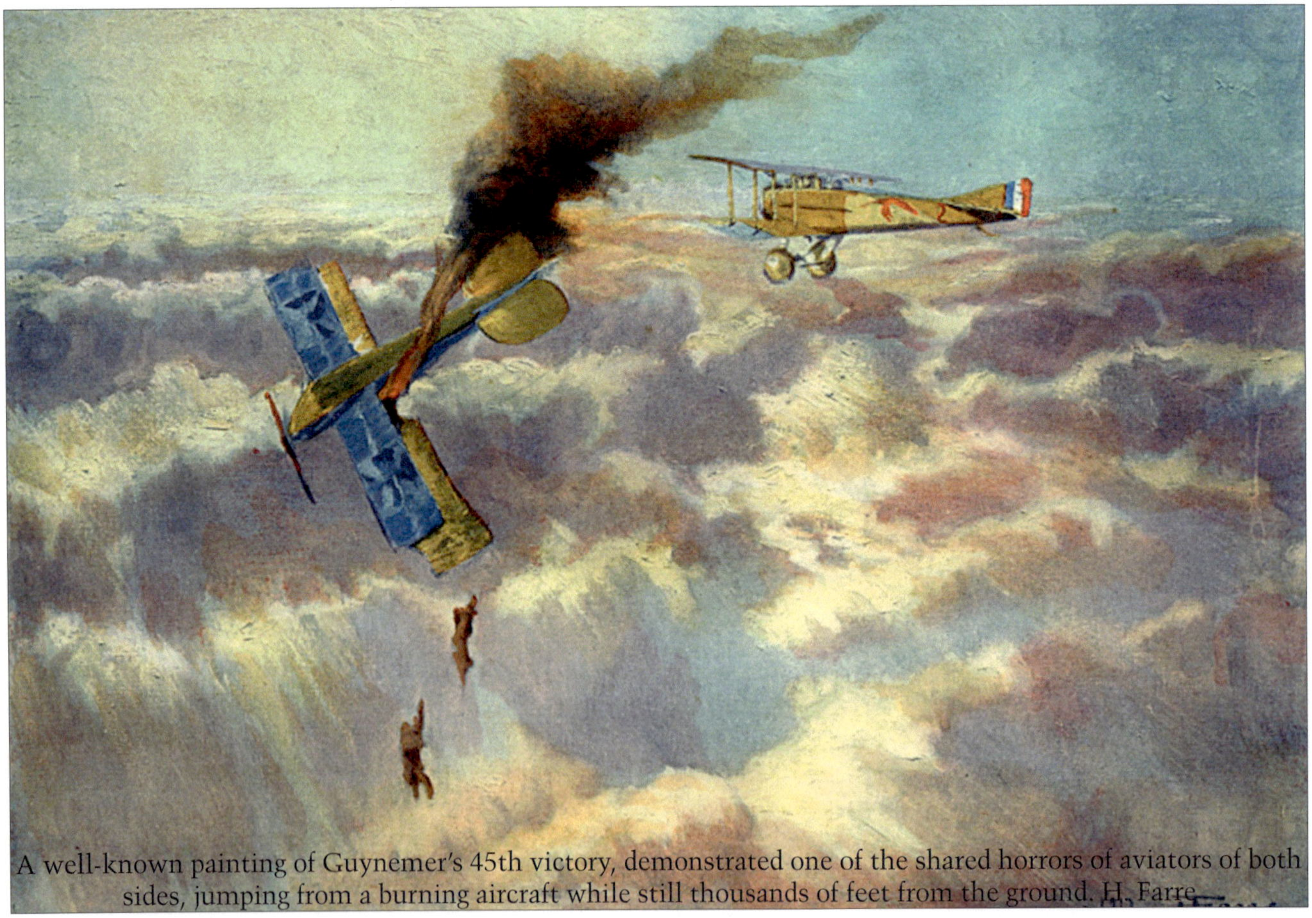

A well-known painting of Guynemer's 45th victory, demonstrated one of the shared horrors of aviators of both sides, jumping from a burning aircraft while still thousands of feet from the ground. H. Farre

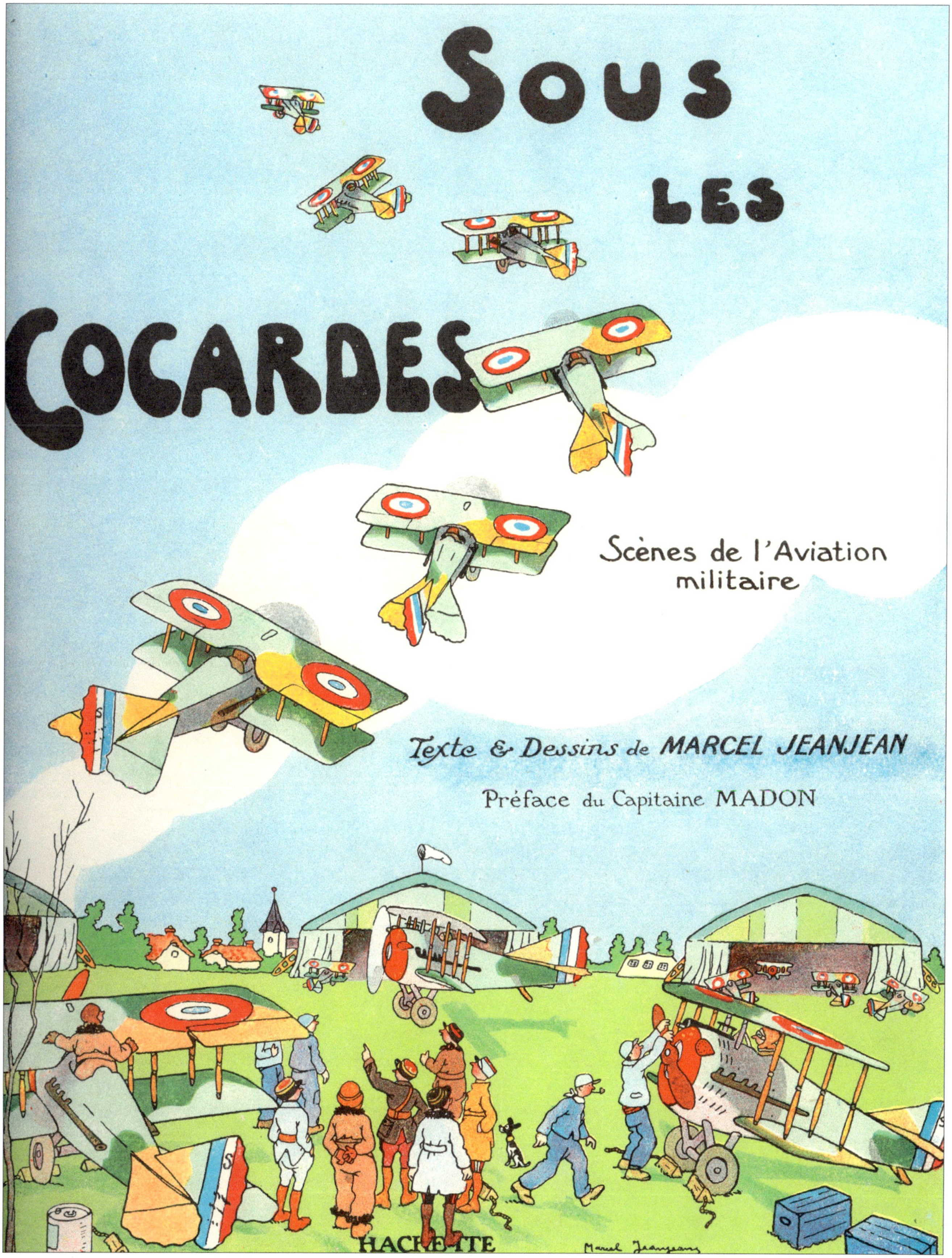

Above: Title page of Marcel Jeanjean's wonderful book of illustrations, published after the war.

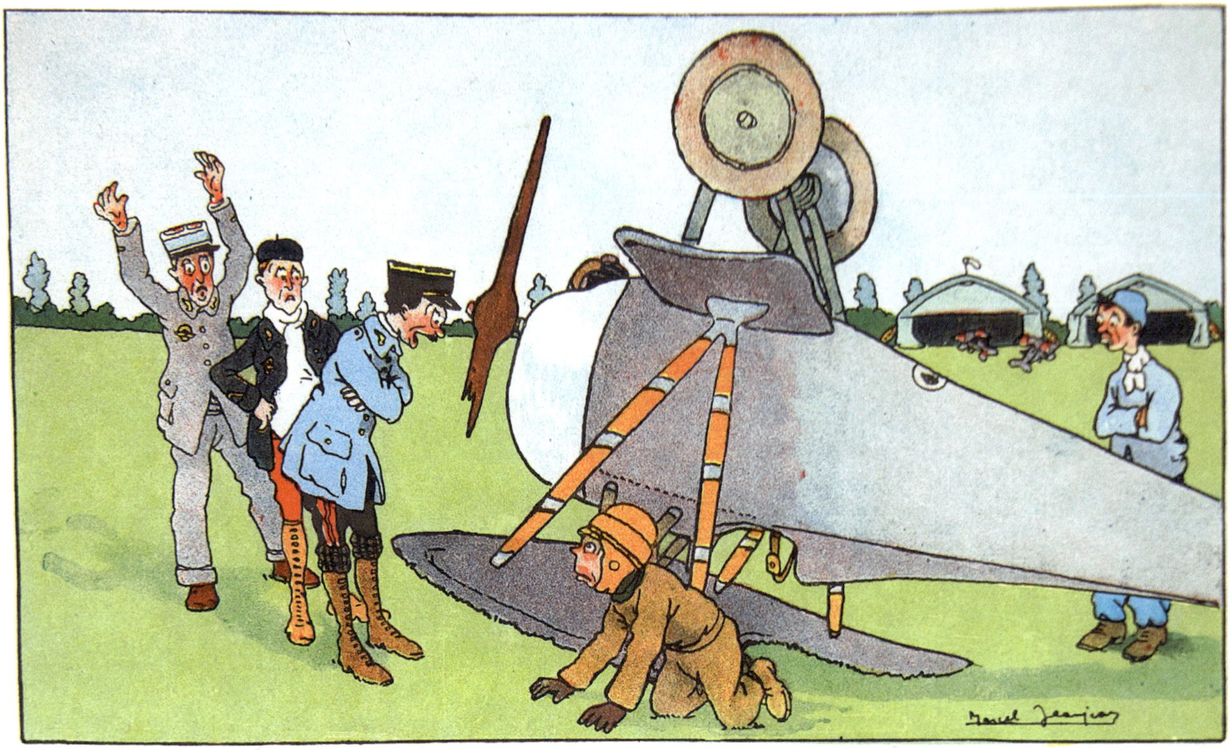

Le Capotage du Caporal.

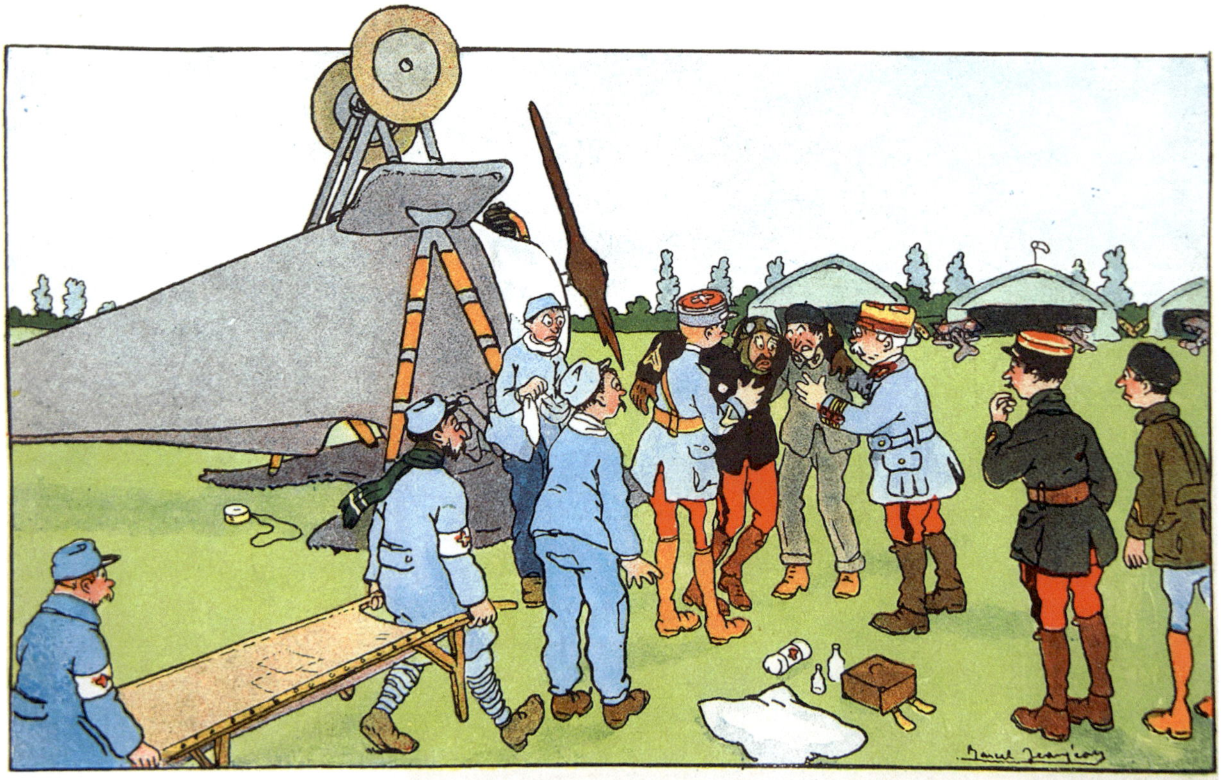

Le Capotage du Capitaine.

Above: Jeanjean's humor is shown at illustrating the difference an NCO pilot would face after a crash landing and the reception for an officer.

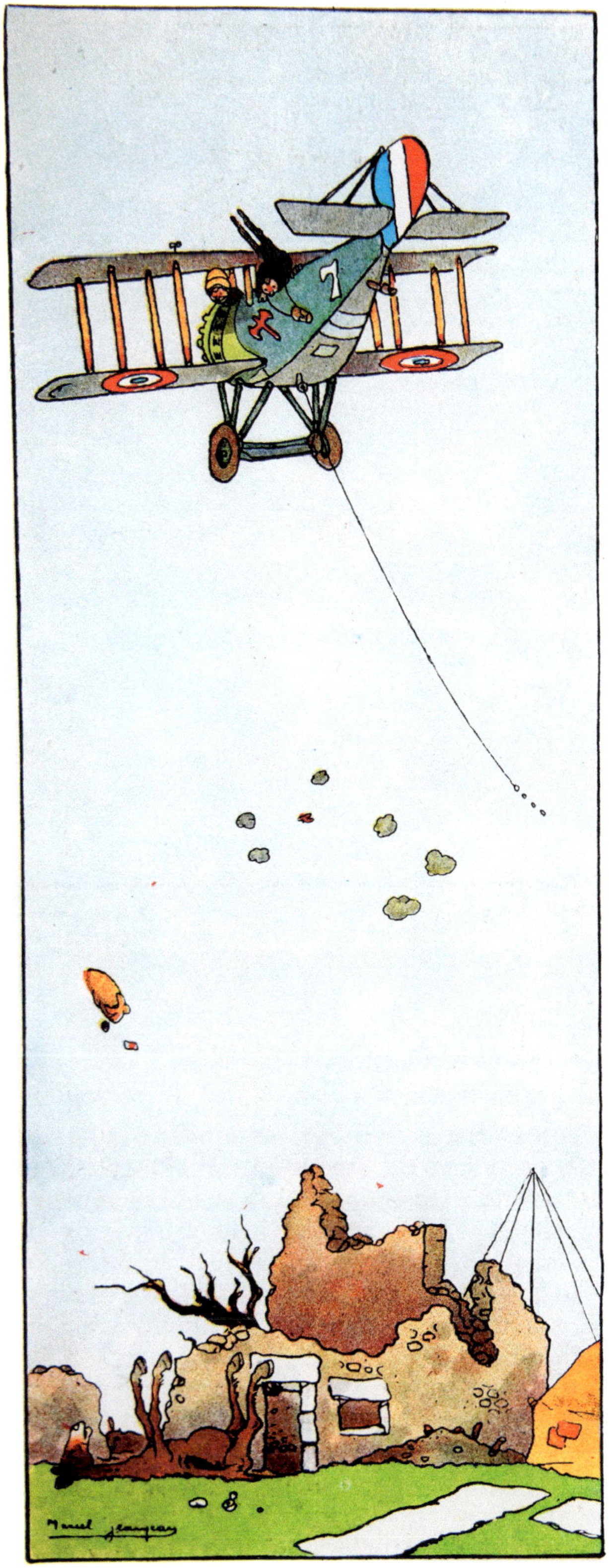

Le Réglage d'Artillerie.

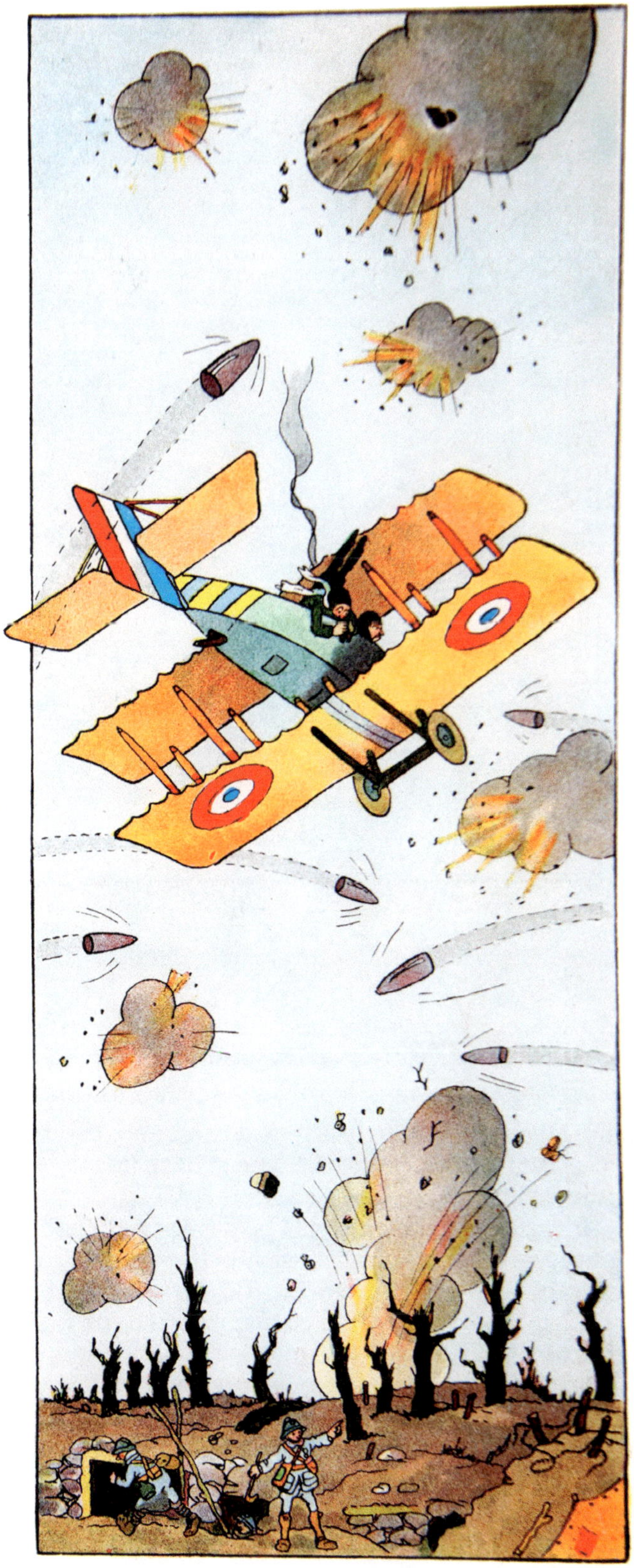

La Liaison d'Infanterie.

Above: Jeanjean flew many frontline missions and here he depicts two common missions he would fly: directing an artillery fire – note the trailing antenna for the aircraft's wireless – and the very dangerous ground support flight, note that the aircraft is at the same altitude of numerous canon shells.

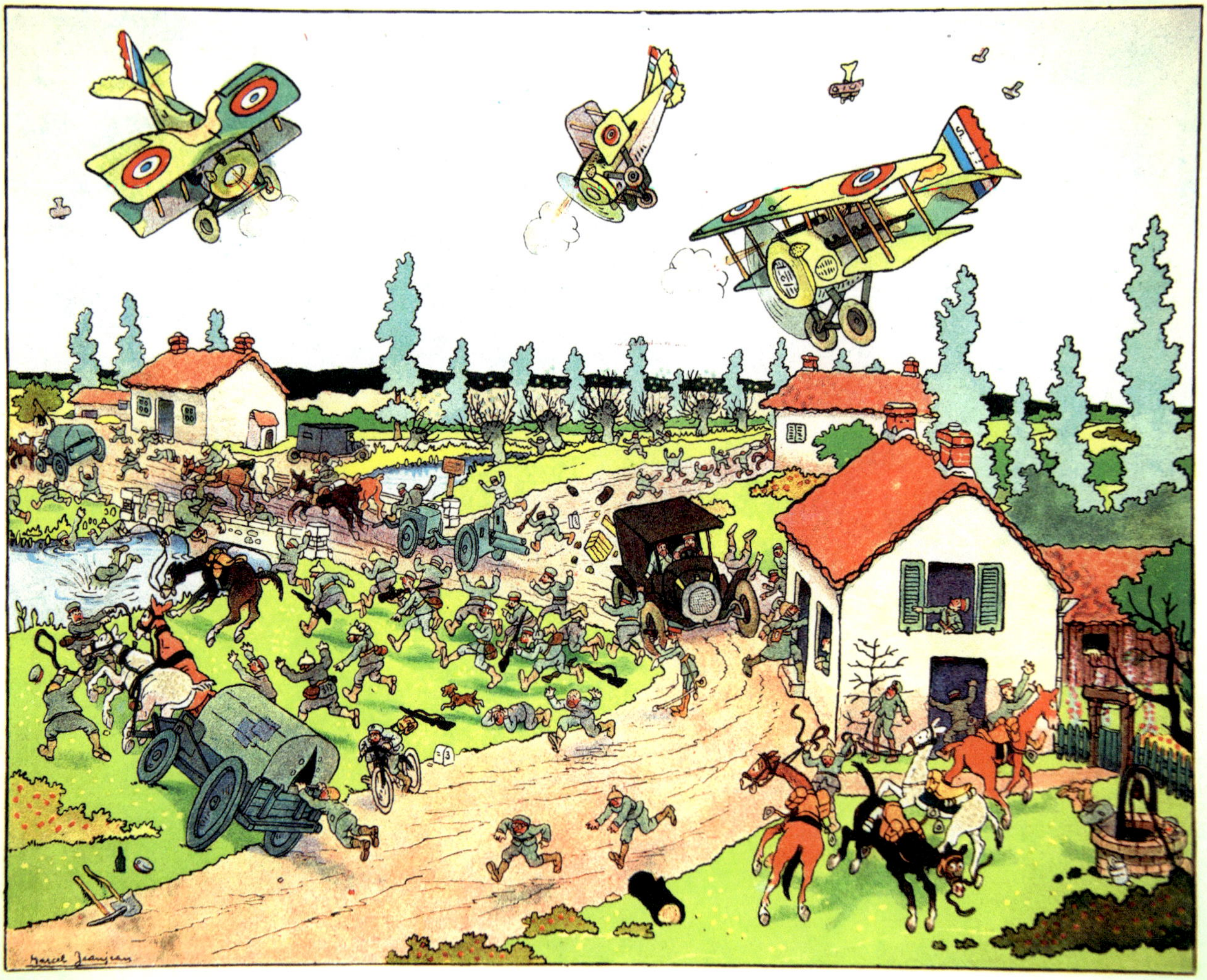

La Patrouille s'amuse.

Above: A flight of French Spads terrorize a German unit during a ground attack mission. M. Jeanjean

Facing Page: Rene Prejelan receives the Croix de Guerre for his service on the Salonika front.

land. I have no time to lose. I sit down immediately and sketch from memory the scene I have just witnessed."

Indeed, an argument could be put forward that Henry Farre was the first classically trained aviation artist the world has known. In 1918, he was present for a display of many of his paintings at the Art Institute of Chicago. The exhibition of his art was to raise funds for a benefit that had been established in France for the widows and orphans of French airmen killed in the war.

The exhibition was so successful, that Farre toured with his paintings throughout the United States. He also published a book of his work, Sky FIghters of France, in 1919. He married an American heiress and settled into his new life in Chicago, where he would live and paint until his death in 1934.

Although not as generally well known as Farre, Marcel Joseph Jeanjean was a combat aviator and talented artist as well. He was born on 31st January 1893 and joined the 6th Hussars in 1913. Going to war with the Hussars, Jeanjean would eventually transfer to aviation and earn his military wings in the fall of 1917. From October, 1917 until January, 1918 he flew as a pilot in the GDE. From January 1918 until the end of the war, Jeanjean was a pilot in the escadrille 33. He piloted a Salmson 2A2 two-seater aircraft and his "cartoon" artworks depict his career in the French AIr Service in the First World War. He would go on to a very successful career after the war, however despite attempting to discover more about him and his wartime career I was unsuccessful: in fact, i cannot locate a photo of him despite trying for over 20 years and with some excellent assistance by outstanding French aviation historians! Jeanjean published a wonderful book, Sous Les Cocardes in 1919, which is a classic work concerning the aviation art of the First World War.

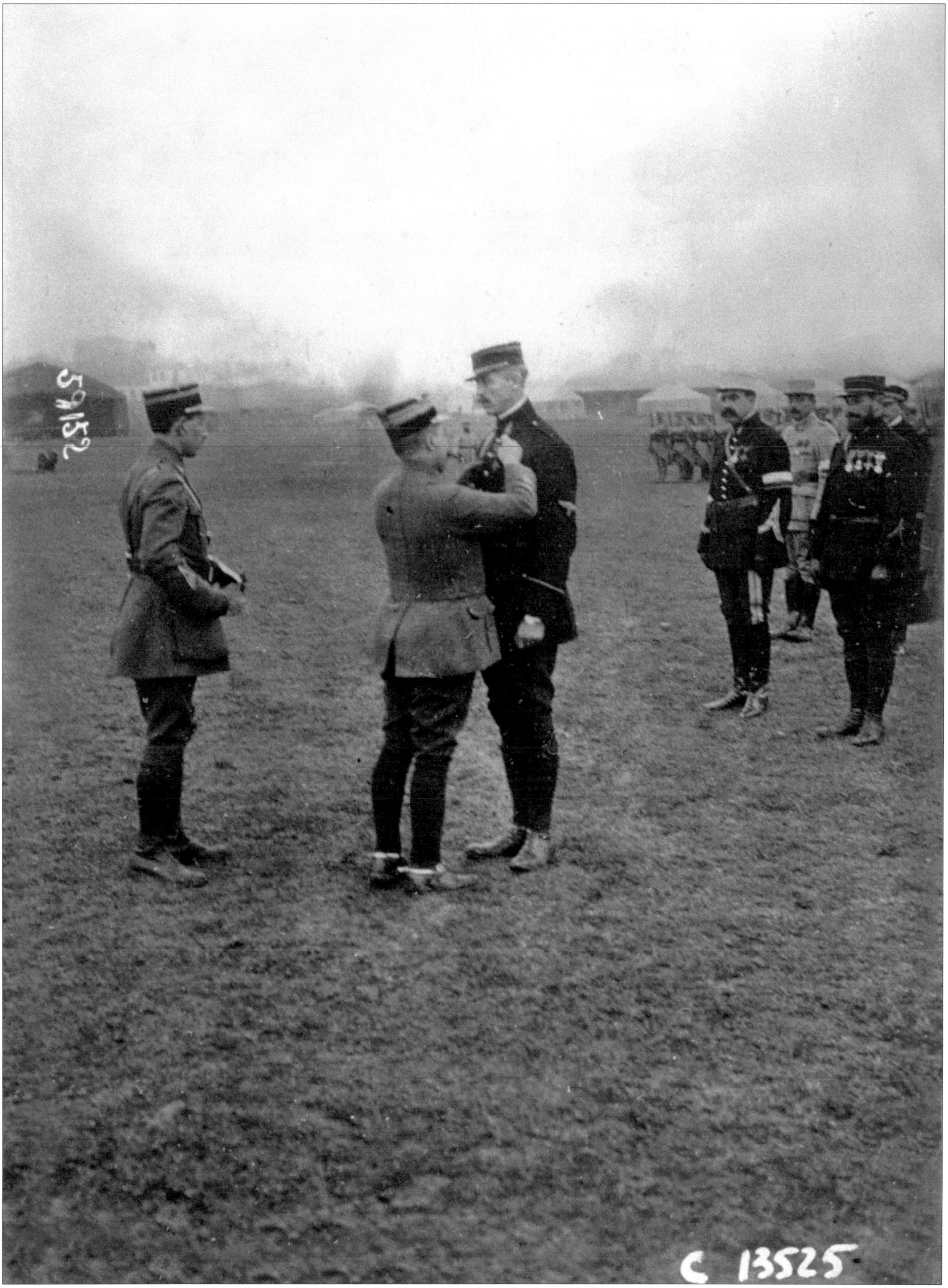
59152
C 13525

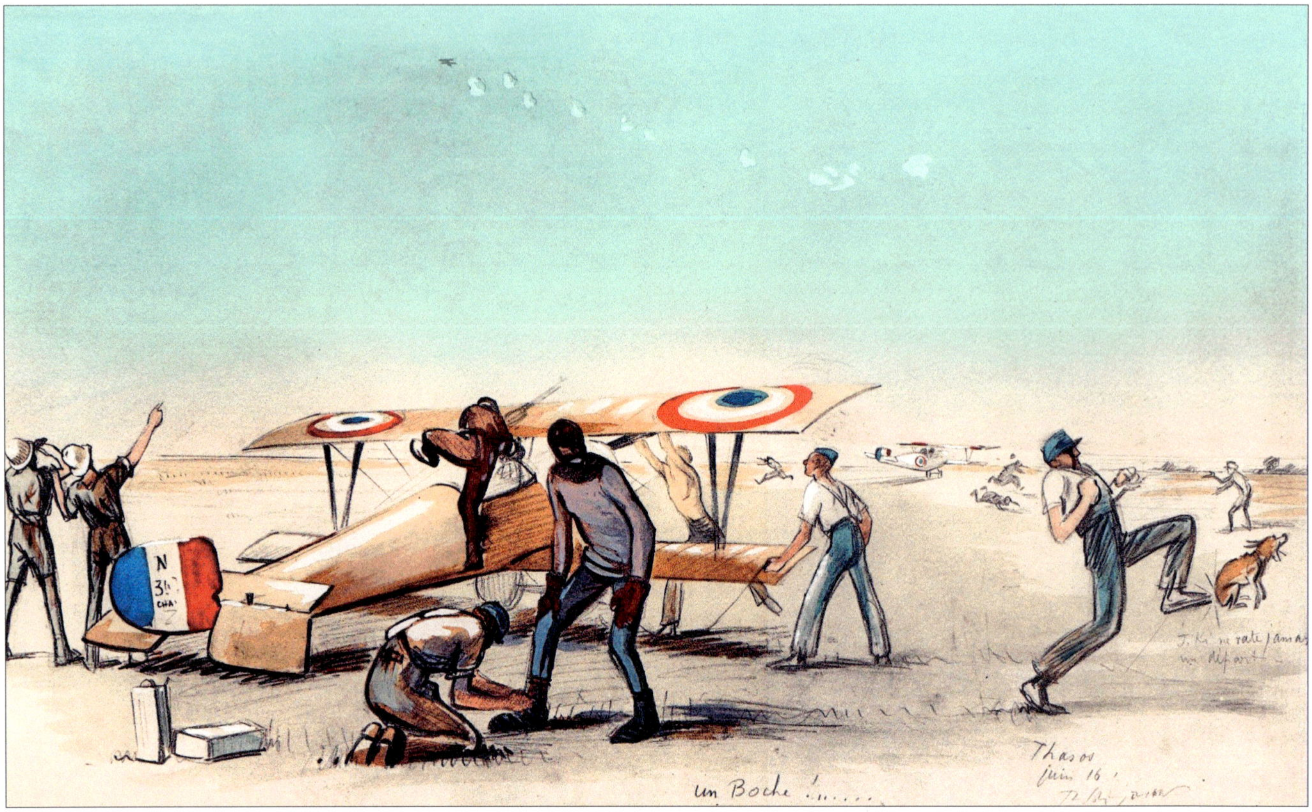
ne rate jamais
un départ
Thasos
juin 16
un Boche !……

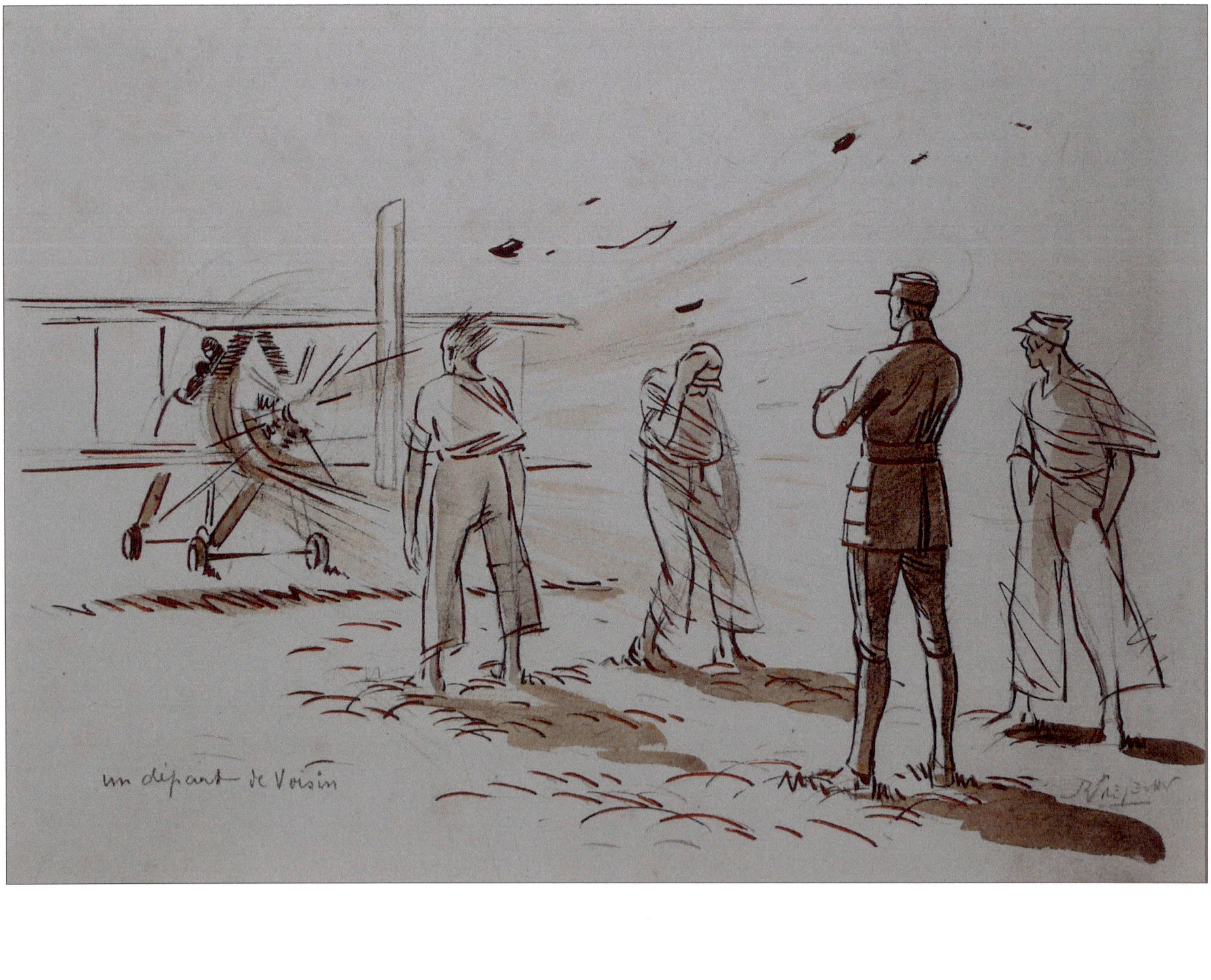
un départ de Voisin

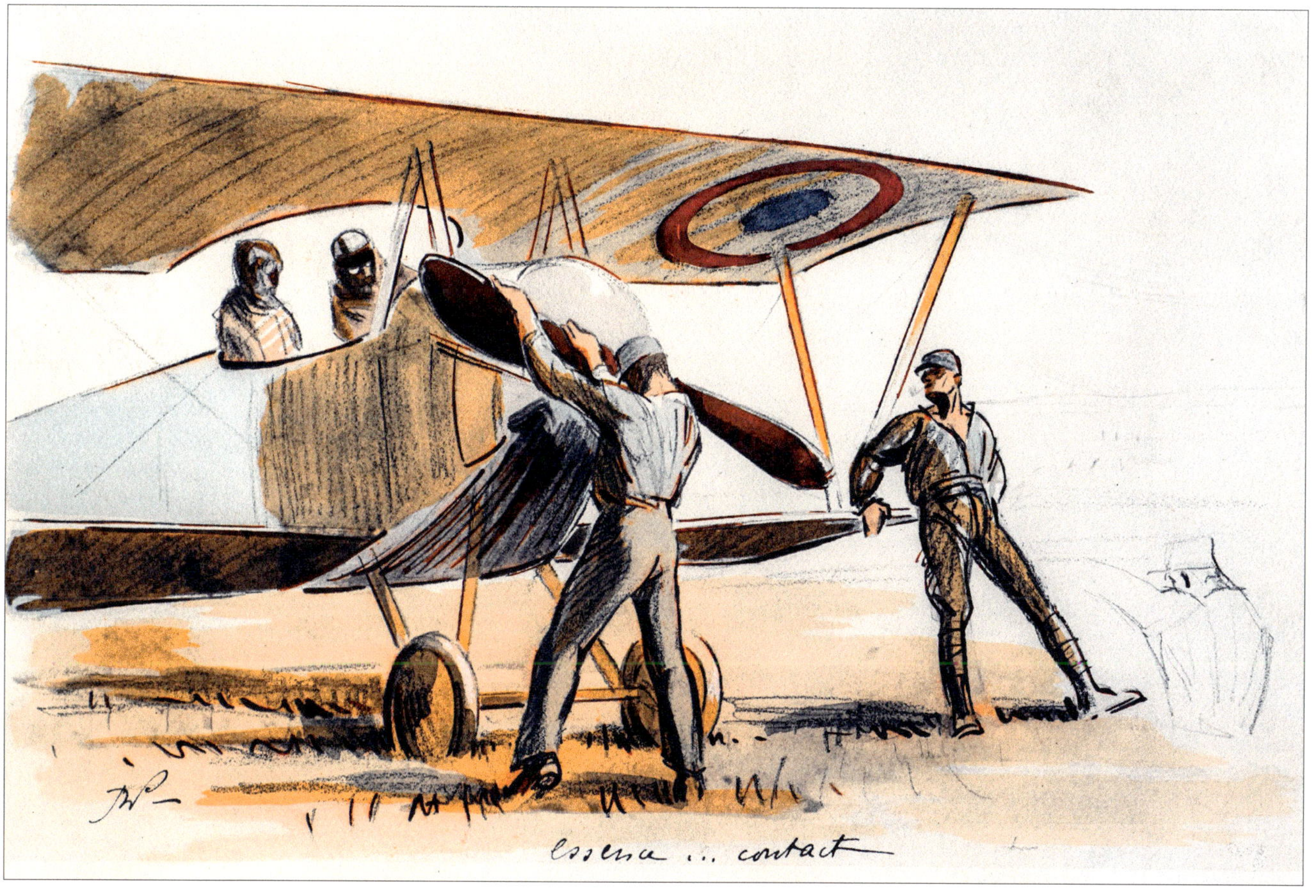

Above: A Nieuport is in the process of starting its engine. R. Prejelan

Facing Page, Above: An enemy aircraft is spotted and highlighted by AA fire. French pilots race to pursue the enemy and an unfortunate mascot needs to be moved aside in the haste. R. Prejelan

Facing Page, Below: A Voison takes off. R. Prejelan

Right: The squadron commander's nightmare. Rene Prejelan

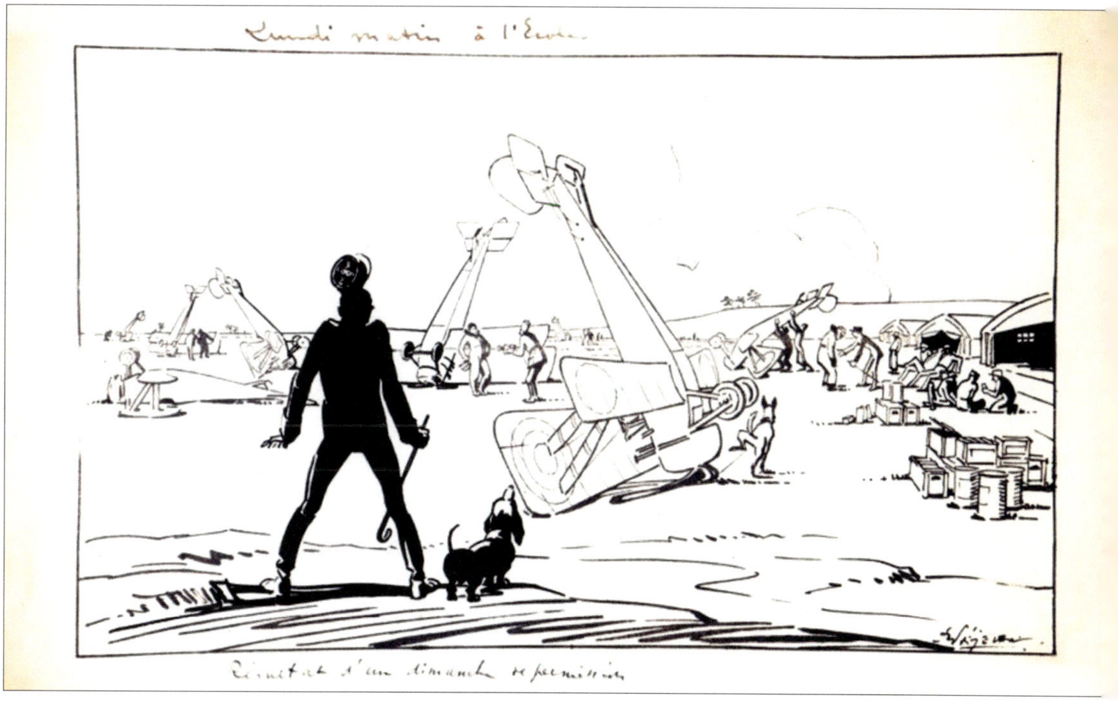

Jeanjean died in 1973.

Another French combat aviator artist who has been elusive to my research is Rene Prejelan. Born in 1877 he had a successful career as a painter, illustrator and humorist. He served in a French unit that saw service on the Salonika front. His portfolio of work from that time is quite a rare find, however his illustrations are outstanding. Prejelan died in 1968.

From the standpoint of a combination of successful combat pilot and talent as an artist, it would be hard to find a person to replace German

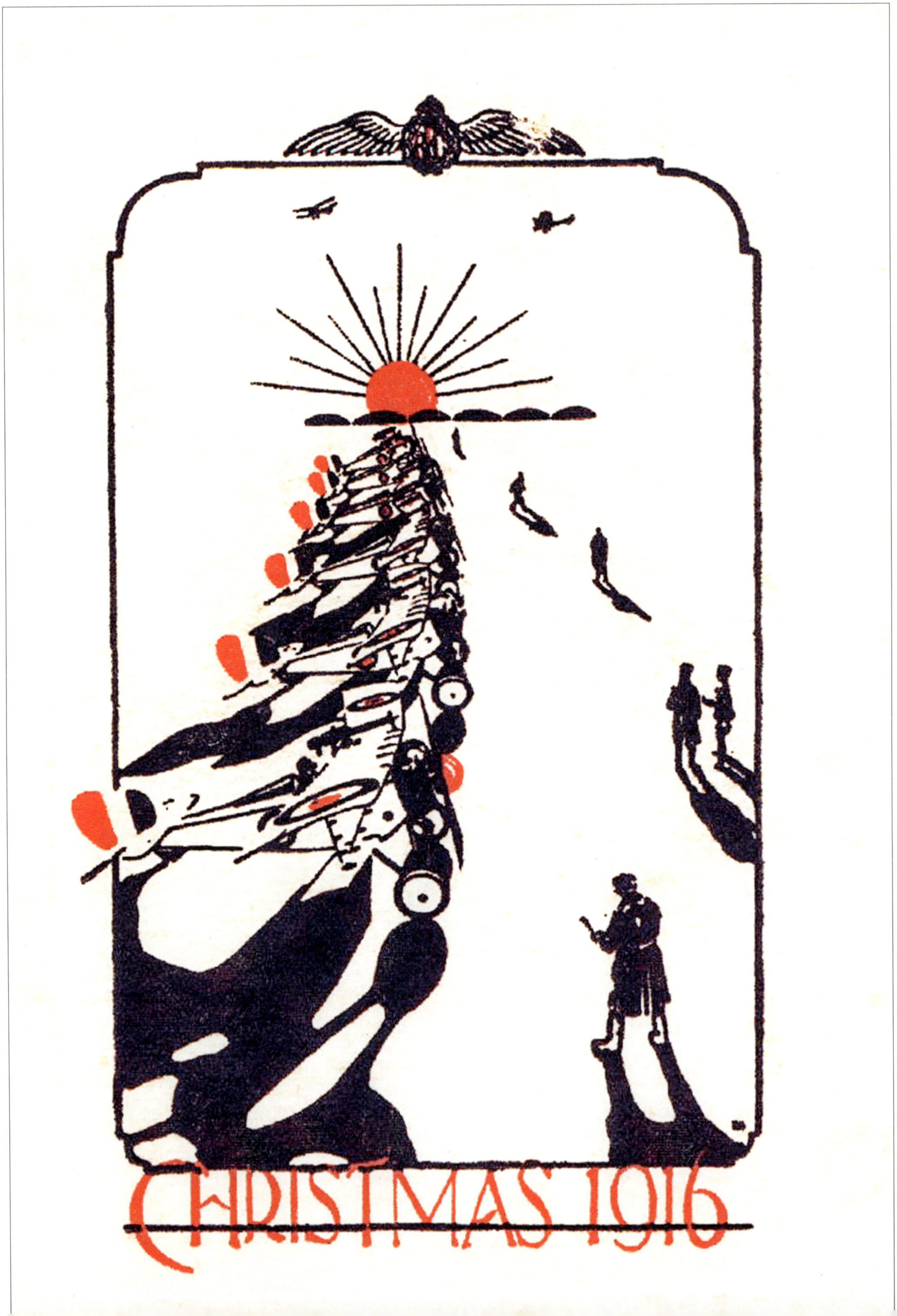
CHRISTMAS 1916

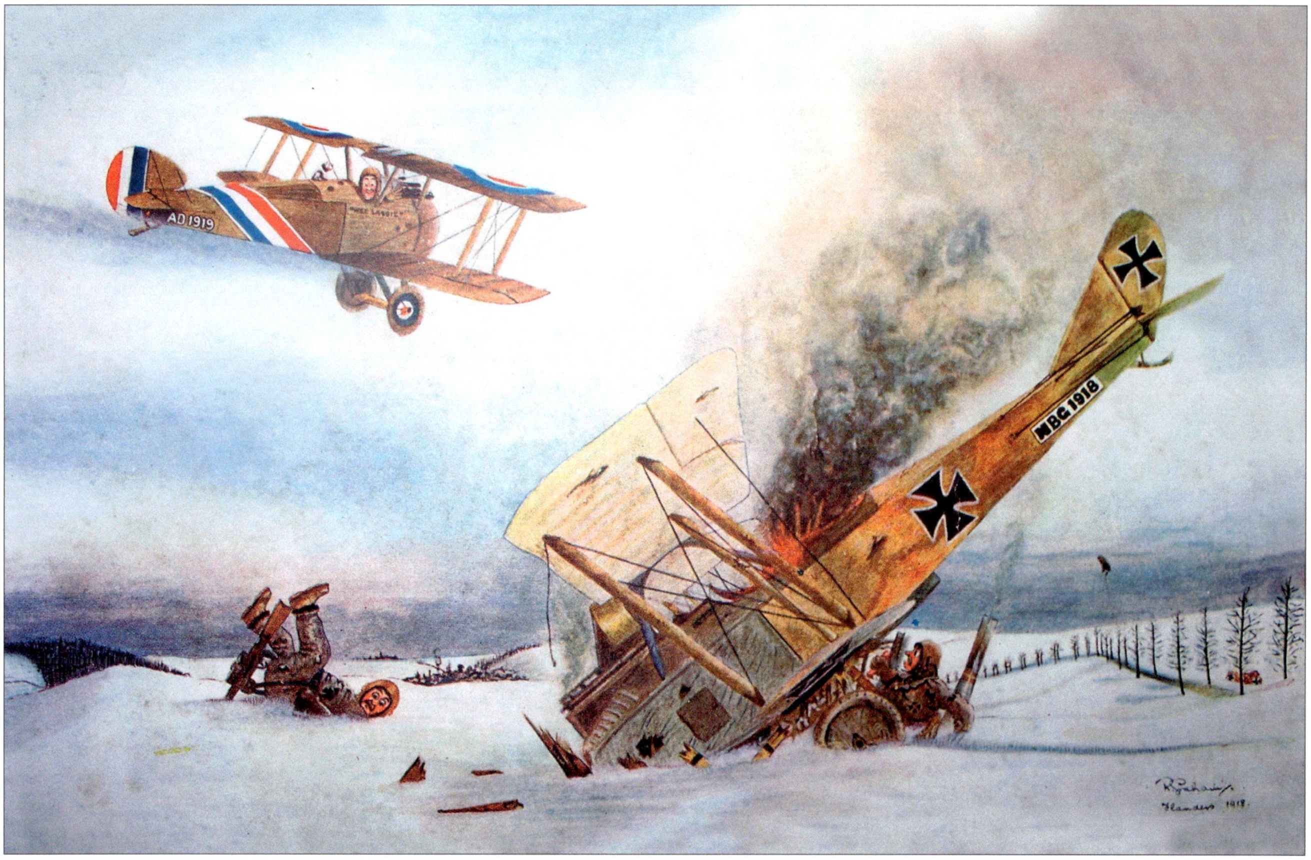

Above: Another RAF Christmas card, this time from No. 213 squadron and it is notable since it is for Christmas 1918, the war is over and takes on a more humorous and less ominous tone. Note the downed German aircraft has the NBG 1918 marking- for no bloody good! Artist unknown

Facing Page: Roderick Hill's wonderful Christmas card for No. 60 sqdn. For 1916. The nearest Nieuport in the flight line is immediately recognizable as it is Albert Ball's with its distinctive red spinner.

fighter ace, Rudolph Stark. His artwork can be seen in the chapter, Unsere Heimat.

There can be no doubt that there were many amateur artists to be found amongst the various aviation units during the First World War. Among the British squadrons in both the RFC and the RNAS (continuing when the two combined into the RAF) there grew a tradition of sending home thematic Christmas Cards. One of the most striking of these is the Christmas card from No. 60 sqdn,, RFC in 1916. It is the work of Roderic Maxwell Hill, born 1894, and his artistic talent is clearly beyond amateur. Hill was fascinated by flight at an early age. Along with his brother Geoffrey, they designed, built and successfully flew a glider in 1913!

After being wounded in the Battle of Loos, Hill transferred to the RFC and was posted to No. 60 squadron in time to participate in the Battle of the Somme in November, 1916. Hill was an excellent pilot and after promotion to the rank of Captain, he was transferred to be the leader of experimental flying at the Royal Aircraft Factory at Farnborough. He would have a lifetime career in the RAF, and was the Commander in Chief of Fighter Command between 1943-45.

These aviator artists pioneered the concept of aviation themed art. In their work can be seen the themes that reflect the feelings and beliefs of the first group of military aviators the world has ever known.

The concepts of individuality, a certain sang froid to be found in their fellow airmen, are mixed into the elements of the vastness of the sky during their missions and the shared lives they felt with each other at their respective aerodromes. Thus, the works of these artists give us all a very unique perspective, an eye witness to what aviation was truly like during the First World War.

Blitzkrieg Birth

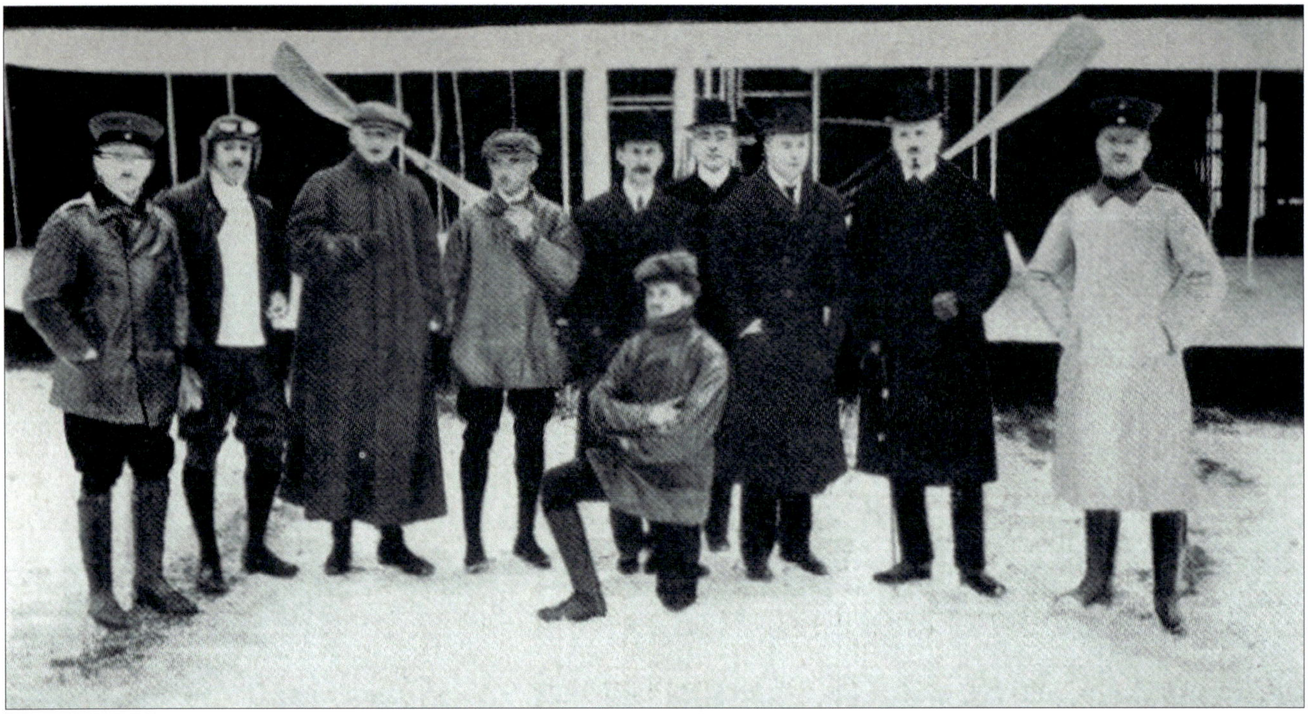

Above: Helmuth Wilberg is one of the group known as Die Alten Adler, the old eagles, German aviators who earned their wings before the start of the First World War, earning license number 26 in 1910. Wilberg (far R) during a demonstration by the Wright brothers in Germany.

"The object of the battle flights is to shatter the enemy's nerve by repeated attacks in close formations."

Any reader of military history will likely have read about the blitzkrieg (lightning war) that Germany unleashed upon the countries of Western Europe in the spring of 1940. However, many of those same readers may not realize that this innovative type of warfare actually began in Flanders and then was effectively used in the German offensive of 1918.

The British Army first deployed tanks in September 1916 without very much success. There were too few tanks used and of those used, many very quickly broke down. Their usage at this time now gave the Germans time to develop a change in their tactics in order to deal with this new foe.

It should be noted that the Allies would ramp up tank production during the war with the British building about 2,600 and the French 4,500 tanks before the end of the war. The German HIgh Command swiftly realized that they did not have the industrial capacity to "out-build" the Allies in numbers of tanks, so they developed their tactics in an appropriate manner.

In fact, when the Germans launched their Spring offensive in 1918, they utilized more captured Allied tanks than German built tanks in that attack. An important response to the Allied tanks was the mounting of a Becker cannon to some of their low flying ground support aircraft. The Allied tanks had been built with enough armor to withstand machine gun fire, but not the 20 mm shells from this gun. Thus, some of the low-flying two seater aircraft were now "tank-busters" decades before the P-47s, Typhoons and Stukas of WWII.

By the time of the Battle of Cambrai in November 1917 , the Allies could field more than 400 tanks in their attack. However, the new German tactics of blitzkrieg were ready to be deployed in a tremendously successful counter-attack. There were several components to their newly developed warfare. First, infantry infiltration movements around strong points and seeking to exploit weak points were honed by General Oskar von Hutier from his experiences on the Eastern Front. New approaches to artillery fire by General Bruchmuller were also employed. But the key to making these changes succeed was the development of new ground-attack methods, units and equipment

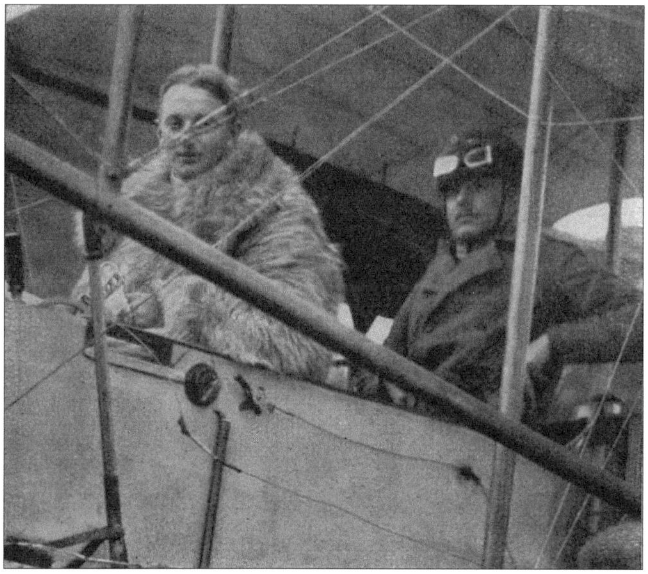

Above: Wilberg took part in the army maneuvers in 1912, during which two aviators died, highlighting the dangers of early aviation. Wilberg on the right, with goggles.

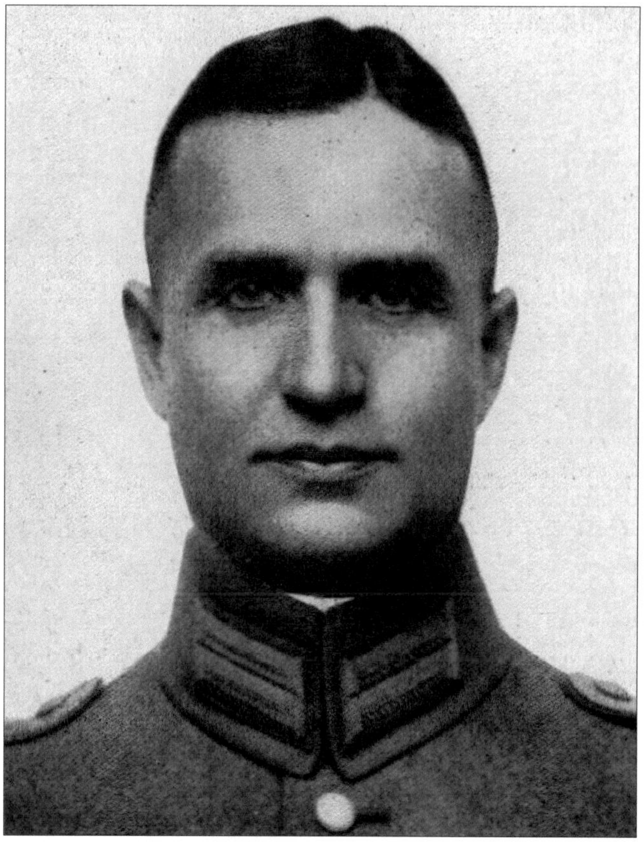

Above: During the course of the war, Wilberg would be credited with the development of the tactics that would develop into the concept of blitzkrieg by developing the needed close support, low altitude attack aircraft units known as schlachstaffeln.

by the German Air Force.

The key figure in these innovations was Captain Helmuth Wilberg. Wilberg would not only develop the schlachtstaffeln (battle squadrons) and help promote the new aircraft designs: small, light two-seater aircraft; he would prove to be the key conncction from thc first world war German Air Force to the Luftwaffe's blitzkrieg tactics in World War II.

Below: Wilberg became the KOFL or commander of aviation units for the German 4th Army on the Western Front. One of the units under his command was the famous JG1. Here Wilberg is seen with Manfred von Richthofen at the time of a formal review of the unit.

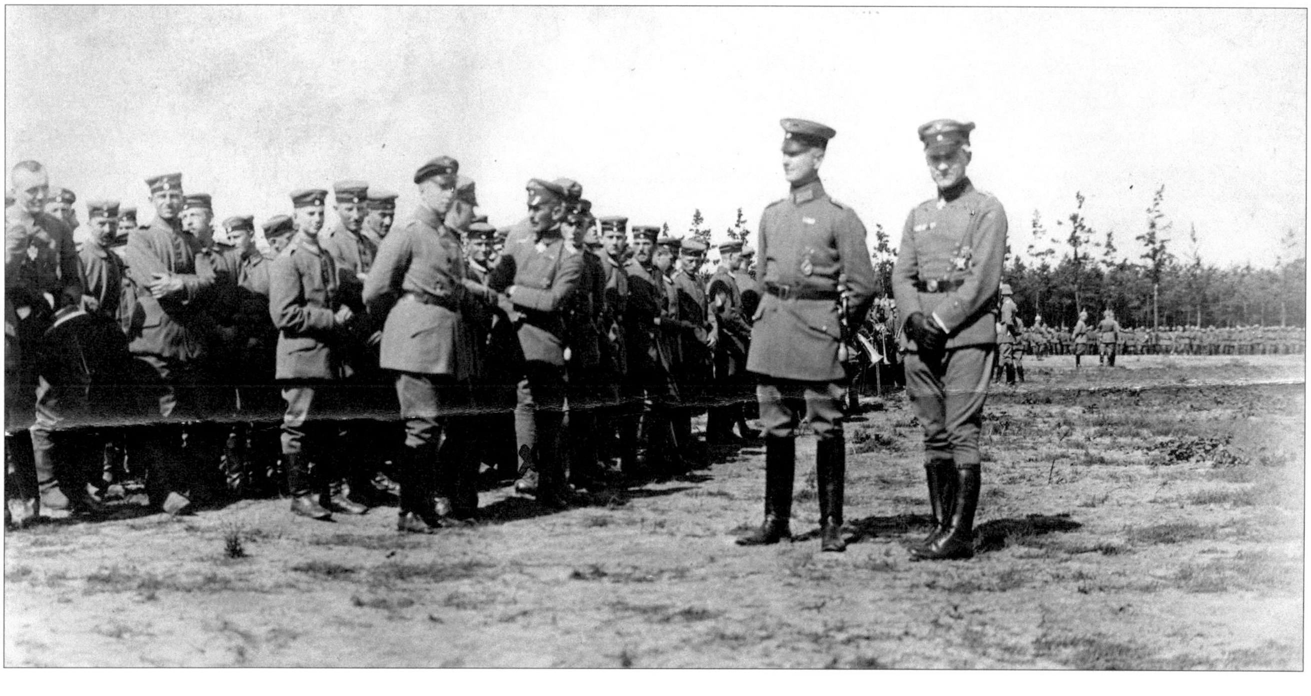

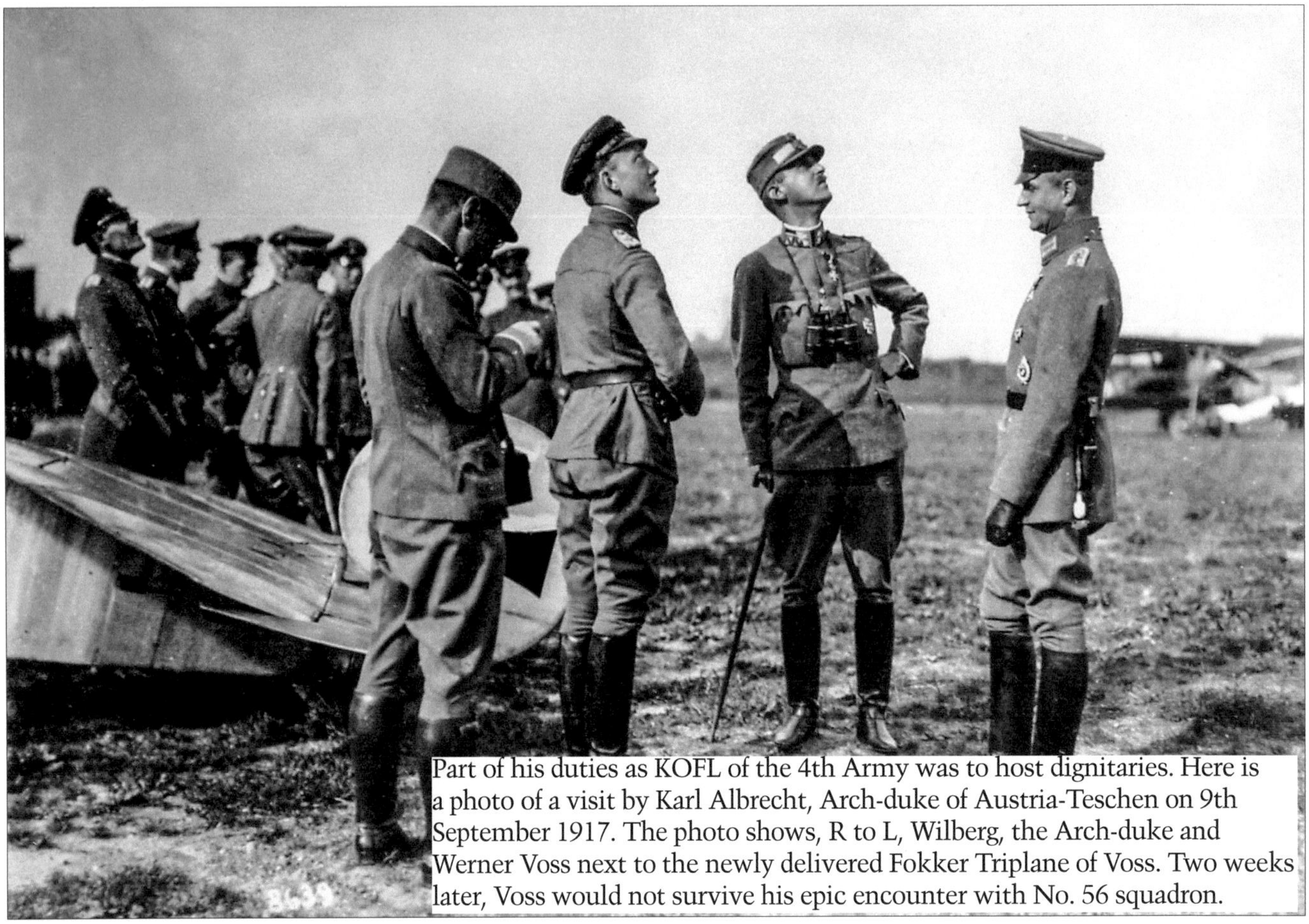

Part of his duties as KOFL of the 4th Army was to host dignitaries. Here is a photo of a visit by Karl Albrecht, Arch-duke of Austria-Teschen on 9th September 1917. The photo shows, R to L, Wilberg, the Arch-duke and Werner Voss next to the newly delivered Fokker Triplane of Voss. Two weeks later, Voss would not survive his epic encounter with No. 56 squadron.

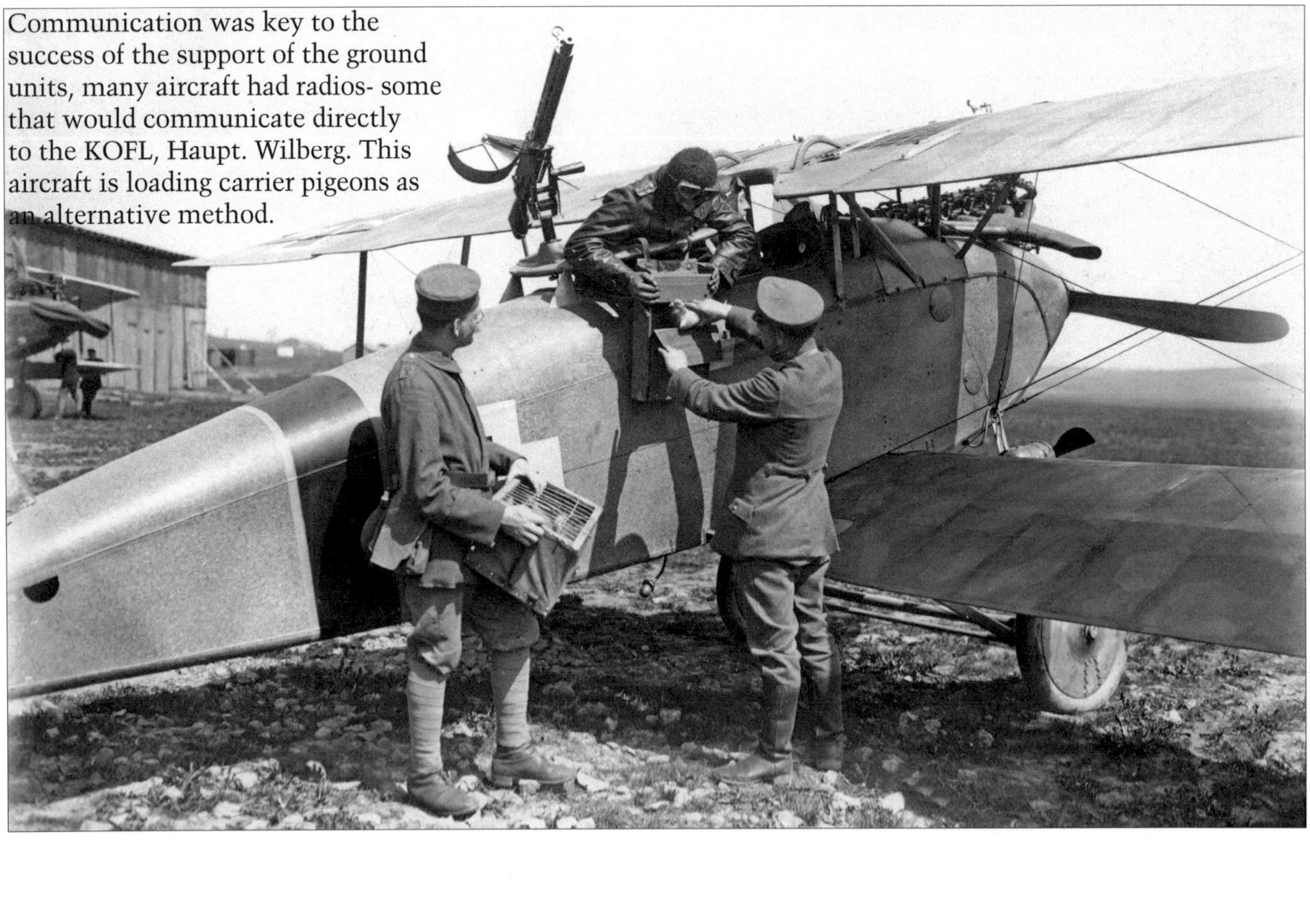

Communication was key to the success of the support of the ground units, many aircraft had radios- some that would communicate directly to the KOFL, Haupt. Wilberg. This aircraft is loading carrier pigeons as an alternative method.

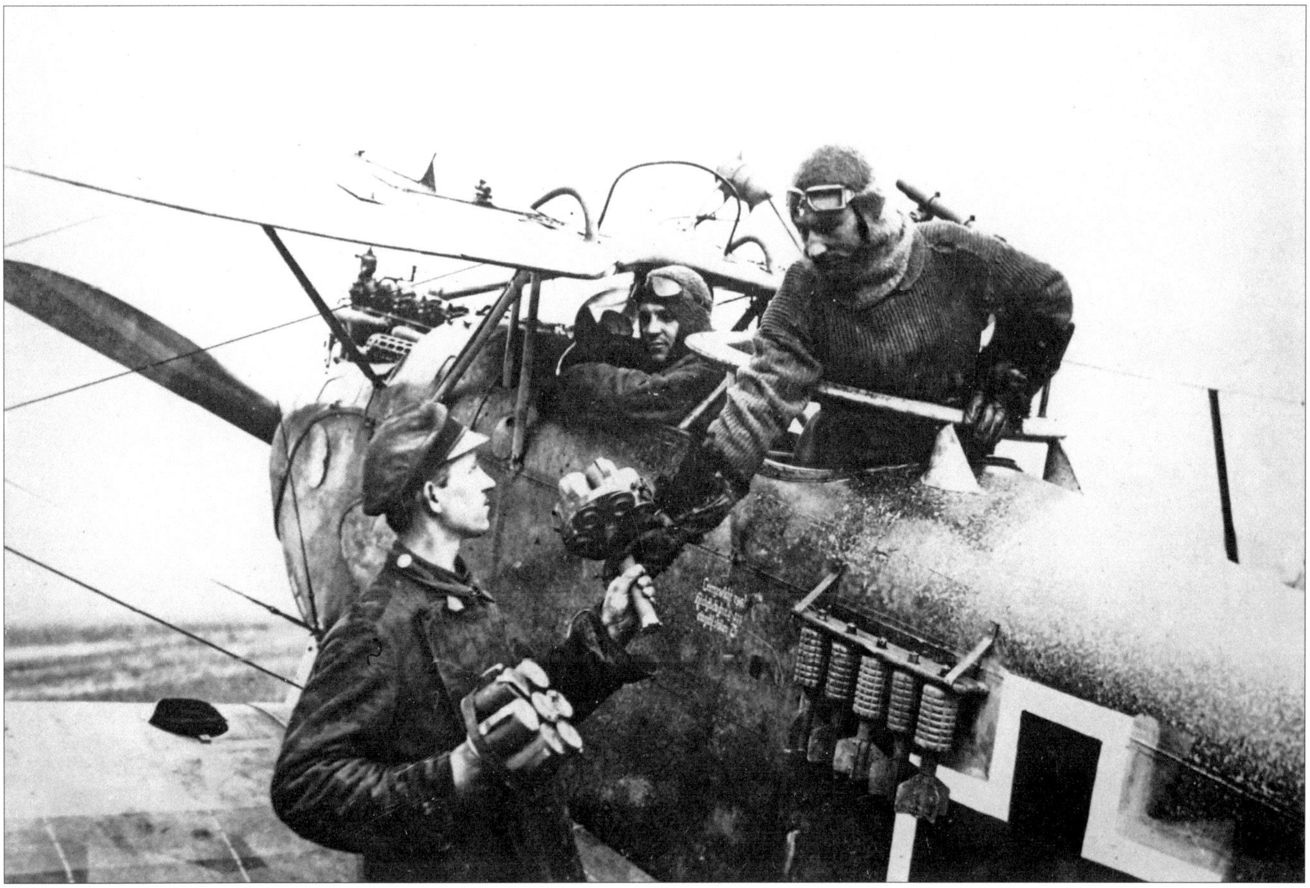

Above: The aircraft were heavily armed for their mission: the machine gun, the grouped hand grenades and the small bombs were part of their weaponry.

Wilberg was born in 1880 to a Jewish mother and a Prussian protestant father. His father was a prominent portrait painter who gave art lessons to the crown princess. His father died when Helmuth was only 4 years old and it was at this point that his life began its unusual course. Because of his father's connection to the royal family, the empress mother decided to look after his family. This help included supporting young Helmuth's attendance at the officer academy. He later spent two years as the military tutor to the royal family. In 1910, he was selected to attend the prestigious General Staff Academy. At this time, he also discovered his passion for flying. Wilberg became a member of Die Alten Adler (German aviators before the first world war) when he earned license number 26 in 1910.

He would be very active in the embryonic German AIr Force, including flying in the 1912

Suddeutsche Rundflug which revealed the dangers of military flying when two of the sixteen aviators were killed.

He proved to be a very capable, intelligent and popular officer within the German Air Force. Thus, Wilberg found himself as commander of the KOFL to the fourth army along the Western Front. General Freiherr von Bulow believed that Wilberg: " [had] the deepest understanding of the employment of the aircraft in the ground battle." As the British began their Flanders campaign in the summer of 1917, that would culminate in the Battle of Cambrai in November 1917. Wilberg prepared his aviation units for the onslaught.

By the end of July 1917, Wilberg was in command of 600 aircraft and approximately 70 squadrons. It was now that the first jagdgeschwader (JGI) was formed under the command of Manfred von Richthofen. The concept of the fighter wing was able to aggregate the experienced fighter pilots into a concentrated, coordinated effort. With sorties of up to fifty aircraft at a time the units were to achieve air superiority over small, designated areas along the front lines.

Wilberg also was instrumental in the development of specialized close air support groups. In this case the units were actually small squadrons of 4-6 aircraft and it was usual for several of these smaller units to group on a given mission to neutralize very specific targets.

"Some radios were also mounted in the ground attack aircraft, and Captain Wilberg used

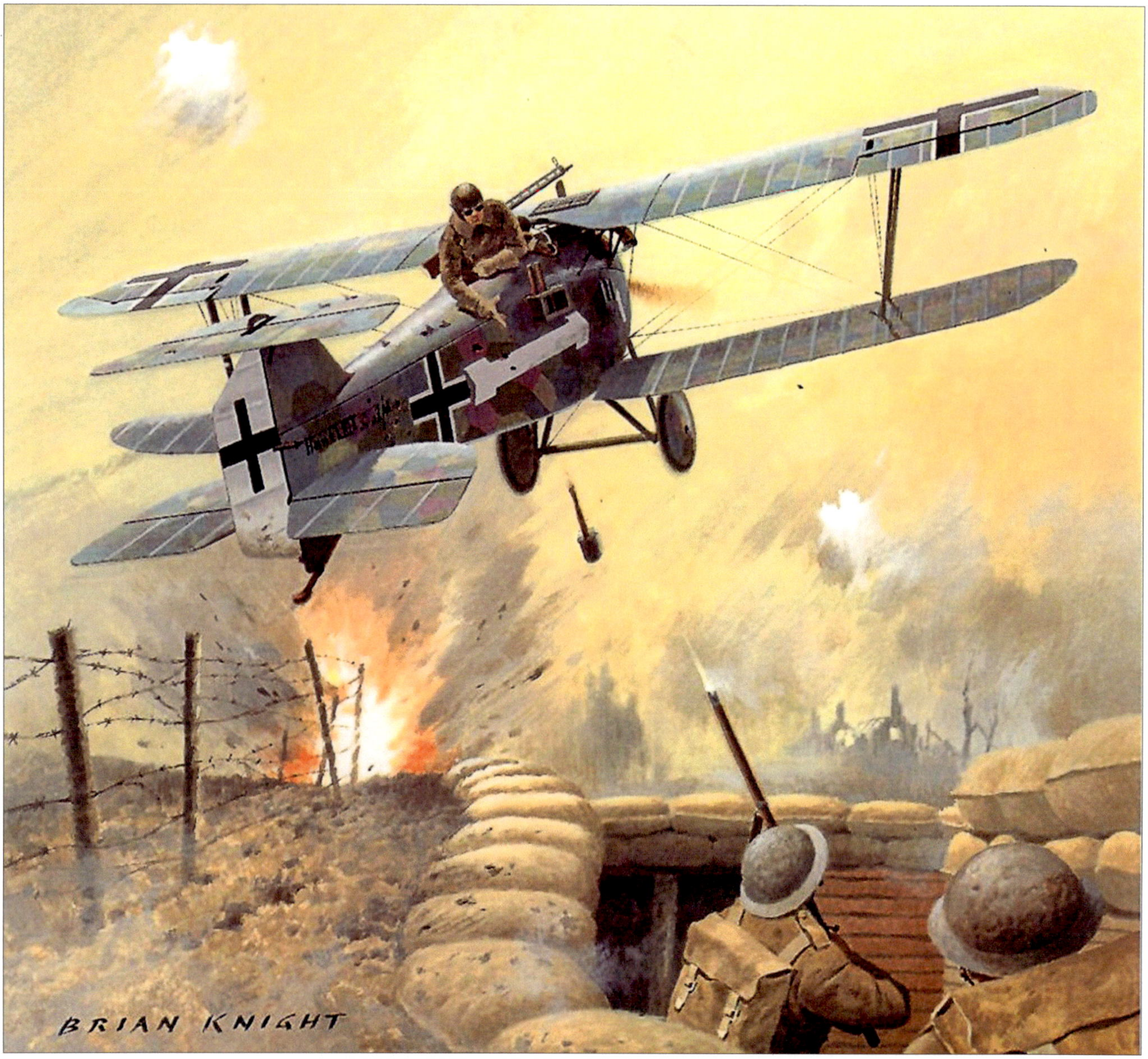

Above: A Hannover CL II engages the British infantry in a front line trench, the often extremely low altitudes these aircraft flew at is seen in this painting. Brian Knight

the radio to direct his ground attack squadrons at the height of the battle. The German Schutzstaffel were able to react quickly, and the employment of his ground attack squadrons at Messines Ridge on 6-14 June, helped slow and control the British offensive. Wilberg also used his artillery observation aircraft to support the ground battle by dropping ammunition, food and medical supplies on forward German infantry units that were hard-pressed by British attacks and unable to get ammunition forward. Captain Wilberg gained a reputation as the Luftstreitkrafte's premier expert on ground attack aviation." James S. Corum

German military planners were now preparing the Kaiserschlacht (Spring 1918 offensive). WIlberg and other aviation leaders now planned their portion of the offensive. They worked very hard to have their superior doctrines and tactics overcome the obvious facts of being outnumbered and "out-resourced".

The key to the offensive's blitzkrieg efforts would be speed of the infantry, co-ordinated and concentrated artillery fire as well as endurance by the smaller groups of stormtrooper units as they bypassed strongpoints and encircled strategic points of the offensive plan. Key to this warfare were the low-flying aircraft that would support communication between the forward units, and between those units and their artillery support.

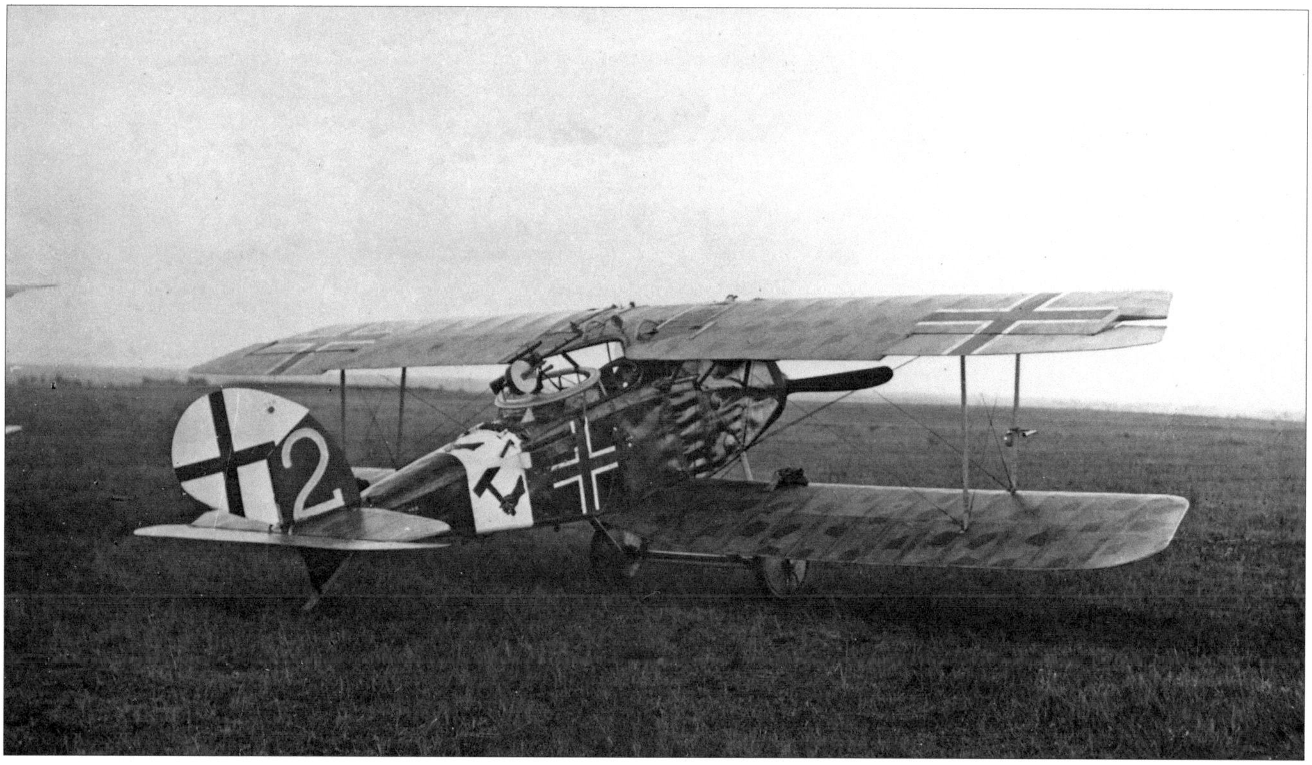

Above: One of the principal aircraft in the attack squadrons was the Halberstadt CL.II through IV series. A small, nimble and lethal aircraft, its crews were very fond of it and also enthusiastically decorated their aircraft.

Below: Their was a high level of morale amongst the crews in the schlachstaffeln, the two-seater crews would fly into combat in close quarters in their shared cockpit. Note the extensive flame decorations on this aircraft.

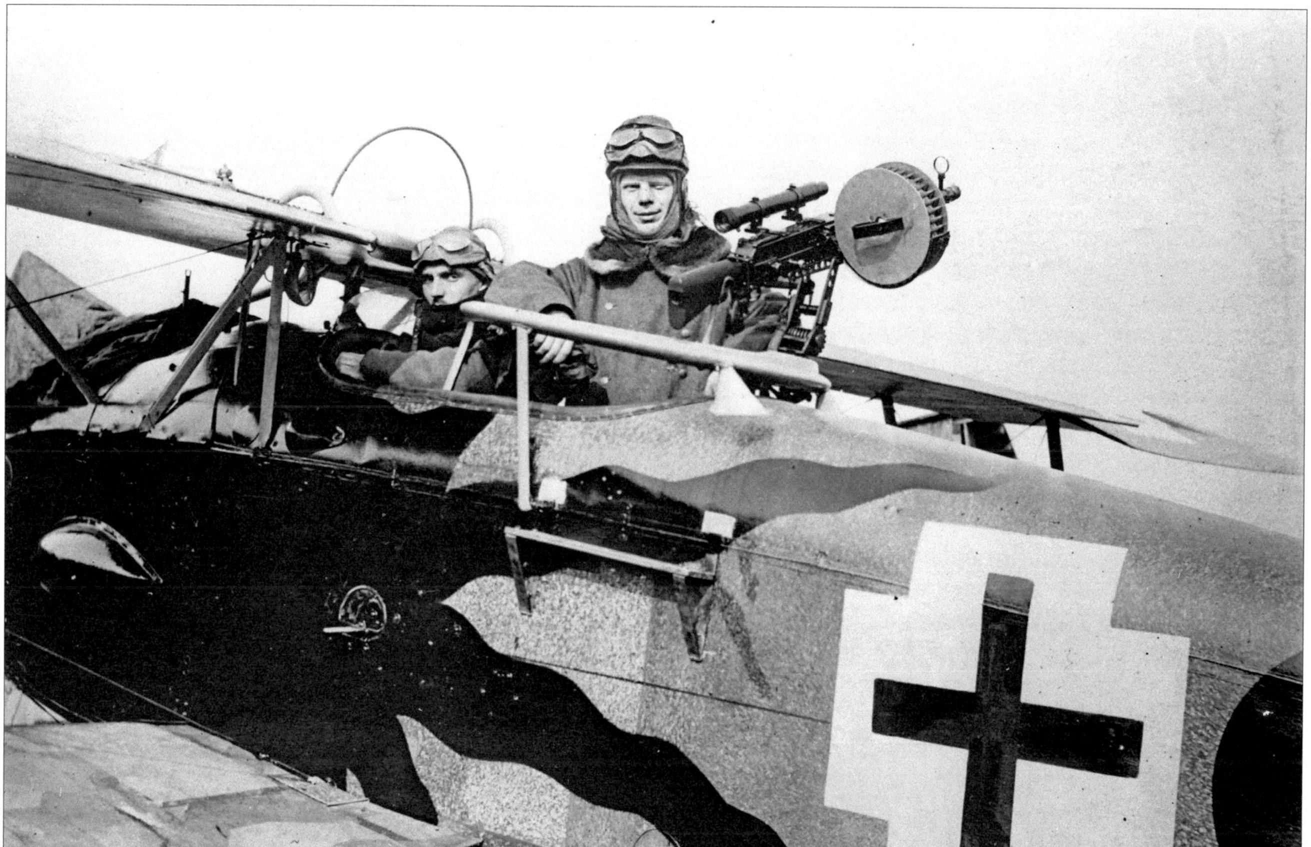

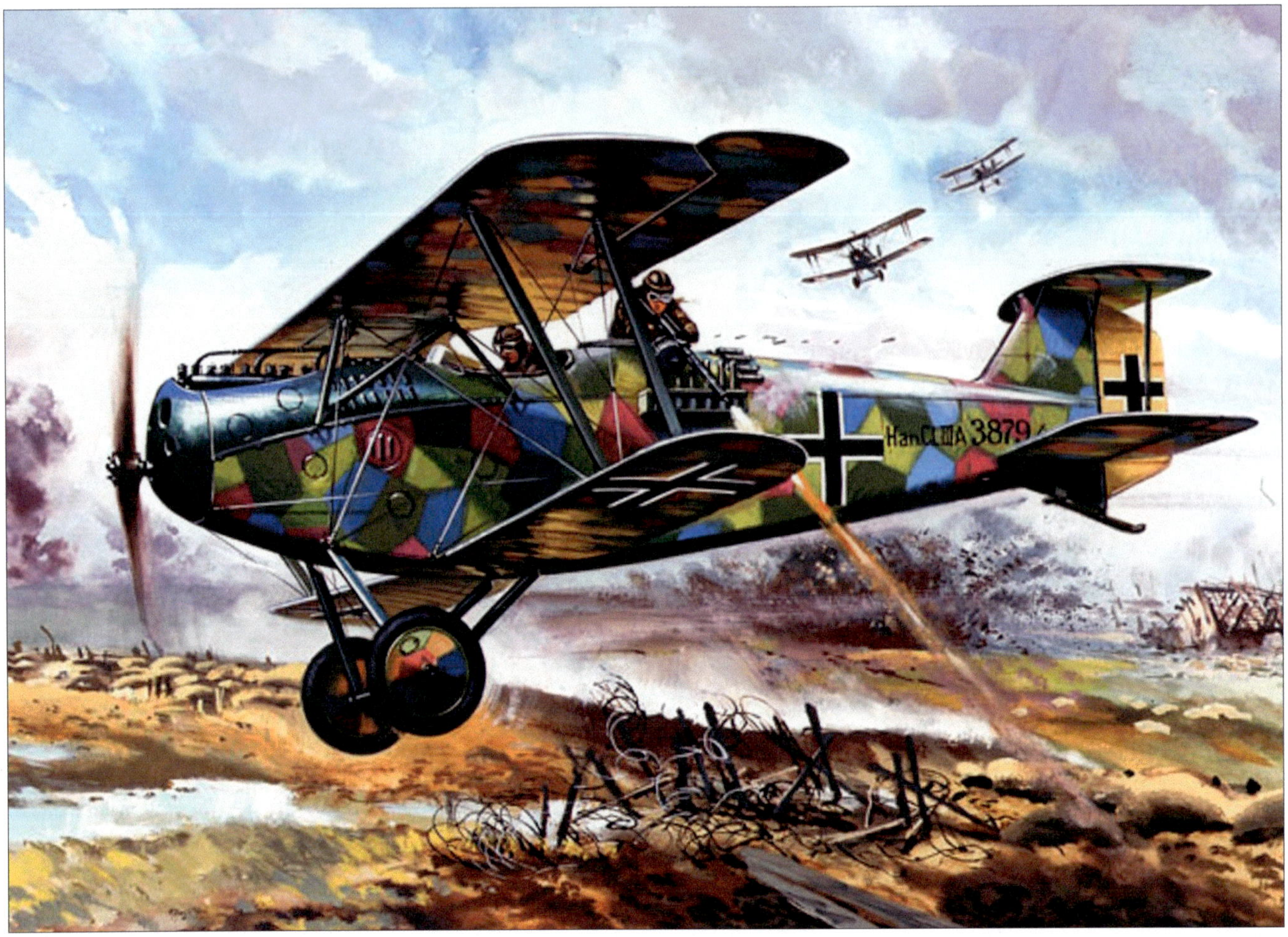

Above: Another Hannover CL II at low altitude in a classic example of support to the German infantry in the most contested spots on the battle field. Roy Cross

WIlberg organized the schlachstaffel into small units of 4-6 aircraft that could group into larger formations as needed to neutralize specific targets. These aircraft were fitted with armor to protect vital components of the aircraft: engine, fuel tank and the aviators themselves. The aircraft were fitted with machine guns for both pilot and observer and could carry small bombs and even hand grenades to drop onto the enemy troops.

The aircraft would fly in tight formations at an altitude of less than 100 feet and would attack the enemy that would be found just in front of the advancing ground troops. After expenditure of their ammunition, the aircraft would fly back to their small, advanced airfields to re-arm and would then immediately take off to hit their new target.

The aircraft had small, shared cockpits for the crew which facilitated the communication between pilot and observer/gunner.

The Halberstadt and Hannover CL (L-leicht, light) designs would prove to be very effective designs for their mission and would also prove to be very popular with their aircrew, who felt themselves to be in superior aircraft and performing vital tasks at a critical point in the war. Thus, the aviators in the schlacht staffeln carried themselves with a certain pride and elan during the 1918 Spring offensive.

A radical new aircraft design was the Junkers J.I, designed by Dr. Hugo Junkers, it was an all metal aircraft with noticeably thick-sectioned cantilever wings that was quite large in size. It's crews would love the aircraft because it provided them with incredible protection and was so robust in its strength, that no Junkers aircraft was lost in air combat during the war!

An incredible statistic considering that the Junkers crews mission was to fly very slowly and at very low altitude over the hottest spots of the battlefield in order to identify where friendly troops were and where the enemy was located. This mission was often accomplished by the simple fact that the Junkers crew would fly over portions of the battlefield until they were shot at, thus finding the location of sheltering enemy troops.

Principally this information would be utilized by

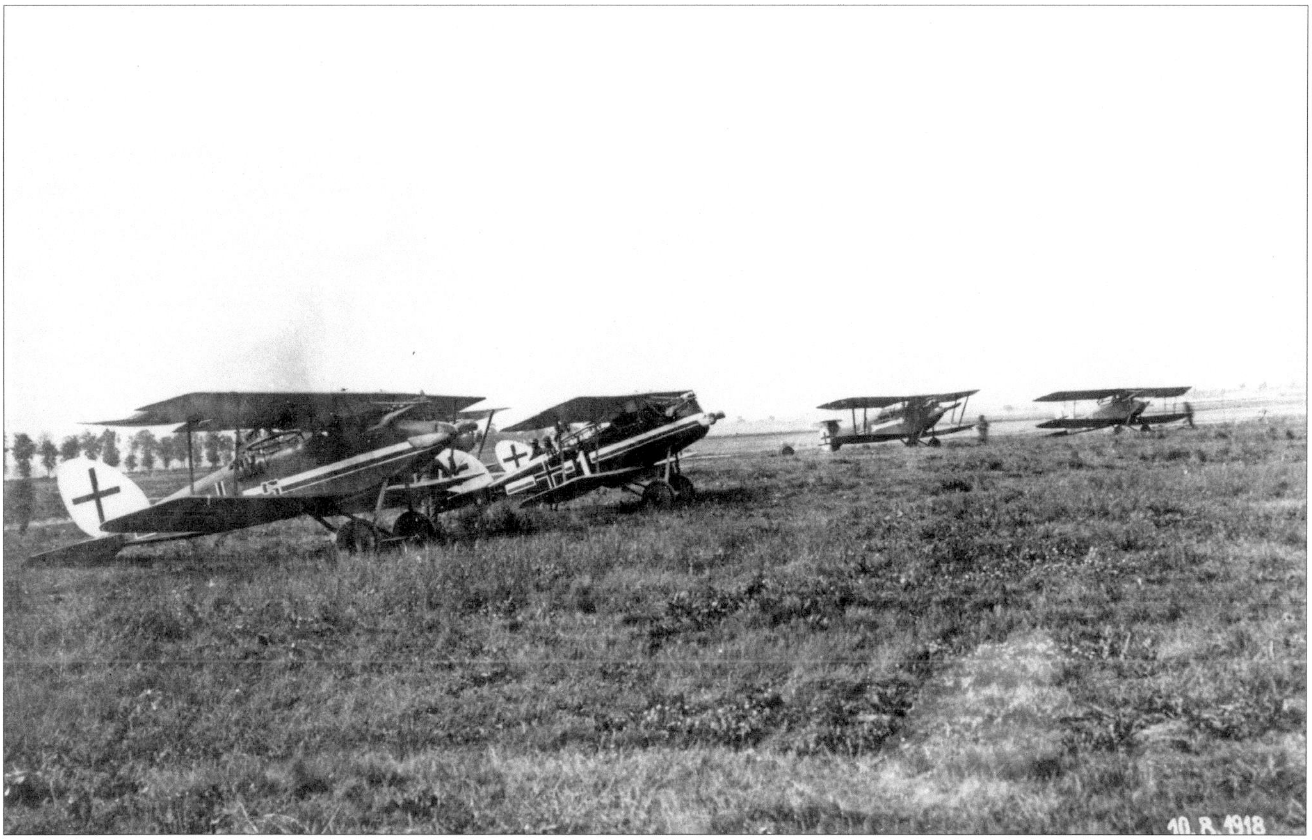

Above: Each Schlachtstaffel was composed of 4-6 aircraft and several units were put together as needed during the battle, here several Halberstadt CL aircraft are joined by another unit that was flying Hannover CL II aircraft.

the artillery units to fire at enemy targets, without hitting their own advancing troops.

The Junkers crew affectionately nicknamed the Junkers J.I the Mobelwagen (furniture van).

American fighter pilot, Charles Biddle, had first flown as a member of the Lafayette Flying Corps (LFC) and had learned very sound combat tactics while flying in escadrille Spa. 73 under the tutelage of the seasoned French ace, Capt. Deullin. Biddle would become an ace in the war and would actually rise to the rank of Major and placed in command of the 2nd pursuit group USAS before the war's end.

He was temporarily flying in the 103rd aero squadron of the USAS in May, 1918. It was during a sortie at that time that Biddle learned how tough a target the Mobelwagen actually was:

"On the morning of May 15 at about 9:30, Baker[Hobey], Lt. Baer and myself sallied forth, in response to a telephone call saying that there were a great many Huns on the line...I noticed a peculiar two-seater circling very low down between the trenches, he could not have been more than 600 meters up...he seemed to circle into the Boche lines with remarkable impunity considering his very low altitude, so decided to investigate...I said that this Hun was flying between the trenches as he was, but in this most terrible of all the battlefields that I have seen, you cannot distinguish the trenches from above, and in many places they consist of simply shell holes joined together...it was the slowest bus you ever saw and I think I could go two miles to his one. All this leads me to believe that it was a new type of German armored plane which they call "Junkers" and which i have read about...They are built especially for this low infantry liaison work and are heavily armored about the fuselage to protect them from fire from the ground, In consequence of their great weight they cannot go very high and are extremely slow. This fellow must have been a squadron leader or something, for he had four big streamers attached to his wings...He certainly got the best of me... I dove down after him and made up my mind this time to get at a good close range... but I made one bad mistake...I had not fired more than one or two shots when I got caught in the back draught from his propeller...so stopped shooting for a second to get in better position...He started to swerve from side to side to get me out from under him so that the machine-gunner could shoot...The Boche and I were at this time about twenty yards apart...that was sufficient for the observer, who proceeded to do the quickest and most accurate bit

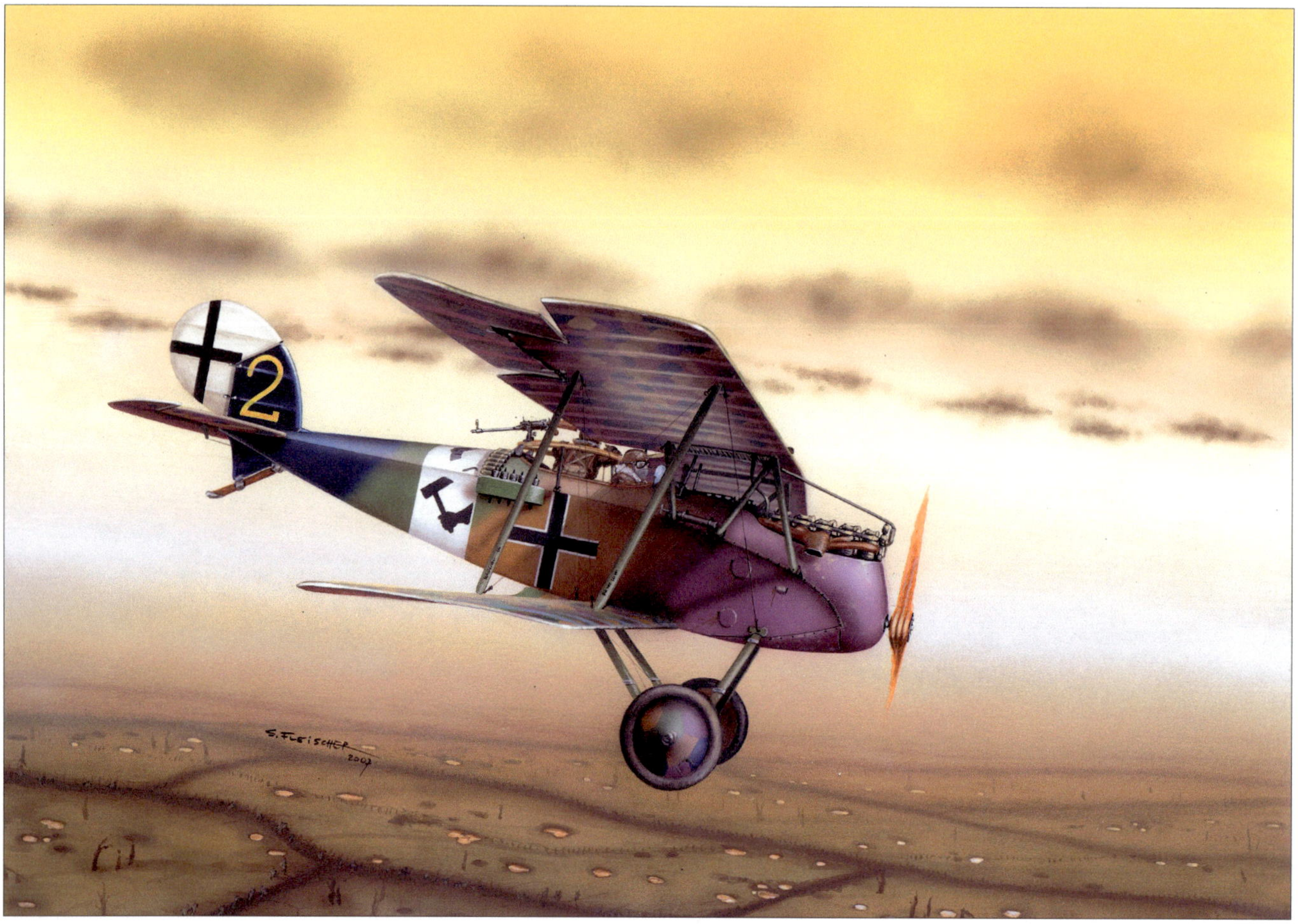

Above: The Halberstadt CL. II was an excellent design for its role in close ground support targeting precise spots to help advancing infantry, the essence of blitzkrieg tactics. Seweryn Fleischer

of shooting that I have yet run up against...his very first shot came smashing through the front of my machine above the motor and caught me just on top of the left knee." Charles Biddle, FIGHTING AIRMAN

Biddle's opponents that day were indeed seasoned veterans, pilot Feldwebel Ernst Schafer and observer Lt. Wilhelm Paul Schreiber (who would earn the Pour le Merite).

Under Wilberg's command, the ground attack units led the way as the infantry followed from one success to the next during the opening days of the German offensive in the Spring of 1918.

Eduard Wolfgang Zorer was an experienced pilot who was given command of one of these new staffels:

"I trained my Staffel in methods to use against enemy infantry, later artillery and other targets, and I led them in three types of engagements-frontal assault, attacking along a line, and assaulting a broad area, wherein the basic principle was to fly close together, with all aircraft unquestioningly and relentlessly following the leading aircraft through its movements at altitude and by flare signals from the beginning of the battle until it was broken off. Black, white and red streamers on the end of the wings and tail of the aircraft were our recognition sign, and, at the sight of them, our infantry were to display white panels that would enable us up above to determine the course of the most advanced lines."

The aircrew of the schlacht staffels obviously were exposed at low altitudes above the areas of the fiercest fighting along the battle front. Zorer, who was an observer/gunner, describes one of his missions with his pilot Vzfw. Schleiffer:

"We were at about 600 meters altitude. We made one pass along the Front in order to orient ourselves precisely in terms of the hard-pitched battle going on. In all of my life I had never seen such a murderous fire of all calibers.Up here we were threatened by death every second by shells[artillery] that flew right by us-first one side, then the other-with such turbulence that we were almost drawn into the abyss below...Now Schleiffer with his fixed machine gun and I with my movable machine gun alternately began to shoot at those fellows...

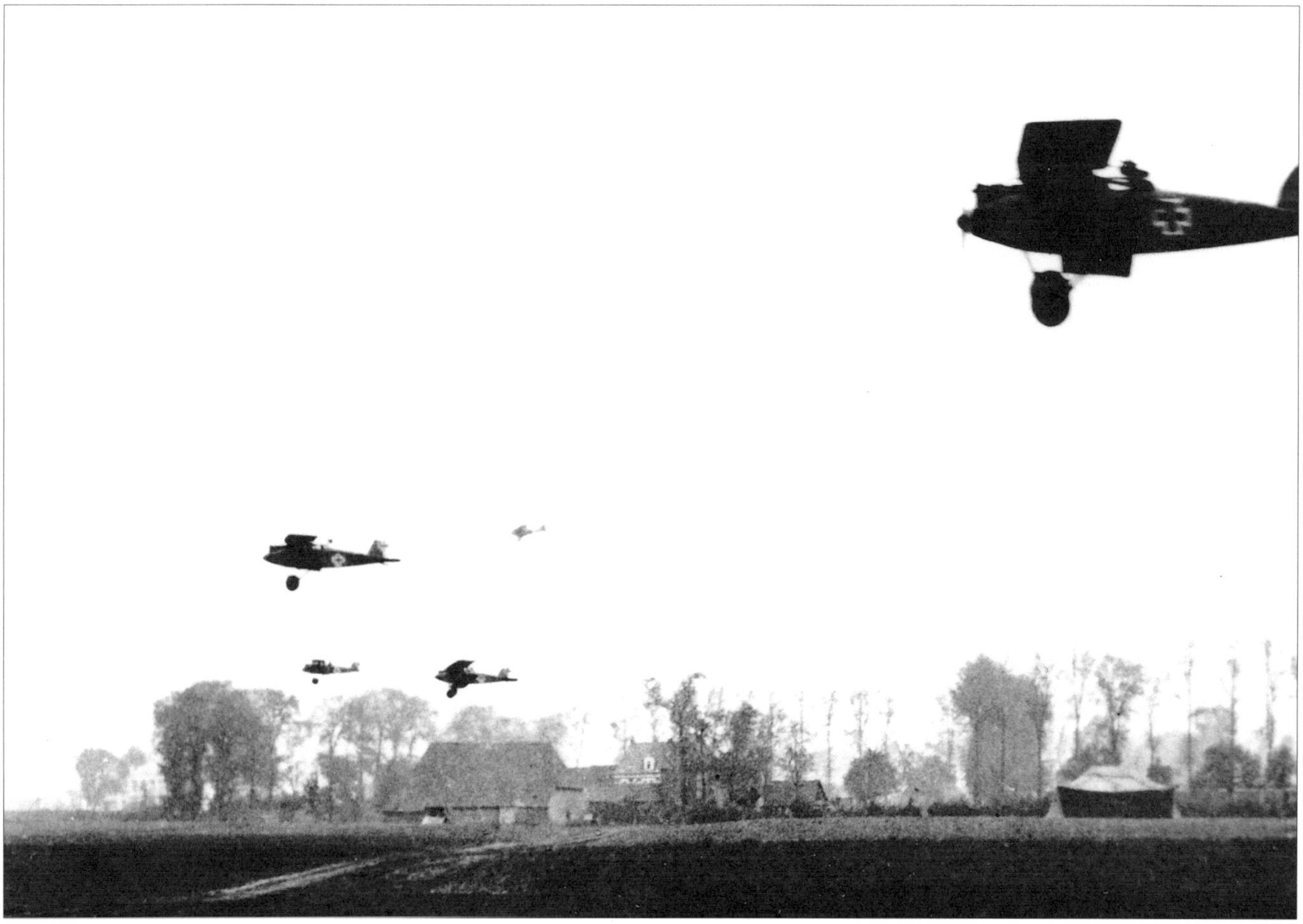

Above: The German attack aircraft flew very low, often only a couple hundred feet off the ground during their missions. Here a flight is headed towards the front lines.

Below: Flying so low naturally put the aircraft and aircrews at great risk from ground fire; here a crew shows their extensively damaged aircraft. That they were able to return to their airfield is a testament to the ruggedness of the Hannover CL type.

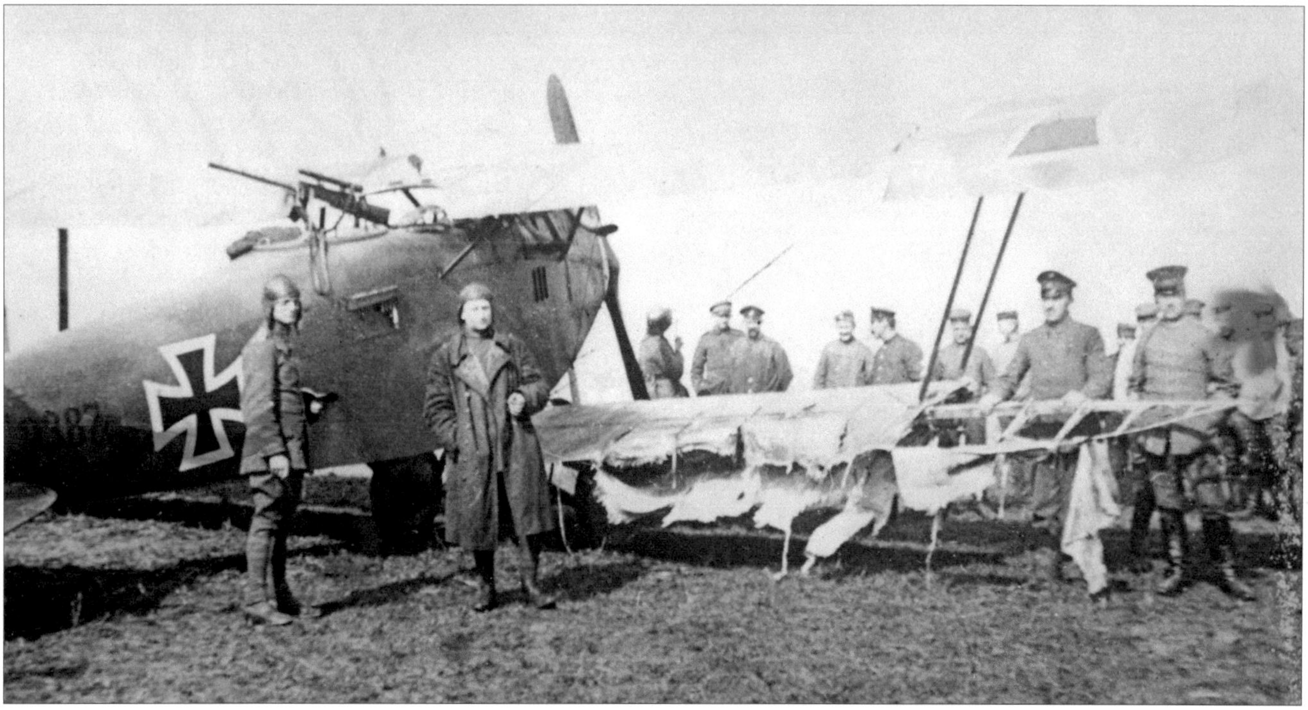

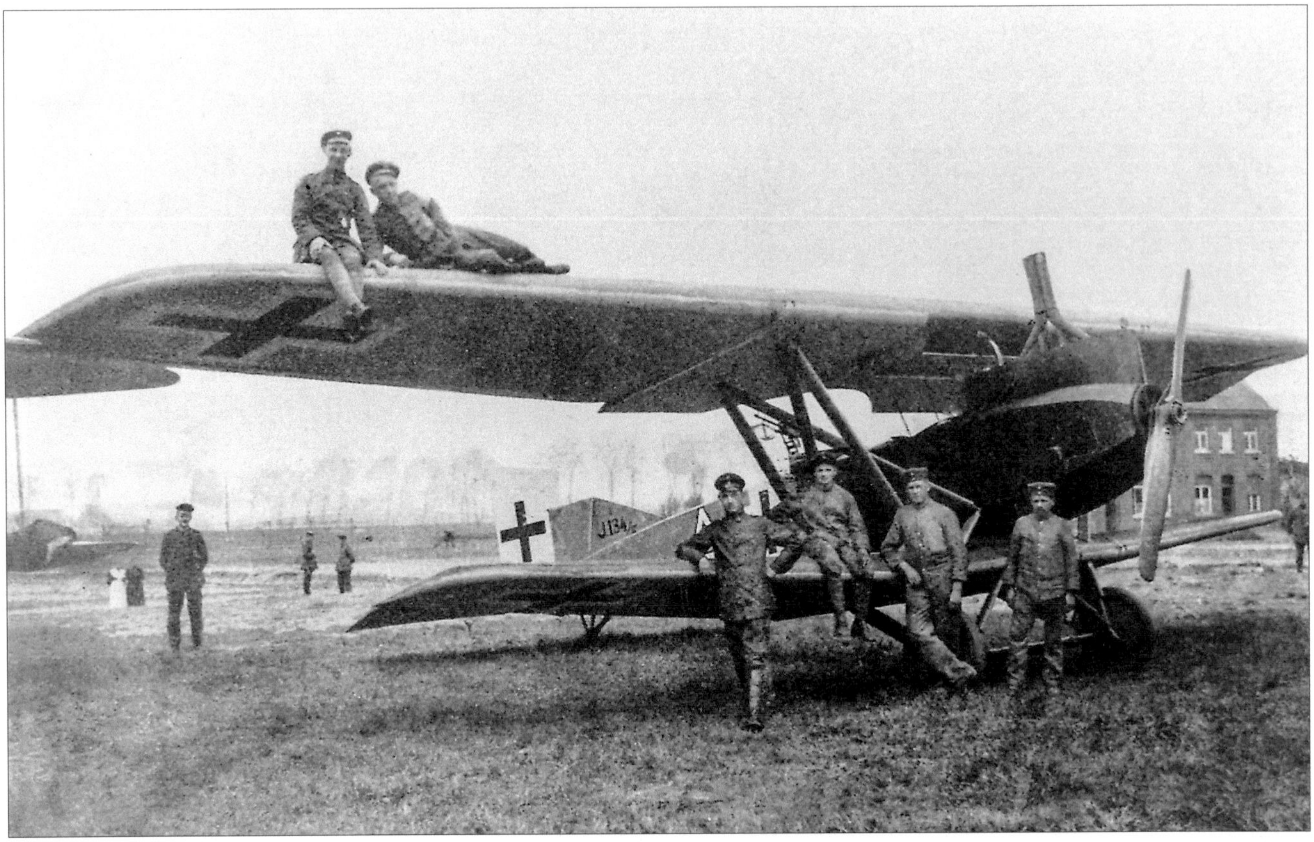

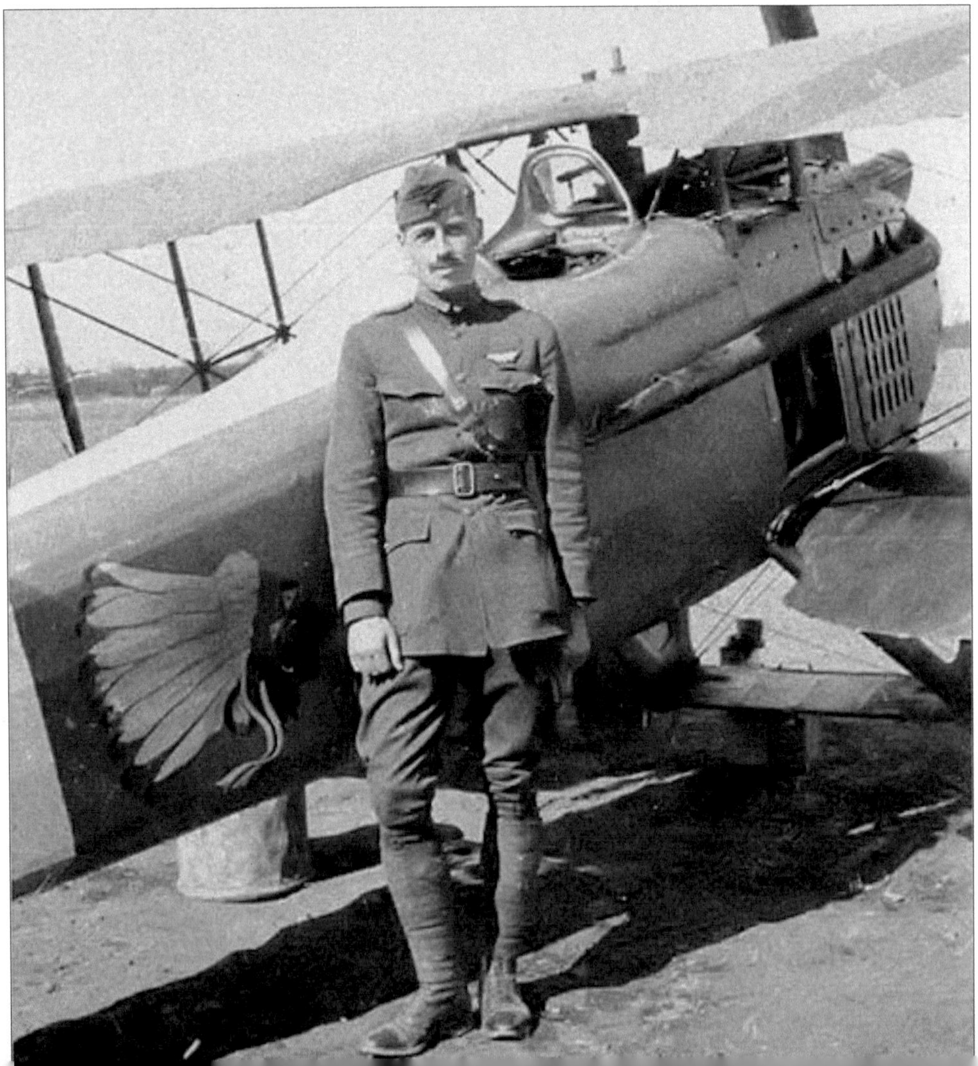

Above: A big improvement in the attack aircraft came with the all-metal Junkers J. I, nicknamed the Mobelwagen (furniture van) by its crews who were fond of it since it provided such excellent protection to them. This photo shows the very aircraft that American Charles Biddle tangled with and its aircrew is seen on the upper right wing: pilot-Ernst Schafer and observer-Wilhelm Schreiber.

Left: Charles Biddle while with the 103rd aero squadron of the USAS, which had formally been the S.124 escadrille Lafayette in the French Air Service, note the unit insignia on the aircraft.

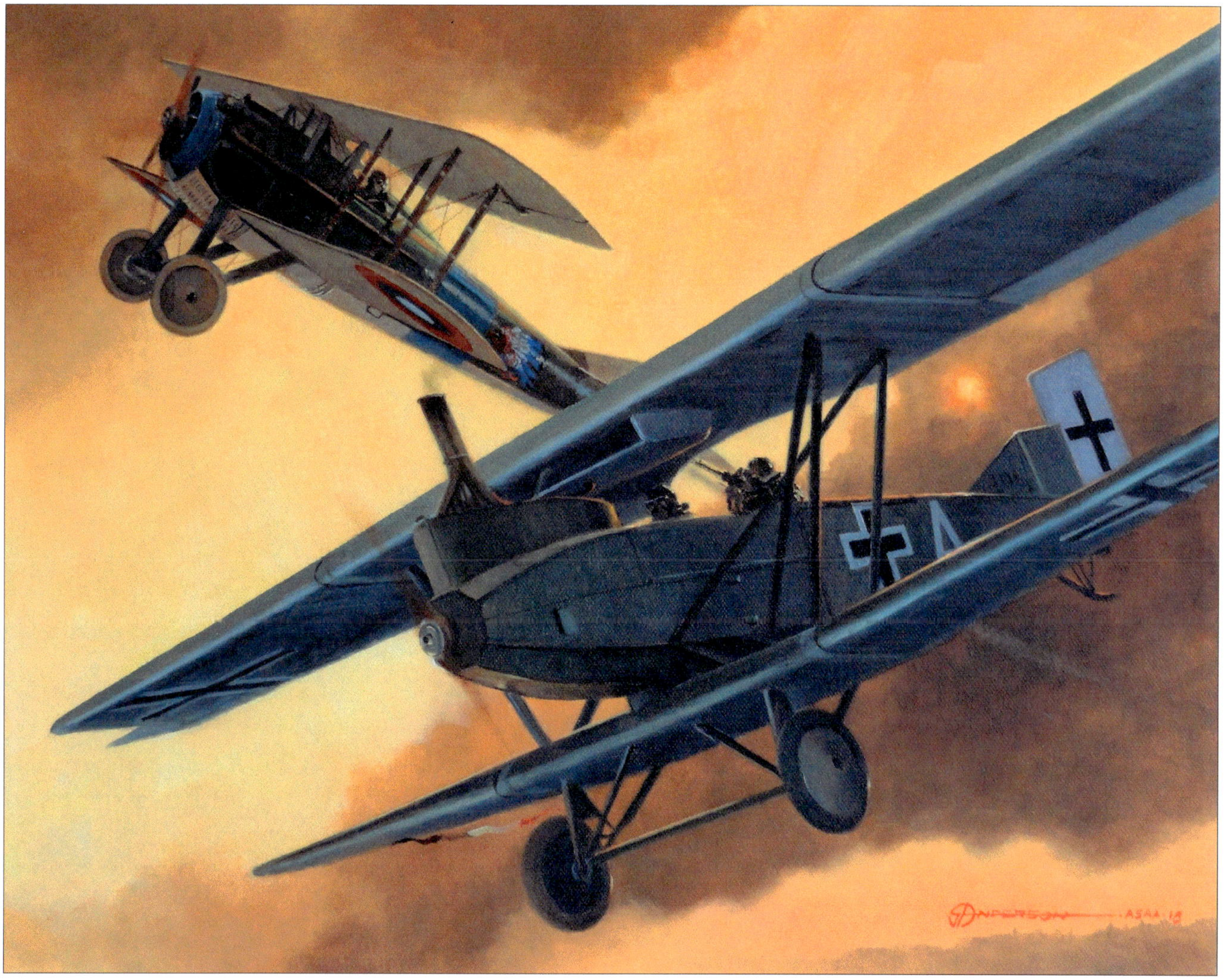

Above: American fighter pilot Charles Biddle finds out how difficult a Junkers J.I can be as he is shot down by its observer-gunner Ltn. Wilhelm Schreiber, who was later awarded the Pour le Mérite. Steve Anderson

But, gradually, the situation for us became ever less pleasant and ever more dangerous. Round after round from rifles and machine guns whistled by our ears...Our oil line broke and covered us both with hot oil, thus requiring us to break away and make an emergency landing."

It is not an exaggeration to state that by 1918 the German AIr Force's principal offensive weapon were the ground attack units. Under the command of WIlberg, they proved to be an extremely cohesive combat force and would fight in an effective manner to the very end of the war.

Despite their best efforts, the spring offensive failed and the German military was forced into defensive mode until the war's eventual end in November, 1918.

However, their efforts proved the capacity of air power to effectively not only support, but actually lead in a rapidly evolving war of maneuver that would be known as blitzkrieg.

Among the stipulations in the Treaty of Versailles, was that the German Air Force was to be severely restricted and essentially disbanded. Col. von Seeckt tasked Capt. Wilberg to draft the German proposal for their post-war air service. It was rejected by the Allies, but von Seeckt was impressed by the professionalism and capability of Wilberg. So impressed that he named Wilberg to be the air chief of the Truppenamt in 1919.

Among the small group of German military leaders in the 1920's and early 1930's, Wilberg was considered the leading military air theorist. In 1925, he led a delegation that traveled extensively throughout the United States visiting air bases and reviewing the first line military aircraft of the day.

With the growth of the Air Force, WIlberg was a leading candidate to be named chief of air staff. That would be impossible in Nazi Germany, because of

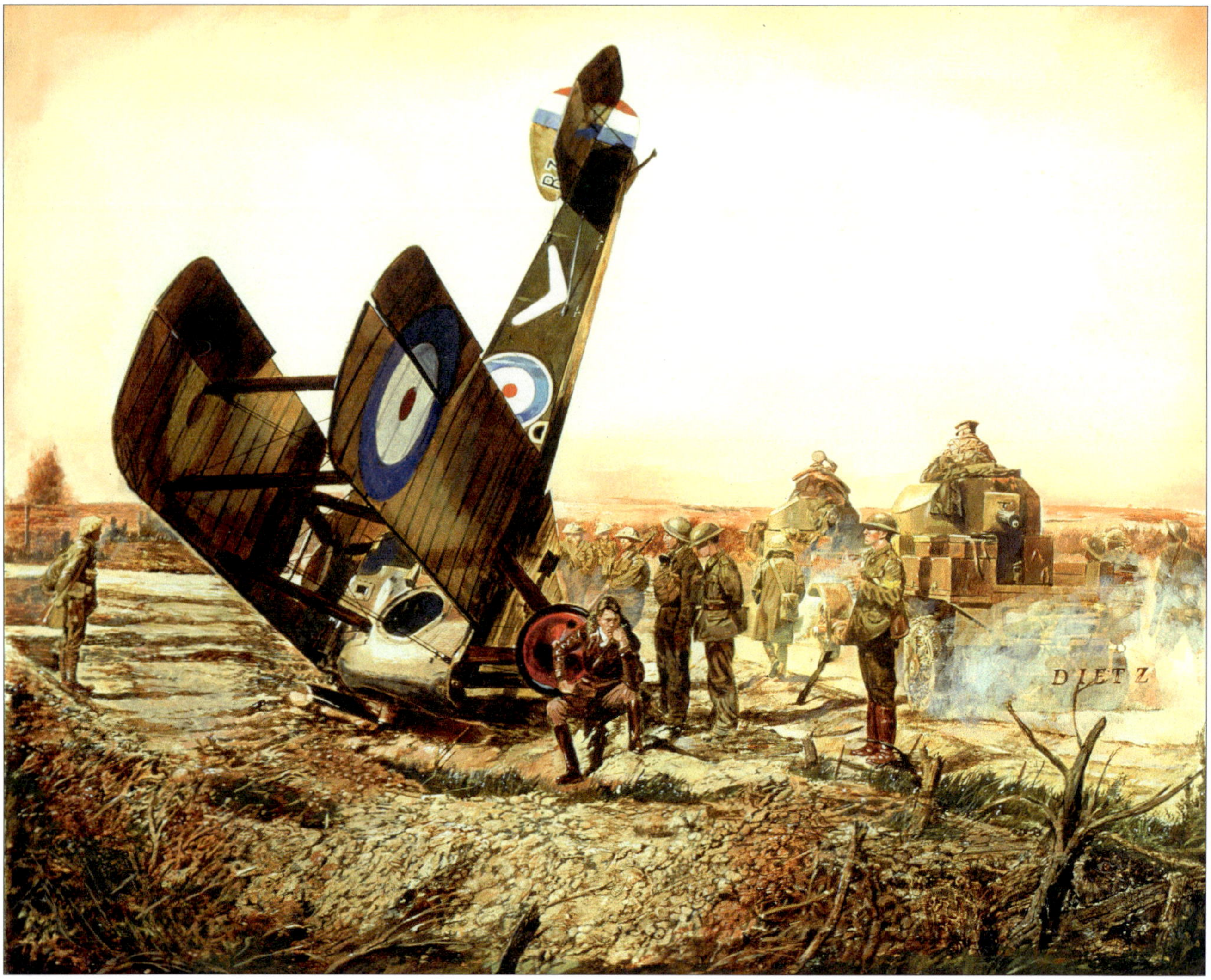

Above: Surprisingly, the RAF was slow to develop their own low-altitude, close support aircraft despite the German's success. Instead, they pressed existing squadrons into that role, much to the intense complaints of the airmen who had previously been fighter pilots. Loss rates in the squadrons went way up because of the dangers of the low flight. Here an Australian airman has made it back, just, to his side of the lines after such a mission. One of the small Cooper bombs that were rigged to the Sopwith Camels is still seen on the underside of the airplane. The pilot and airplane will be picked up be a tender from his squadron and returned, both to fight another day. Jim Dietz

the fact that his mother was Jewish. Goring himself, not desirous of losing such an experienced and talented officer to the nascent Luftwaffe, signed a special order that declared WIlberg and his family to be Aryan.

In 1935, Wilberg was a senior officer and played a key role in the development of the new Luftwaffe's strategy and tactical doctrines. He was the committee chief and editor of the manual, Regulation 16: the conduct of aerial war. This book would become the foundation of Luftwaffe plans for WWII.

On the 26th July 1936, Lt. General Wilberg was ordered to form a special staff with the purpose of organizing Luftwaffe assistance to the fascists in Spain. Special staff W (for Wilberg) was thus created. Within only four days, aircraft, pilots and ground crew were embarked to Spain.

In history's first major airlift, Junker JU-52 transport aircraft were used to bring 13,000 Moroccan troops and their equipment from North Africa to the Spanish mainland.

General der Flieger Helmuth Wilberg formally retired in 1938. During WWII, he was called back to service and was placed in charge of an advanced training/piloting school. He died in the air crash of a Junkers Ju-52 transport aircraft while on his way to attend the funeral of the German fighter ace Werner Molders.

Alpine Eagles

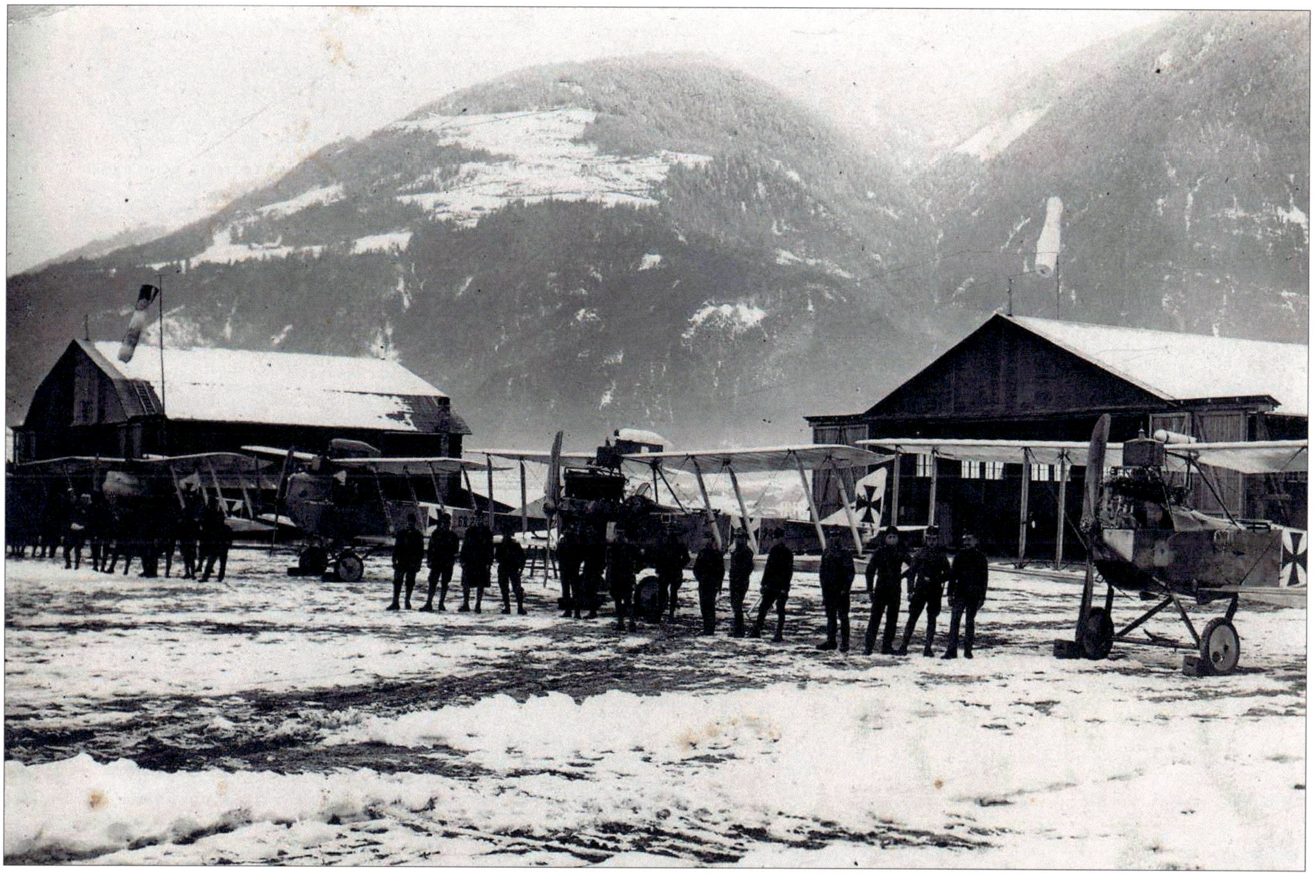

Above: Almost all of the Austro-Hungarian aces flew the durable and ever reliable Hansa-Brandenburg C.I at some point in the war. Most aviators would fly the two-seater at the front before being transferred to a jager(hunter/fighter) unit. This photo also shows well the ever present mountains and inhospitable terrain that the Alpine Eagles would face.

"It is amazing they did as much as they did, all things considered. The flyers of the Austro-Hungarian Empire had to surmount the end products of an empire and a system that were severely flawed and doomed to eventual failure."

Dr. Martin O'Connor

It is one of history's ironies that the First World War was essentially started by the Austro-Hungarian empire and that the war eventually destroyed that very same empire. It had been a critical part of European history and crucial to the politics of the prior 300 years or more. It was the remaining vestige of the Holy Roman Empire and in its borders the empire optimistically harbored what was considered a multi-ethnic collaboration that in reality represented an ever fraying grouping of antagonistic, nationalistic minded multiethnic regions of people. What "on paper" seemed to be one of the largest empires in Europe in the autumn of 1914 was actually an empire teetering on the brink of collapse, internally frayed by its multiple, distinct ethnic populations of people often within overlapping and contested geographic regions.

The Austro-Hungarian armed forces were woefully unprepared for a modern war and yet they were rapidly forced to contend with four separate fronts at the same time: Italy in the west, Serbia in the south, Russia in the east and then another separate front opened when the Allies supported Macedonia with sizable numbers of men and material.

The citizens of the empire accounted for 14 separate languages and they all lived under a very strict, archaic social system that was based on an eastern feudal system only minimally changed during the previous centuries. Which mattered, because as the Austro Hungarian army was sent to these fronts, there would be numerous soldiers within that army that actually felt more kinship for

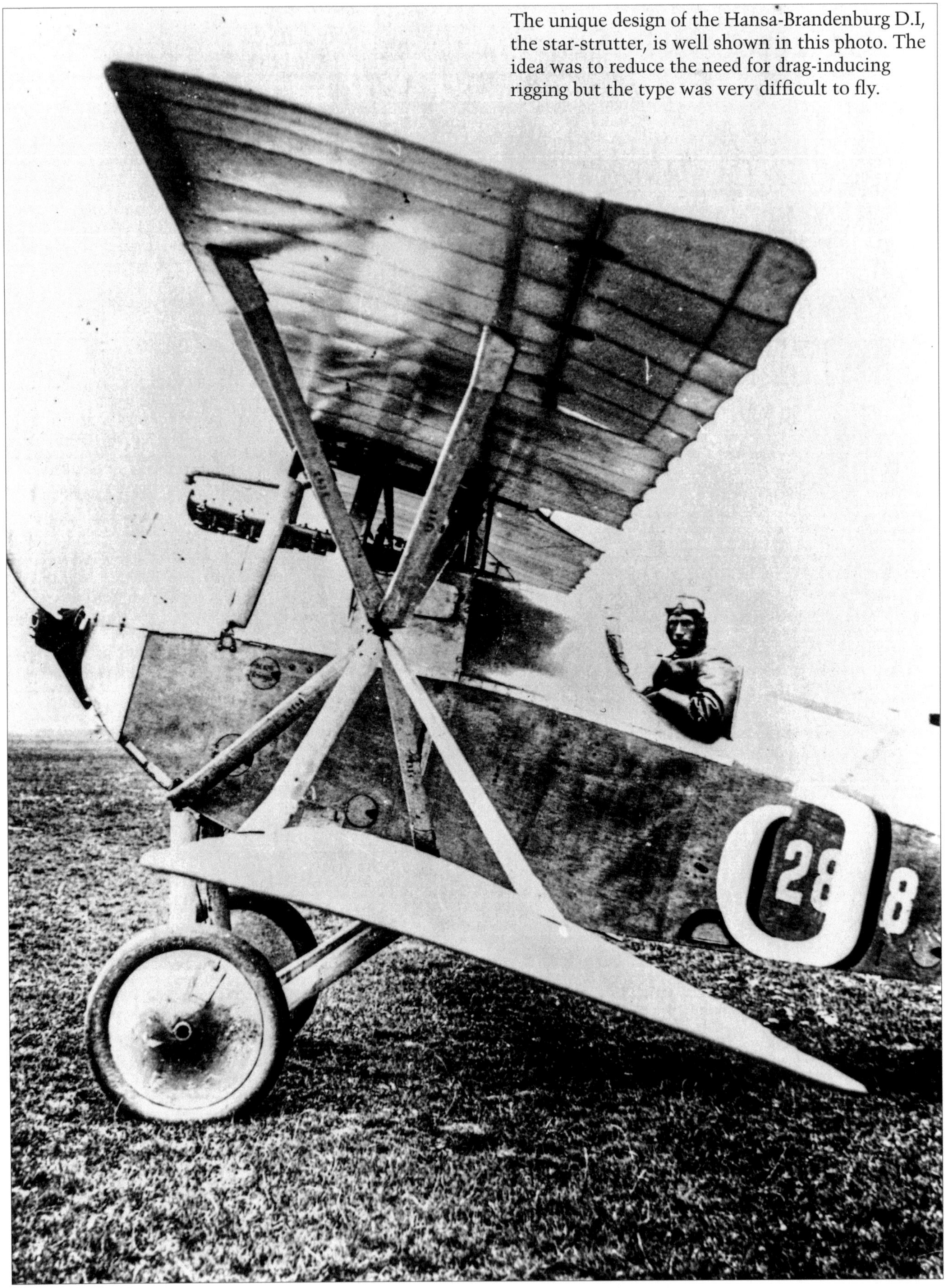

The unique design of the Hansa-Brandenburg D.I, the star-strutter, is well shown in this photo. The idea was to reduce the need for drag-inducing rigging but the type was very difficult to fly.

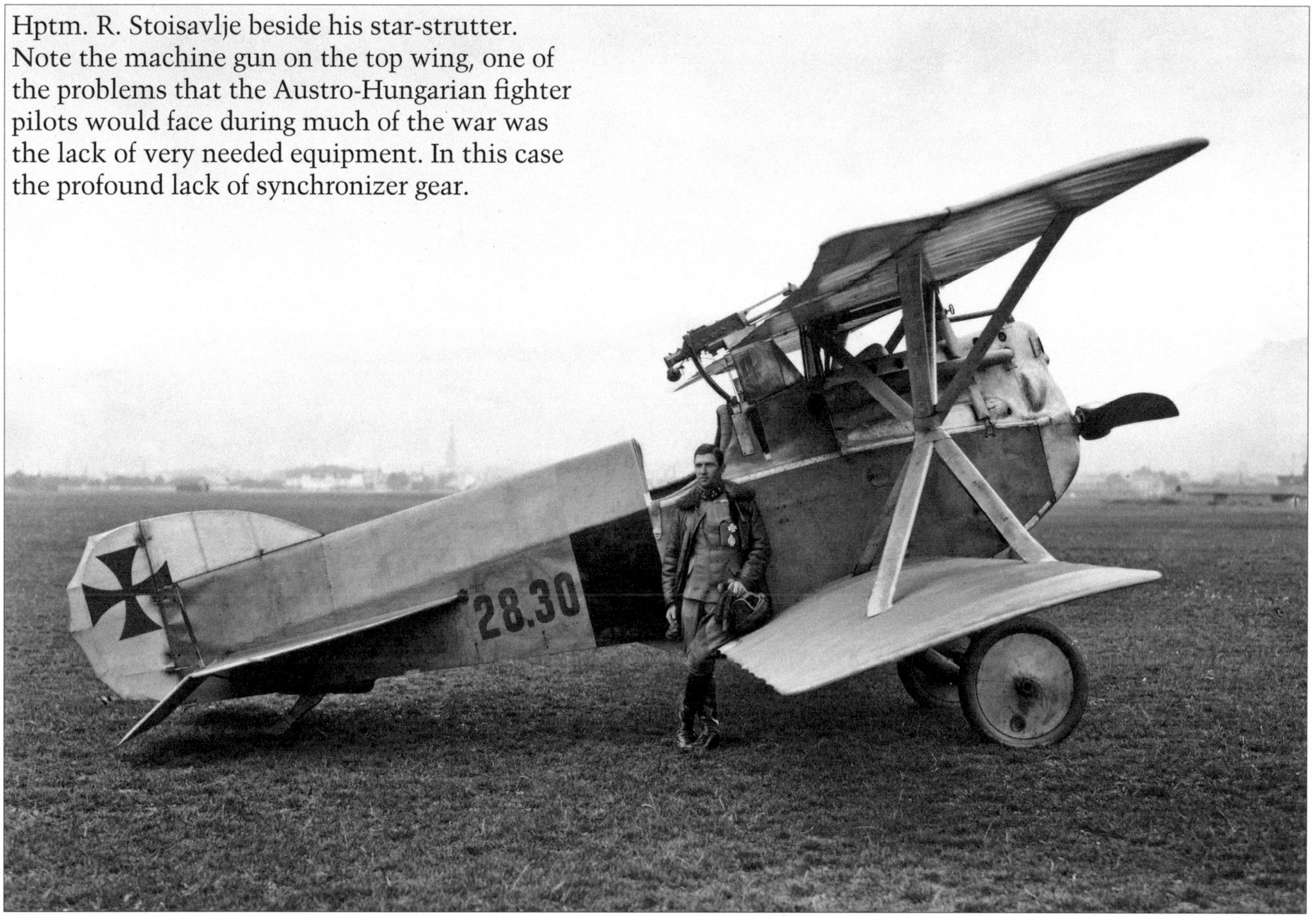

Hptm. R. Stoisavlje beside his star-strutter. Note the machine gun on the top wing, one of the problems that the Austro-Hungarian fighter pilots would face during much of the war was the lack of very needed equipment. In this case the profound lack of synchronizer gear.

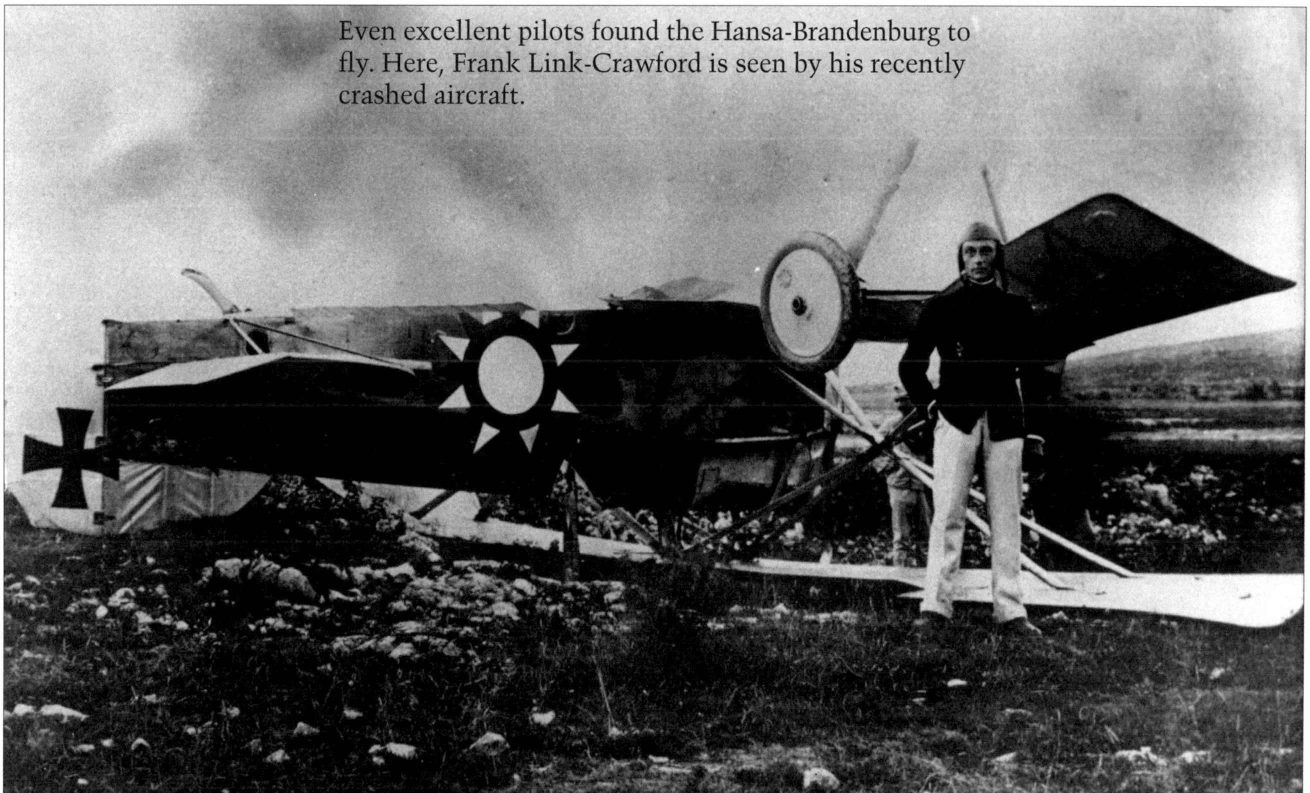

Even excellent pilots found the Hansa-Brandenburg to fly. Here, Frank Link-Crawford is seen by his recently crashed aircraft.

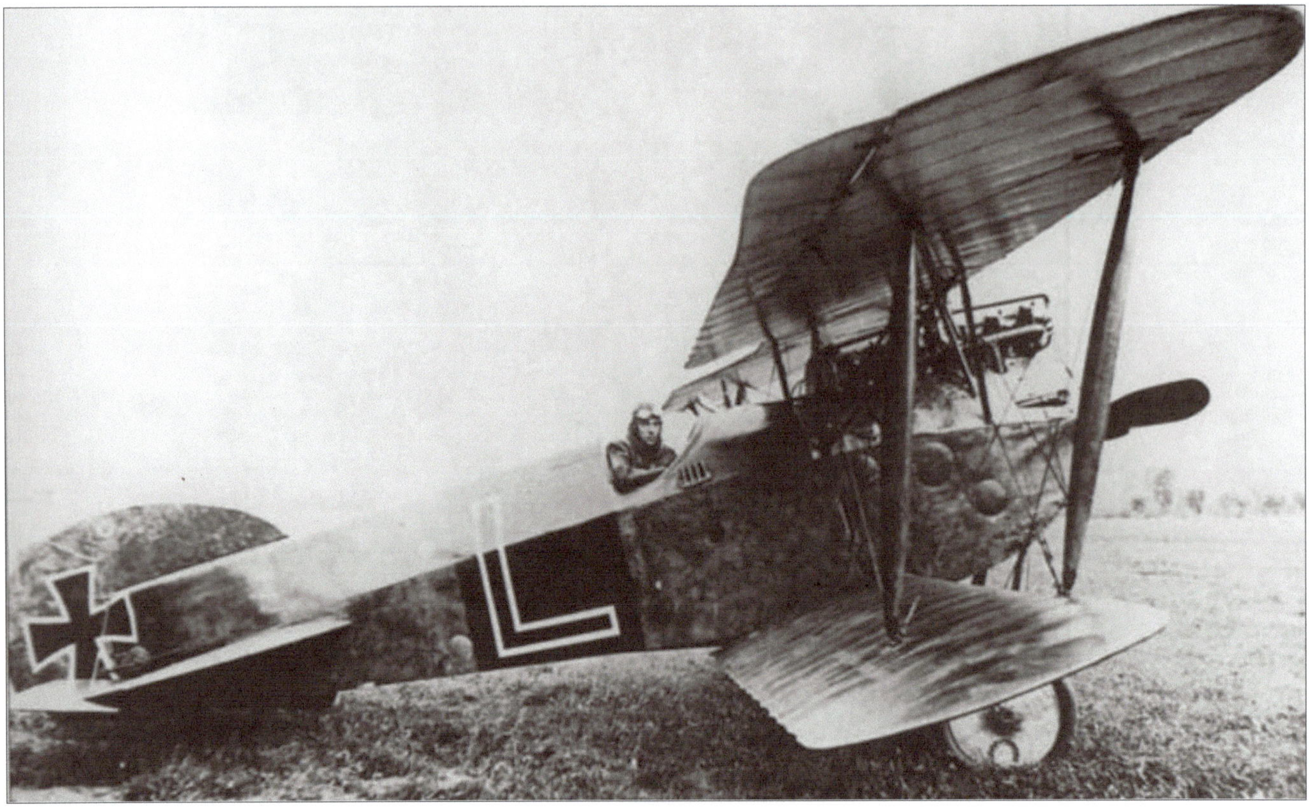

Above: The Phönix D.II and D.IIa fighters were based on the Hansa-Brandenburg but were a much better design and one that found favor with the frontline fighter pilots of the LFT.

Below: Josef Kiss takes off in his Phönix D.I fighter, note the mountain sides awaiting a crash if anything goes wrong. Al Forbes

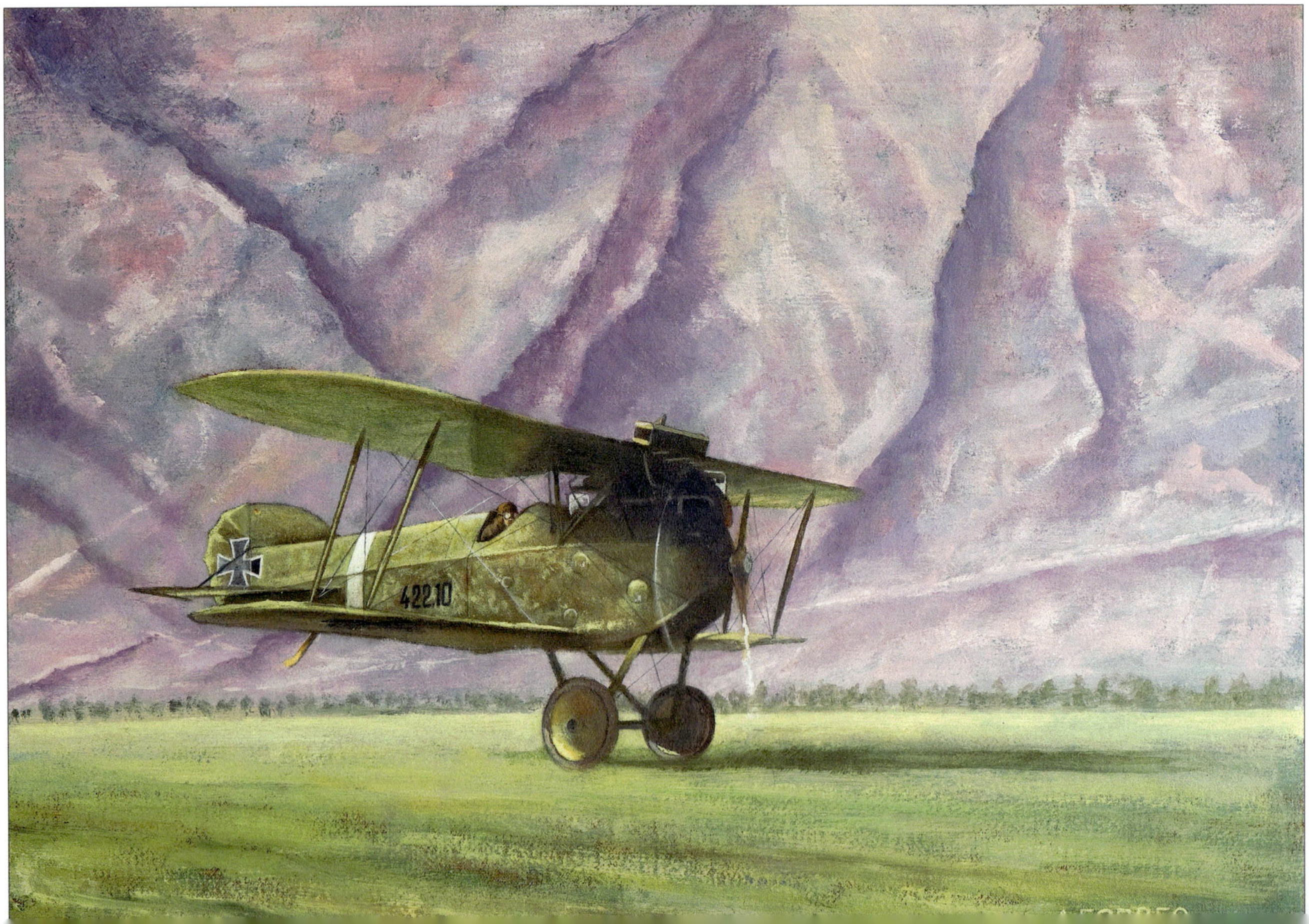

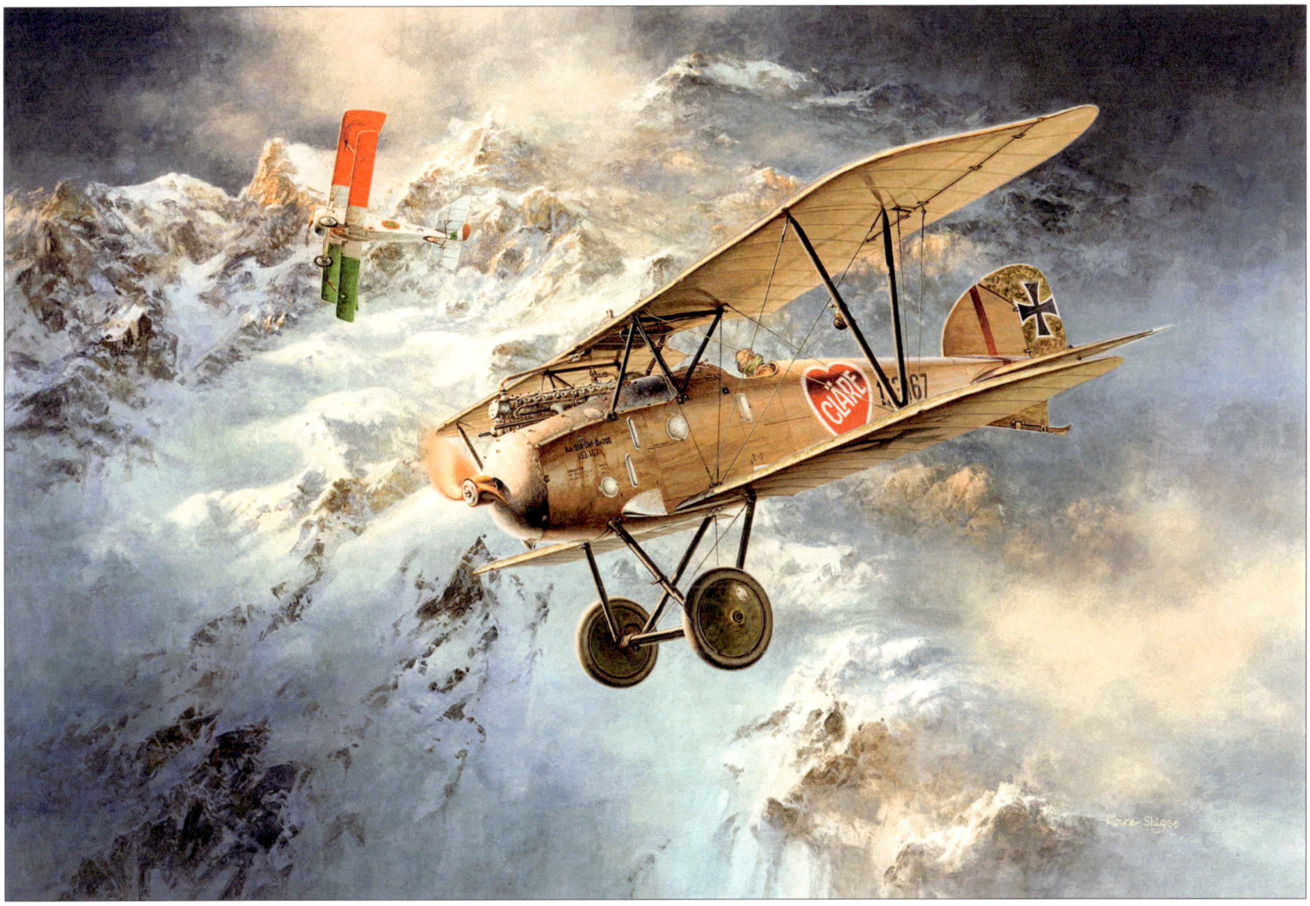

Above: An epic dogfight above the inhospitable terrain of the Alps. Shigeo Koike

the troops they were sent to fight. With the official language of the military being German, and with the stubbornly strict caste-like distinctions separating the officers from the men: many units could not understand commands given by their officers without translation and in essence, the men felt very little loyalty to those same officers.

To further complicate the matters faced by the Austro-Hungarian military, the empire was still primarily agrarian in nature and had very little overall industry to support their expanded military needs. During the course of the entire war, Austro-Hungary could manage the production of only 5,000 aircraft and approximately 4,000 aero engines. Italy alone produced 20.000 aircraft and nearly 38,000 engines in the war! At its maximum effort, the Austro-Hungarian Air Service (the Luftfahrtruppen, LFT, aviation troops) had 500 operational aircraft to cover the four different fronts it found itself contesting.

The aviators of the Austro-Hungarian would answer their call of duty with devotion and courage. They made the sacrifices asked of them until the very end of the war without wavering in the slightest. They paid the price for their service: approximately 38 % of them were killed in the war. These Alpine Eagles were undoubtedly brave pilots and unfortunately for them their efforts were severely handicapped by their leaders in so many ways.

One of the most important impediments to their success was Camillo Castiglioni, one of Europe's richest individuals and a man that would come to control much of the aviation industry in the Austro-Hungarian empire. Before the war began, Castiglioni bought the Lohner aircraft company which was the principal provider of aircraft to the LFT and then Castiglioni merged all of Lohner into his Hansa-Brandenburg company. This action then made certain that he had complete control over the provision of aircraft to the LFT. During the war he would expand his control over the aviation industry within the Austro-Hungarian empire when he acquired the UFAG and Phonix firms as well. Castiglioni had incredible political power and used that power to help dictate which companies got contracts, even if the aircraft designs were not deserving of production.

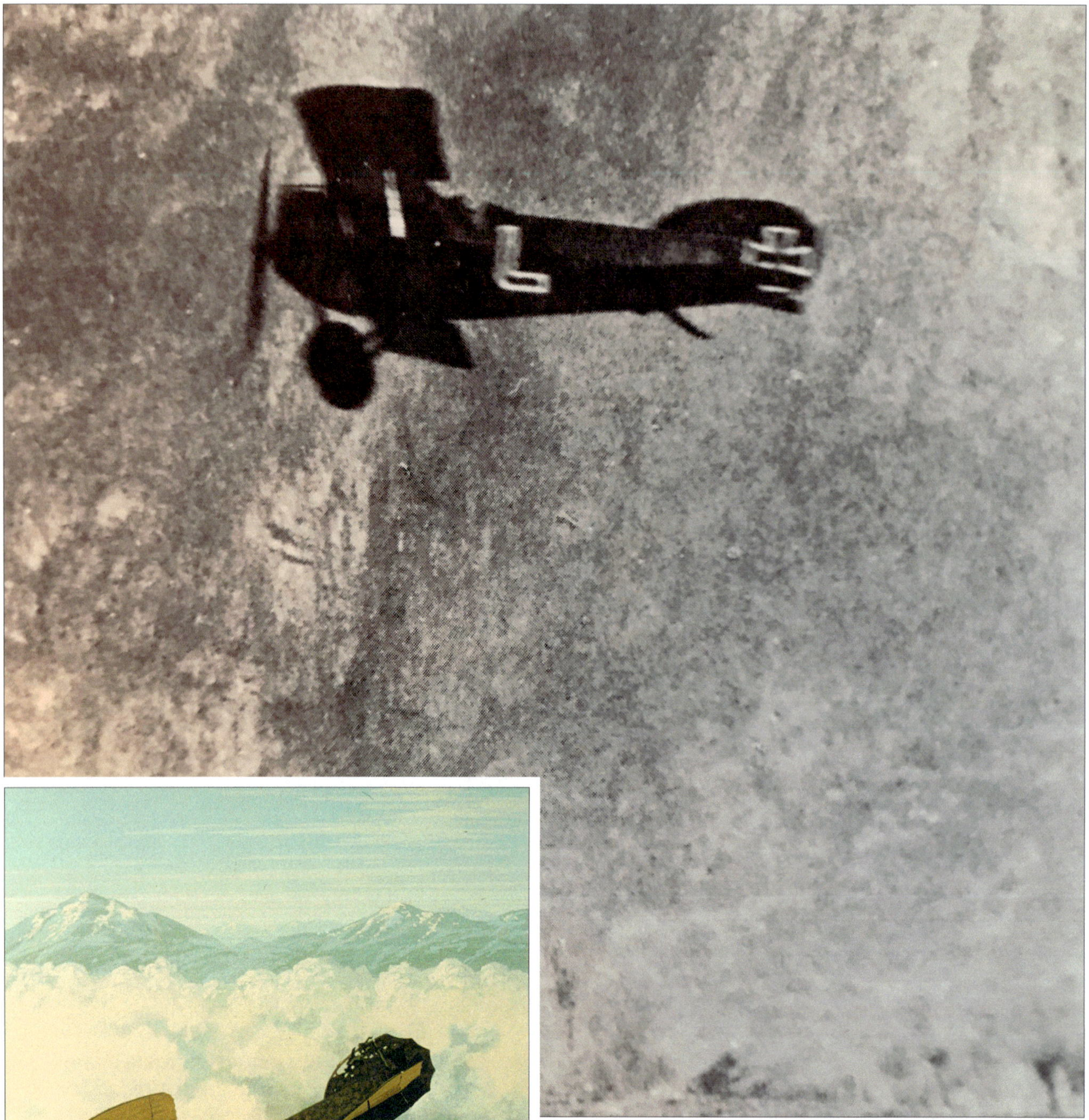

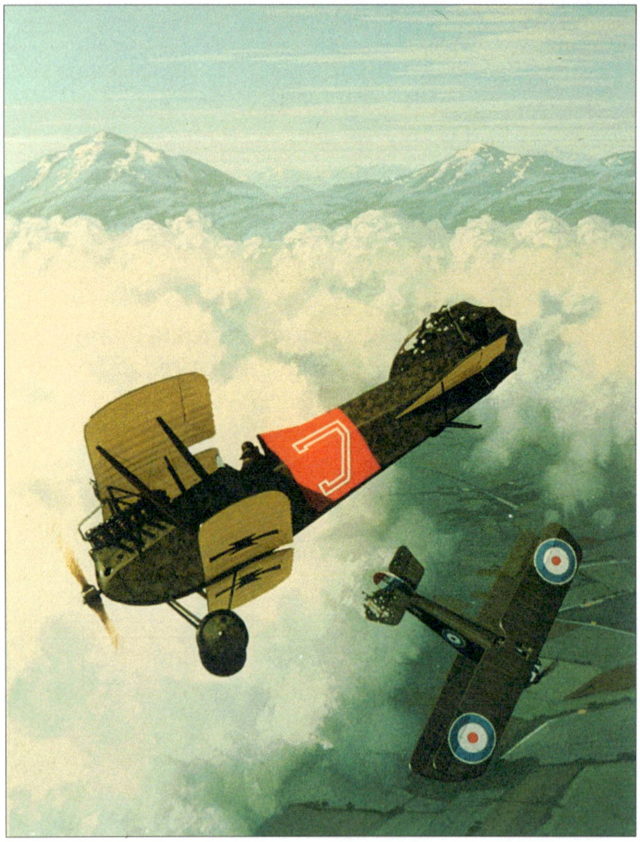

Above: The prior photo showed Link-Crawford in his Phonix D.II , here he is in flight, note the steep mountain terrain near his airfield.

Left: Linke-Crawford in an intense dogfight with an English Sopwith Camel. Keith Woodcock.

Another factor was simply the number of aircraft companies: Aviatik, MAG, OEFFAG and Lloyd in addition to the ones owned by Castiglioni. The government was keen on getting some contracts to all the companies and to ensure that these production facilities were spread out throughout the

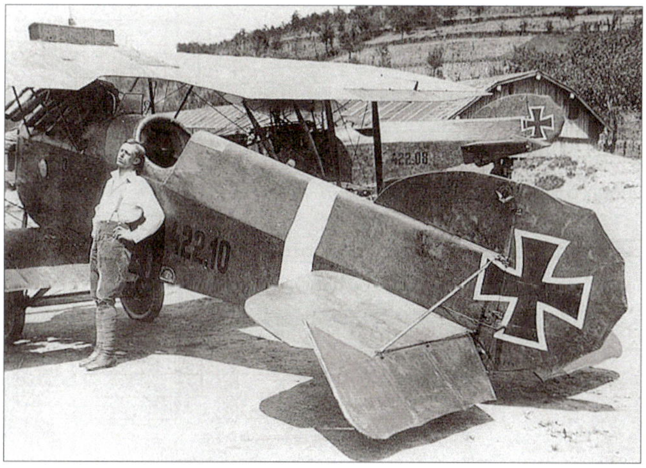

Above: Josef Kiss, enjoying the summer sunshine, the single white band was his personal insignia.

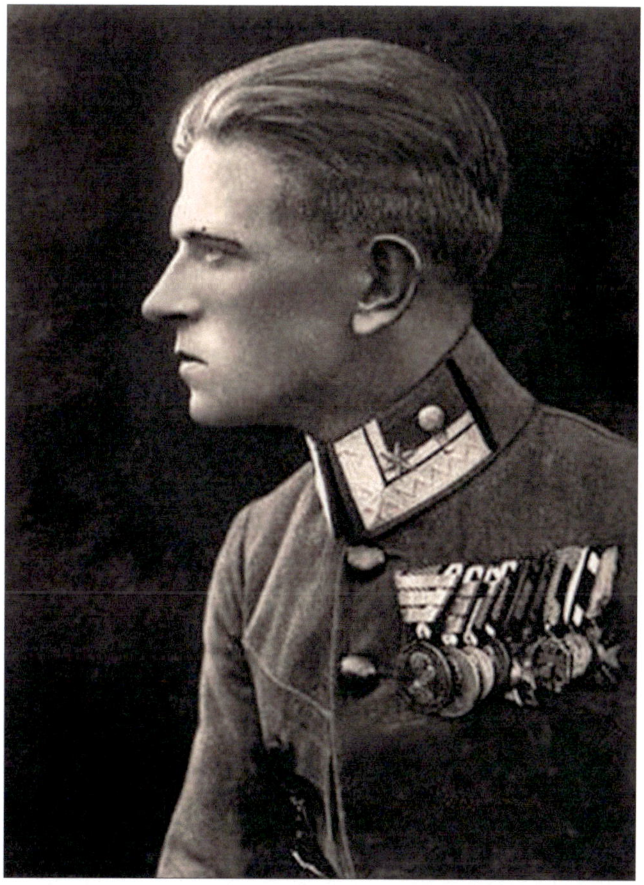

Right: Julius Arigi was one of the members of the well-known Kaiser Staffel.

Below: Line-up of the Phoenix D.I fighters of the Kaiser-staffel. The aircraft on the bottom left was flown by Alexander Kasza, a very good friend of Josef Kiss. Kasza added the heart to his aircraft after Kiss was shot down and killed.

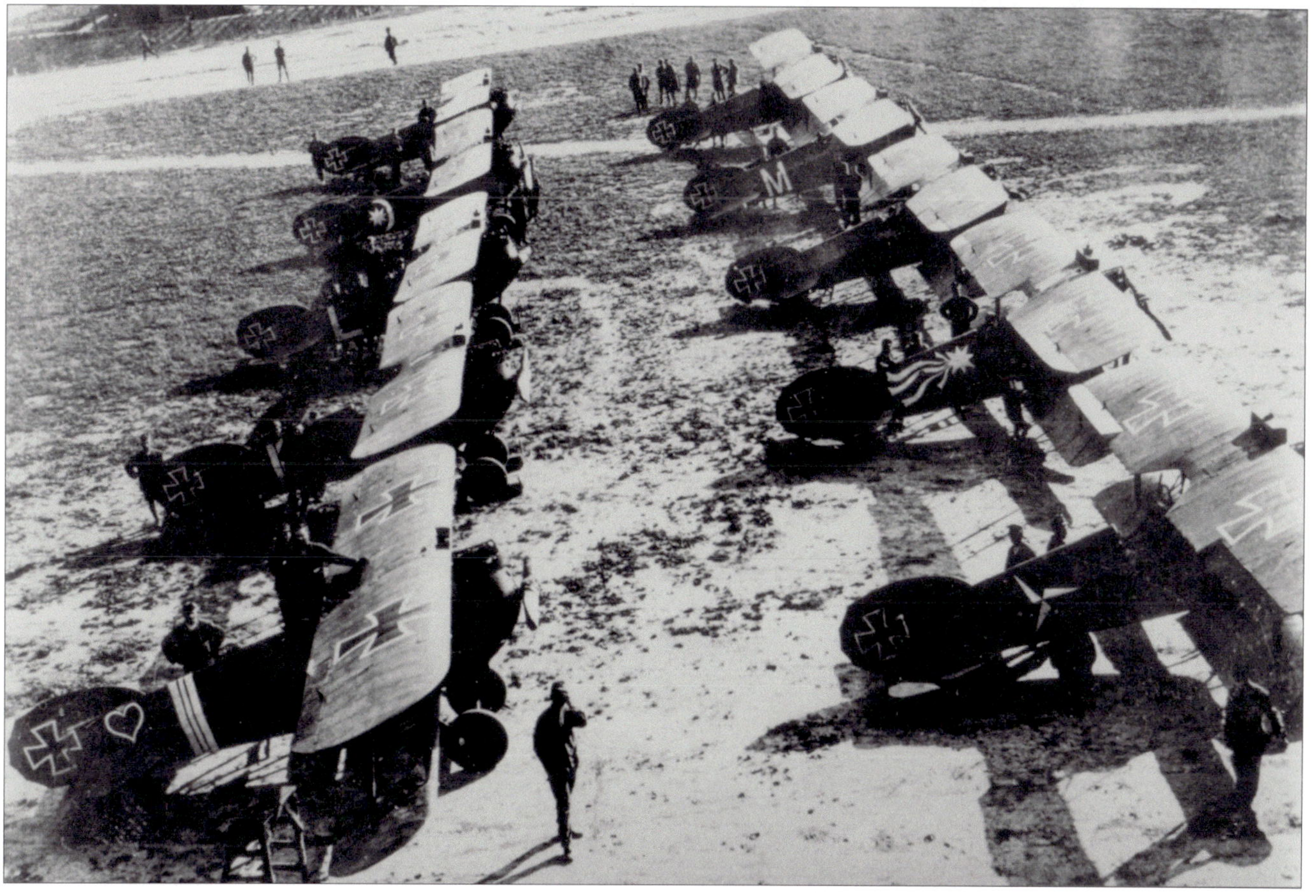

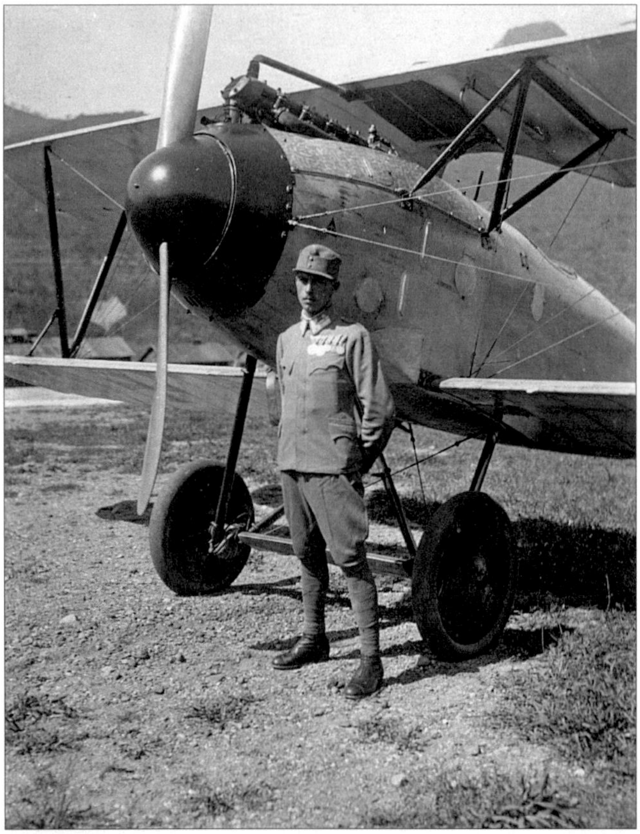

empire in an attempt to appease all of the different ethnicities in their vast empire: so that money and contracts flowed in a manner to appease the various groups that were otherwise contemplating independence from Viennese control. These choices would ultimately lead to a very diluted capability that lead to a chronic undersupply of capable combat aircraft for the frontline aviators.

Germany recognized the profound inability for its ally to provide enough sound aircraft designs and noted its production problems. Therefore, during the war several German aviation firms bought Austro-Hungarian firms and ran them as subsidiaries to produce aircraft of German design: Albatros bought Phonix, DFW bought Lloyd and Aviatik opened a manufacturing facility near Vienna. Germany could not simply send aircraft that it produced to help

Left: Josef Kiss beside his Oeffag Albatros D.III at Pergine airfield.

Below: Kaiser Karl I visited Pergine airfield on 14th September 1917 to review Flik 55J, "his" Kaiser-Staffel". Note the narrow mountain valley that created a very tight space for an airfield. At the end of the valley was a mansion, Villa Giulia della Rosen, which was used as a billet by the aviators.

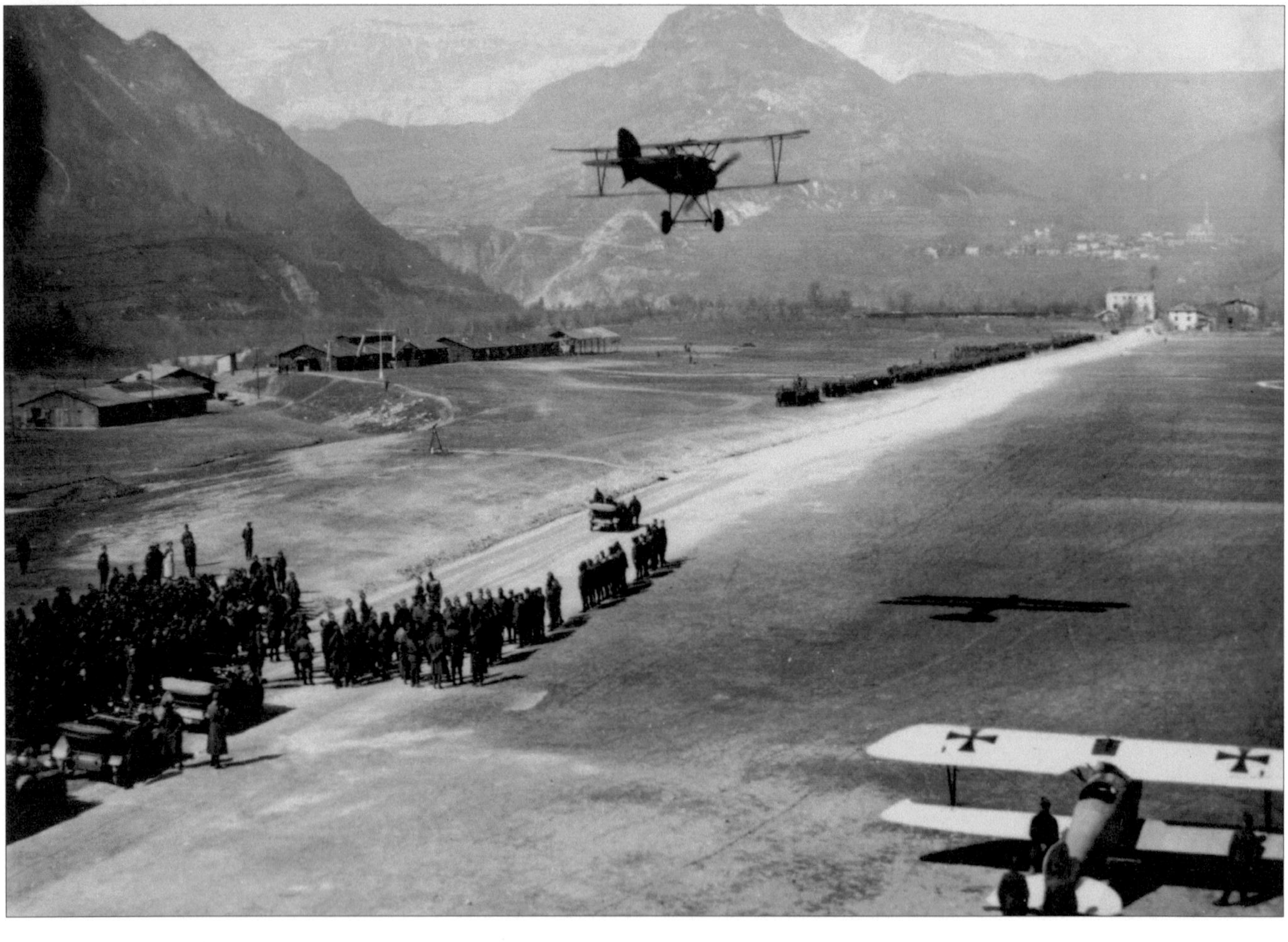

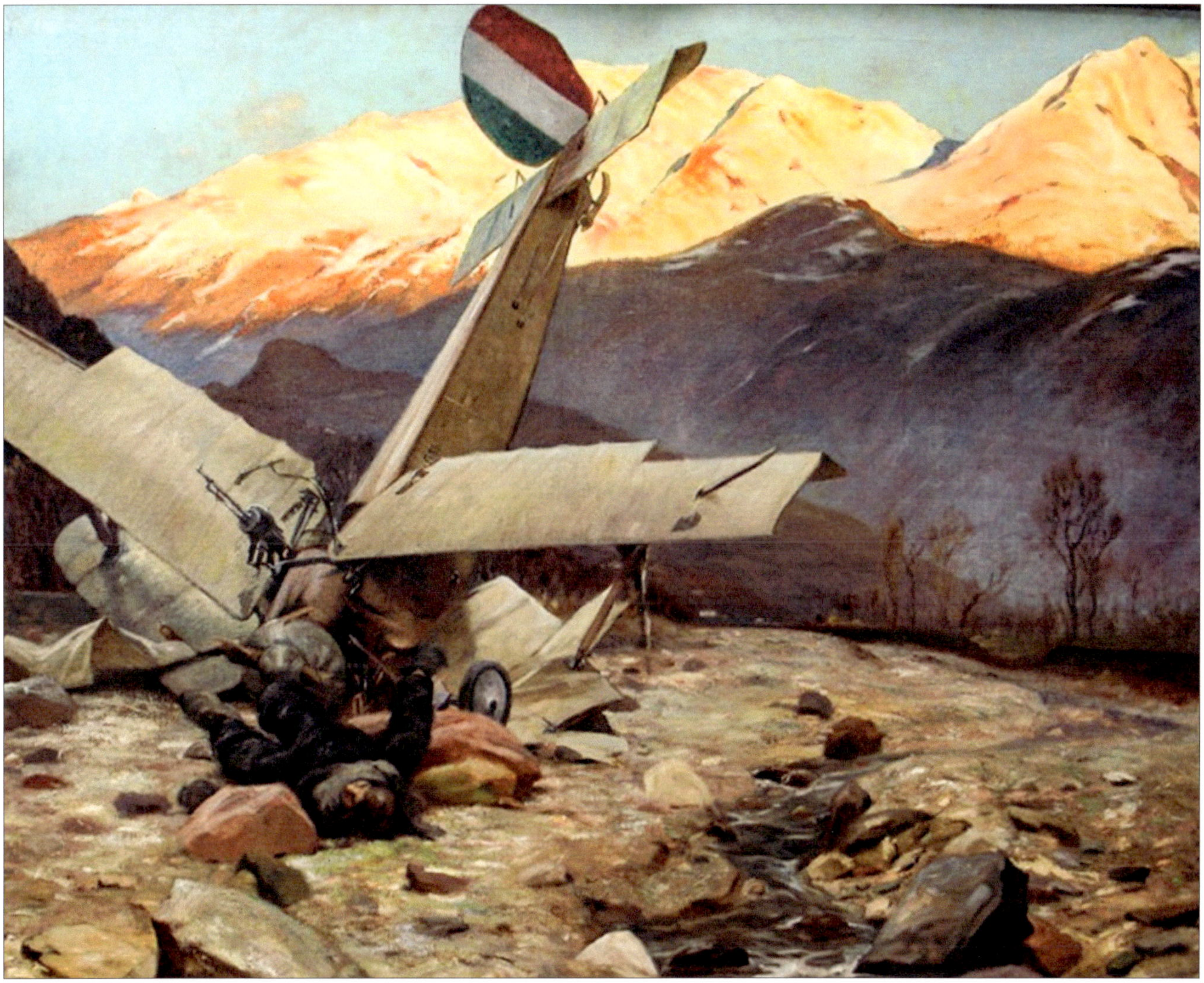

Above: The dangers and the remoteness of the Alpine areas are highlighted in this contemporary painting of a downed Italian airman and his Nieuport 11. Max von Poosch, 1917.

Austro-Hungary as it needed its own aircraft on the Western and Eastern fronts for their own armies.

In late 1915 and early 1916 in the knowledge that their Austro-Hungarian allies were in desperation for a fighter aircraft, Germany sent a small number of Fokker monoplane fighters to their Austrian allies.The Fokkers were becoming outdated and over-matched on the Western Front. The few Fokkers did not make much of an impact at all because the Italians already had Nieuport fighters and there were too few of the Fokker monoplanes.

In November of 1916 a new fighter, a design of Ernst Heinkel at Hansa-Brandenburg began to appear in units of the LFT. It was referred to as the Hansa Brandenburg D.I or as the KD (KampfDoppeldecker, biplane fighter) star-strutter because of its very unique strut design. It was not long before the type was given the name Fliegende Sarg (flying coffin) or Totschläger (killer).

Because of the power and influence of Castiglioni, the Hansa-Brandenburg D.I was put into production and followed by the little improved D.II. A review of the aircraft from a frontline unit in May 1917 stated:

"In the pilots' unanimous opinions, they cannot give full attention to the combat at hand if they are so totally occupied with controlling the aircraft."

Another assessment from Flik 41J echoed their sentiments:

"The KD is absolutely useless...the best pilots, and only they can fly the type, are shackled, ruin their nerves and perish in crashes over the airfield, without their expert skill achieving anything."

In fact the reports were so consistently

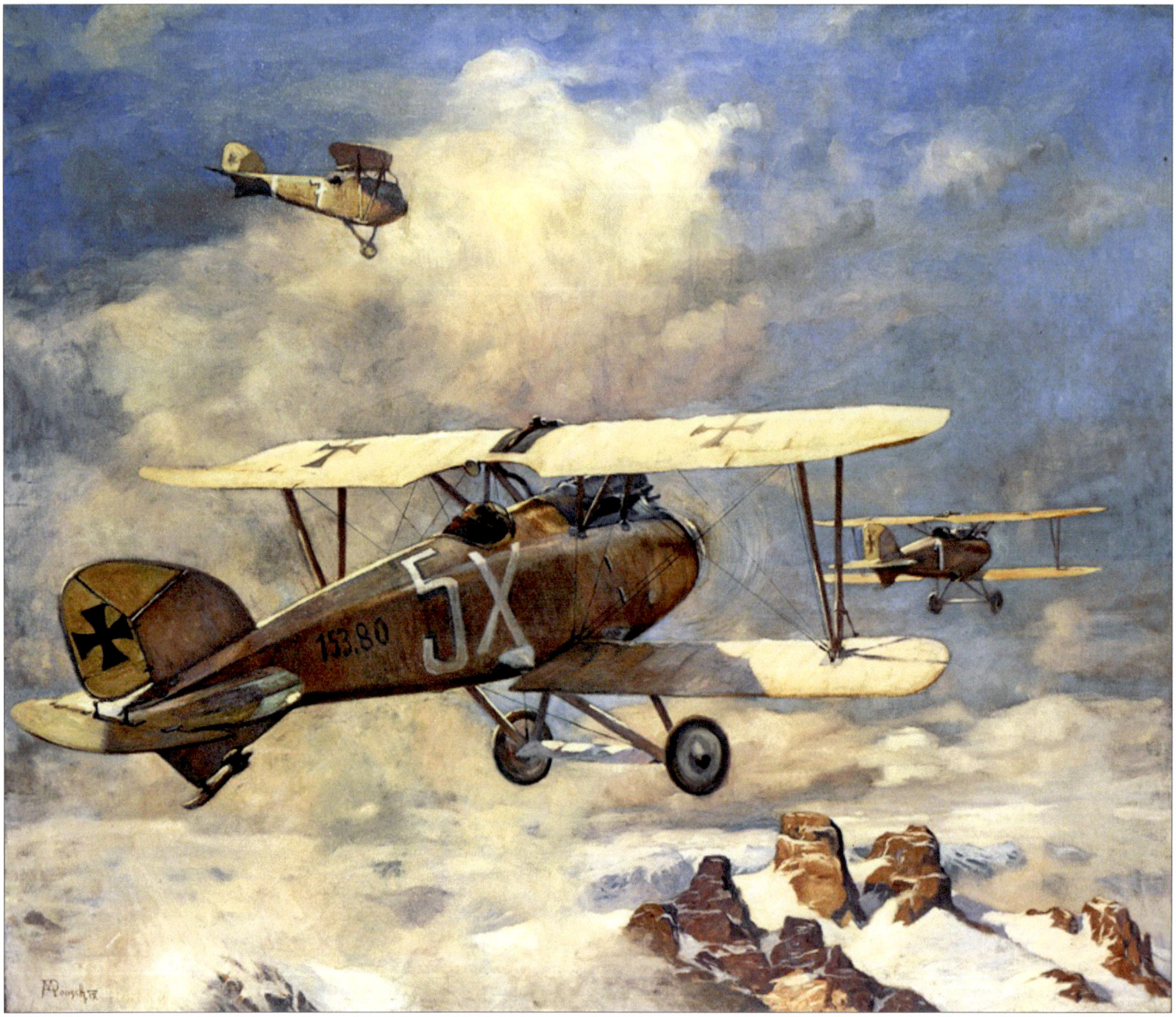

Above: The famous Kasier-Staffel's most successful flight. The leader is Hptm. Josef von Maier (number 1), Arigi (5X), and Kiss (7). Max von Poosch, 1917.

horrendous, that Oberst. Emil Uzelac, the commander of the LFT, personally went to the front lines on a surprise inspection to see for himself. Uzelac was an excellent pilot himself and was known to periodically test fly aircraft to see how they would perform for his aviators. He was much beloved by his command.

He arrived at the aerodrome of Flik 19 in November 1916 and he then test flew one of the KD fighters there. He promptly crashed the airplane and was lucky to emerge with only a concussion!

Phonix had been ordered to produce the KD under license, however some of the engineers at the company took the basic KD design and reworked it into the Phonix D.I fighter. It proved to be much more successful and was enthusiastically received by the frontline fighter pilots. Aces such as Frank Linke-Crawford flew the type and submitted positive impressions of the type. Phoenix fighters would then go on through several variations and continue to be used in service until the end of the war.

Another aircraft that went into production was the Aviatik D.I which was as fast as the Italian Hanroit fighters and British Sopwith Camels that it would face. It had a good rate of climb and was quite maneuverable. Unfortunately it was plagued by structural fragility and tended to start to shed its wing fabric if the aircraft was put into too steep or too prolonged of a dive, as Link-Crawford would tragically discover on the 30th July 1918.

It was not only the lack of a successful fighter design that hampered the LFT in its efforts to combat their enemy. Too many of the commanding officers of the LFT flicks(squadrons) happily

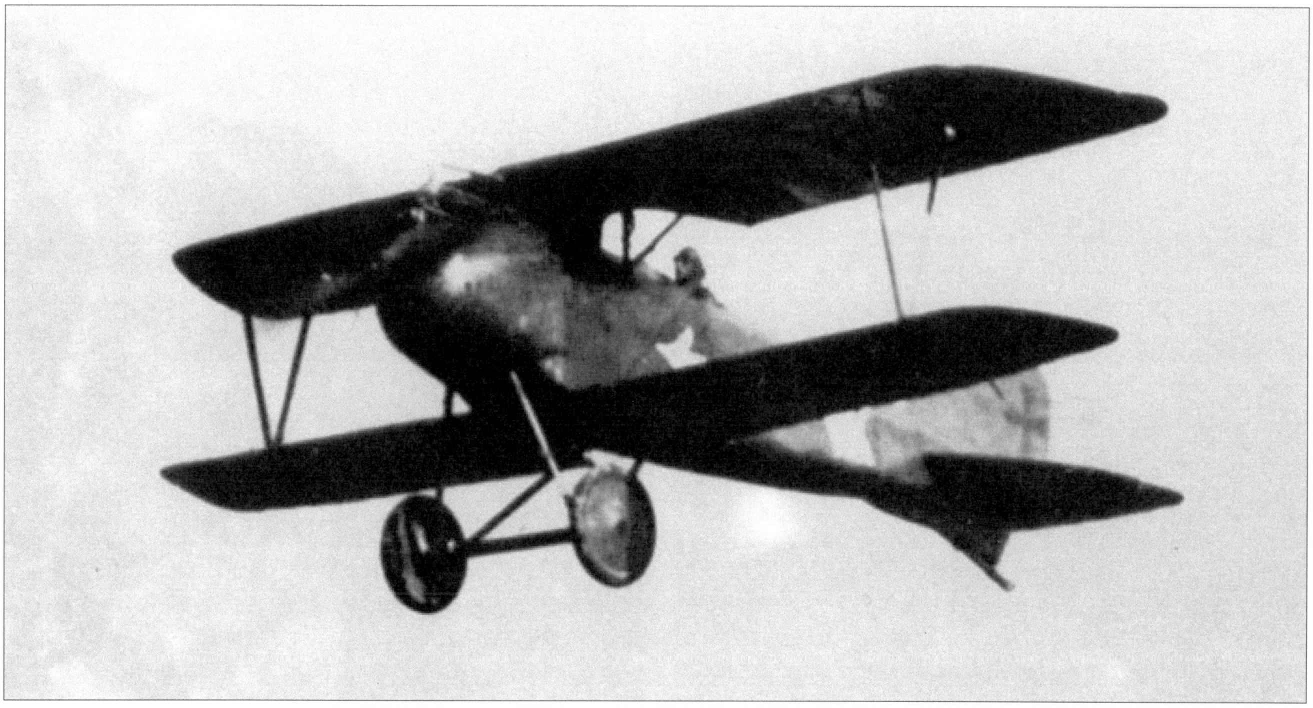

Above: The most successful of the Austro-Hungarian fighter designs was the Oeffag Albatros fighter, although based on the basic design of the German Albatros fighter it had many improvements and a better power plant which led to a superior performance than the German built aircraft.

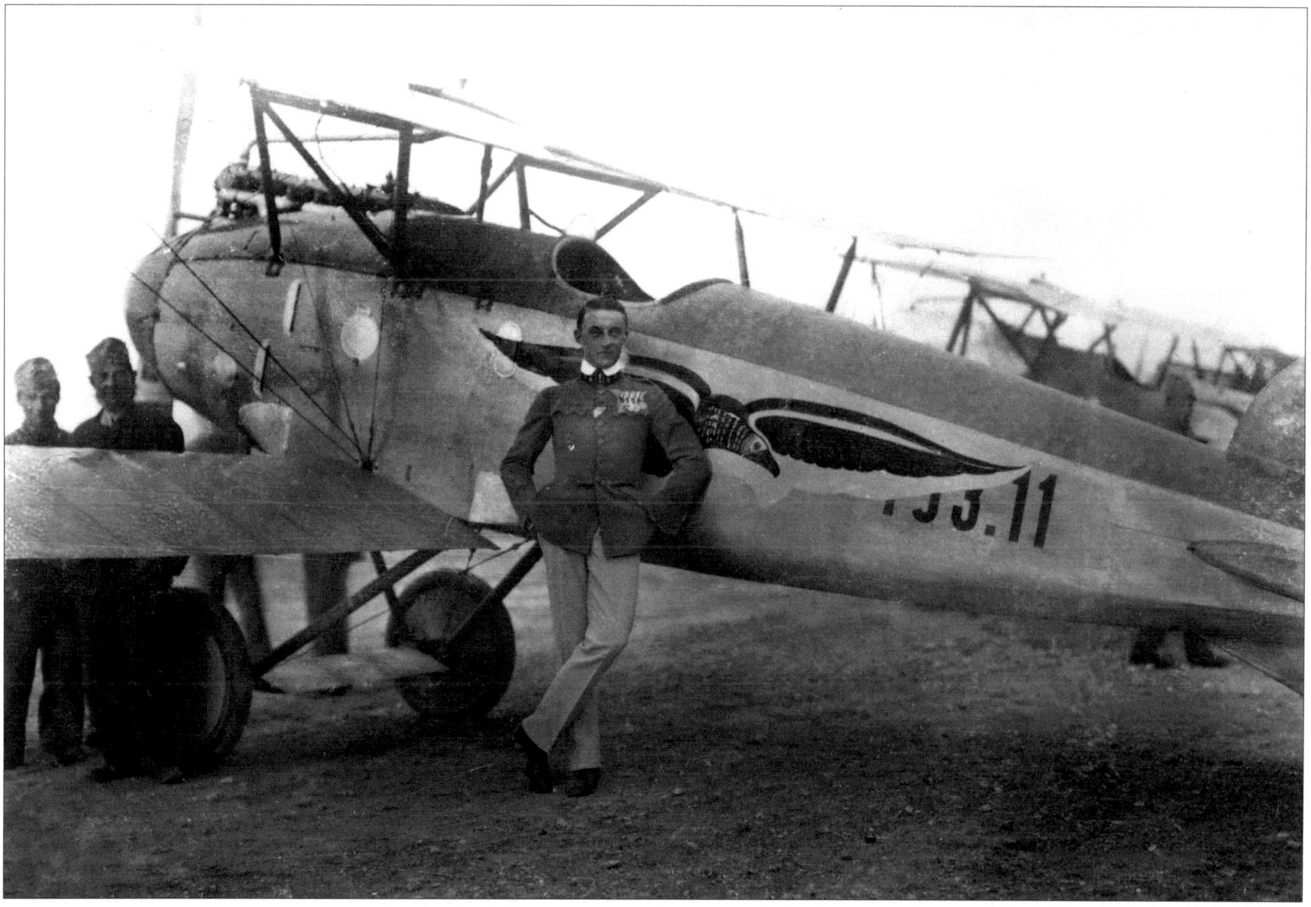

Above: Linke-Crawford beside his Oeffag Albatros. Note the prominent falcon marking on his aircraft, which would help give rise to his being known as the Falcon of Feltre.

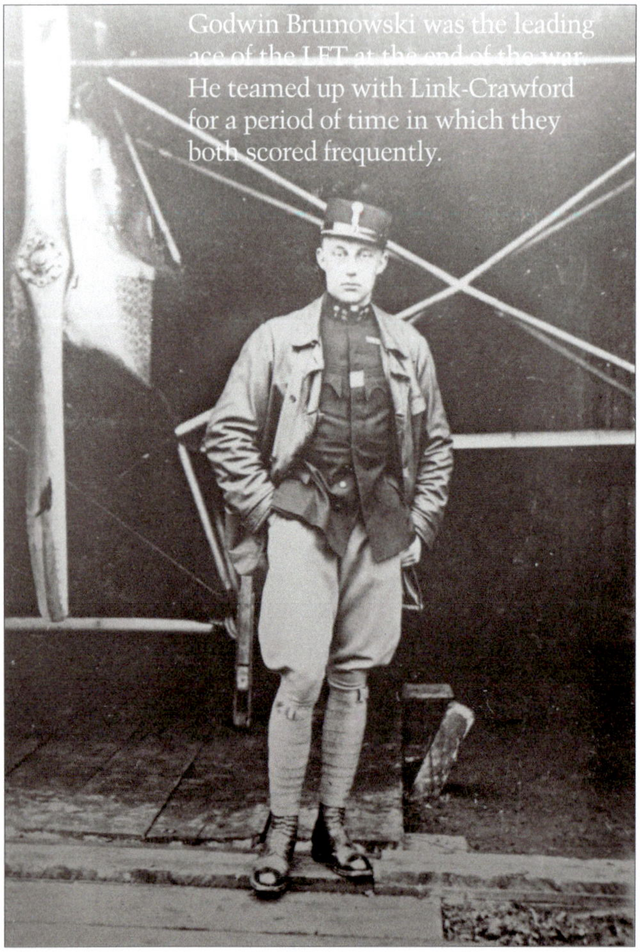

Godwin Brumowski was the leading ace of the LFT at the end of the war. He teamed up with Link-Crawford for a period of time in which they both scored frequently.

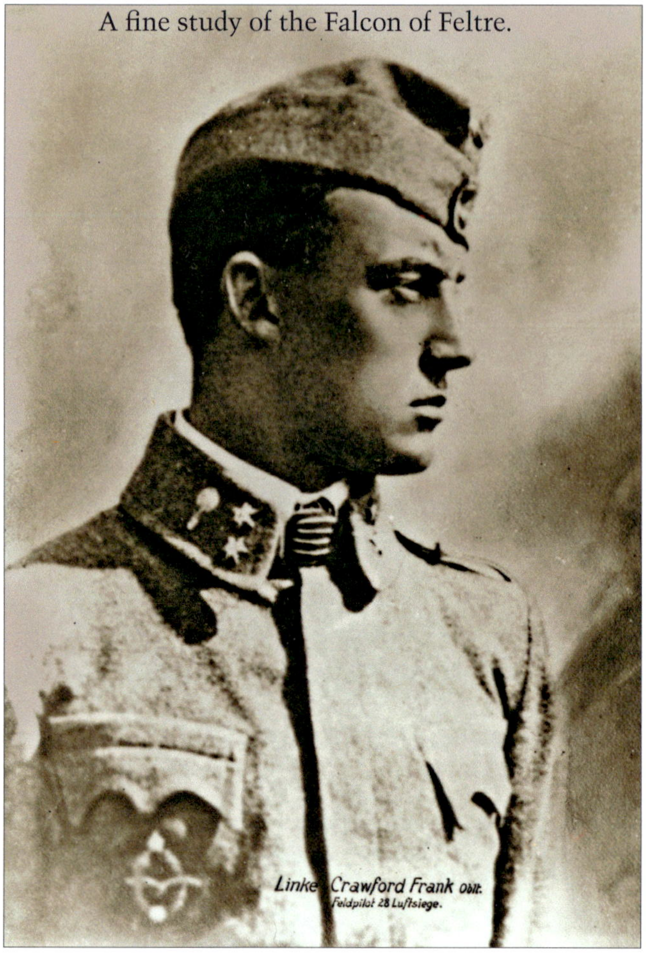

A fine study of the Falcon of Feltre.

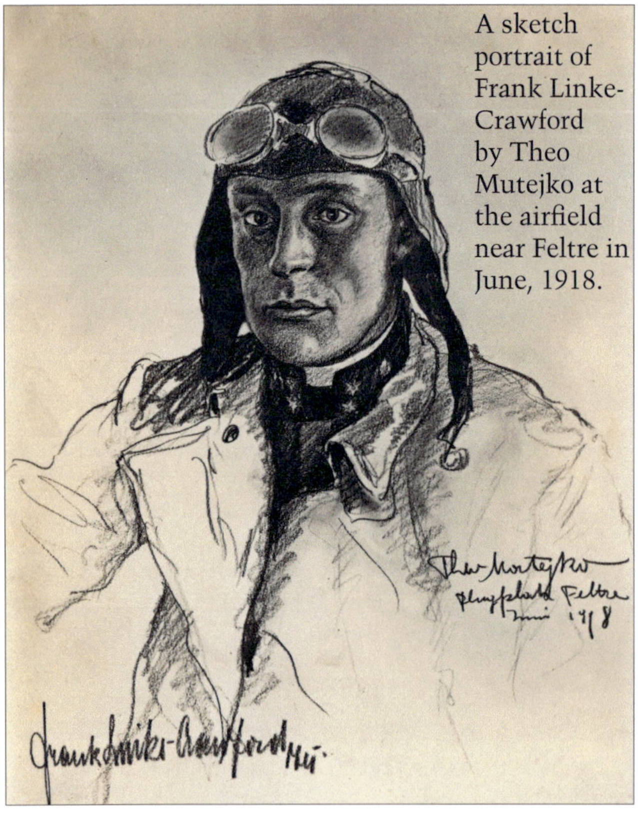

A sketch portrait of Frank Linke-Crawford by Theo Mutejko at the airfield near Feltre in June, 1918.

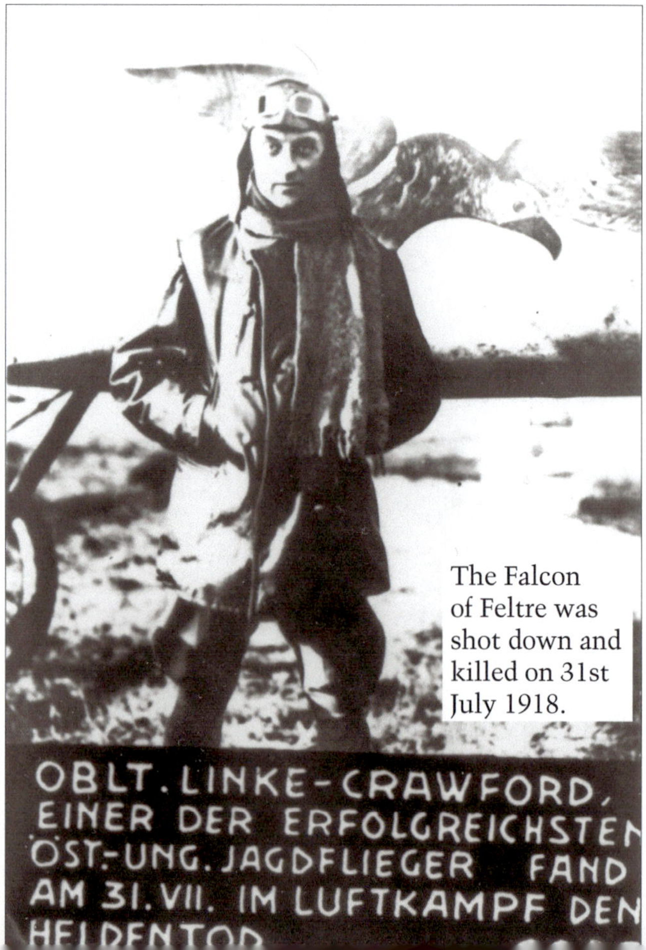

The Falcon of Feltre was shot down and killed on 31st July 1918.

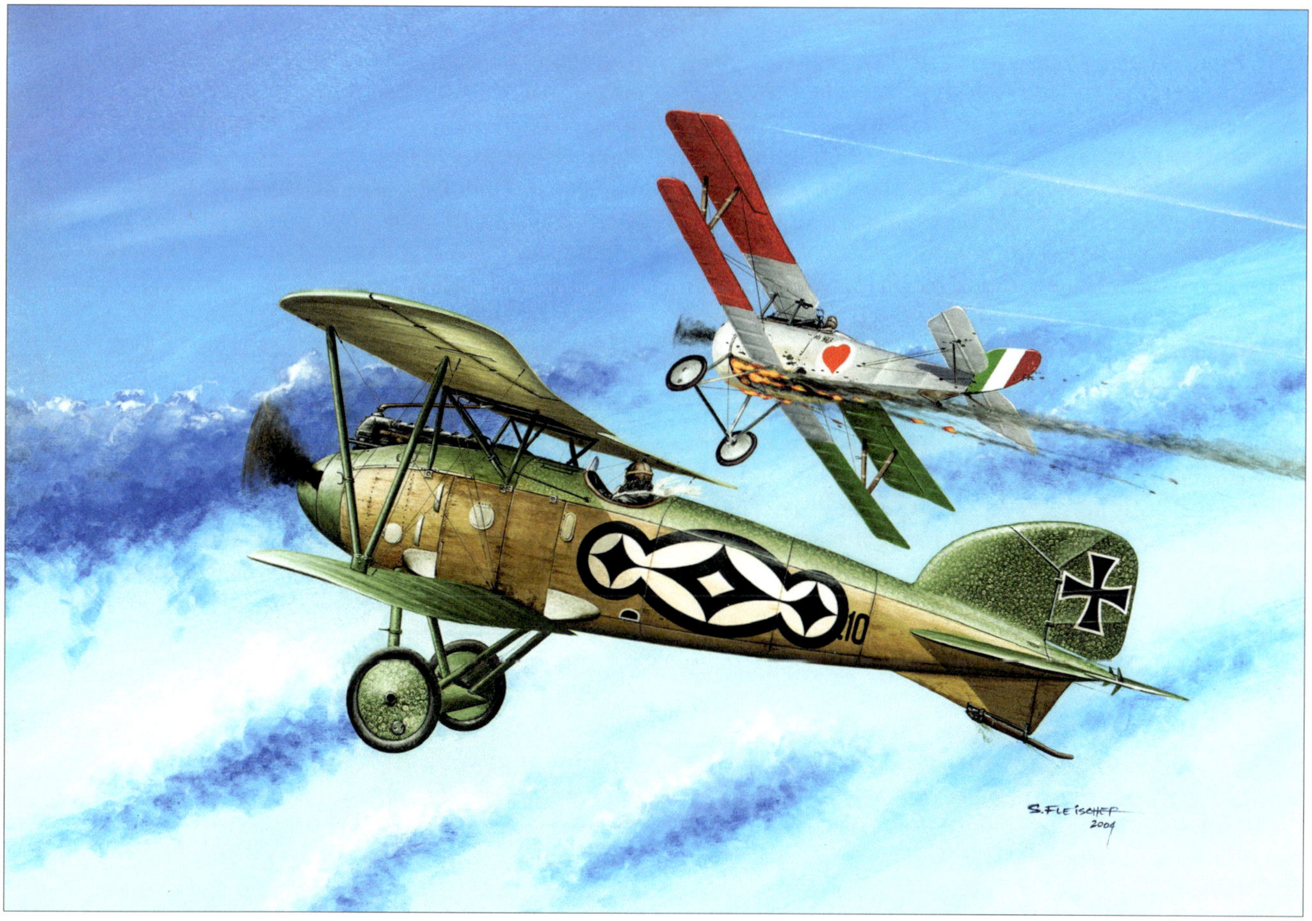

Above: Godwin Brumowski flew this Oeffag Albatros D.III between February and March 1918. Seweryn Fleischer.

promoted the rigid and stifling "norms" of the social strata in how they conducted the war. Many of them completely ostracized the NCO pilots and the cohesiveness of the units in action suffered for that reason. To many of the higher commanders in the LFT, the NCO pilot was nothing more than a chauffeur. Offiziersstellvertreter Julius Arigi would eventually become the second leading ace in the LFT with 32 confirmed victories. He began his score in a spectacular fashion.

On the 22nd August 1916 while he was an NCO pilot in flik 6 "chauffeuring" his officers/observers in a Hansa-Brandenburg C.I Arigi learned of reports that six Italian Farman aircraft were flying towards the naval base at Durazzo in order to bomb the base. Arigi asked permission to take off and attack them, but his request was refused by his commander since there were no officers available and it would be unthinkable to engage in a mission without an officer in command. Frustrated beyond belief, Arigi fumed. Finally, he coerced another NCO to join him and together they took off to intercept the Italians. They succeeded in downing five of the six aircraft, successes all witnessed by multiple ground observers! Instead of congratulations, Arigi was reprimanded because his own commander had been scolded for allowing the aircraft to take off without an officer!

Lt.(posth.) Josef Kiss ultimately became the fourth most successful fighter ace with a tally of 19 victories. Like Arigi, he was an NCO pilot throughout his career. Hungarian Kiss was born in humble circumstances but was a true patriot. When the war began, he quit school and immediately enlisted in the infantry. By not finishing his education and gaining a matriculation certificate, Kiss was prevented from becoming an officer in the Austro-Hungarian armed services. Despite this formal restriction, it would become his fervent desire to somehow become an officer.

He was wounded on the Russian front and while recovering form his injuries, he successfully applied for a transfer to the LFT. In April, 1916 he was assigned to flik 24 flying the Hansa-Brandenburg C.I. In a short time span he scored three confirmed victories, two of them being Caproni bombers. During one of those combats, Kiss's aircraft was hit 70 times by return fire of the Caproni's three gunners!

His successes were noticed and he earned multiple awards for bravery. He also earned a transfer to a newly formed, fighter unit flik 55J. This unit was led by Hptm. Josef von Maier, himself a successful fighter ace and an admired, respected commander. He did not let the rigid caste-like system interfere in his command. In fact, he formed the flight which would make flik 55J famous as the Kaiser Staffeln. Arigi and Kiss as NCO pilots joined their commander von Maier. Success followed success for the three of them as they flew together and gained fame for their exploits. On the 15th November the trio shot down three Caproni bombers over the course of a single day.

On the 26th January 1918 Kiss shot down his 19th confirmed victory which made him the leading ace of the LFT at the time.

On the 27th January 1918, Kiss became engaged in an uneven dogfight. He was trapped by himself as he fought off three enemy aircraft. To make matters worse, his machine guns had jammed and were impossible to clear. After many minutes of superb flying simply to avoid his enemies lethal fire, he was hit in the abdomen and severely wounded. Kiss spiralled down and out of the fight. Somehow he made it back to his airfield and was taken to a local field hospital for care.

When his good friend and comrade Arigi heard about this, he immediately drove to the hospital. Upon arrival, Arigi found that Kiss had not been treated yet, the surgeon was eating his lunch and did not wish to be distrubed.

Arigi disturbed him in no uncertain terms and did not stand for the surgeon's response that since Kiss was not even an officer there was no rush. Faced with the imminent risk of significant bodily harm from Arigi, the surgeon interrupted his lunch and he no doubt made sure to perform the needed surgery as skillfully as possible.

However, Kiss had been severely injured and the surgery required the removal of part of his bowels. But he survived. By late March he was back with his unit, now flying the Phonix D.II fighter.

He had come back too quickly however, he was still weakened by his old wounds and struggled to keep going. The end came on the 24th May 1918 when he was shot down and killed. The next day, Josef Kiss was promoted to the rank of an officer, the only NCO in the Austro-Hungarian air service to achieve that promotion in the entire war.

On the 27th May Pergine aerodrome was a busy place. Hundreds of aviators were gathered from multiple units to come and pay respect as flik 55J were ceremoniously burying their beloved comrade, the now Lt. Josef Kiss. As the bustling funeral preparations were underway there was a sudden commotion. A low flight of multiple Allied aircraft, combined Italian, English and French aircraft appeared overhead and flew past. As they flew past, a single large bundle was dropped. In it, a large memorial wreath was found with banners in several languages: "Our last salute to our courageous foe,"

The best fighter design from Austro-Hungary during the war was its version of the Albatros DIII design. Following its success on the Western Front, the Albators company was directed to begin producing its design in Austria as well for the LFT. Since both the Hansa-Brandenburg company and the Phonix company were already building their own fighter aircraft and because, finally the Austro-Hungarians were sensing the need to introduce some competition to the firms owned by Castiglioni, the contract was given to the Oesterreichische Daimler Motoren firm (Oeffag).

Oeffag was already building the outstanding Austro-Daimler aircraft engine at the time and so now had to develop the means to manufacture an aircraft as well. Although similar in outward appearance, the Oeffag-Albatros D.III fighter was its own design. Any of the problems with lower wing flutter were solved by building a more robust attachment of the lower wing and the engines produced by Oeffag gave the type a faster top speed and better rate of climb than the German built Albatros fighters. In addition, the Oeffag types were built with a high degree of quality control. A successful fighter ace, Friedrich Hefty had this to say:

"...an aeroplane of an excellent design, perfectly balanced and especially fit for aerobatics... It has the pilot's complete trust. Because of its excellent handling and performance it is preferred over every other fighter."

Finally, the Austro-Hungarian fighter pilots had a capable aeroplane they could trust as they flew into combat. However, their Italian opponents were now flying Hanroits and Spads and were far more numerous in the skies. In addition, there were French units and British units flying the superior Sopwith Camels and Bristol F2b fighters. As a result:

"So the Austro-Hungarian airmen based on the Southwest Front had to face an increasingly numerically superior enemy who also had better machines. Nonetheless, they fought thousands of times in unequal combats and accomplished admirable things, thus deserving the highest praise." Paolo Varriale

Perhaps one of the finest examples of these courageous Alpine Eagles was Frank Linke-Crawford

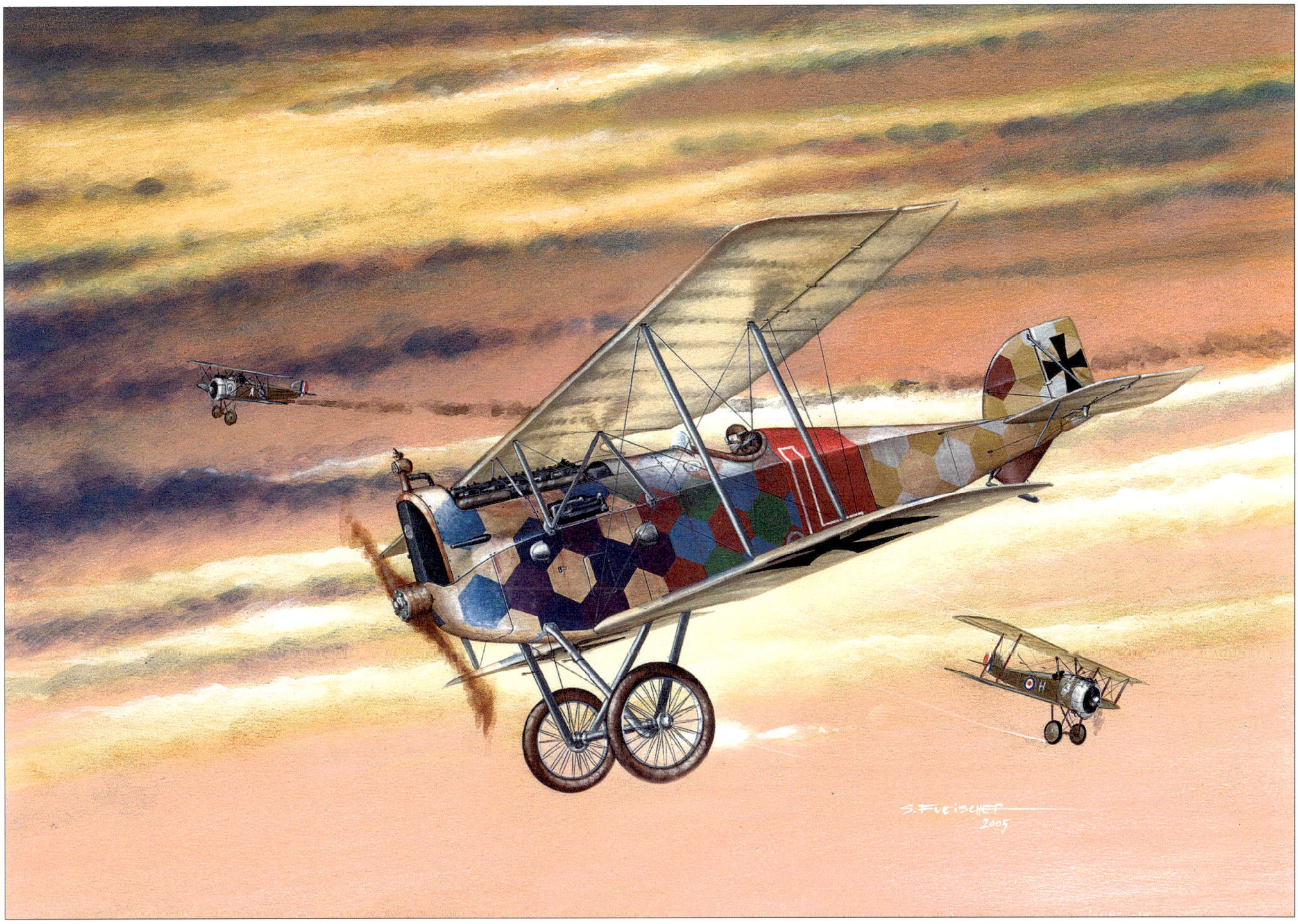

Above: Despite having his choice of aircraft, Linke-Crawford continued to fly the Aviatik D.I which had structural problems, although he was often successful as he is here in downing a British Sopwith Camel. Seweryn Fleischer

who would become known as the Falcon of Feltre. He was born in Krakow(now Cracow, Poland) on the 18th August 1893 to Major Albert Linke and his English wife Lucy Crawford. Linke was a career military man and so it was not surprising that his son attended the top military academy in the Austro-Hungarian empire as a means to follow in his father's footsteps. The outbreak of war found Linke-Crawford serving in a crack cavalry unit on the Eastern Front. His courage from the outset was exemplary and he received numerous awards for his efforts. However, repetitive medical problems induced the young warrior to apply for transfer to the airs service and this was granted in December, 1915.

After training, he was initially assigned as an observer and after multiple missions and again receiving several commendations for his bravery, Linke-Crawford was admitted to pilot school in September of 1916.

By January, 1917 he was assigned to Flik 12 on the Isonzo battlefront. He flew both two-seater aircraft on reconnaissance and bombing missions as well as flying as an escort aircraft on other missions. His forthright and even aggressive flying began to be noticed by his comrades.

He persevered in his duty on 1st May, 1917 and successfully completed a long range reconnaissance mission despite the repeated attacks from an Italian Spad. Upon inspection, his force-landed Hansa-Brandenburg C.I was found to have 68 bullet holes in it!

Unknown to Linke-Crawford, the Italian fighter pilot had been Francesco Baracca who would become the top Italian fighter ace of the war. Baracca's Spad fighter was found to have a few bullet holes as well and he described his adversary as abilissimo (very clever).

Linke-Crawford was flying an Aviatik two-seater as an escort fighter on 2nd August 1917 when he was downed by Pier Ruggero Piccio who would eventually score 26 confirmed victories in the war.

Having survived these and other aerial combats, in August 1917 Linke-Crawford's piloting skills and his combat experience earned him a transfer to a newly formed unit Flik 41 J(jager, hunter). This new unit was commanded by Godwin Brumowski: who would survive the war as the leading ace of the Austro-Hungarian empire. Brumowski and Linke-Crawford formed an excellent team and soon began to score.

By September Linke-Crawford had scored his fifth victory and was now an ace. During this time he began to wear a red flying helmet and so earned the nickname "red head" from his comrades. He also was awarded more medals, including the rarely given Gold Bravery Medal for officers.

By December his score stood at 13 and he was given his own command. Although on paper the strength of a Flik was supposed to be much more, the newly formed Flik 60J was composed of Linke-Crawford and six NCO pilots. They were issued Phonix D.I aircraft which was generally well-received by the pilots for its sturdiness and predictable flying characteristics.

In March, 1918 the unit was moved to an aerodrome near Feltre, which along with the markings on Linkd-Crawford's Albatros fighter now earned him the sobriquet "The Falcon of Feltre". It was here that the unit also received a new fighter aircraft, the Aviatik D.I which was sometimes referred to as the Berg fighter in reference to its designer Julius von Berg.

During the month of May 1918 alone, Flik 60J lost four Aviatik fighters due to wing/structural failures and forced the grounding of the remaining types.

Despite this, Linke-Crawford apparently preferred the Aviatik over the Phonix D.II as he continued to fly it in combat. He must have appreciated its excellent maneuverability and felt that as an accomplished pilot he could avoid the dangers of the aircraft's design. On May 10, 1918 Linke-Crawford flew his Berg fighter while downing a Bristol F2b in the morning and then a Sopwith Camel in the afternoon.

By downing two of the best fighter aircraft the British had, Linke-Crawford clearly showed his skills and his ability to get the best out of the Aviatik design.

Unfortunately, his best was not good enough on the 31st July 1918. Again flying in his Aviatik fighter, Linke-Crawford fought two Sopwith Camels of No. 45 squadron to a standstill, but then later in the same sortie he was involved in a turning melee of a dogfight with two Italian Hanroits when his aircraft was seen to go into a involuntary spin. The Aviatik finally came out of the spin, but then flew straight and level. The Italian pilots also noted that there were pieces of fabric shedding off the wings of the Austro-Hungarian fighter. Closing in they hit the helpless Linke-Crawford's fighter and set it ablaze. He crashed into the ground just outside of the town of Guia, he was 24 years old. The renowned aviation historian Dr. Martin O'Connor wrote this:

"Frank Linke-Crawford was a fine pilot and a fine man. His even-handed leadership inspired devotion in those who flew under him. His rapport with his men was superb, particularly with non-commissioned personnel, quite an unusual state of affairs in the strictly-stratified forces of the Austro-Hungarian empire."

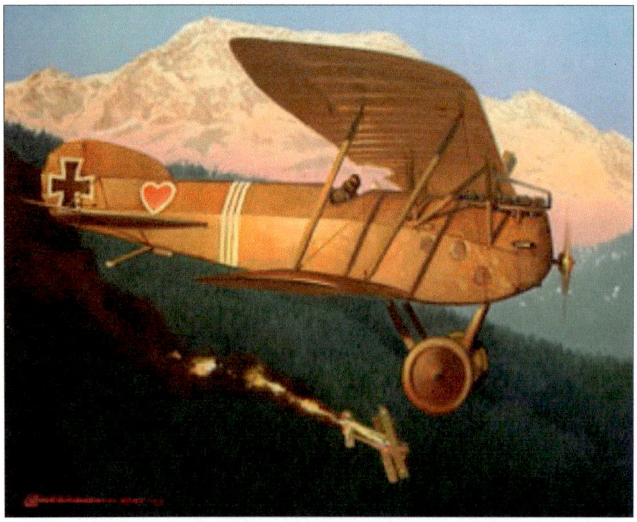

Above: Alexander Kasza in his Phonix D.IIa is shown bringing down an Italian Caproni bomber. Note the personal insignia of 3 stripes on the fuselage and the additional red heart on the tail, added after his good friend Josef Kiss was killed in combat.

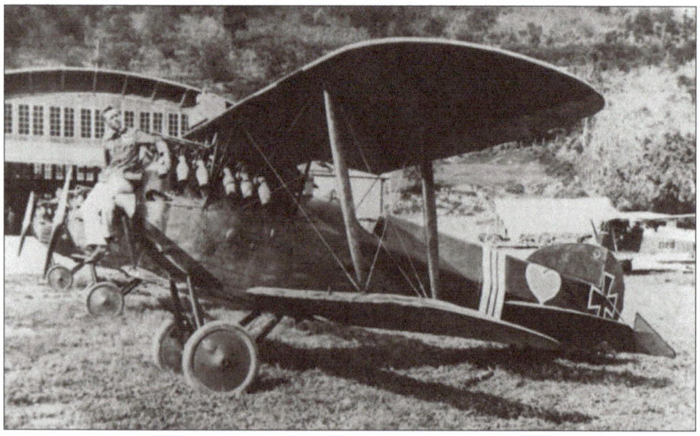

Above: Alexander Kasza perches atop his Phonix D.IIa fighter as a member of Flik 55J at Peringe airfield in June, 1918. He would score six victories in the war.

Critical Side Show

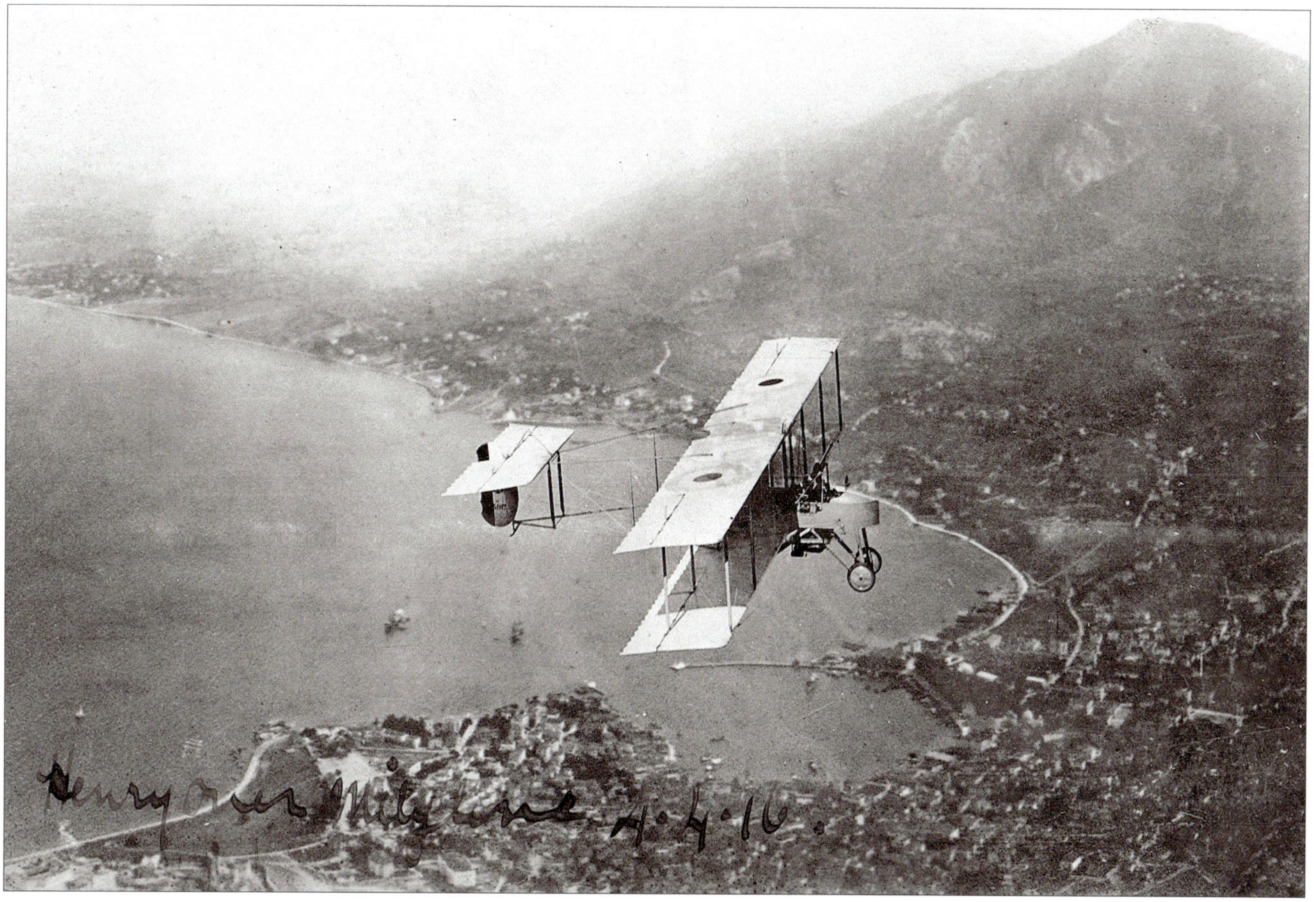

Above: A British Henri Farman bomber headed towards a Turkish target during the Dardenelles. The British had no resistance initially, then the German ace Buddeke arrived and made things much more difficult for the British air units.

Right: For the Dardanelles campaign one of the principal air bases was the RNAS base at Imbros. Note the open air tents and the Union-Jack insignia on the lower wing of the Sopwith Tabloid.

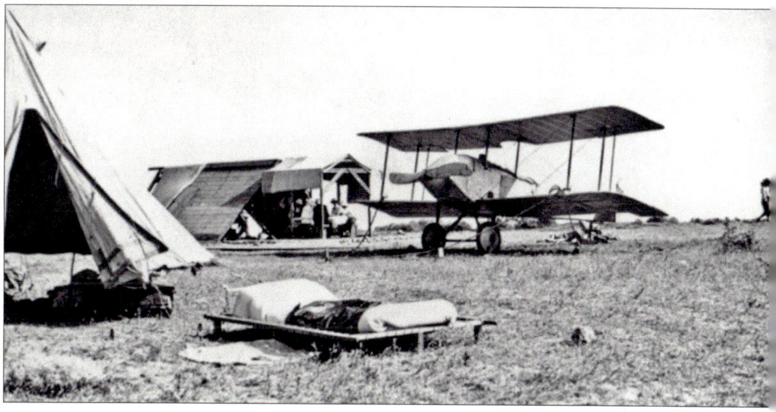

The Middle East / Palestine theatre during the First World War is generally relegated to the concept of a "side show" : naturally considered as being of secondary importance to the Western Front. However, for the Allies and especially for the British there were aspects of this area of fighting that were of critical importance to its efforts in the war.

Foremost was the protection of the Suez Canal. It was imperative for Britain to keep the Suez Canal open to allow shipping of material and troops from its colonies of India, Australia and New Zealand Without the "shortcut" through the canal, the ships would have to go around the horn of Africa.

These troops would be especially important in the fight against Turkey. And the merchant ships bringing needed war materials and other supplies could more quickly get to ports in southern France and/or Italy by spending much less time on the journey and also reduce their exposure to the U-boat menace.

A third and no less important aspect to the Middle East sector was its increasingly important source for crude oil in order to provide petroleum to

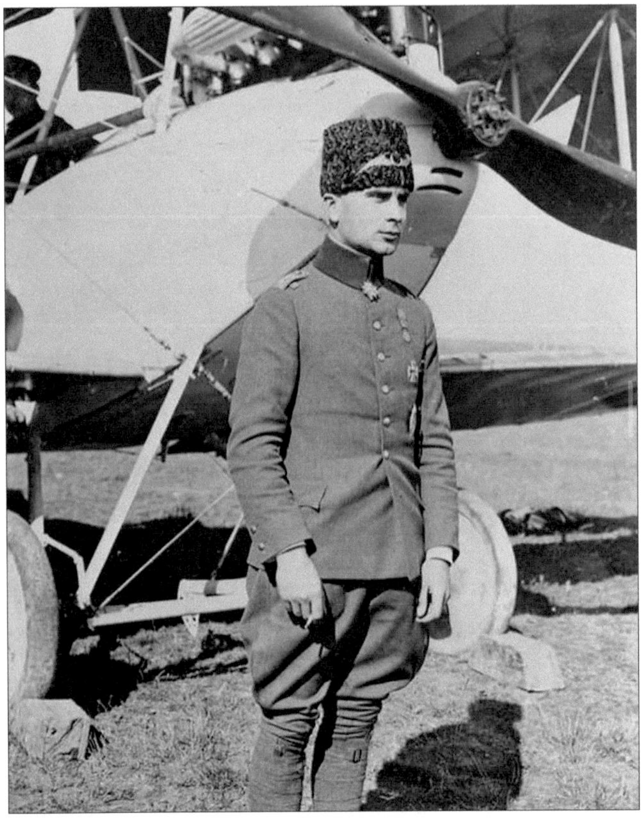

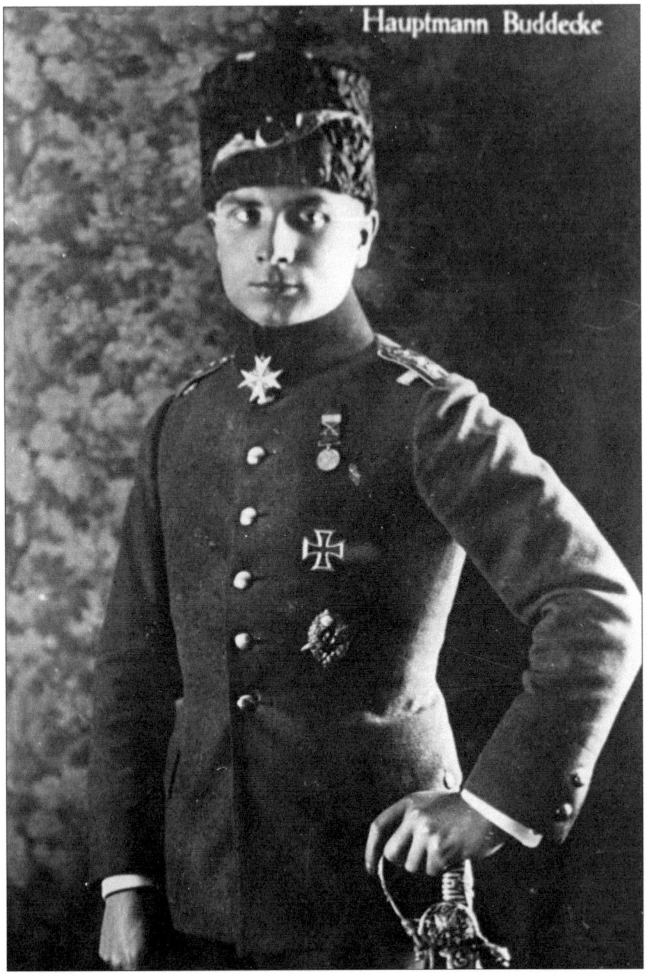

Above: Soon after his arrival. El Schahin [Hunting Falcon], made his arrival known by shooting down several of the British Farman bombers while flying a Fokker E. III fighter. Buddecke arrived in Jan 1916 but was transferred back to the Western Front to command the newly formed Jasta 4 in August 1916.

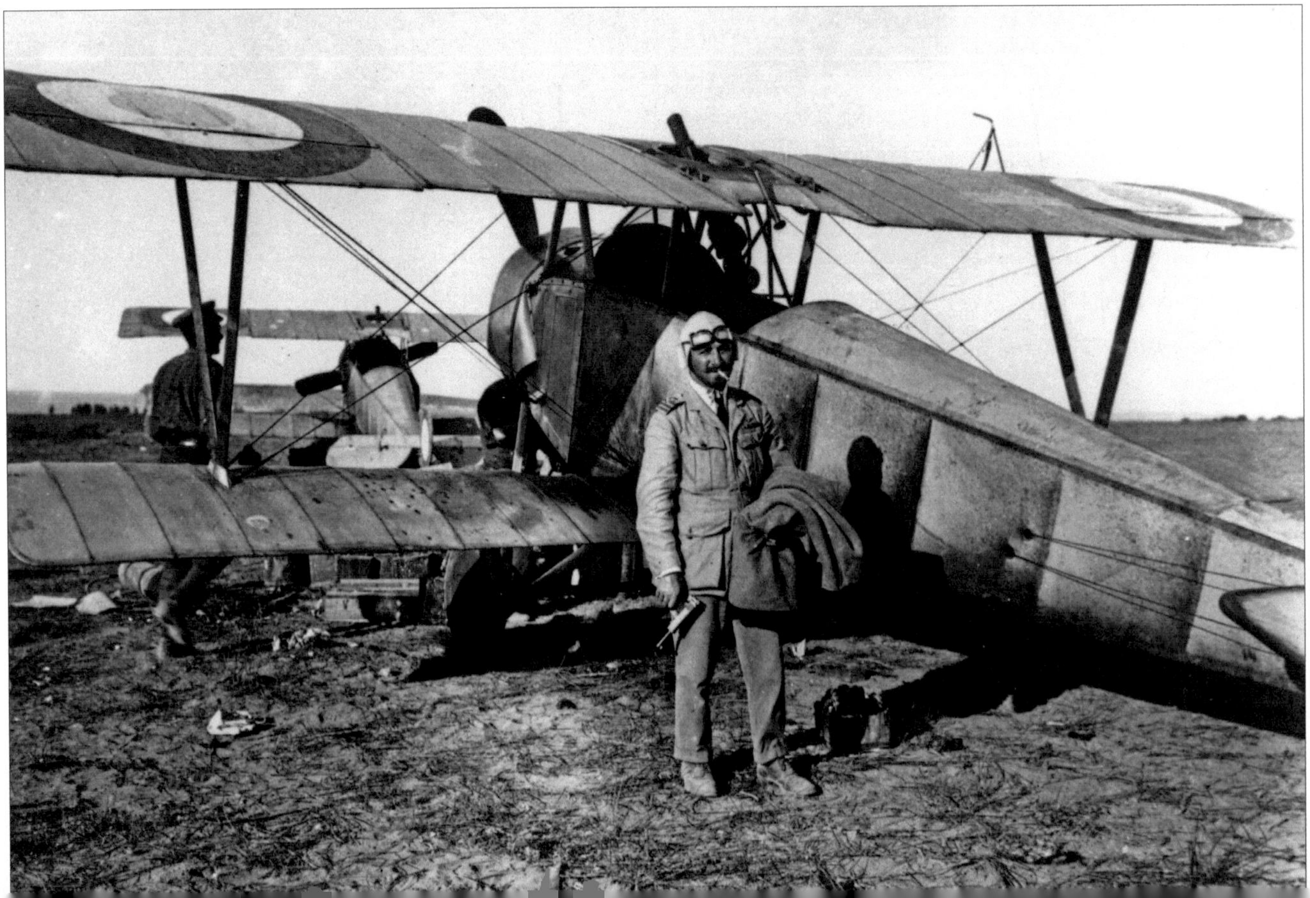

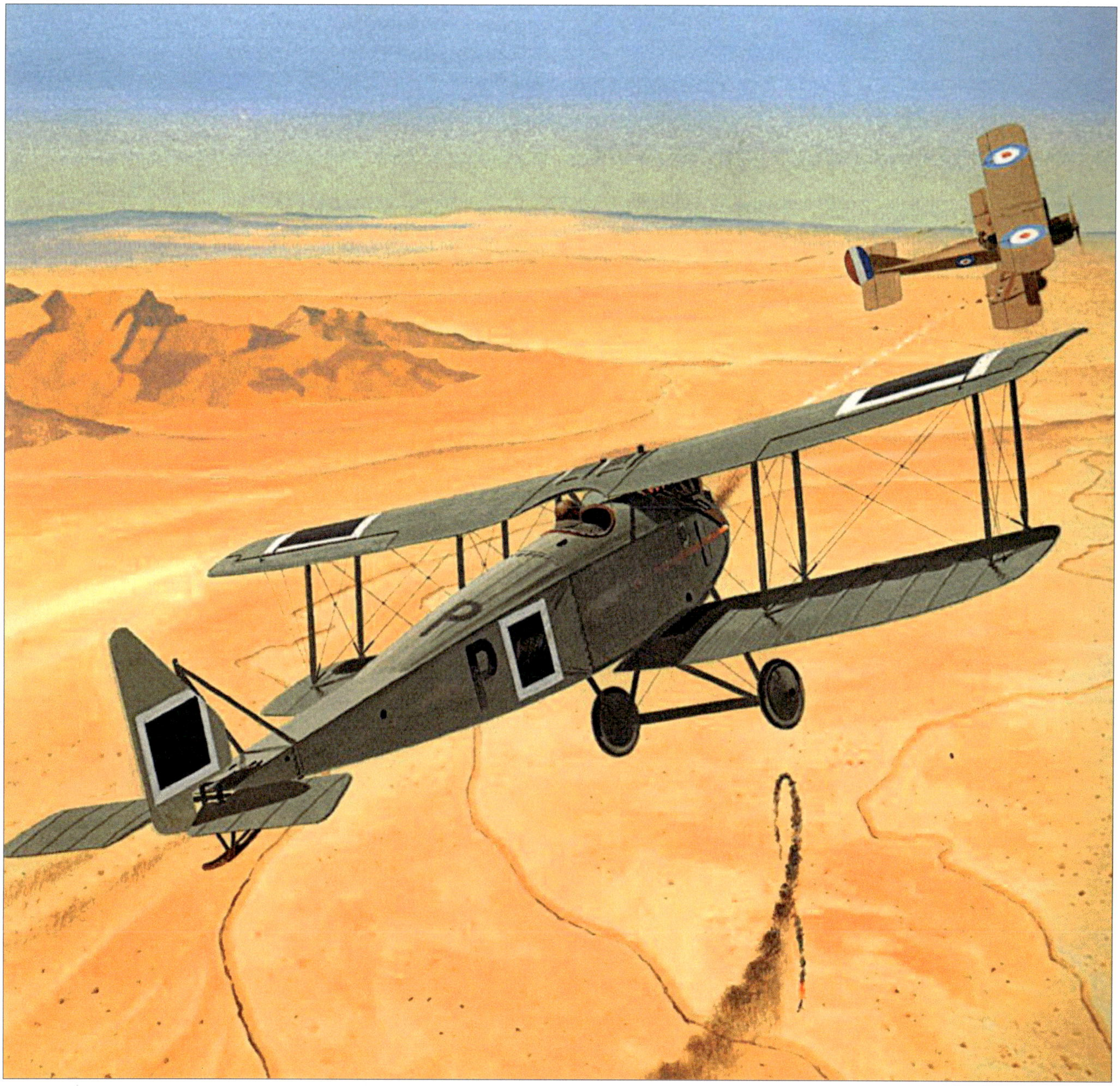

Above: For a long time, the German air units in Turkey and Palestine had superior aircraft. Although inferior in total number, the Fokker fighters and the Halberstadt fighters as shown here in Turkish colors would rule the dogfights in the skies. The Turkish insignia of a black square was likely do to painting of the Black Crosses that the aircraft were initially decorated with-the cross not being welcome since it was a Crusader symbol. Brian Knight

Facing Page, Above Right: Buddecke would quickly return to Turkey, and he would command his small unit, FA 5, which he called Der Wanderzirkus [wandering circus] from December 1916 until January 1918. His orders were to protect the majority of the West coast of Turkey which forced Buddecke to move his small number of aircraft up and down the coast to where the action was most intense for the moment.

Facing Page, Below: The enigmatic Cmdr.Ralph Samson was put in command of the RNAS units in the Dardenelles and Palestine. Note the Lewis gun positioned to fire through a cut-out opening in the top wing of the Nieuport.

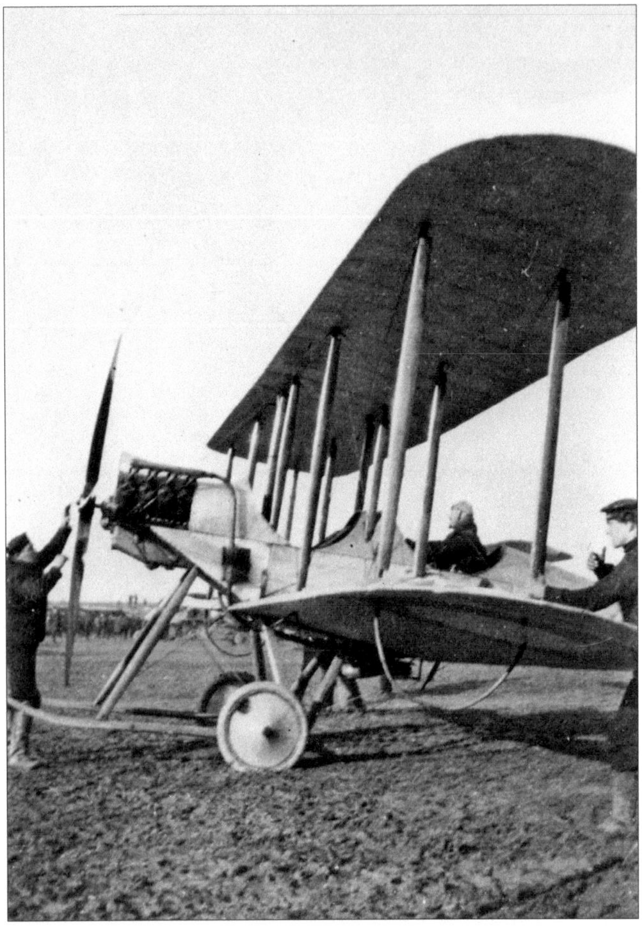

Above: Samson was an excellent pilot, here he is at the controls of an early Be2. He was a much beloved leader and a very offensive -minded one as well.

the Allies. Although approximately 80% of fuel for the Allied war effort would originate from the USA, the remainder would come from the Middle East and its fuel would be an even more important supply in its own region of the war.

Both the US Navy and the Royal Navy had begun to convert their fleets to fuel from coal (the German fleet had not). Now with the surge in the machinery of the war: trucks, cars, tanks, airplanes, etc.; the control of the infant oil fields in the Middle East became important for the first time in history.

Germany had developed its large scale strategy based on coal-powered trains, getting men and supplies as close to the front and then depending on the infantry marching and horses pulling guns, and supplies to the battlefront itself.

The Allies would develop a strategy dependent much more on hundreds to thousands of gasoline powered trucks to bring troops and materials to the battle. Which ultimately gave them more flexibility and a better capability of supplying what was needed in a much faster and much more precise fashion.

A prime example of this difference could be found in the delivery of needed gasoline to the FA 300 unit in Palestine. The German squadron ultimately depended on 30 camels in a six day round trip, carrying fuel in 25 liter cans from the nearest railroad station at Hafir el Auja. Whereas the Allied units built roads in order to deliver supplies by truck convoy.

Yet, the Germans also were in need of refined oil

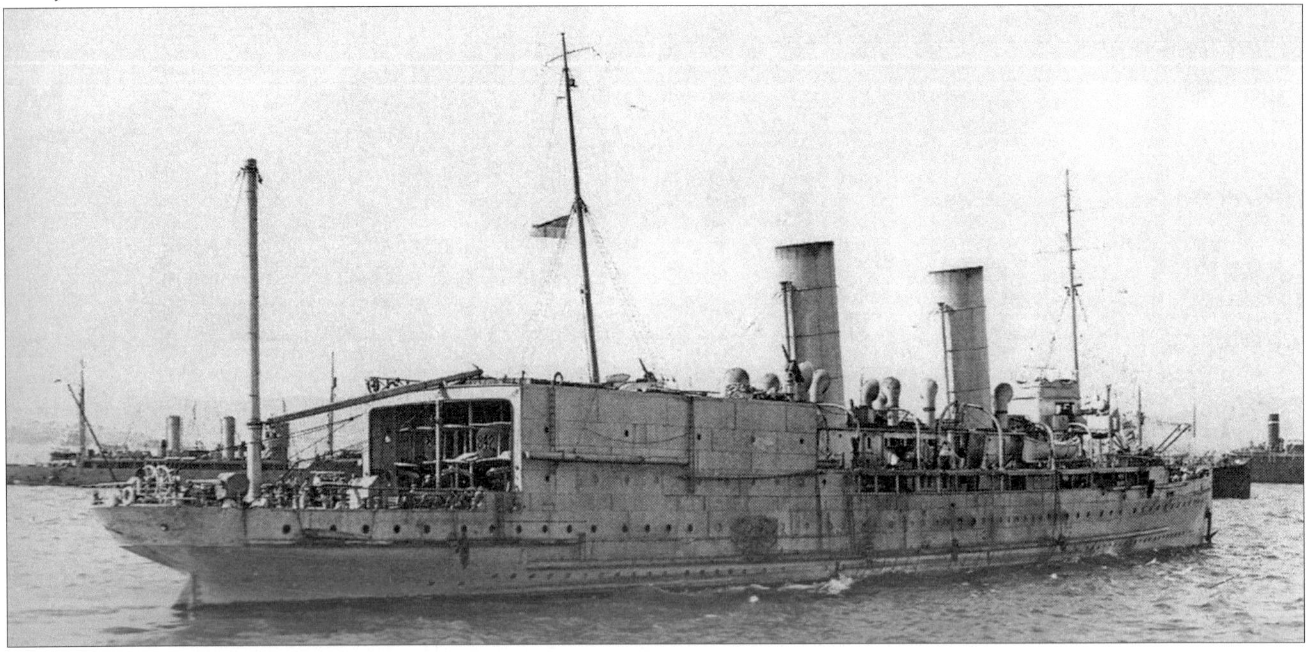

Above: The seaplane carrier *BEN-MY-CHREE* [Gaelic, meaning The Woman of my Heart] was a crucial factor in the ability of the EIESS squadron to effectively move and deploy its limited air assets in the Middle East. It was involved in the Sea Battle of El Arish airfield.

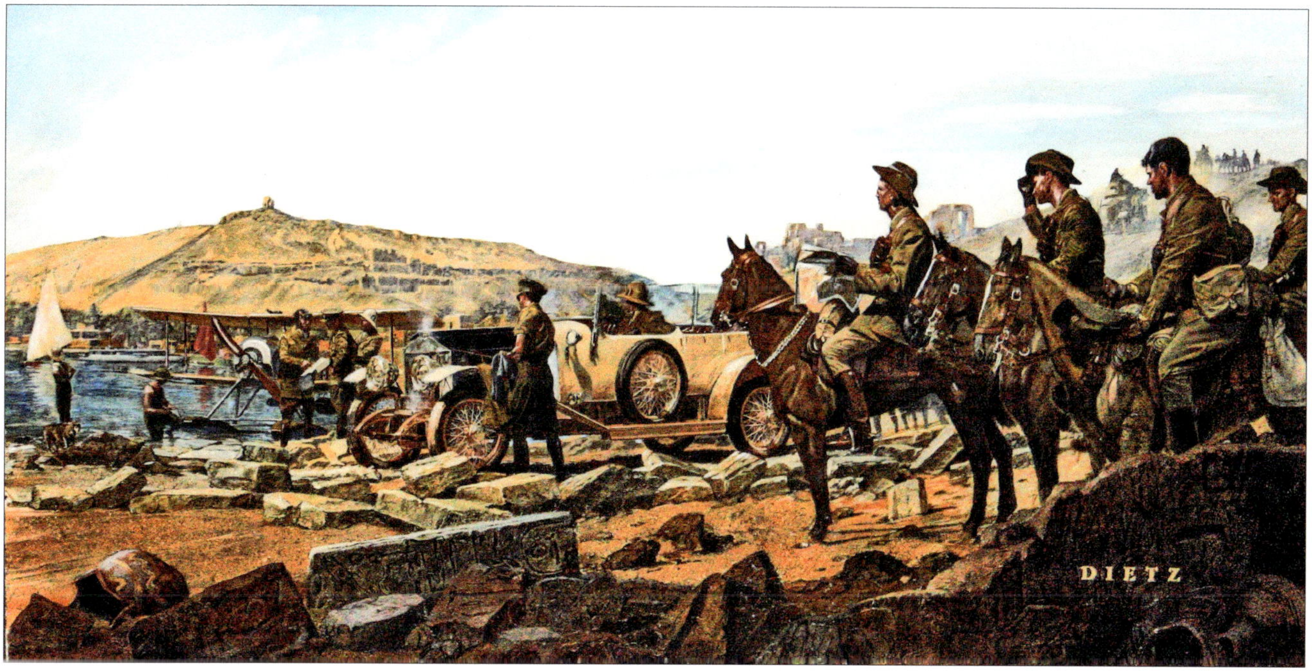

Above: The floatplanes of the EIESS could quickly dispatch orders, communications all along the coast of Palestine. Jim Dietz

if they were to be able to continue to send U-boats out to sea and be able to send their airplanes into the sky. As the war went on year after year, fuel shortages began to hamper Germany's capacity to effectively wage war.

The Germans knew that they were deficient in the realm of oil supplies/ oil refining capacity and so they were eager to support their Turkish allies in pushing the British out of the Sinai and to successfully overtake the Middle East.

The Germans of course could not rely on America as a source of its oil, thus the Middle East and its nascent oil fields were that much more important to them. It should also be noted that the Germans attempted to obtain the oil fields in Romania during the war, without success.

Therefore, the stage was set for the Middle Eastern theatre of war, a sideshow that nonetheless was of crucial importance to the belligerent nations. For Britain, to hold on to the Suez Canal and the oil supplies to be found in the Middle East whereas the Germans via their Turkish allies meant to

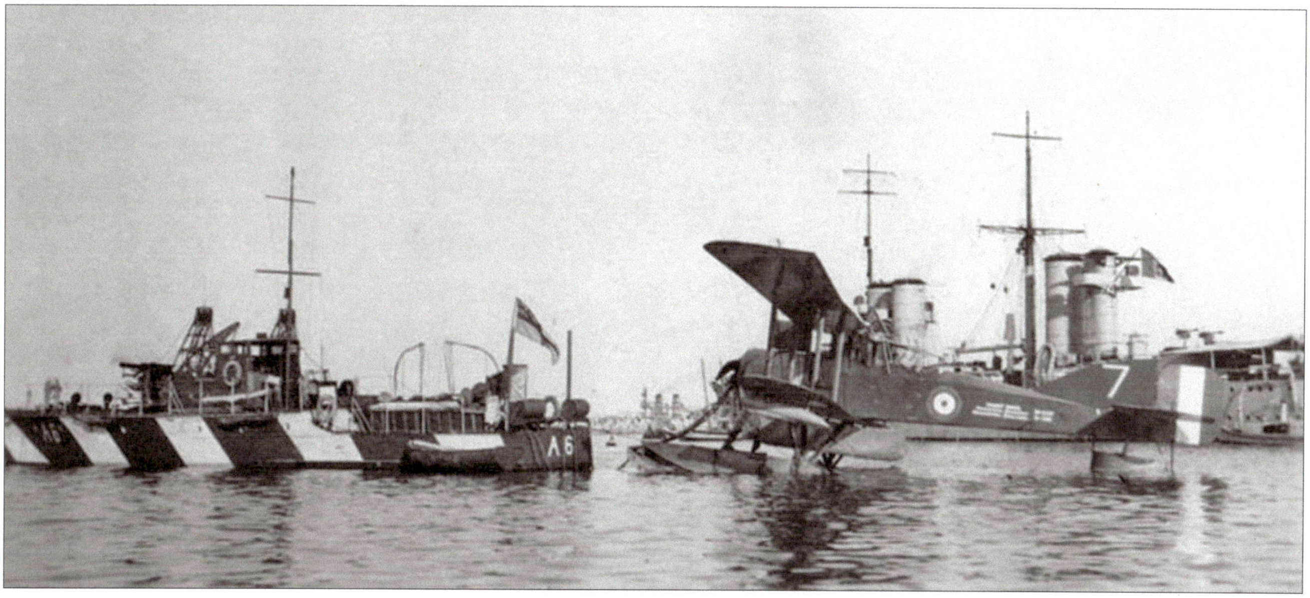

Above: The Short Seaplane was one type of the aircraft that was assigned to the *BEN-MY-CHREE*; this particular aircraft is seen in service in Italy.

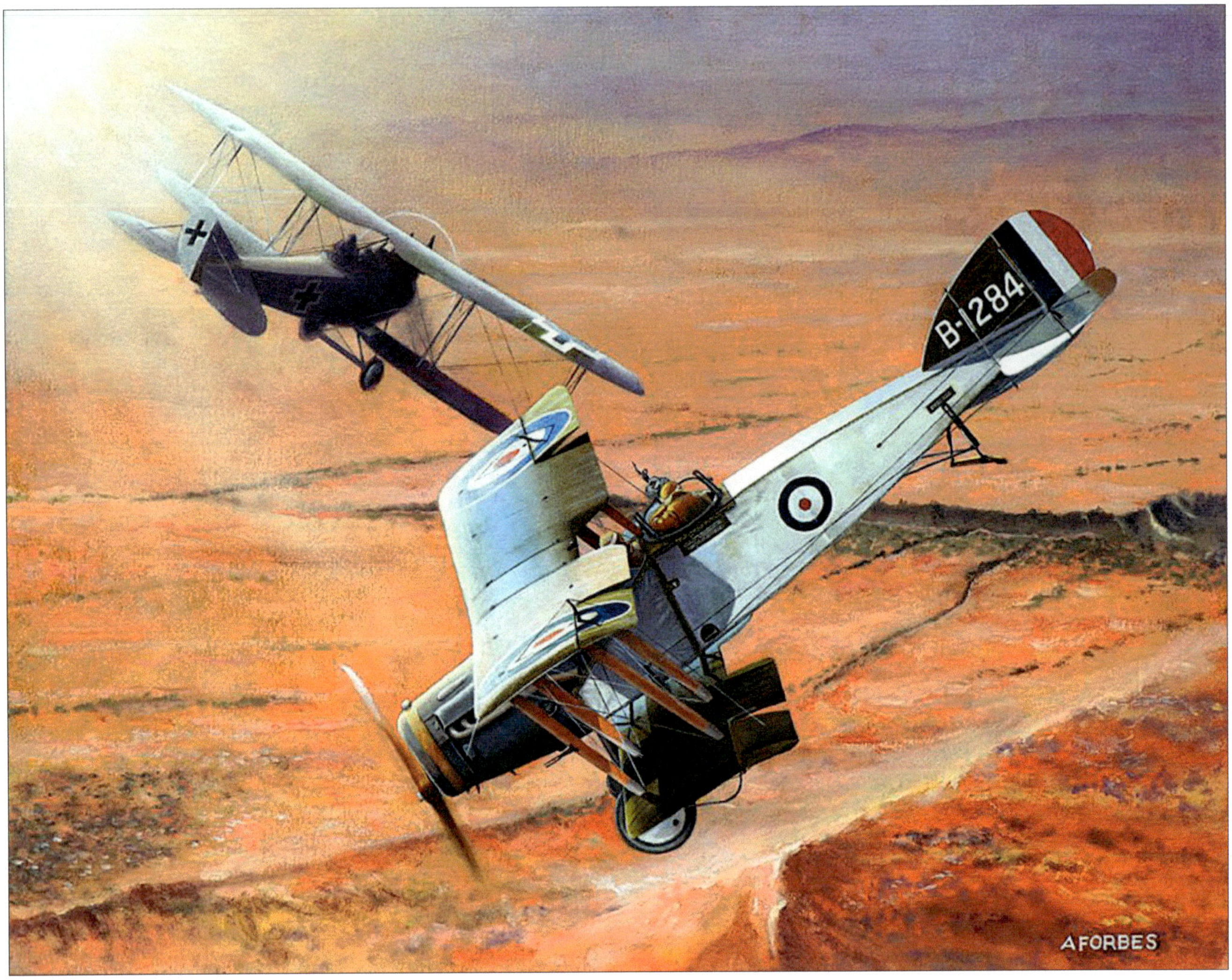
B-1284
AFORBES

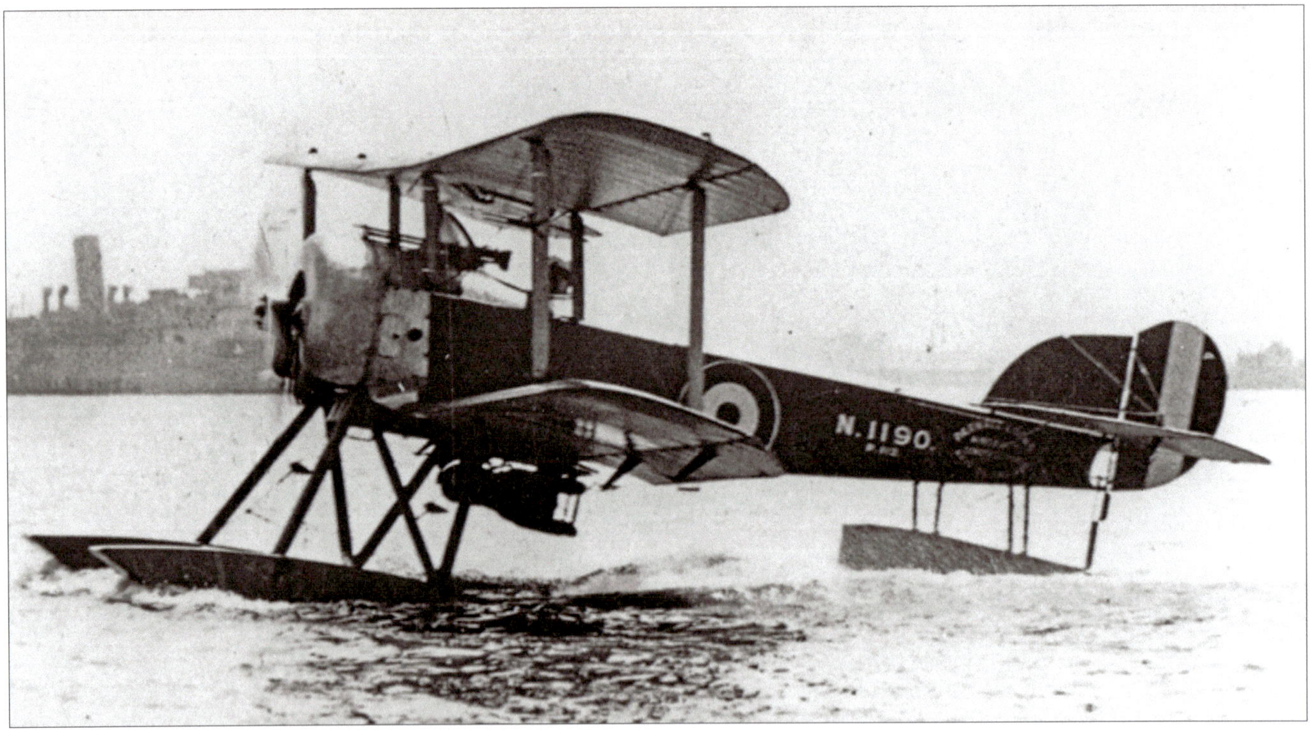
N.1190.

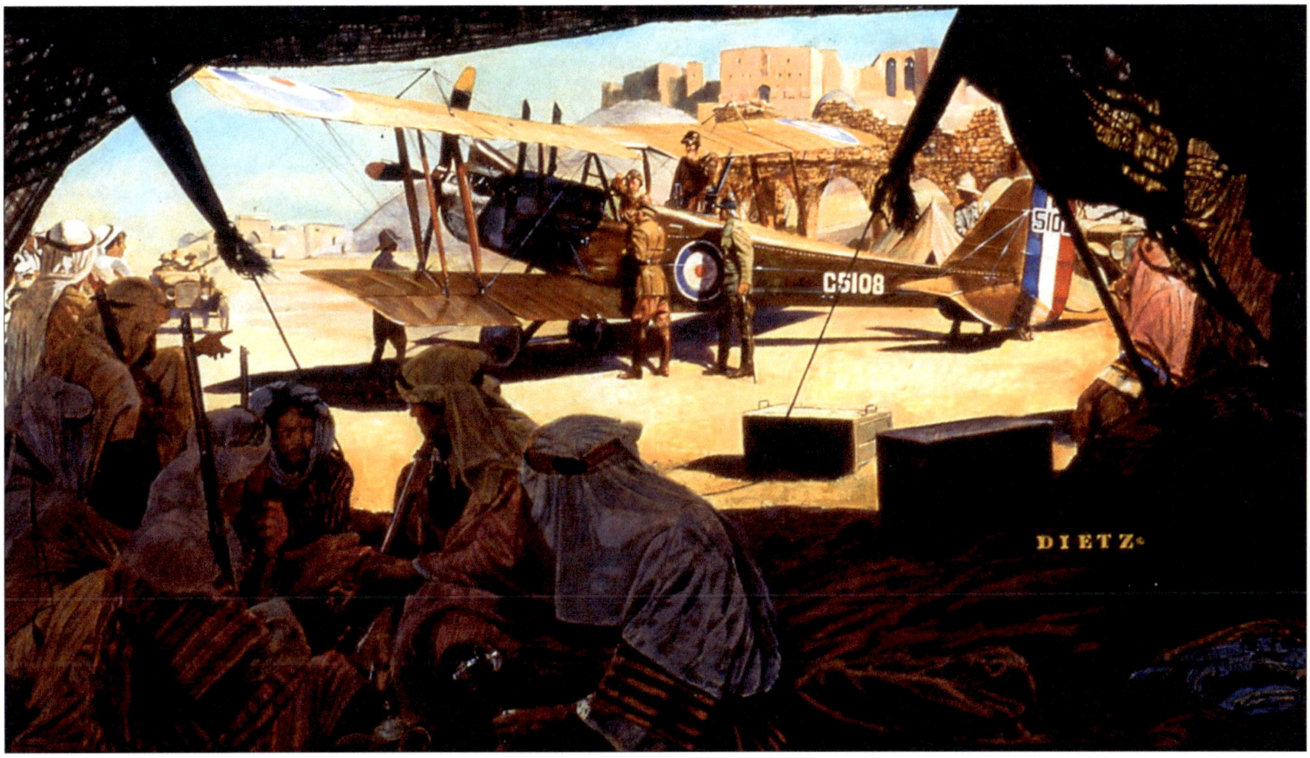

Above: The Middle-East would prove to be a assignment filled with adventure and sometimes dangerous liaison missions to support T.E. Lawrence's Arab uprising. Jim Dietz

Facing Page, Above: Because of the variety of missions that were needed and the relatively fewer number of fighter planes in Palestine, air combat was often between two-seater aircraft. At first the excellent German Rumpler C.I was more than enough to out perform any of the aircraft that the Australian No. 1 squadron had available, but that dominance ended upon the arrival of the Bristol Fighter. Alan Forbes

Facing Page, Below: The much modified Sopwith Baby floatplane also was on the roster of the BEN-MY-CHREE, note the Lewis gun to fire through the propeller and the bombs, this is likely how the Sopwith floatplanes that attacked the German airfield at El Arish would have appeared.

accomplish the opposite.

In addition, the Allies (primarily the British) would heavily invest men and material into the Gallipoli / Dardanelles campaign in order to secure a water route to connect with Russia. Those efforts of course would end in disaster. However, the offensive that would successfully drive Turkey to an armistice would come from the desert, from Palestine.

Germany was well aware of their Turkish Allies deficiency in air power and made significant efforts to overcome that weakness. The Turkish air service would eventually be involved in Mesopotamia, the Caucasus, and battling the T.E. Lawrence driven Arab revolt, while a special German unit, FA 300, would provide the air support in Palestine.

As a result of geography, the British forces fighting in the Middle East were primarily Indian or ANZAC(Australian, New Zealand) troops. Because the RFC was pre-occupied on the Western Front, initially the air service in the Middle East was No.1 sqdn, AFC (Australian Flying Corps) along with several units of the RNAS.

The tragic Dardanelles campaign was supplied via the British bases in Egypt and the capacity of bringing the ANZAC troops through the Suez Canal. Therefore, both the Germans and the Turkish military leaders planned an offensive through the Sinai Peninsula with the hope of breaching the defenses at the canal in order to sink ships in the canal or somehow destroy its functionality in some manner. With the loss of the Suez Canal, the Gallipoli campaign would lose a vital link in its supply chain.

The German command pushed their Turkish allies into mounting an offensive in the summer of 1916 after learning that the British had moved a significant amount of their forces from the Egyptian/ Palestinian area to the Western Front to support their offensives in France(in fact 8 of 12 divisions had been re-deployed). Understanding that Turkey did not have sufficient air resources, Flieger-

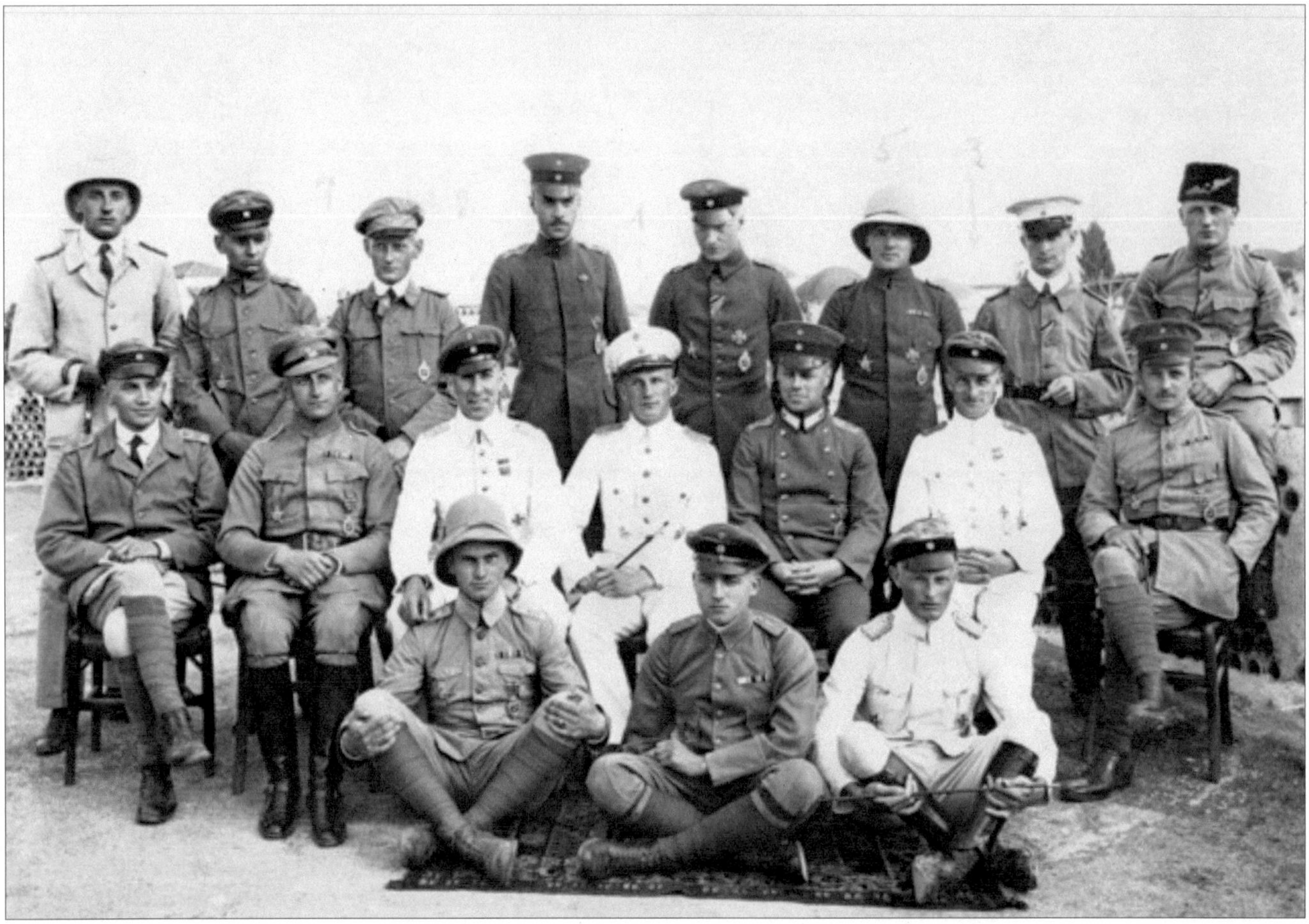

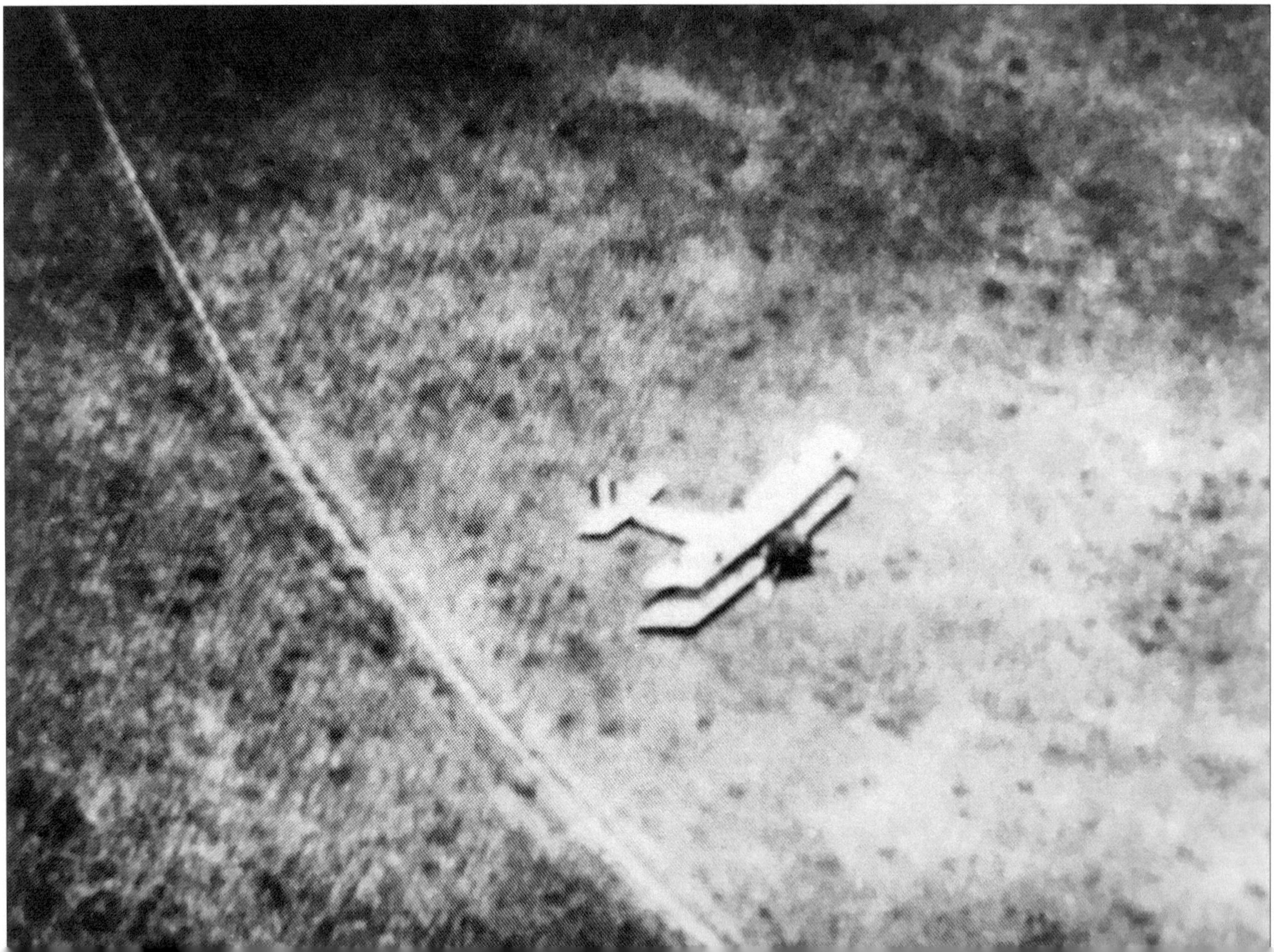

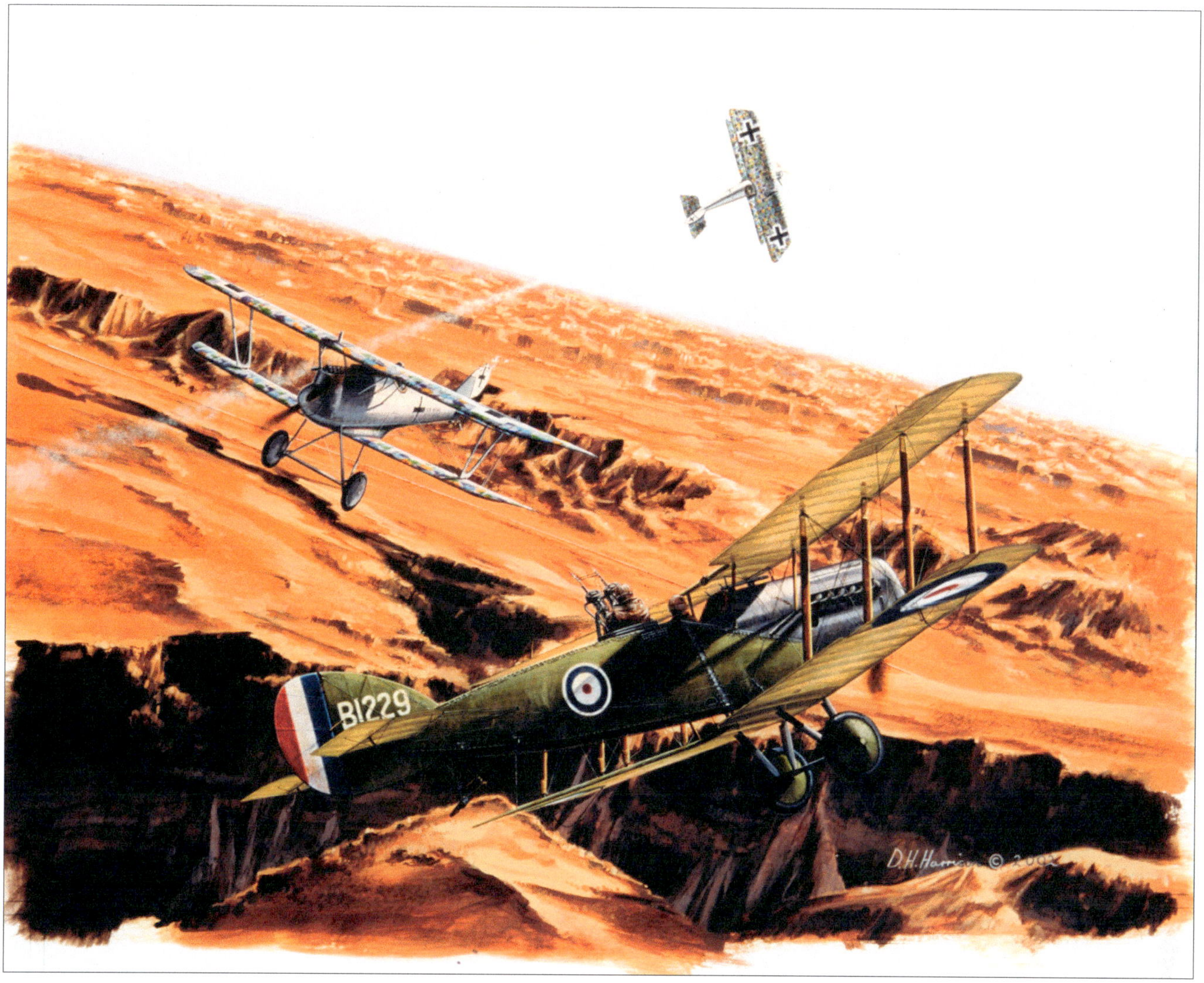

Above: On the 22nd September 1918, Capt. Ross Smith and gunner Edward Mustard shot down two of three Pfalz fighters in combat. The two Pfalz fighters force landed onto the desert below and then proceeded to taxi back to a friendly outpost. Smith and Mustard then proceeded to strafe the two Pfalz fighters on the ground in the hopes of making them unserviceable. D.H. Harrison.

Facing Page, Above: This group photo of FA 300 is from early summer 1917 at El Ramleh airfield. In the middle row, from the left: (3) Falke (4) Felmy (6) Obltn. G. Felmy. This photo was included in the message bag that Felmy dropped on a British airfield on 13th July along with his photo with 2/Lt. Vautin and his hand-written note to his "sporty adversaries".

Facing Page, Below: This photo was taken by Falke of the Matinsyde F.1 that he and Felmy shot down on 16 May 1917. The British pilot, Capt. Murray Jones was wounded but managed to walk back to British lines. Of note, the same pilot, in a different Martinsyde G.100, was shot down by Felmy and Falke on 6th of April! Thus, Capt. Murray Jones was both the first and second of Felmy's victories at FA 300.

Abteilung 300 Pascha was established and sent to the region.

The name Pascha was employed to emulate a Turkish word pasa, which signifies a person of high rank. Thus, the Germans were attempting to indicate their seriousness in the support and their respect of their ally in the naming of the assigned air unit.

Initially 10 pilots and 6 observers were sent along with 6 Rumpler C.I two-seaters and 2 Pfalz monoplane fighters. The aircraft were sent in crates by railroad as far as possible, and then had to be brought by caravan the remainder of the way. Despite the long and difficult journey, men and machines were ready in January, 1916.

Thc "stage" was set from the beginning, with the very experienced and capable German aviators having much fewer aircraft than their

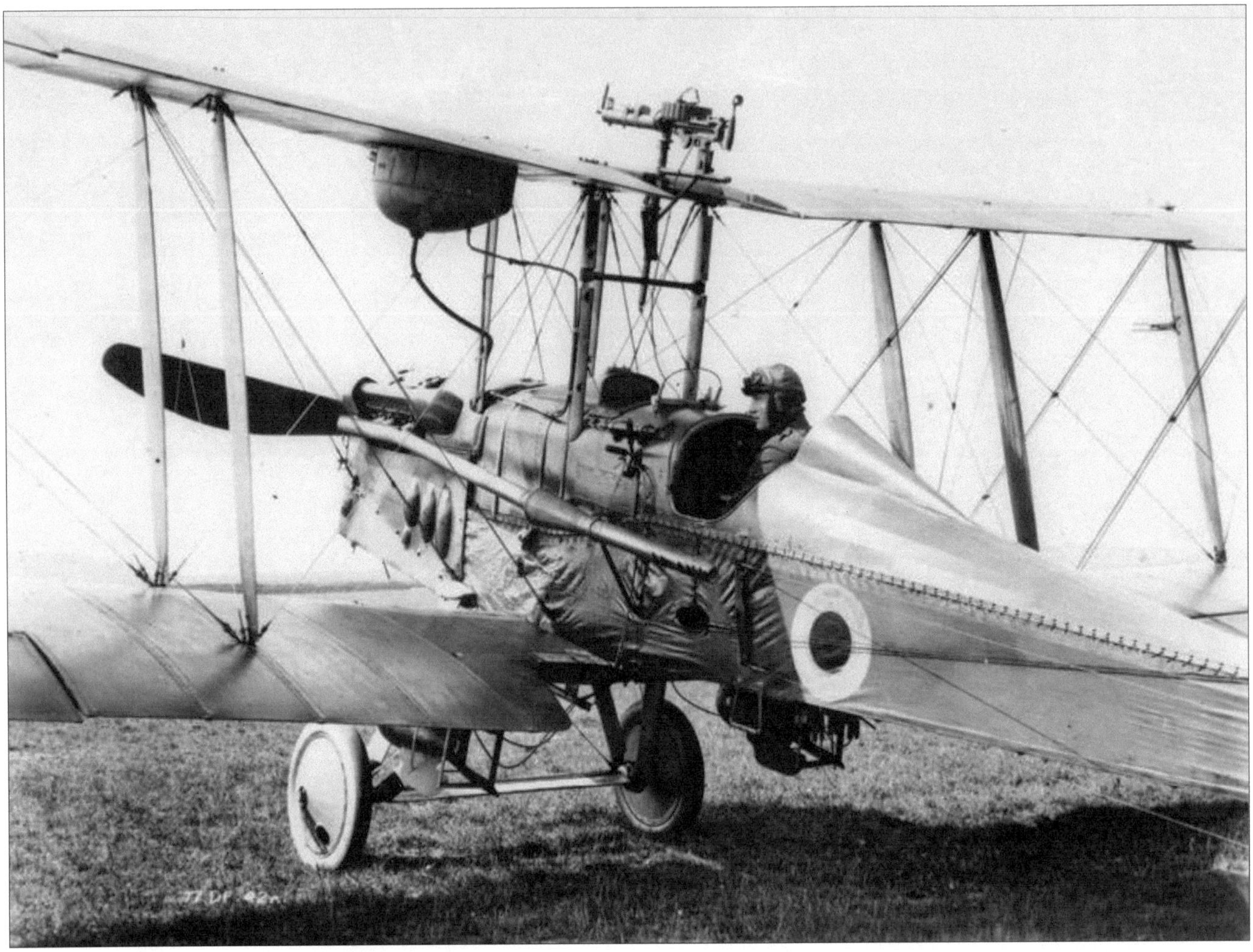

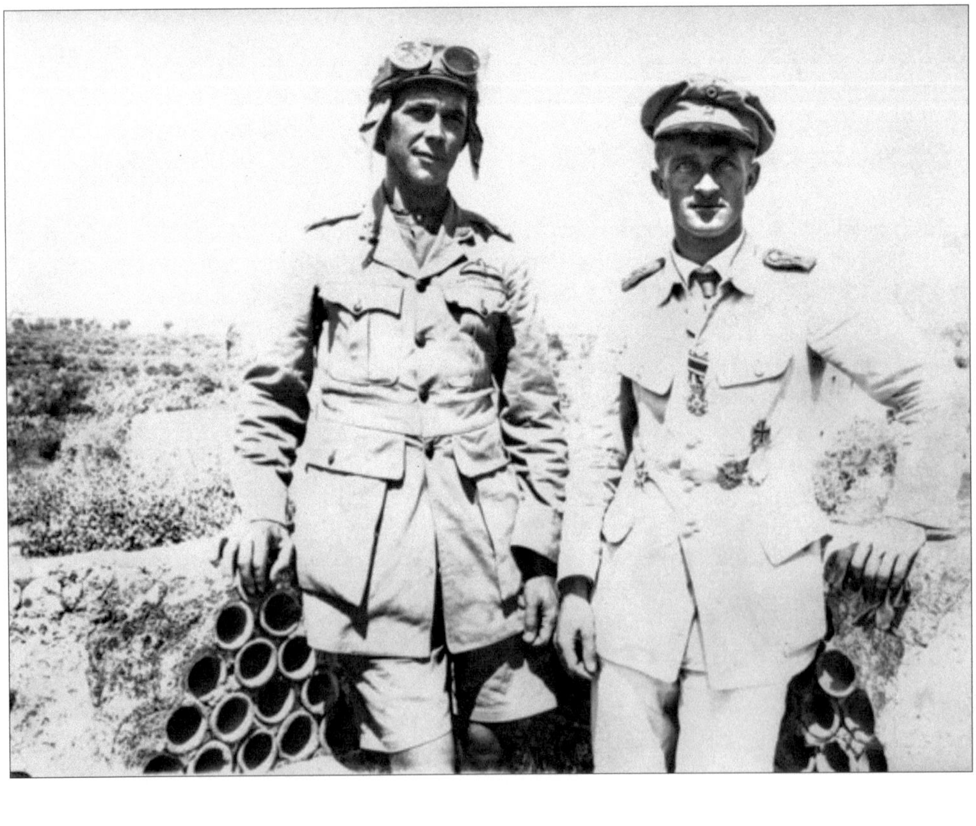

Above: 2/Lt. Claude Vautin was flying this type of aircraft, the hapless BE12, when he was brought down by Obltn. Felmy for his fourth victory.

Left: Vautin and Felmy, the Germans hosted their captive for several days at their airfield, where he voiced incredulity at the few aircraft on hand. This photo was then dropped at a British airfield to let them know that Vautin had survived.

principal Australian adversaries in No. 1 sqdn. AFC. Depending on availability, the Australians could have between 20-30 aircraft at any one time. However, their aircraft were markedly inferior in performance and consisted of the ubiquitous BE 2 , the marginally capable Martinsyde S.1 "Elephant" [not a nickname that indicated fondness for the aircraft by its pilots] and several D.H.1 pusher aircraft (long relieved of any combat duty elsewhere).

In addition to the AFC squadron, in May 1916 the RNAS was present in the form of the East India and Egypt Seaplane Squadron (EIESS) which was led by the fearless W.Commander C.R. Samson. The EIESS was composed of several seaplane tenders such as the HMS Ben-my-Chree and other similar vessels. Their aircraft were variants of Short Seaplanes and Sopwith Baby type single-seat floatplanes. This unit would often undergo joint operations with British monitors and other vessels which could bring significant firepower onto the Turkish land forces and did just that on several critical occasions.

One of the pilots in FA 300 was Lt. Hans Henkel who wrote this remark which was translated by the renowned First World War German aviation historian Peter Kilduff:

"The Turks sit here and the British there, and both form up troops, slowly advancing their camps, roads and railway lines for the coming battle. But in the middle lies the desert, silent and peaceful, aside from minor skirmishes by patrols, which make determined efforts to acquire temporary possession of one small oasis or another."

This quote accurately describes the lulls in fighting that would be followed by an offensive by one side or the other that would soon become stalled, and then fighting would return back to the lull, these "rhythms" that so characterized the fighting in the desert during the First World War.

Initially the Germans established their main aerodrome at Beersheba, inland from Gaza and this base would remain their main center of operations until the end of the war. With initial advances made by the Turkish forces in later 1916 and into early 1917, FA 300 also established an airbase at El Arish.

The initial Turkish advance in the early summer of 1916 was able to advance to the Suez Canal, but there it stalled and was not able to disrupt the Allied usage of the canal at all. During the slow withdrawal of its infantry, the British aircraft from both the AFC and the EIESS unit bombed the Turkish ground forces when possible. FA 300 would put their Rumpler C.I aircraft into the air and were usually effective at clearing the skies. The Rumpler was a far superior aircraft to any type the British had.

Therefore the British mounted a large scale raid on the German aerodrome near El Arish on the early morning of 18 June 1916. The raid was carried out by BE 2 aircraft manned only with a pilot so as to carry bombs and several Martinsyde G.I as escorts. The aggressive British attack was successful at destroying several German aircraft on the ground and wounding several of the aviators. All in all, a big set-back for FA 300 as they did not have that many aircraft or personnel to begin with.

Hence, despite the inferior aircraft the British were able to significantly hamper the German aviation unit's capabilities with this attack. Throughout the campaign, the German aerodrome of FA 300 was a prime and frequent target. Another example of this unwanted attention would come to be called the "Sea Battle of El Arish"(although it was not on the coast!). On the morning of 17 September 1916 several Short Seaplanes escorted by Sopwith Baby floatplanes of EIESS attacked the airfield with bombs and machine guns.

Via wireless transmission, the British airmen also directed naval gunfire against the aerodrome. A couple of the Rumplers managed to get off the ground and chased the British aircraft back to sea and then dropped several bombs against the seaplane tender without hitting it, but causing it to steam further out to sea in retreat. One of the Rumpler crews, Bulow/Halser successfully shot down two of the Sopwith Baby floatplanes!

However, the continued British attacks coupled with the approaching winter weather which promised heavy rain and few flying days caused the German unit to retreat back to its main base at Beersheba.

However, the fortitude of the members of FA 300 was undaunted by this move. On the 13th November Ltn. Schultheiss and Obltn. Falke undertook a memorable mission. Taking off from Beersheba, they landed at the old aerodrome of El Arish to take on more fuel and from there they flew to Cairo to drop several bombs onto the heart of the British efforts in Egypt. The flight was 5 hrs 30 mins, and went without any problems. They then landed at El Arish and took on more fuel for the return to Beersheba for a total of 7 hrs 30 mins in their mission!

The British returned the favor in November bombing the Beersheba aerodrome and the railroad station near it.

Both British and German aviators in the Middle East faced unique dangers as they performed their missions. All the aviators suffered through illnesses such as dysentery and malaria. They all had to face horrible weather such as sudden sand storms and also had to contend with uninviting terrain which made emergency landings often become disastrous.

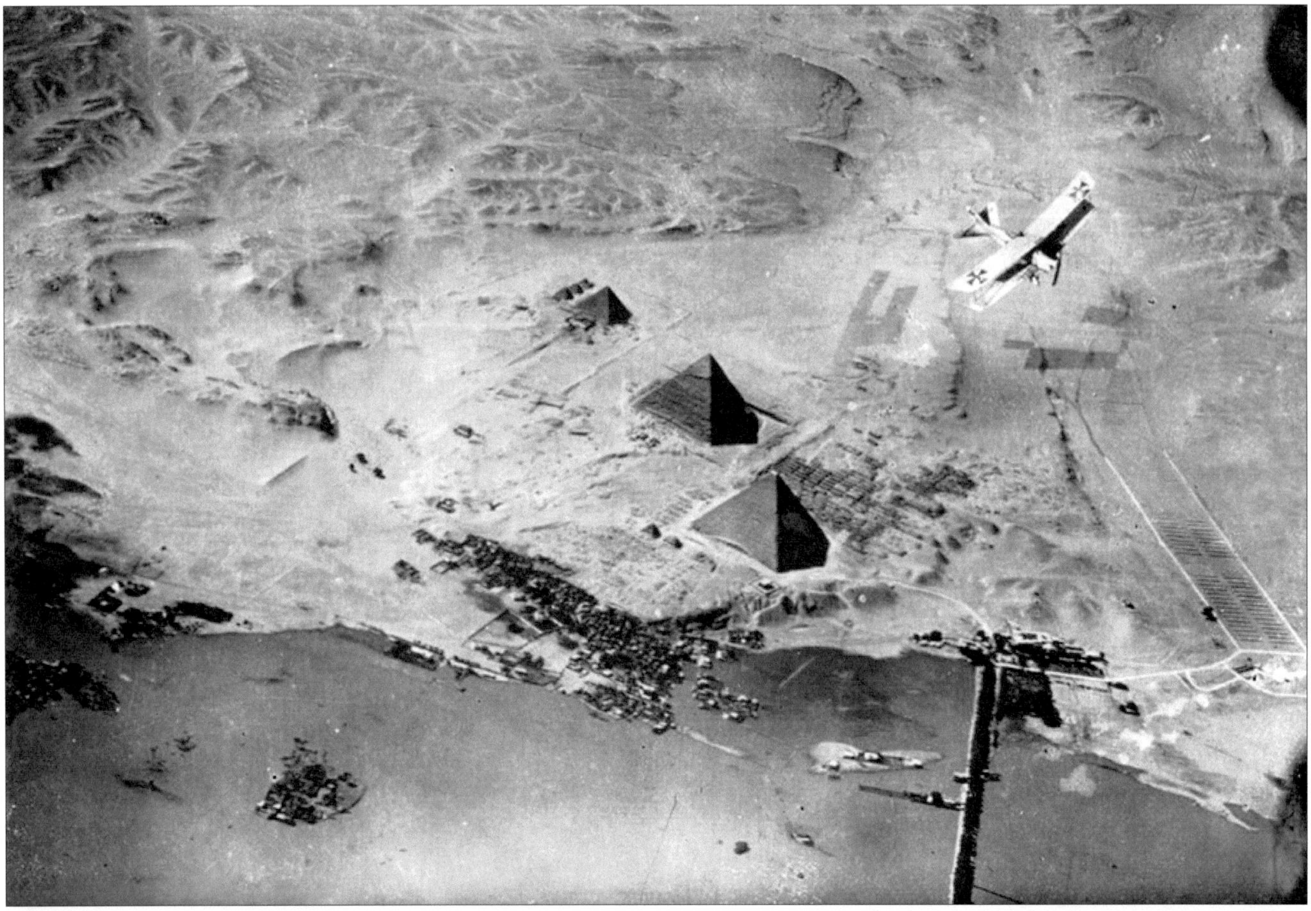

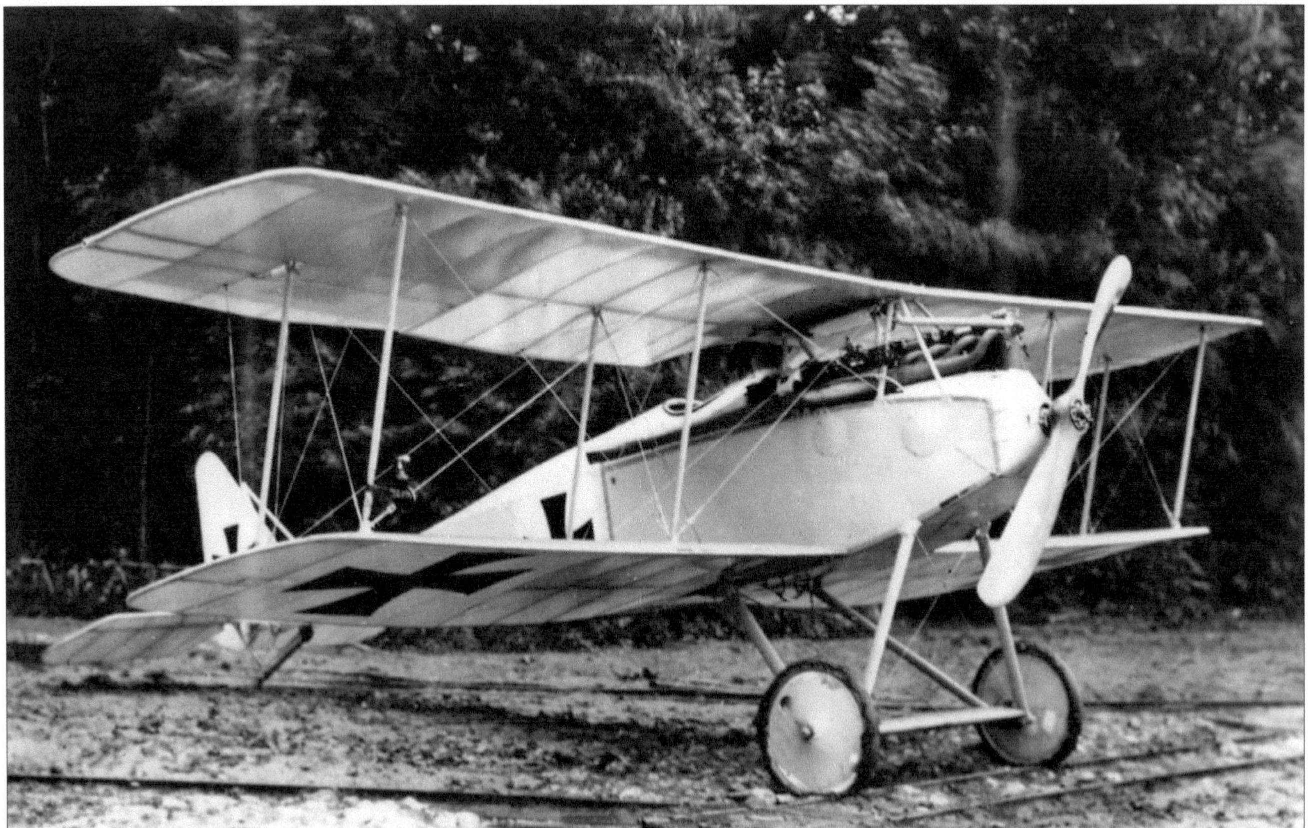

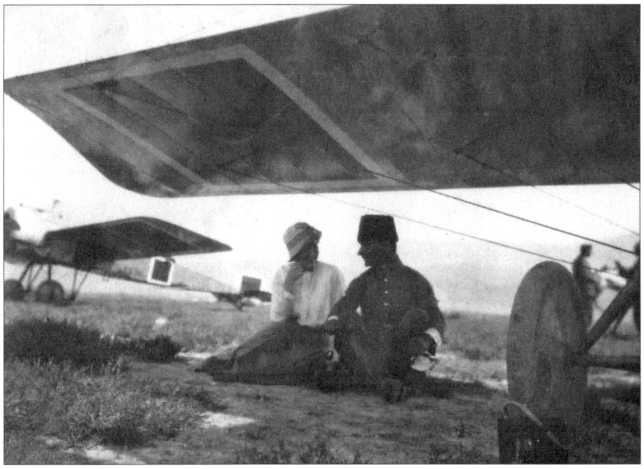

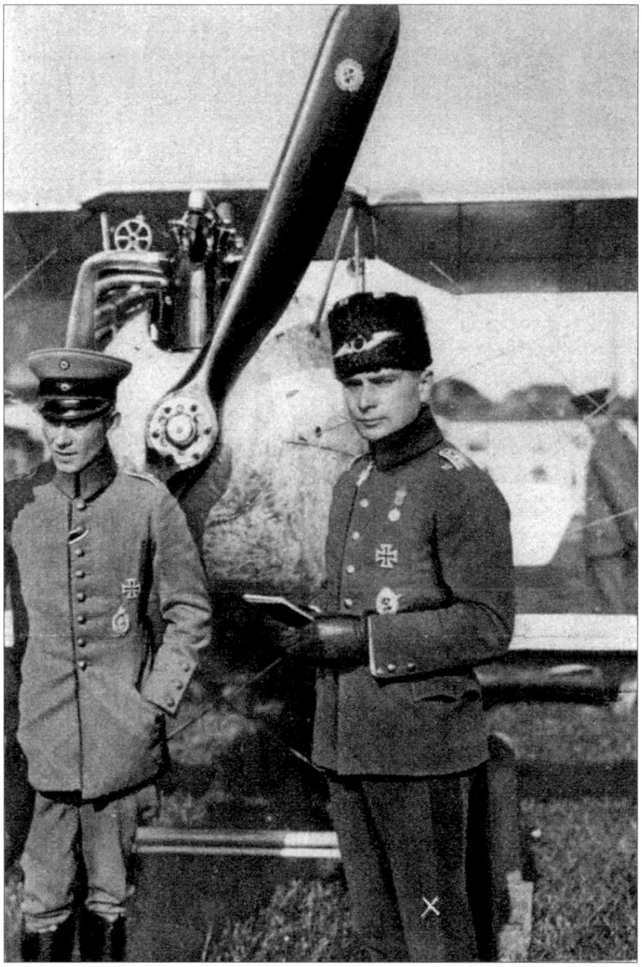

Above: Buddeke in a quiet moment with some pleasant company. Note the Fokker EIII monoplanes in Turkish markings.

Facing Page, Above: A composite photo meant to depict Ltn. Schultheiss and Obltn. Falke during their epic flight to Cairo and back.

Right: Buddeke in front of a Halberstadt fighter; note the single machine gun.

Facing Page, Below: A sturdy, maneuverable fighter, the Halberstadt found a new operational significance in the Middle East after becoming obsolete on the Western Front.

Below: Aircraft types not deemed suitable for the Western Front would find operational status in the Middle East. Here are two Vickers Bullets, less than fifty were built in the war, despite being a fast aircraft and of capable performance.

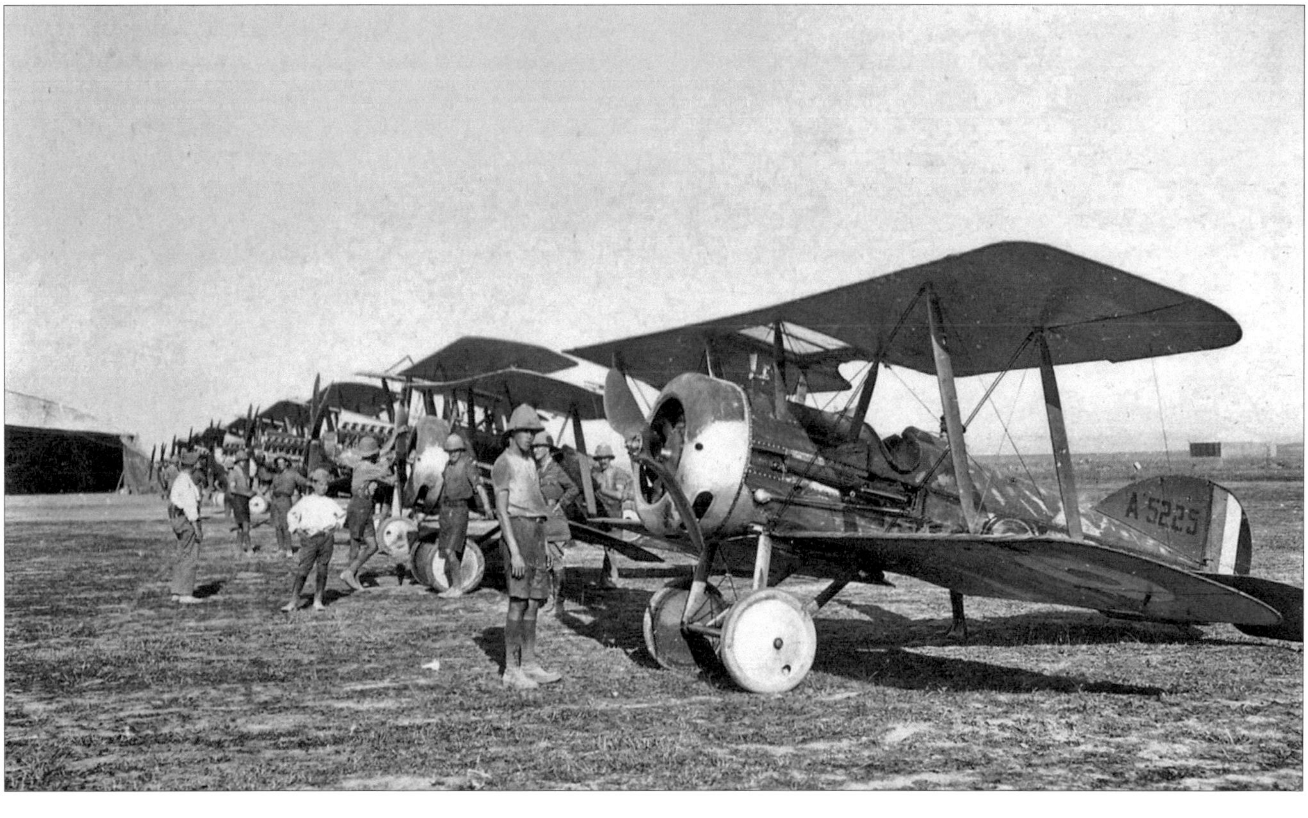

If they survived the crash landing, they would then have to face hostile natives- Arabs or Bedouins who could be cruel to any captives and often did not care which side of the war the aviator was fighting for. They would also face the relentless desert: its heat, and other dangers as they tried to walk back to friendly ground forces.

These shared dangers and hardships fostered a certain camaraderie and respect that the aviators held for each other, despite being adversaries in the war. The relatively few numbers of airmen involved in the air war in the Middle East also meant that they came to know of each other as stories were told in their respective officer's mess as well as reading of each other's exploits in newspapers and magazines.

One of the first, recognized "stars" of the Middle East was Hptm. Hans-Joachim Buddeke who was the third German aviator to be a recipient of the Pour le Merite. Surprisingly, he was actually in Chicago, Illinois at the outbreak of the First World War. Having taught himself how to fly in a modified Nieuport monoplane with the thought of beginning a career in aviation, he immediately changed all of his plans and using an assumed name, Buddeke made his way back to Germany, where he offered his services as a pilot.

He would score three victories flying one of the new Fokker monoplane fighters as a member of FFA 23 between September and November of 1915. He learned of a planned expansion of German aviation units in Turkey in December of 1915, including a Fokkerstaffel. Buddeke may have found the chance to go and see another part of the world to his liking and so applied for and was accepted as part of the contingent.

Eventually over 400 German aviators would serve in the Turkish air service and over 450 aircraft were sent to the region. Buddeke would continue to successfully down Allied aircraft and seemed to be invincible in the skies. He would earn the nickname El Schahin(The Hunting Falcon) before being recalled to take command of Jasta 4 on the Western Front in August, 1916.

One of the pilots who came in a group of replacement aviators in the autumn of 1916 was Obltn. Gerhard Felmy, whose older brother Helmuth was the commander of FA 300. Felmy would make quite a name for himself before he was compelled to return to Germany in August, 1917 in order to recover from a severe bout of malaria.

The Allied forces were expanded with new RFC units as well as some new aircraft, although the BE 12 fighter was not very effective while the Re8 observation aircraft was a definite improvement over the seriously outdated Be2 types that would still carry on at times.

The Germans brought more Rumpler C.I aircraft and now also had several Albatros D.III fighters which were clearly superior to any of the Allied aircraft, although the Germans remained far outnumbered.

Felmy flew several bold missions that earned him significant respect amongst his comrades and his Australian and British opponents. On the 24th May 1917 Felmy flew with his observer Falke in a Rumpler C.i to a position about 100 kms behind the British front lines. Landing , the two proceeded to place explosives and managed to blow up a significant rail line as well as destroying a water pipe line and nearby telegraph wire lines. They returned without incident and the two performed several similar missions causing significant hampering of the Allied supply lines.

In addition, Felmy flew both the Rumpler and the Albatros D.III in combat and achieved 4 victories. Amazingly, his first and third victories were over the same pilot! Alan Murray-Jones was shot down by Felmy on both 6th April and the 16th May, 1917. Both times Murray-Jones was able to successfully force land his Martinsyde G.I and make his way back to British outposts, but both times his aircraft was destroyed.

On the 8th July, Felmy was flying an Albatros D.III fighter when he brought down 2/Lt. Claude Vautin who was flying the horrendously inferior Be12 at the time. Felmy brought Vautin to FA 300's aerodrome where he was entertained for several days before being sent to a POW camp. Vautin was reported to have said, "Where are the others?", when he saw the few German aviators and the relatively few number of aircraft, such was the reputation of the FA 300 aviators amongst their Australian counterparts.

Felmy flew over an Allied airbase and dropped a photo and note from Vautin to allow his comrades (and family back home) know that he was alive. In addition, Felmy arranged so that a British aircraft could fly over the FA 300 airbase at an arranged time to drop a canvas bag that included clothes, kit and even letters for Vautin as he headed off to captivity! Later in the campaign Felmy would drop off a two page letter to his Allied adversaries that he signed: "with sportly respects, your G. Felmy, Obtln."

Despite the best efforts of the German airmen in FA 300, by the end of 1917 the air war was turning against them. When Felmy was sent back to Germany in August, the commanding officer of the German forces , General von Kress, wrote in his memoirs:"...with Felmy's departure our luck departed too; after his leaving we were no longer

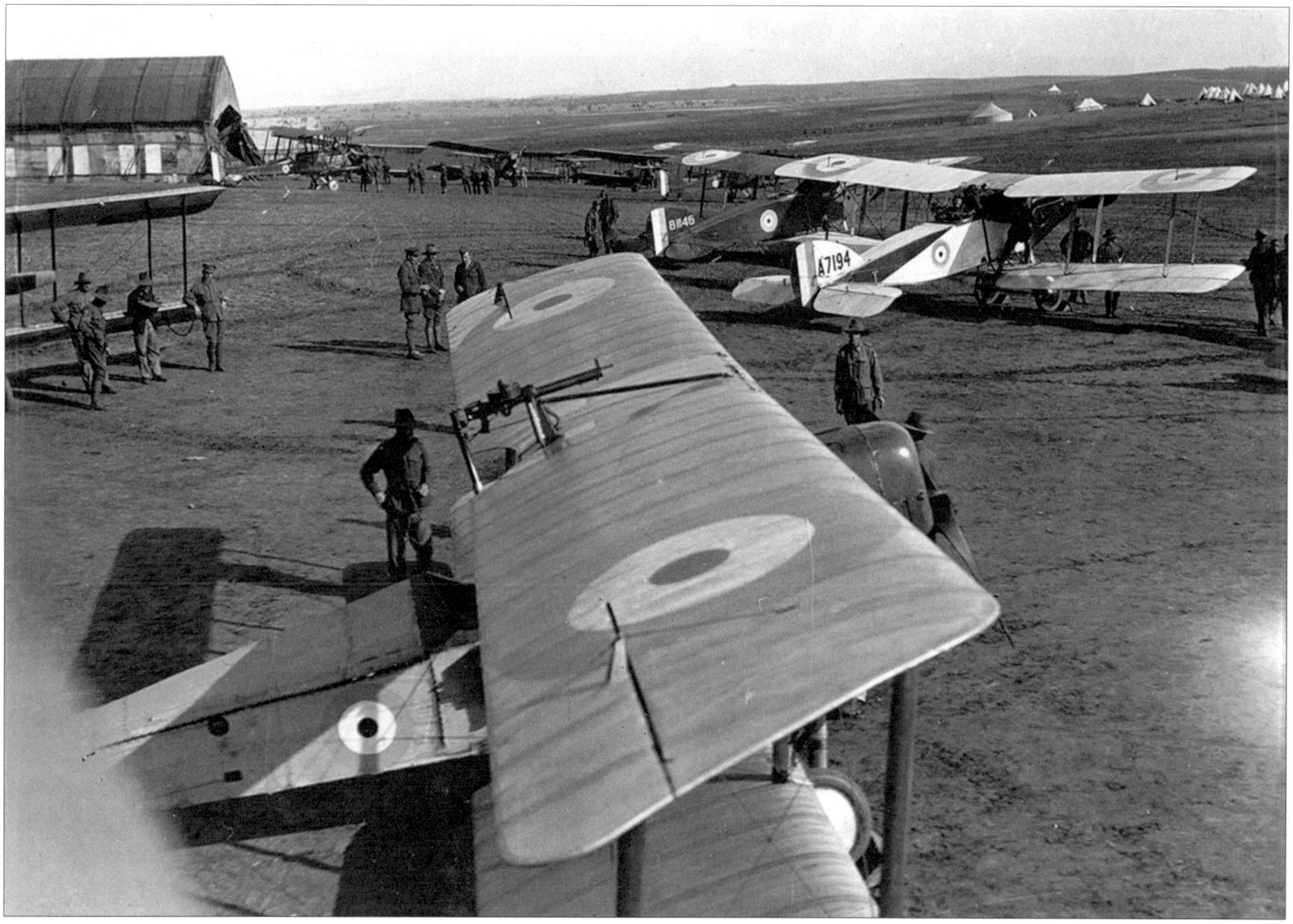

Above: The beginning of Allied air supremacy in Palestine. Note the Martinsyde G.100 in the foreground, now replaced by the superlative Bristol Fighter which would be much more effective against the German Rumpler and Albatros aircraft.

masters of the air."

As 1917 ended, even more Allied aircraft arrived in the region to further the imbalance in numbers of aircraft against the Germans. However, more importantly, the new aircraft were also of much improved quality, such as the Bristol F2b fighter. Now, the Australians had not only more aircraft, but aircraft types that were equal if not even better than the aircraft flown by the Germans.

The ground war also changed from an Allied defensive posture to protect the Suez canal to an offensive, attacking attitude that would drive any Turkish ground forces from Egypt, then from the Sinai peninsula and then retreating back into Palestine towards Damascus. The tide of battle had irrevocably shifted.

On the 19th September 1918 the Allies were successful at the Battle of Megiddo and forced the Turkish forces to retreat. On the 21st September Bristol fighters of No.1 sqdn, AFC discovered a large force of Turkish soldiers retreating along the road at Wadi El Fara. It was a narrow gorge with steep cliffs lining the road on either side. The Australian aircrew transmitted their discovery back to their airbase and flights of aircraft then proceeded to pound the Turkish forces relentlessly. One of the aircrew described their effective use of airpower as "sheer butchery". In the most effective air to ground attack during the entire war, by the end of the day, the Turkish seventh army was decimated to the point of no longer being a functional fighting force.

On the 22nd September 1918, T.E. Lawrence requested help because his Arab fighters were being bombed and repeatedly attacked by German aircraft. Opposing the Arabs were several DFW two-seater aircraft and three Pfalz D.IIIa fighters. A Handley-Page bomber filled with spare parts, fuel cans and other supplies also brought Lawrence back to his troops. In addition, two Bristol F2b fighters flew into the makeshift airfield near Um es Surab. As the Australian airmen were sitting down for a late breakfast, a lookout called out that he had spotted several German aircraft approaching. Pilot Captain Ross Smith took off with his experienced gunner Edward "Pard" Mustard.

They approached the enemy flight which

consisted of a DFW escorted by two Pfalz D.III fighters. They dove to the attack and quickly dispatched the DFW in flames near Mafraq. The two German fighters departed and so Smith and Mustard returned to their field to finish their meal.

Again their repast was interrupted and so the duo raced back to their Bristol fighter and took off once again to attack a newly reported flight of German aircraft. As they approached the enemy flight:

"This time their foe was all three of the Pfalz scouts, and the combination of Smith's aggressive flying and Pard Mustard's accurate shooting soon forced two of the aircraft to the ground near the railway, where they taxied at high speed to Turkish outposts. The third scout scurried away towards Deraa and with the throttle wide open the Australians followed. Unable to gain enough ground for effective action to be taken, they eventually abandoned the chase and turned for home. On the way back they visited both of the machines that had been forced down earlier and from close range peppered each aircraft with fifty rounds. When they returned to Um es Surab they expressed their regret at having to return to the squadron duties because there they had complete mastery of the sky, and the enemy was not seen again." author, Felix Noble

Turkish forces formally surrendered on 31st October 1918. The British Commander-in-chief, General Sir Edmund Allenby had this to say of his Australian airmen:

"This squadron(No.1, AFC) played an important role in making this achievement possible. You gained for us absolute mastery of the air, thereby enabling my cavalry, artillery and infantry to carry out their work on the ground practically unmolested by hostile aircraft. This undoubtedly was a factor of paramount importance in the success of our arms here."

Facing Page: Lt. Malcolm G. Begg was an experienced, valued ,and honored RFC pilot. He had already received the M.C. and a bar to the M.C. for "conspicuous gallantry...came down to 700 feet to obtain valuable information under a heavy fire from the ground...great courage and determination." He was a member of No. 63 squadron as it was sent to the Middle East in August, 1917. They were assigned to the 1st Indian Army that was holding the Tigris front. After being unloaded, Capt. J.E. Philpott and Lt. M.G. Begg flew two assembled RE8 aircraft over 400 miles to land at Samarra on the 10th September 1917. On the 25th September, during their first sortie in Mesopotamia, the pair spied a Turkish Halberstadt fighter at a lower altitude not far from the enemy's airfield at Tekrit. Begg did not hesitate; he put his RE8 into a sudden and steep dive to attack his enemy. He managed to shoot a burst of fire from his machine guns at the Halberstadt before the "top plane extensions folded [ripped off] at about 5,000 feet... the speed indicator had displayed 115 mph". Somehow Begg was able to regain control of his plunging aircraft and managed a controlled crash-landing with the result that both he and his observer Cpl. O.W. Grant were captured. The aircrew of the other RE8 were forced to also land nearby because of engine failure and were captured as well (Capt. J.R.Philpott who died of dysentery while in captivity and Cpl. Grant). The unidentified pilot of the Halberstadt was forced to land because of damage to his engine by Begg's accurate burst of fire; the Halberstadt pilot requested confirmation of two victories over the RE8 aircraft. Story courtesy of Robin Begg and the family archive. Artwork by Darryl Legge

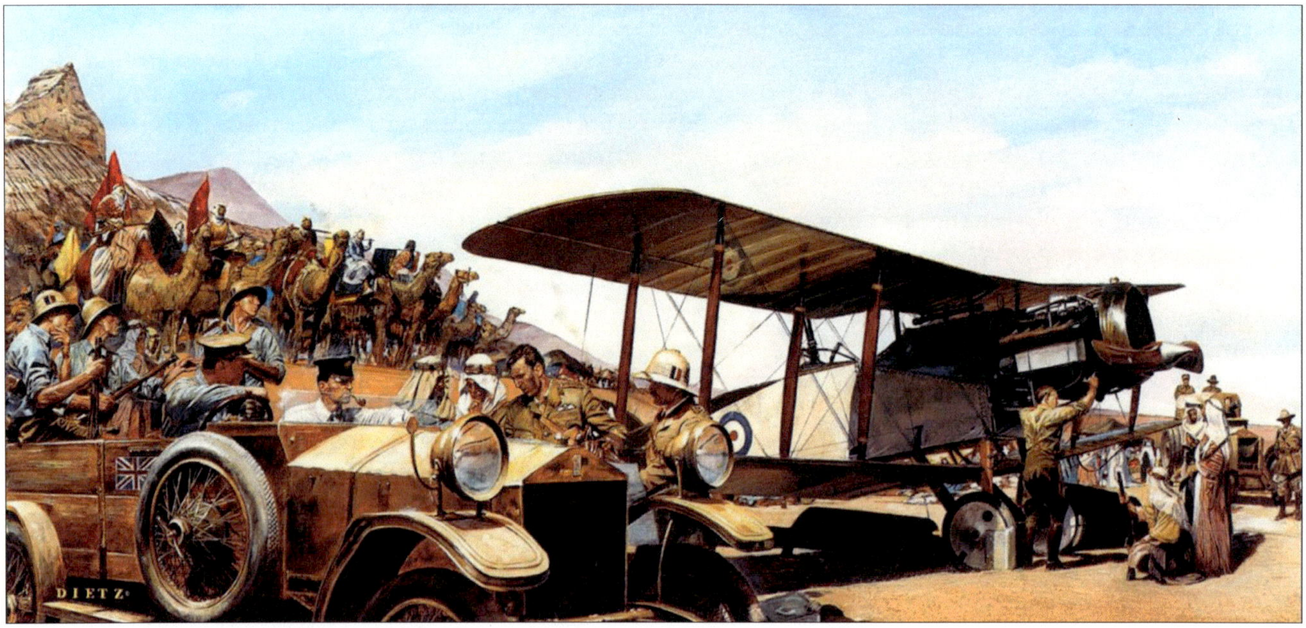

Above: The modern equipment of the British forces combined with the T.E. Lawrence led Arab fighters eventually led to success in the Middle East. Jim Dietz

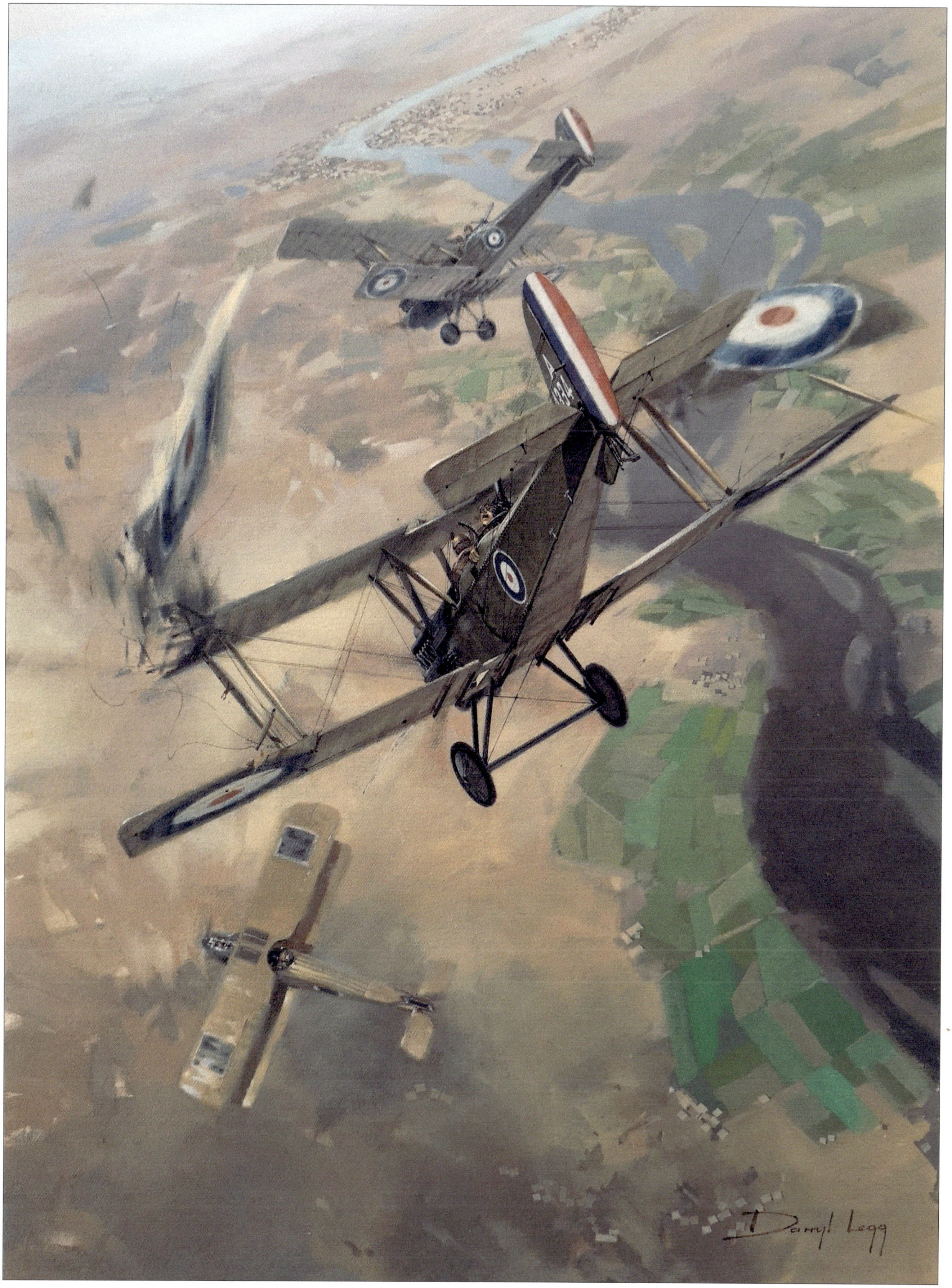
A
4334
Darryl Legg

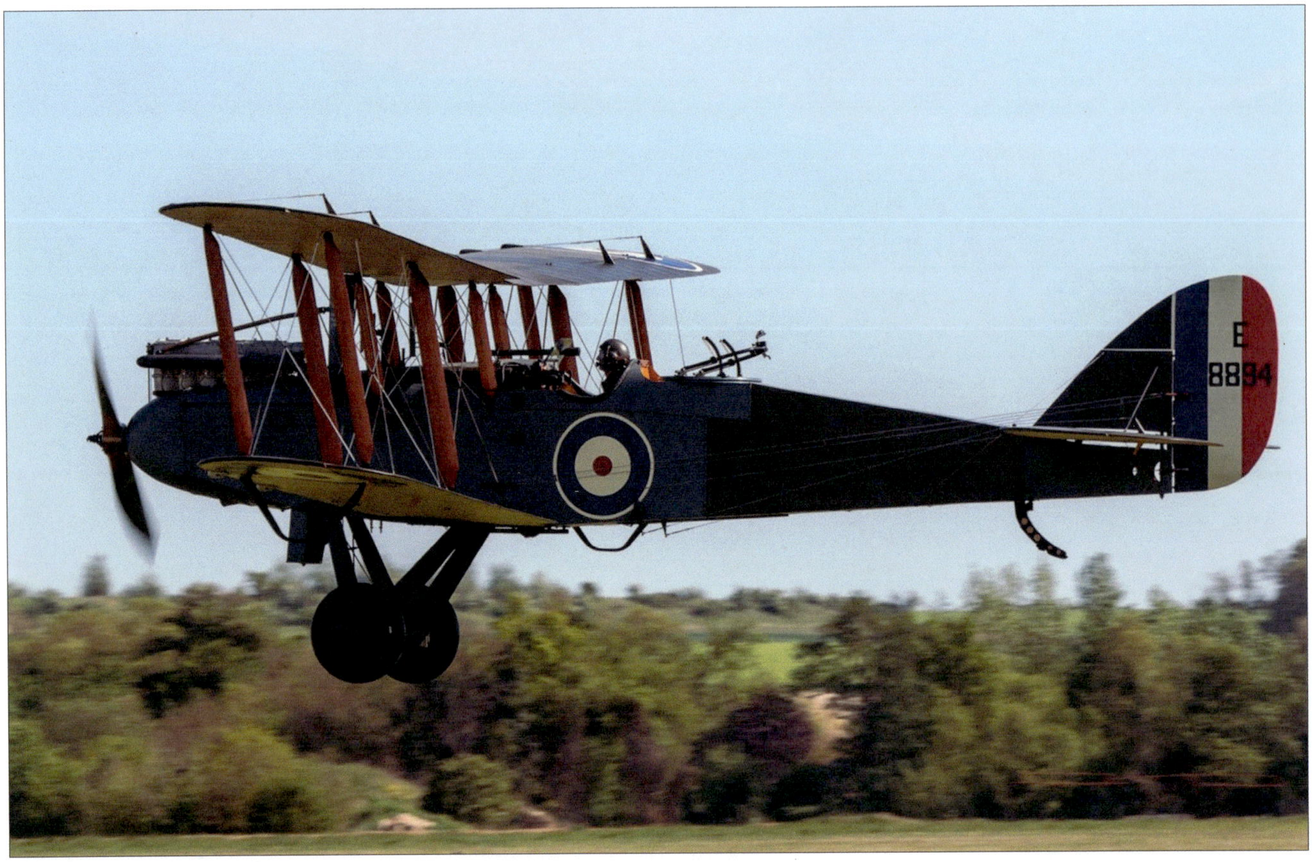

Above: A modern replica of a D.H. 9 aircraft as used so effectively by the Australian aircrew during their "blitzkrieg" attack at Wadi Fara.

Facing Page, Above: The Battle of Megiddo ended the war in the Middle East, a significant component of the British victory was the destruction of the Turkish 7th Army by air attacks alone. The Australian No. 1 squadron dropped three tons of bombs and expanded more than 24,000 machine gun bullets while flying 90 sorties in a four-hour time frame.

"But the climax of the air attack, and the holocaust of the miserable Turks, fell in the valley by which the Escdraelon drained to the Jordan [river] by Beisan. The modern motor road, the only way of escape for the Turkish divisions, was scalloped between cliff and precipice in a murderous defile. For four hours our aeroplanes replaced one another in series above the doomed columns: nine tons of small bombs and grenades and fifty thousand rounds of S.A.A. were rained upon them. When the smoke cleared it was seen that the organization of the enemy had melted away. They were a dispersed horde of trembling individuals, hiding for their lives in every fold of the vast hills. Nor did their commanders ever rally them again. When our cavalry entered the silent valley the next day they could count ninety guns, fifty lorries, nearly a thousand carts abandoned with all their belongings. The R.A.F. lost four killed. The Turks lost a corps." T. E Lawrence. Art by Steve Anderson

Below: No. 1 Squadron AFC had equipped several of their Bristol Fighters with radio communications and these aircraft effectively directed the flights of other aircraft so as to attack the Turkish 7th army forces in waves and in a near continuous fashion.

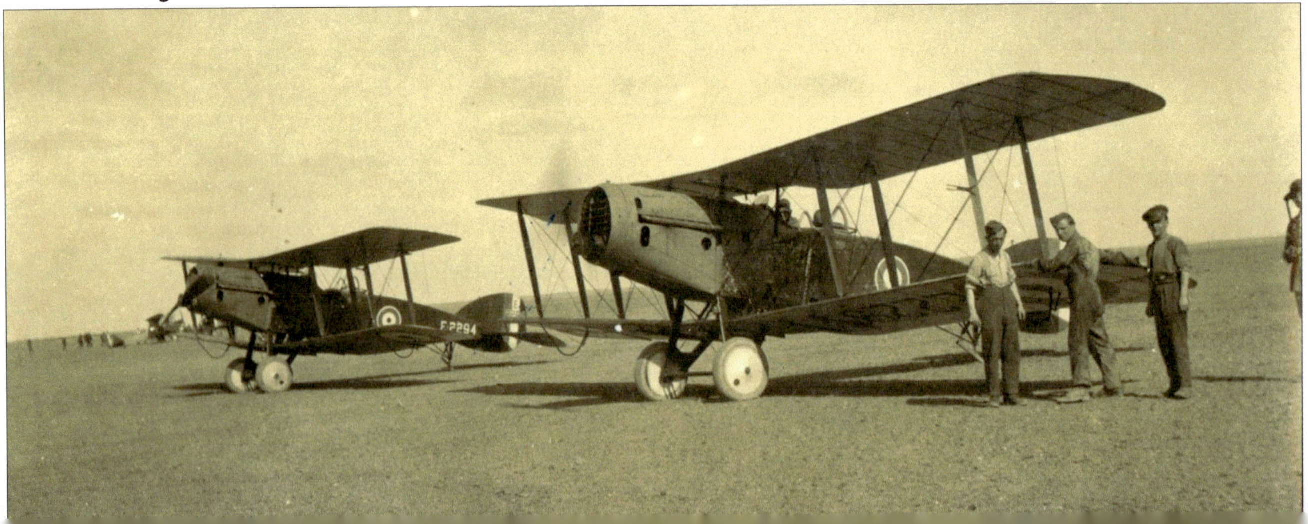

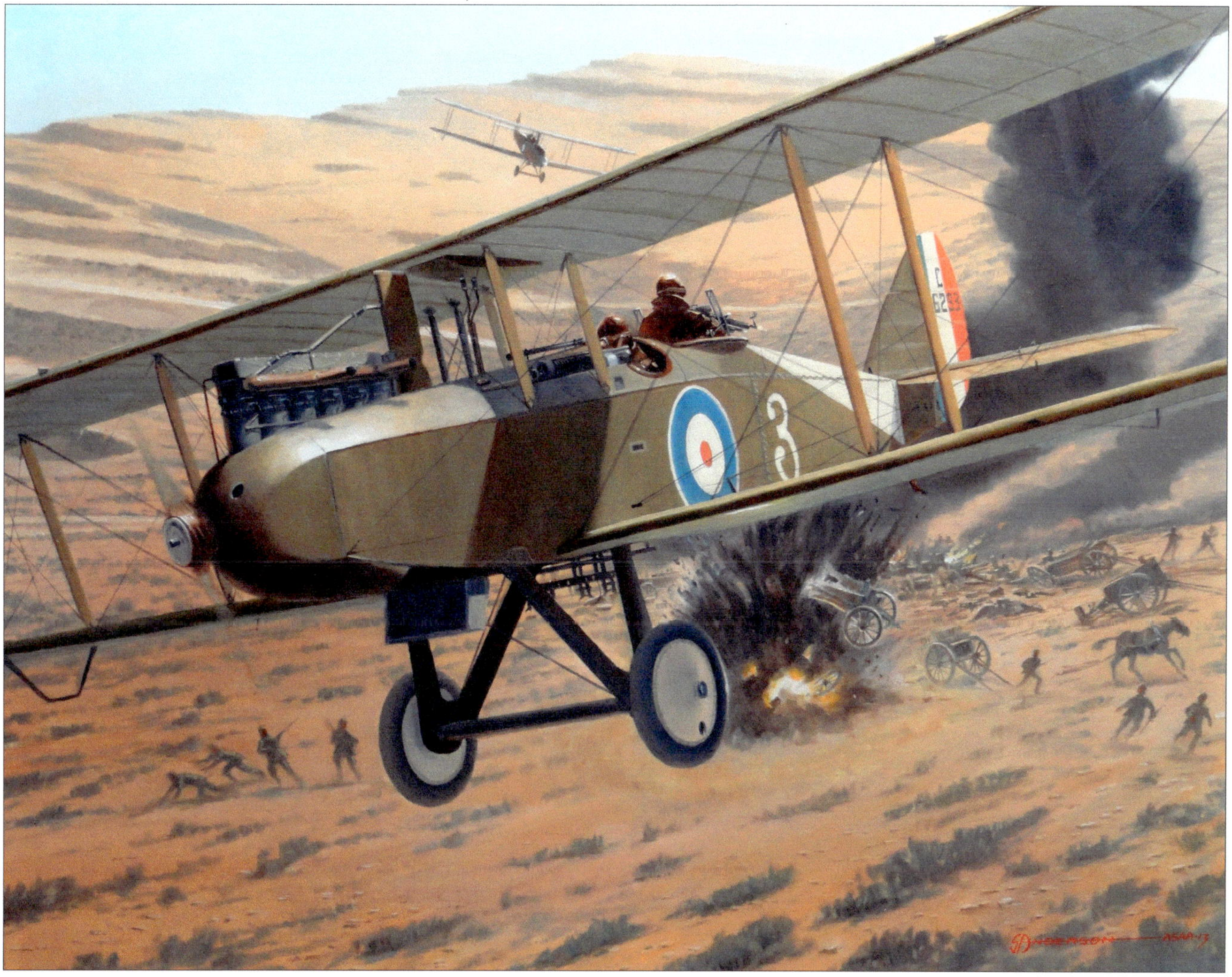

Below: "The panic and the slaughter beggared all description. The long, winding, hopeless column of traffic was so broken and wrecked...that the bombing machines gave up all attempts to estimate the losses under the attack, and were sickened of the slaughter." F. M. Cutlack.

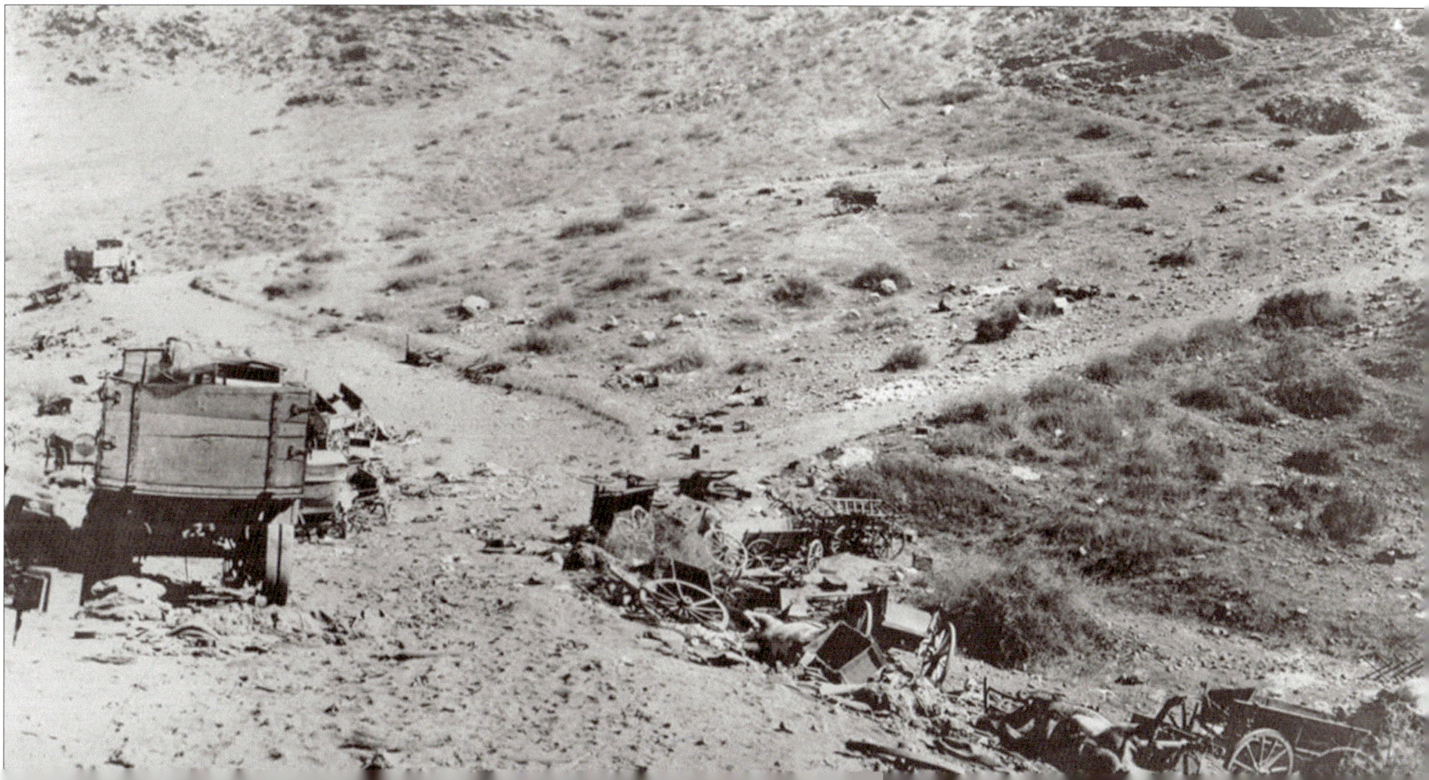

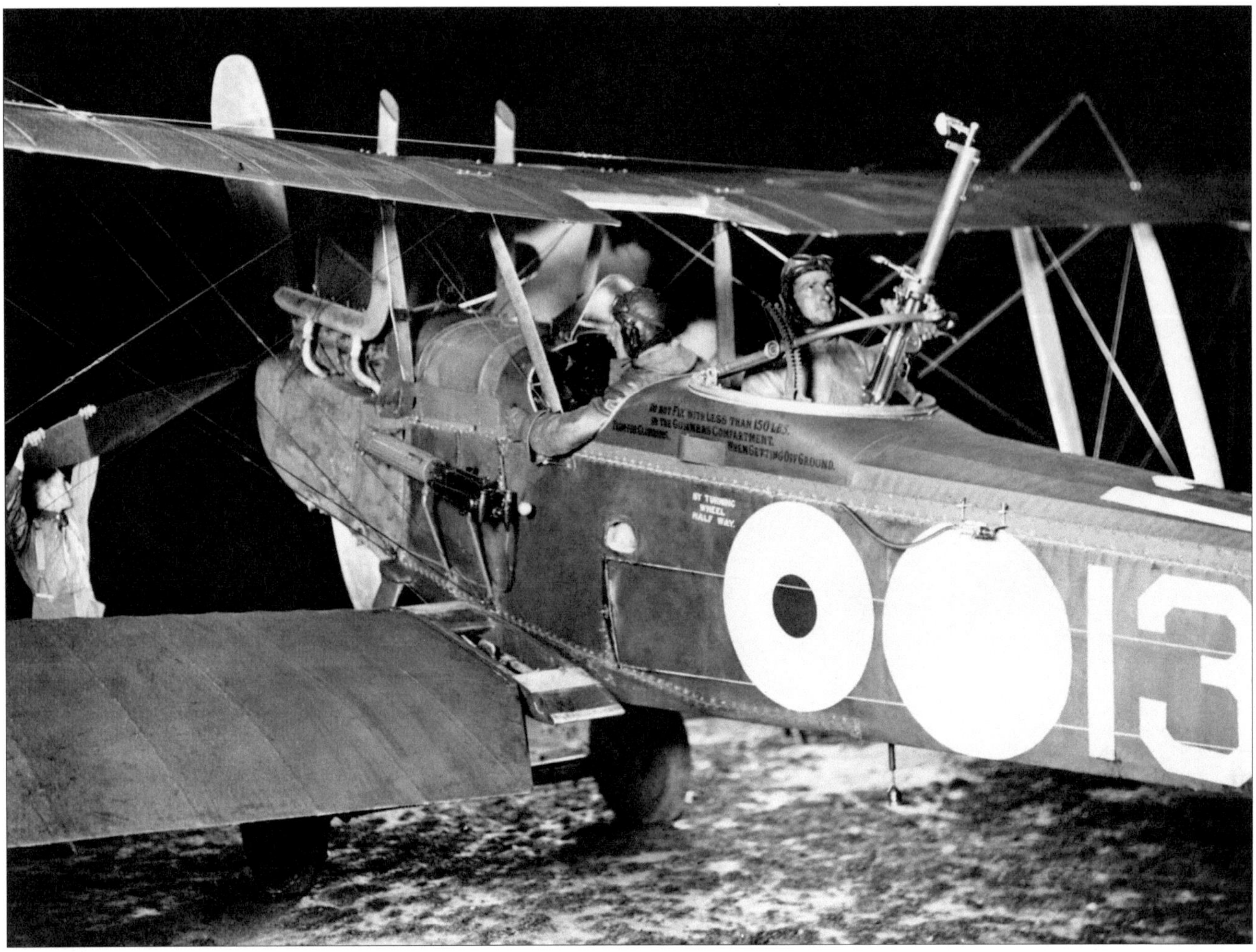

Above: A good close up of the RE8, note the pilot's Vickers machine gun location.

Below: An RE8 of 63 squadron during its service in Mesopotamia. Note the personal insignia of a skull and cross-bones at the front of the aircraft

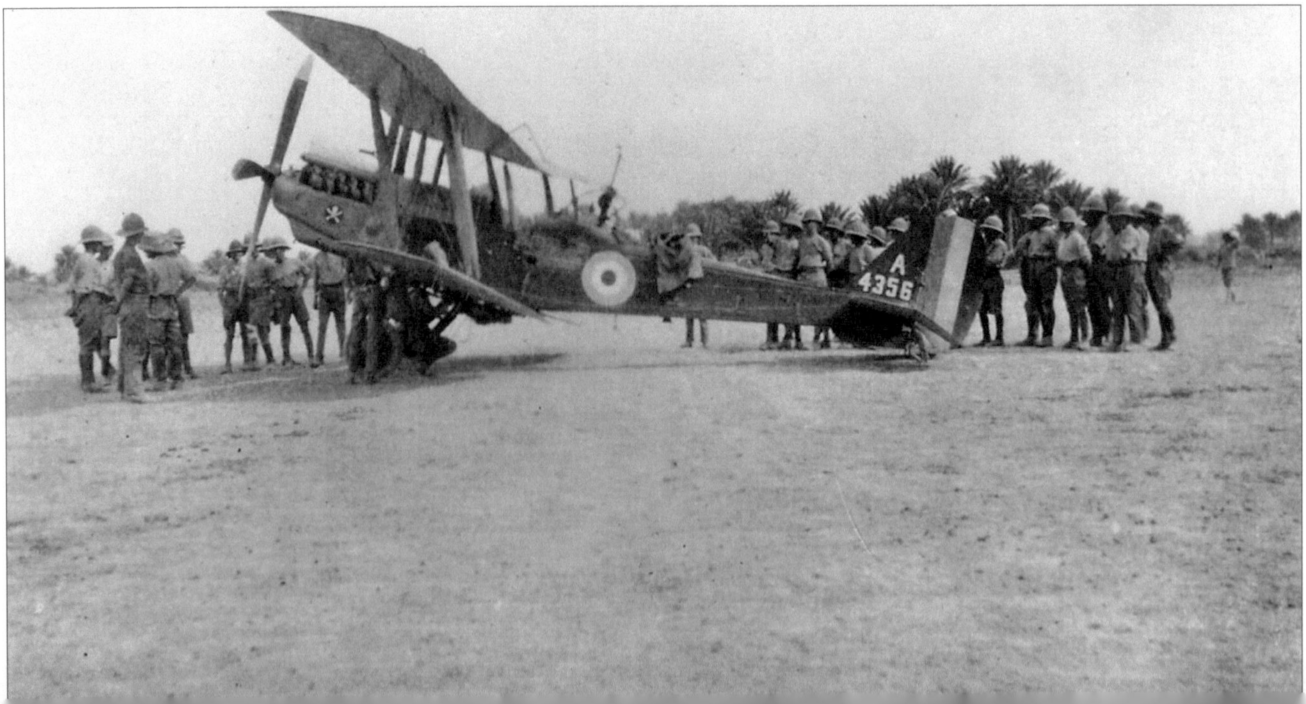

Artists For Volume 2

The reader should remember that all artwork in the *Art of the Dogfight* series is copyrighted and cannot be copied for any reason, in any format without the permission of the artist or the appropriate copyright holder. At the end of volume 4 there is an expanded artist biography section. The page numbers to find an artist's work is for the convenience of the reader. The artwork has been permitted by the artist unless otherwise noted.

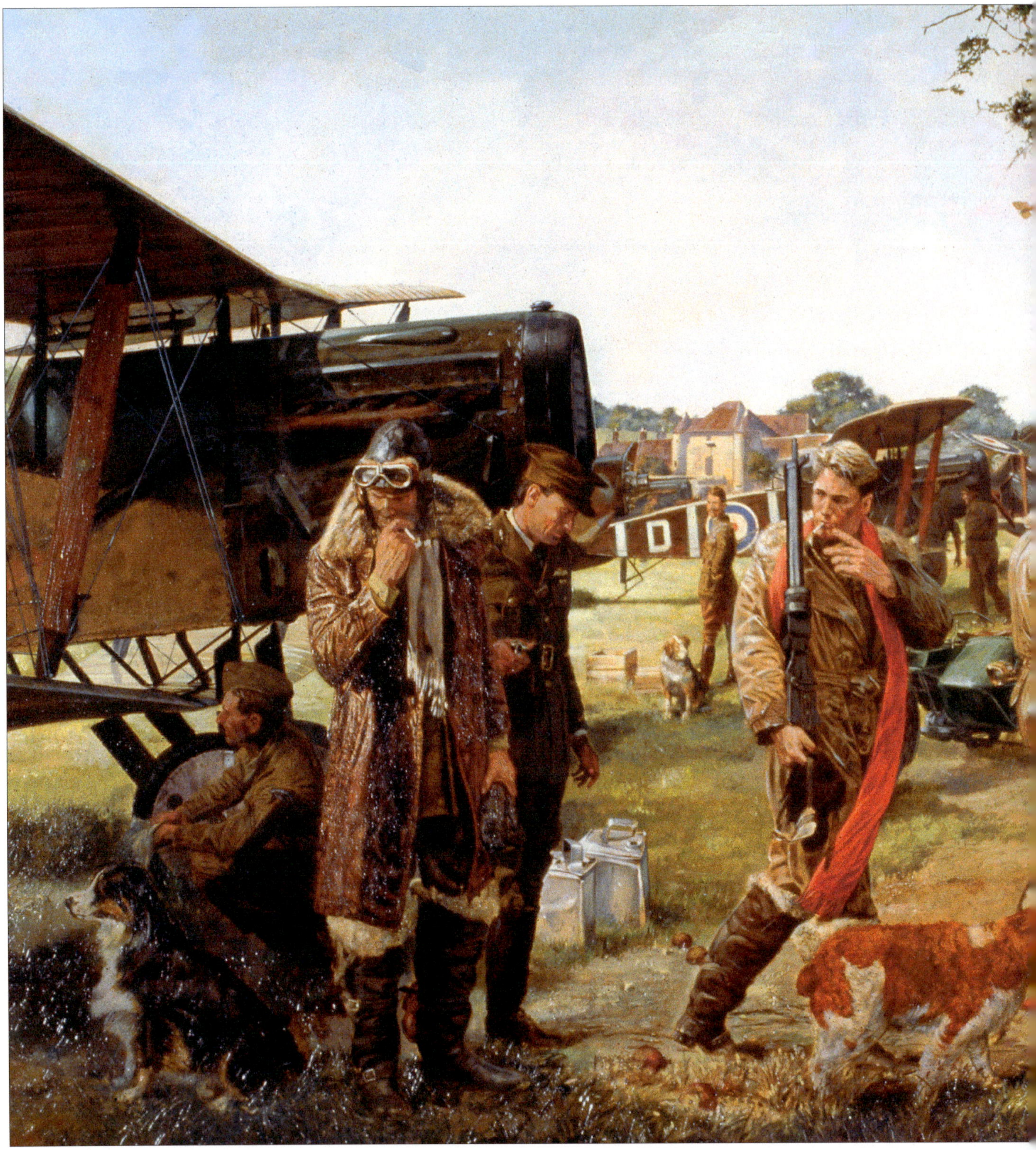

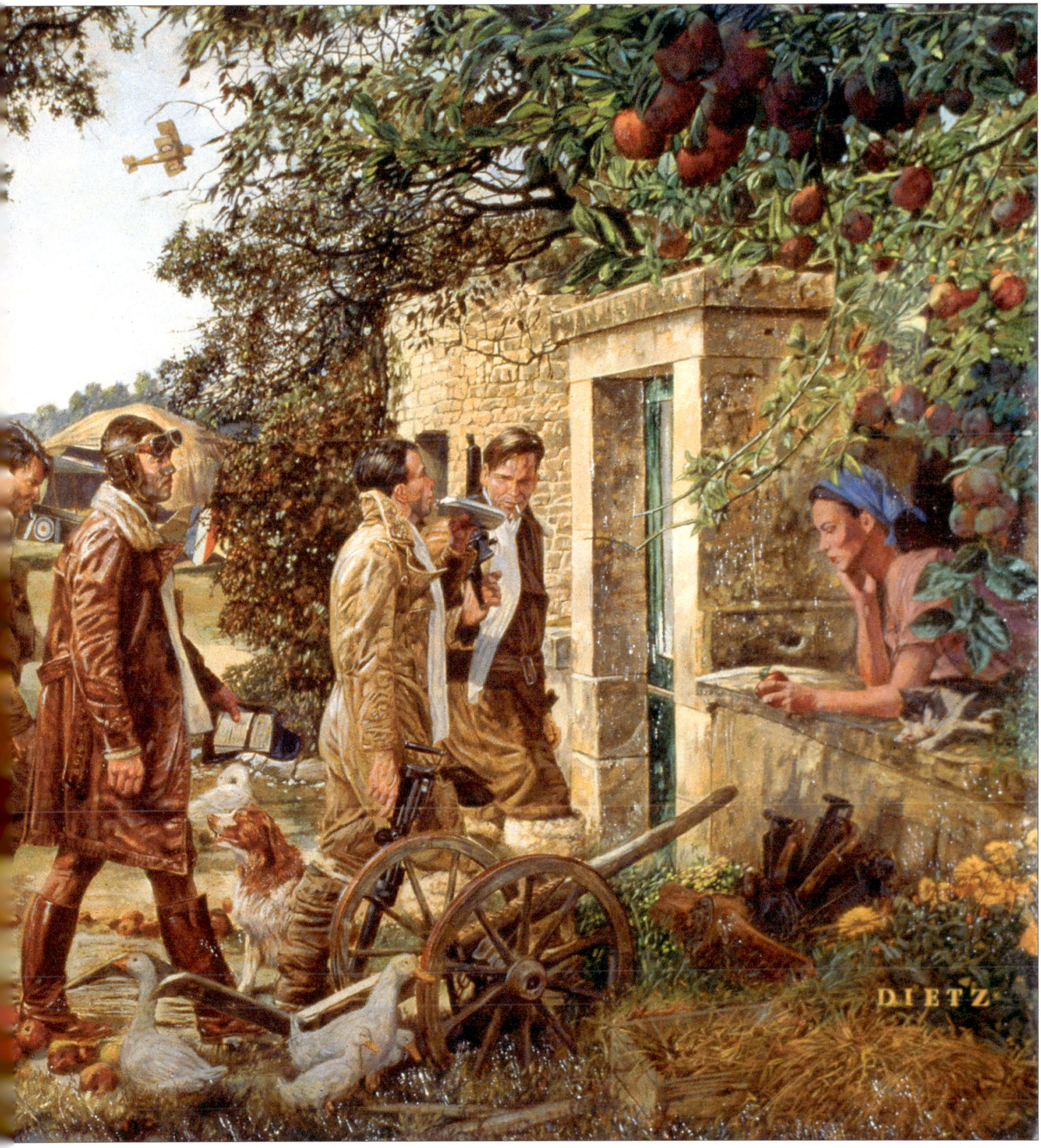
DIETZ

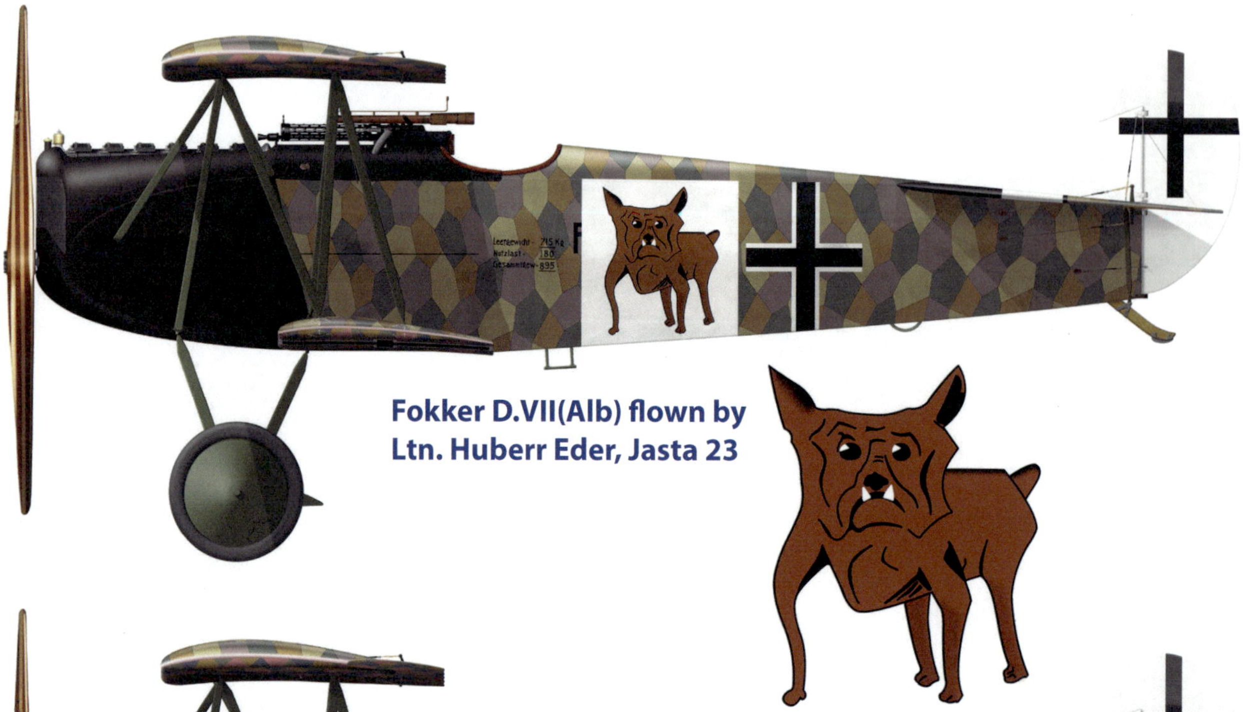

Fokker D.VII(Alb) flown by Ltn. Huberr Eder, Jasta 23

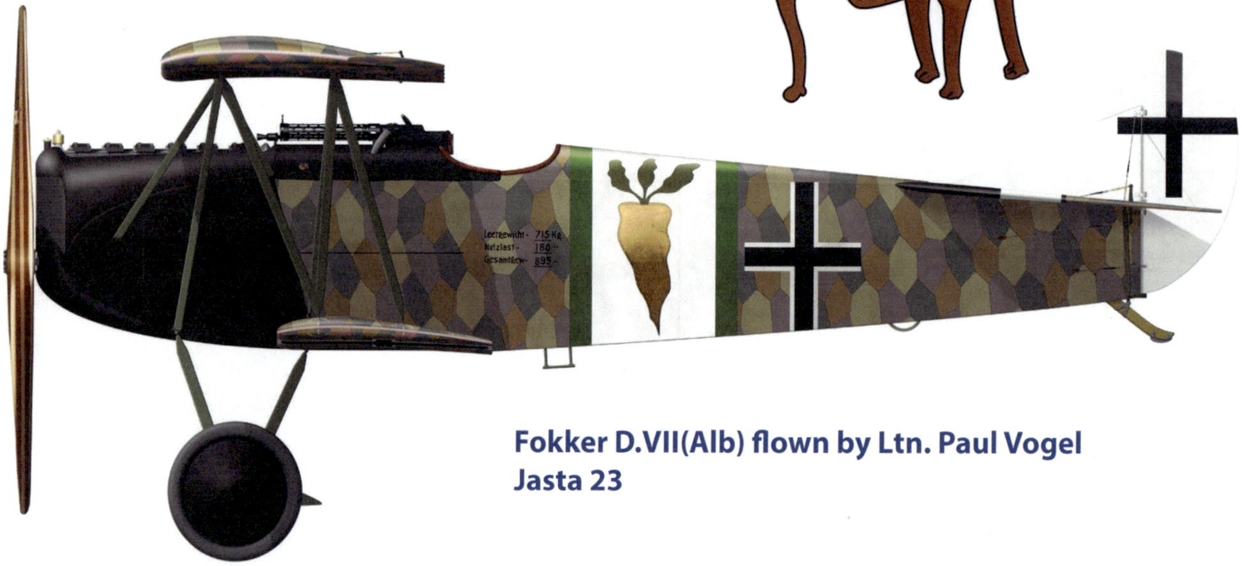

Fokker D.VII(Alb) flown by Ltn. Paul Vogel Jasta 23

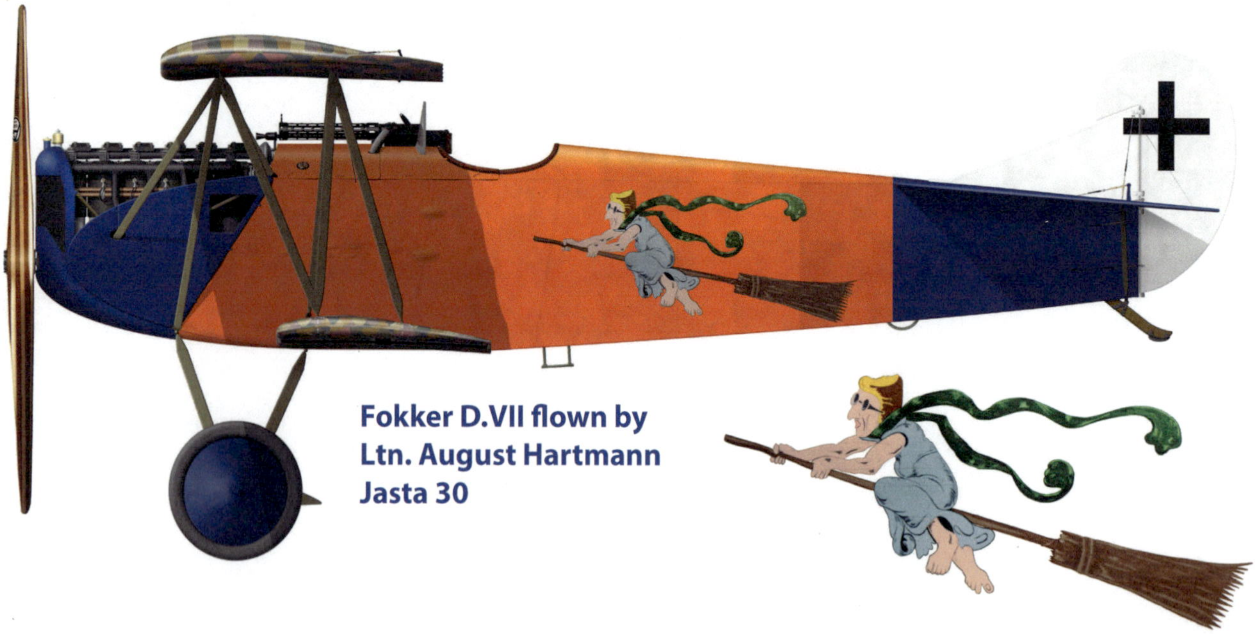

Fokker D.VII flown by Ltn. August Hartmann Jasta 30

Made in the USA
Las Vegas, NV
03 November 2026

57421648R00128